ICSA Book Series in Statistics

Series editors

More information about this series at http://www.springer.com/series/13402

Ding-Geng (Din) Chen · John Dean Chen
Editors

Monte-Carlo Simulation-Based Statistical Modeling

Springer

Editors
Ding-Geng (Din) Chen
University of North Carolina
Chapel Hill, NC
USA

and

University of Pretoria
Pretoria
South Africa

John Dean Chen
Risk Management
Credit Suisse
New York, NY
USA

ISSN 2199-0980 ISSN 2199-0999 (electronic)
ICSA Book Series in Statistics
ISBN 978-981-10-3306-3 ISBN 978-981-10-3307-0 (eBook)
DOI 10.1007/978-981-10-3307-0

Library of Congress Control Number: 2016960187

Printed on acid-free paper

This Springer imprint is published by Springer Nature
The registered company is Springer Nature Singapore Pte Ltd.
The registered company address is: 152 Beach Road, #22-06/08 Gateway East, Singapore 189721, Singapore

Preface

Over the last two decades, advancements in computer technology have enabled accelerated research and development of Monte-Carlo computational methods. This book is a compilation of invited papers from some of the most forward-thinking statistical researchers. These authors present new developments in Monte-Carlo simulation-based statistical modeling, thereby creating an opportunity for the exchange ideas among researchers and users of statistical computing.

Our aim in creating this book is to provide a venue for timely dissemination of the research in Monte-Carlo simulation-based statistical modeling to promote further research and collaborative work in this area. In the era of big data science, this collection of innovative research not only has remarkable potential to have a substantial impact on the development of advanced Monte-Carlo methods across the spectrum of statistical data analyses but also has great promise for fostering new research and collaborations addressing the ever-changing challenges and opportunities of statistics and data science. The authors have made their data and computer programs publicly available, making it possible for readers to replicate the model development and data analysis presented in each chapter and readily apply these new methods in their own research.

The 18 chapters are organized into three parts. Part I includes six chapters that present and discuss general Monte-Carlo techniques. Part II comprises six chapters with a common focus on Monte-Carlo methods used in missing data analyses, which is an area of growing importance in public health and social sciences. Part III is composed of six chapters that address Monte-Carlo statistical modeling and their applications.

Part I: Monte-Carlo Techniques (Chapters "Joint Generation of Binary, Ordinal, Count, and Normal Data with Specified Marginal and Association Structures in Monte-Carlo Simulations"–"Quantifying the Uncertainty in Optimal Experiment Schemes via Monte-Carlo Simulations")

Chapter "Joint Generation of Binary, Ordinal, Count, and Normal Data with Specified Marginal and Association Structures in Monte-Carlo Simulations" presents a unified framework for concurrently generating data that include the four major types of distributions (i.e., binary, ordinal, count, and normal) with specified marginal and association structures. In this discussion of an important supplement to existing methods, Hakan Demirtas unifies the Monte-Carlo methods for specified types of data and presents his systematic and comprehensive investigation for mixed data generation. The proposed framework can then be readily used to simulate multivariate data of mixed types for the development of more sophisticated simulation, computation, and data analysis techniques.

In Chapter "Improving the Efficiency of the Monte-Carlo Methods Using Ranked Simulated Approach", Hani Samawi provides an overview of his development of ranked simulated sampling; a key approach for improving the efficiency of general Monte-Carlo methods. Samawi then demonstrates the capacity of this approach to provide unbiased estimation.

In Chapter "Normal and Non-normal Data Simulations for the Evaluation of Two-Sample Location Tests", Jessica Hoag and Chia-Ling Kuo discuss Monte-Carlo simulation of normal and non-normal data to evaluate two-sample location tests (i.e., statistical tests that compare means or medians of two independent populations).

Chapter "Anatomy of Correlational Magnitude Transformations in Latency and Discretization Contexts in Monte-Carlo Studies" proposes a general assessment of correlational magnitude changes in the latency and discretization contexts of Monte-Carlo studies. Further, authors Hakan Demirtas and Ceren Vardar-Acar provide a conceptual framework and computational algorithms for modeling the correlation transitions under specified distributional assumptions within the realm of discretization in the context of latency and the threshold concept. The authors illustrate the proposed algorithms with several examples and include a simulation study that demonstrates the feasibility and performance of the methods.

Chapter "Monte-Carlo Simulation of Correlated Binary Responses" discusses the Monte-Carlo simulation of correlated binary responses. Simulation studies are a well-known, highly valuable tool that allows researchers to obtain powerful conclusions for correlated or longitudinal response data. In cases where logistic modeling it used, the researcher must have appropriate methods for simulating correlated binary data along with associated predictors. In this chapter, author Trent Lalonde presents an overview of existing methods for simulating correlated binary response data and compares those methods with methods using R software.

Chapter "Quantifying the Uncertainty in Optimal Experiment Schemes via Monte-Carlo Simulations" provides a general framework for quantifying the sensitivity and uncertainty that result from the misspecification of model parameters in optimal experimental schemes. In designing life-testing experiments, it is widely accepted that the optimal experimental scheme depends on unknown model parameters, and that misspecified parameters can lead to substantial loss of efficiency in the statistical analysis. To quantify this effect, Tony Ng, Yu-Jau Lin, Tzong-Ru Tsai, Y.L. Lio, and Nan Jiang use Monte-Carlo simulations to evaluate the robustness of optimal experimental schemes.

Part II: Monte-Carlo Methods for Missing Data (Chapters "Markov Chain Monte-Carlo Methods for Missing Data Under Ignorability Assumptions"–"Application of Markov Chain Monte-Carlo Multiple Imputation Method to Deal with Missing Data from the Mechanism of MNAR in Sensitivity Analysis for a Longitudinal Clinical Trial")

Chapter "Markov Chain Monte-Carlo Methods for Missing Data Under Ignorability Assumptions" presents a fully Bayesian method for using the Markov chain Monte-Carlo technique for missing data to sample the full conditional distribution of the missing data given observed data and the other parameters. In this chapter, Haresh Rochani and Daniel Linder show how to apply these methods to real datasets with missing responses as well as missing covariates. Additionally, the authors provide simulation settings to illustrate this method.

In Chapter "A Multiple Imputation Framework for Massive Multivariate Data of Different Variable Types: A Monte-Carlo Technique", Hakan Demirtas discusses multiple imputation for massive multivariate data of variable types from planned missingness designs with the purpose to build theoretical, algorithmic, and implementation-based components of a unified, general-purpose multiple imputation framework. The planned missingness designs are highly useful and will likely increase in popularity in the future. For this reason, the proposed multiple imputation framework represents an important refinement of existing methods.

Chapter "Hybrid Monte-Carlo in Multiple Missing Data Imputations with Application to a Bone Fracture Data" introduces the Hybrid Monte-Carlo method as an efficient approach for sampling complex posterior distributions of several correlated parameters from a semi-parametric missing data model. In this chapter, Hui Xie describes a modeling approach for missing values that does not require assuming specific distributional forms. To demonstrate the method, the author provides an R program for analyzing missing data from a bone fracture study.

Chapter "Statistical Methodologies for Dealing with Incomplete Longitudinal Outcomes Due to Dropout Missing at Random" considers key methods for handling longitudinal data that are incomplete due to missing at random dropout. In this chapter, Ali Satty, Henry Mwambi, and Geert Muhlenbergs provide readers with an overview of the issues and the different methodologies for handling missing data in longitudinal datasets that result from dropout (e.g., study attrition, loss of follow-up). The authors examine the potential strengths and weaknesses of the various methods through two examples of applying these methods.

In Chapter "Applications of Simulation for Missing Data Issues in Longitudinal Clinical Trials", Frank Liu and James Kost present simulation-based approaches for addressing missing data issues in longitudinal clinical trials, such as control-based imputation, tipping-point analysis, and a Bayesian Markov chain Monte-Carlo method. Computation programs for these methods are implemented and available in SAS.

In Chapter "Application of Markov Chain Monte-Carlo Multiple Imputation Method to Deal with Missing Data from the Mechanism of MNAR in Sensitivity Analysis for a Longitudinal Clinical Trial", Wei Sun discusses the application of Markov chain Monte-Carlo multiple imputation for data that is missing not at random in longitudinal datasets from clinical trials. This chapter compares the patterns of missing data between study subjects who received treatment and study subjects who received a placebo.

Part III: Monte-Carlo in Statistical Modellings and Applications (Chapters "Monte-Carlo Simulation in Modeling for Hierarchical Generalized Linear Mixed Models"–"Bootstrap-Based LASSO-type Selection to Build Generalized Additive Partially Linear Models for High-Dimensional Data")

Chapter "Monte-Carlo Simulation in Modeling for Hierarchical Generalized Linear Mixed Models" adds a discussion of Monte-Carlo simulation-based hierarchical models, taking into account the variability at each level of the hierarchy. In this chapter, Kyle Irimata and Jeffrey Wilson discuss Monte-Carlo simulations for hierarchical linear mixed-effects models to fit the hierarchical logistic regression models with random intercepts (both random intercepts and random slopes) to multilevel data.

Chapter "Monte-Carlo Methods in Financial Modeling" demonstrates the use of Monte-Carlo methods in financial modeling. In this chapter, Chuanshu Ji, Tao Wang, and Leicheng Yin discuss two areas of market microstructure modeling and option pricing using Monte-Carlo dimension reduction techniques. This approach uses Bayesian Markov chain Monte-Carlo inference based on the trade and quote database from Wharton Research Data Services.

Chapter "Simulation Studies on the Effects of the Censoring Distribution Assumption in the Analysis of Interval-Censored Failure Time Data" discusses using Monte-Carlo simulations to evaluate the effect of the censoring distribution assumption for interval-censored survival data. In this chapter, Tyler Cook and

Jianguo Sun investigate the effectiveness and flexibility of two methods for regression analysis of informative case I and case II interval-censored data. The authors present extensive Monte-Carlo simulation studies that provide readers with guidelines regarding dependence of the censoring distribution.

Chapter "Robust Bayesian Hierarchical Model Using Monte-Carlo Simulation" uses Monte-Carlo simulation to demonstrate a robust Bayesian multilevel item response model. In this chapter, Geng Chen uses data from patients with Parkinson's disease, a chronic progressive disease with multidimensional impairments. Using these data, Chen illustrates applying the multilevel item response model to not only deal with the multidimensional nature of the disease but also simultaneously estimate measurement-specific parameters, covariate effects, and patient-specific characteristics of disease progression.

In Chapter "A Comparison of Bootstrap Confidence Intervals for Multi-level Longitudinal Data Using Monte-Carlo Simulation", Mark Reiser, Lanlan Yao, and Xiao Wang present a comparison of bootstrap confidence intervals for multilevel longitudinal data using Monte-Carlo simulations. Their results indicate that if the sample size at the lower level is small, then the parametric bootstrap and cluster bootstrap perform better at the higher level than the two-stage bootstrap. The authors then apply the bootstrap methods to a longitudinal study of preschool children nested within classrooms.

Chapter "Bootstrap-Based LASSO-Type Selection to Build Generalized Additive Partially Linear Models for High-Dimensional Data" presents an approach to using a bootstrap-based LASSO-type selection to build generalized additive partially linear models for high-dimensional data. In this chapter, Xiang Liu, Tian Chen, Yuanzhang Li, and Hua Liang first propose a bootstrap-based procedure to select variables with penalized regression and then apply their procedure to analyze data from a breast cancer study and an HIV study. The two examples demonstrate the procedure's flexibility and utility in practice. In addition, the authors present a simulation study that shows, when compared with the penalized regression approach, their variable selection procedure performs better.

As a general note, the references for each chapter are included immediately following the chapter text. We have organized the chapters as self-contained units so readers can more easily and readily refer to the cited sources for each chapter.

To facilitate readers' understanding of the methods presented in this book, corresponding data and computing program can be requested from the first editor by email at DrDG.Chen@gmail.com.

The editors are deeply grateful to many who have supported the creation of this book. We thank the authors of each chapter for their contributions and their generous sharing of their knowledge, time, and expertise to this book. Second, our sincere gratitude goes to Ms. Diane C. Wyant from the School of Social Work, University of North Carolina at Chapel Hill for her expert editing and comments of this book which substantially uplift the quality of this book. We gratefully acknowledge the professional support of Hannah Qiu (Springer/ICSA Book Series coordinator) and Wei Zhao (associate editor) from Springer Beijing that made publishing this book with Springer a reality.

We welcome readers' comments, including notes on typos or other errors, and look forward to receiving suggestions for improvements to future editions of this book. Please send comments and suggestions to any of the editors listed below.

October 2016

Ding-Geng (Din) Chen
University of North Carolina, Chapel Hill, USA
University of Pretoria, South Africa

John Dean Chen
Credit Suisse, New York, NY, USA

The original version of the book frontmatter was revised: For detailed information please see erratum. The erratum to the book frontmatter is available at DOI 10.1007/978-981-10-3307-0_19

About the Book

This book brings together expert researchers engaged in Monte-Carlo simulation-based statistical modeling, offering them a forum to present and discuss recent issues in methodological development as well as public health applications. It is divided into three parts, with the first providing an overview of Monte-Carlo techniques, the second focusing on missing data Monte-Carlo methods, and the third addressing Bayesian and general statistical modeling using Monte-Carlo simulations. The data and computer programs used here will also be made publicly available, allowing readers to replicate the model development and data analysis presented in each chapter, and to readily apply them in their own research. Featuring highly topical content, the book has the potential to impact model development and data analyses across a wide spectrum of fields, and to spark further research in this direction.

Contents

Editors and Contributors

About the Editors

Prof. Ding-Geng Chen is a fellow of the American Statistical Association and currently the Wallace Kuralt distinguished professor at the University of North Carolina at Chapel Hill. He was a professor at the University of Rochester and the Karl E. Peace endowed eminent scholar chair in biostatistics at Georgia Southern University. He is also a senior consultant for biopharmaceuticals and government agencies with extensive expertise in clinical trial biostatistics and public health statistics. Professor Chen has written more than 150 referred publications and co-authored/co-edited ten books on clinical trial methodology, meta-analysis, causal-inference and public health statistics.

Mr. John Dean Chen is specialized in Monte-Carlo simulations in modelling financial market risk. He is currently a Vice President at Credit Suisse specializing in regulatory stress testing with Monte-Carlo simulations. He began his career on Wall Street working in commodities trading market risk before moving to trading structured notes within the Exotics Interest Rate Derivatives desk at Barclays Capital. He transitioned back to risk at Mitsubishi UFJ working in its model risk group. During his career in the financial industry, he witnessed in person the unfolding of the financial crisis, and the immediate aftermath consuming much of the financial industry. He graduated from the University of Washington with a dual Bachelors of Science in Applied Mathematics and Economics.

Contributors

Rawan Allozi Division of Epidemiology and Biostatistics (MC923), University of Illinois at Chicago, Chicago, IL, USA

Geng Chen Clinical Statistics, GlaxoSmithKline, Collegeville, PA, USA

Tian Chen Department of Mathematics and Statistics, University of Toledo, Toledo, OH, USA

Tyler Cook University of Central Oklahoma, Edmond, OK, USA

Hakan Demirtas Division of Epidemiology and Biostatistics (MC923), University of Illinois at Chicago, Chicago, IL, USA

Shelley Gray Speech and Hearing Science, Arizona State University, Tempe, AZ, USA

Jessica R. Hoag Department of Community Medicine and Health Care, Connecticut Institute for Clinical and Translational Science, University of Connecticut Health Center, Farmington, USA

Yiran Hu Division of Epidemiology and Biostatistics (MC923), University of Illinois at Chicago, Chicago, IL, USA

Gul Inan Department of Statistics, Middle East Technical University, Ankara, Turkey

Kyle M. Irimata School of Mathematical and Statistical Sciences, Arizona State University, Tempe, AZ, USA

Chuanshu Ji Department of Statistics and Operations Research, University of North Carolina, Chapel Hill, NC, USA

N. Jiang Department of Mathematical Sciences, University of South Dakota, Vermillion, SD, USA

James Kost Merck & Co. Inc., North Wales, PA, USA

Chia-Ling Kuo Department of Community Medicine and Health Care, Connecticut Institute for Clinical and Translational Science, University of Connecticut Health Center, Farmington, USA

Trent L. Lalonde Department of Applied Statistics and Research Methods, University of Northern Colorado, Greeley, CO, USA

Yuanzhang Li Division of Preventive Medicine, Walter Reed Army Institute of Research, Silver Spring, MD, USA

Hua Liang Department of Statistics, George Washington University, Washington, DC, USA

Y.-J. Lin Department of Applied Mathematics, Chung Yuan Christian University, Chung-Li District, Taoyuan city, Taiwan

Daniel F. Linder Department of Biostatistics and Epidemiology, Medical College of Georgia, Augusta University, Augusta, GA, Georgia

Y.L. Lio Department of Mathematical Sciences, University of South Dakota, Vermillion, SD, USA

G. Frank Liu Merck & Co. Inc., North Wales, PA, USA

Xiang Liu Health Informatics Institute, University of South Florida, Tampa, FL, USA

Sheng Luo Department of Biostatistics, The University of Texas Health Science Center at Houston, Houston, TX, USA

G. Molenberghs I-BioStat, Universiteit Hasselt & KU Leuven, Hasselt, Belgium

A. Satty Faculty of Mathematical Sciences and Statistics, Alneelain University, Khartoum, Sudan

H.K.T. Ng Department of Statistical Science, Southern Methodist University, Dallas, TX, USA

Levent Ozbek Department of Statistics, Ankara University, Ankara, Turkey

Mark Reiser School of Mathematical and Statistical Science, Arizona State University, Tempe, AZ, USA

Haresh Rochani Department of Biostatistics, Jiann-Ping Hsu College of Public Health, Georgia Southern University, Statesboro, GA, Georgia

Hani Michel Samawi Department of Biostatistics, Jiann-Ping Hsu College Public Health, Georgia Southern University, Statesboro, Georgia

Henry Mwambi School of Mathematics, Statistics and Computer Science, University of KwaZulu-Natal, Pietermaritzburg, South Africa

Jianguo Sun University of Missouri, Columbia, MO, USA

Wei Sun Manager Biostatistician at Otsuka America, New York, USA

T.-R. Tsai Department of Statistics, Tamkang University, Tamsui District, New Taipei City, Taiwan

Ceren Vardar-Acar Department of Statistics, Middle East Technical University, Ankara, Turkey

Tao Wang Bank of America Merrill Lynch, New York, NY, USA

Xiao Wang Statistics and Data Corporation, Tempe, AZ, USA

Jeanne Wilcox Division of Educational Leadership and Innovation, Arizona State University, Tempe, AZ, USA

Jeffrey R. Wilson W.P. Carey School of Business, Arizona State University, Tempe, AZ, USA

Hui Xie Simon Fraser University, Burnaby, Canada; The University of Illinois at Chicago, Chicago, USA

Lanlan Yao School of Mathematical and Statistical Science, Arizona State University, Tempe, AZ, USA

Leicheng Yin Exelon Business Services Company, Enterprise Risk Management, Chicago, IL, USA

Zhigang Zhang Memorial Sloan Kettering Cancer Center, New York, NY, USA

Part I
Monte-Carlo Techniques

Joint Generation of Binary, Ordinal, Count, and Normal Data with Specified Marginal and Association Structures in Monte-Carlo Simulations

Hakan Demirtas, Rawan Allozi, Yiran Hu, Gul Inan and Levent Ozbek

Abstract This chapter is concerned with building a unified framework for concurrently generating data sets that include all four major kinds of variables (i.e., binary, ordinal, count, and normal) when the marginal distributions and a feasible association structure are specified for simulation purposes. The simulation paradigm has been commonly employed in a wide spectrum of research fields including the physical, medical, social, and managerial sciences. A central aspect of every simulation study is the quantification of the model components and parameters that jointly define a scientific process. When this quantification cannot be performed via deterministic tools, researchers resort to random number generation (RNG) in finding simulation-based answers to address the stochastic nature of the problem. Although many RNG algorithms have appeared in the literature, a major limitation is that they were not designed to concurrently accommodate all variable types mentioned above. Thus, these algorithms provide only an incomplete solution, as real data sets include variables of different kinds. This work represents an important augmentation of the existing methods as it is a systematic attempt and comprehensive investigation for mixed data generation. We provide an algorithm that is designed for generating data of mixed marginals, illustrate its logistical, operational, and computational details; and present ideas on how it can be extended to span more complicated distributional settings in terms of a broader range of marginals and associational quantities.

H. Demirtas (✉) · R. Allozi · Y. Hu
Division of Epidemiology and Biostatistics (MC923), University of Illinois at Chicago, 1603 West Taylor Street, Chicago, IL 60612, USA
e-mail: demirtas@uic.edu

G. Inan
Department of Statistics, Middle East Technical University, Ankara, Turkey

L. Ozbek
Department of Statistics, Ankara University, Ankara, Turkey

D.-G. Chen and J.D. Chen (eds.), *Monte-Carlo Simulation-Based Statistical Modeling*, ICSA Book Series in Statistics, DOI 10.1007/978-981-10-3307-0_1

1 Introduction

Stochastic simulation is an indispensable part and major focus of scientific inquiry. Model building, estimation, and testing typically require verification via simulation to assess the validity, reliability, and plausibility of inferential techniques, to evaluate how well the implemented models capture the specified true population values, and how reasonably these models respond to departures from underlying assumptions, among other things. Describing a real notion by creating mirror images and imperfect proxies of the perceived underlying truth; iteratively refining and occasionally redefining the empirical truth to decipher the mechanism by which the process under consideration is assumed to operate in a repeated manner allows researchers to study the performance of their methods through simulated data replicates that mimic the real data characteristics of interest in any given setting. Accuracy and precision measures regarding the parameters under consideration signal if the procedure works properly; and may suggest remedial action to minimize the discrepancies between expectation and reality.

Simulation studies have been commonly employed in a broad range of disciplines in order to better comprehend and solve today's increasingly sophisticated issues. A core component of every simulation study is the quantification of the model components and parameters that jointly define a scientific phenomenon. Deterministic tools are typically inadequate to quantify complex situations, leading researchers to utilize RNG techniques in finding simulation-based solutions to address the stochastic behavior of the problems that generally involve variables of many different types on a structural level; i.e., causal and correlational interdependencies are a function of a mixture of binary, ordinal, count, and continuous variables, which act simultaneously to characterize the mechanisms that collectively delineate a paradigm. In modern times, we are unequivocally moving from mechanistical to empirical thinking, from small data to big data, from mathematical perfection to reasonable approximation to reality, and from exact solutions to simulation-driven solutions. The ideas presented herein are important in the sense that the basic mixed-data generation setup can be augmented to handle a large spectrum of situations that can be encountered in many areas.

This work is concerned with building the basics of a unified framework for concurrently generating data sets that include all four major kinds of variables (i.e., binary, ordinal, count, and normal) when the marginal distributions and a feasible association structure in the form of Pearson correlations are specified for simulation purposes. Although many RNG algorithms have appeared in the literature, a fundamental restriction is that they were not designed for a mix of all prominent types of data. The current chapter is a systematic attempt and compendious investigation for mixed data generation; it represents a substantial augmentation of the existing methods, and it has potential to advance scientific research and knowledge in a meaningful way. The broader impact of this framework is that it can assist data analysts, practitioners, theoreticians, and methodologists across many disciplines to simulate mixed data with relative ease. The proposed algorithm constitutes a comprehensive set of computational tools that offers promising potential for building enhanced computing infrastructure for research and education.

We propose an RNG algorithm that encompasses all four major variable types, building upon our previous work in generation of multivariate ordinal data (Demirtas 2006), joint generation of binary and normal data (Demirtas and Doganay 2012), ordinal and normal data (Demirtas and Yavuz 2015), and count and normal data (Amatya and Demirtas 2015) with the specification of marginal and associational parameters along with other related work (Emrich and Piedmonte 1991; Demirtas and Hedeker 2011, 2016; Demirtas et al. 2016a; Ferrari and Barbiero 2012; Yahav and Shmueli 2012). Equally importantly, we discuss the extensions on nonnormal continuous data via power polynomials that would handle the overwhelming majority of continuous shapes (Fleishman 1978; Vale and Maurelli 1983; Headrick 2010; Demirtas et al. 2012; Demirtas 2017a), count data that are prone to over- and under-dispersion via generalized Poisson distribution (Demirtas 2017b), broader measures of associations such as Spearman's rank correlations and L-correlations (Serfling and Xiao 2007), and the specification of higher order product moments. Conceptual, algorithmic, operational, and procedural details will be communicated throughout the chapter.

The organization of the chapter is as follows: In Sect. 2, the algorithm for simultaneous generation of binary, ordinal, count, and normal data is given. The essence of the algorithm is finding the correlation structure of underlying multivariate normal (MVN) data that form a basis for the subsequent discretization in the binary and ordinal cases, and correlation mapping using inverse cumulative distribution functions (cdfs) in the count data case, where modeling the correlation transitions for different distributional pairs is discussed in detail. Section 3 presents some logistical details and an illustrative example through an R package that implements the algorithm, demonstrating how well the proposed technique works. Section 4 includes discussion on limitations, future directions, extensions, and concluding remarks.

2 Algorithm

The algorithm is designed for concurrently generating binary, ordinal, count, and continuous data. The count and continuous parts are assumed to follow Poisson and normal distributions, respectively. While binary is a special case of ordinal, for the purpose of exposition, the steps are presented separately. Skipped patterns are allowed for ordinal variables. The marginal characteristics (the proportions for the binary and ordinal part, the rate parameters for the count part, and the means and variances for the normal part) and a feasible Pearson correlation matrix need to be specified by the users. The algorithmic skeleton establishes the basic foundation, extensions to more general and complicated situations will be discussed in Sect. 4.

The operational engine of the algorithm hinges upon computing the correlation matrix of underlying MVN data that serve as an intermediate tool in the sense that binary and ordinal variables are obtained via dichotomization and ordinalization, respectively, through the threshold concept, and count variables are retrieved by correlation mapping using inverse cdf matching. The procedure entails modeling the correlation transformations that result from discretization and mapping.

In what follows, let B, O, C, and N denote binary, ordinal, count, and normal variables, respectively. Let Σ be the specified Pearson correlation matrix which comprises of ten submatrices that correspond to all possible variable-type combinations.

Required parameter values are p's for binary and ordinal variables, λ's for count variables, (μ, σ^2) pairs for normal variables, and the entries of the correlation matrix Σ. These quantities are either specified or estimated from a real data set that is to be mimicked.

1. Check if Σ is positive definite.
2. Find the upper and lower correlation bounds for all pairs by the sorting method of Demirtas and Hedeker (2011). It is well-known that correlations are not bounded between -1 and $+1$ in most bivariate settings as different upper and/or lower bounds may be imposed by the marginal distributions (Hoeffding 1940; Fréchet 1951). These restrictions apply to discrete variables as well as continuous ones. Let $\Pi(F, G)$ be the set of cdf's H on R^2 having marginal cdf's F and G. Hoeffding (1940) and Fréchet (1951) proved that in $\Pi(F, G)$, there exist cdf's H_L and H_U, called the lower and upper bounds, having minimum and maximum correlation. For all $(x, y) \in R^2$, $H_L(x, y) = max[F(x) + G(y) - 1, 0]$ and $H_U(x, y) = min[F(x), G(y)]$. For any $H \in \Pi(F, G)$ and all $(x, y) \in R^2$, $H_L(x, y) \leq H(x, y) \leq H_U(x, y)$. If δ_L, δ_U, and δ denote the Pearson correlation coefficients for H_L, H_U, and H, respectively, then $\delta_L \leq \delta \leq \delta_U$. One can infer that if V is uniform in [0, 1], then $F^{-1}(V)$ and $G^{-1}(V)$ are maximally correlated; and $F^{-1}(V)$ and $G^{-1}(1 - V)$ are maximally anticorrelated. In practical terms, generating X and Y independently with a large number of data points before sorting them in the same and opposite direction give the approximate upper and lower correlation bounds, respectively. Make sure all elements of Σ are within the plausible range.
3. Perform logical checks such as binary proportions are between 0 and 1, probabilities add up to 1 for ordinal variables, the Poisson rates are positive for count variables, variances for normal variables are positive, the mean, variance, proportion and rate vectors are consistent with the number of variables, Σ is symmetric and its diagonal entries are 1, to prevent obvious misspecification errors.
4. For B-B combinations, find the tetrachoric (pre-dichotomization) correlation given the specified phi coefficient (post-dichotomization correlation). Let X_1, X_2 represent binary variables such that $E[X_j] = p_j$ and $Cor(X_1, X_2) = \delta_{12}$, where p_j $(j = 1, 2)$ and δ_{12} (phi coefficient) are given. Let $\Phi[t_1, t_2, \rho_{12}]$ be the cdf for a standard bivariate normal random variable with correlation coefficient ρ_{12} (tetrachoric correlation). Naturally, $\Phi[t_1, t_2, \rho_{12}] = \int_{-\infty}^{t_1} \int_{-\infty}^{t_2} f(z_1, z_2, \rho_{12}) dz_1 dz_2$, where $f(z_1, z_2, \rho_{12}) = [2\pi(1 - \rho_{12}^2)^{1/2}]^{-1} \times exp\Big[-(z_1^2 - 2\rho_{12}z_1z_2 + z_2^2)/(2(1 - \rho_{12}^2))\Big]$. The connection between δ_{12} and ρ_{12} is reflected in the equation

$$\Phi[z(p_1), z(p_2), \rho_{12}] = \delta_{12}(p_1 q_1 p_2 q_2)^{1/2} + p_1 p_2$$

Solve for ρ_{12} where $z(p_j)$ denotes the p_j^{th} quantile of the standard normal distribution, and $q_j = 1 - p_j$. Repeat this process for all B-B pairs.

5. For B-O and O-O combinations, implement an iterative procedure that finds the polychoric (pre-discretization) correlation given the ordinal phi coefficient (post-discretization correlation). Suppose $\mathbf{Z} = (Z_1, Z_2) \sim N(0, \Delta_{Z_1 Z_2})$, where $\mathbf{Z}$ denotes the bivariate standard normal distribution with correlation matrix $\Delta_{Z_1 Z_2}$ whose off-diagonal entry is $\delta_{Z_1 Z_2}$. Let $\mathbf{X} = (X_1, X_2)$ be the bivariate ordinal data where underlying $\mathbf{Z}$ is discretized based on corresponding normal quantiles given the marginal proportions, with a correlation matrix $\Delta_{Z_1 Z_2}$. If we need to sample from a random vector (X_1, X_2) whose marginal cdfs are F_1, F_2 tied together via a Gaussian copula, we generate a sample (z_1, z_2) from $\mathbf{Z} \sim N(0, \Delta_{Z_1 Z_2})$, then set $\mathbf{x} = (x_1, x_2) = (F_1^{-1}(u_1), F_2^{-1}(u_2))$ when $\mathbf{u} = (u_1, u_2) = (\Phi(z_1), \Phi(z_2))$, where Φ is the cdf of the standard normal distribution. The correlation matrix of $\mathbf{X}$, denoted by $\Delta_{X_1 X_2}$ (with an off-diagonal entry $\delta_{X_1 X_2}$) obviously differs from $\Delta_{Z_1 Z_2}$ due to discretization. More specifically, $|\delta_{X_1 X_2}| < |\delta_{Z_1 Z_2}|$ in large samples. The relationship between $\delta_{X_1 X_2}$ and $\delta_{Z_1 Z_2}$ can be established via the following algorithm (Ferrari and Barbiero 2012):

   a. Generate standard bivariate normal data with the correlation $\delta^0_{Z_1 Z_2}$ where $\delta^0_{Z_1 Z_2} = \delta_{X_1 X_2}$ (Here, $\delta^0_{Z_1 Z_2}$ is the initial polychoric correlation).
   b. Discretize Z_1 and Z_2, based on the cumulative probabilities of the marginal distribution F_1 and F_2, to obtain X_1 and X_2, respectively.
   c. Compute $\delta^1_{X_1 X_2}$ through X_1 and X_2 (Here, $\delta^1_{X_1 X_2}$ is the ordinal phi coefficient after the first iteration).
   d. Execute the following loop as long as $|\delta^v_{X_1 X_2} - \delta_{X_1 X_2}| > \varepsilon$ and $1 \leq v \leq v_{max}$ (v_{max} and ε are the maximum number of iterations and the maximum tolerated absolute error, respectively, both quantities are set by the users):
   (a) Update $\delta^v_{Z_1 Z_2}$ by $\delta^v_{Z_1 Z_2} = \delta^{v-1}_{Z_1 Z_2} g(v)$, where $g(v) = \delta_{X_1 X_2} / \delta^v_{X_1 X_2}$. Here, $g(v)$ serves as a correction coefficient, which ultimately converges to 1.
   (b) Generate bivariate normal data with $\delta^v_{Z_1 Z_2}$ and compute $\delta^{v+1}_{X_1 X_2}$ after discretization.

   Again, one should repeat this process for each B-O (and O-O) pair.

6. For C-C combinations, compute the corresponding normal-normal correlations (pre-mapping) given the specified count-count correlations (post-mapping) via the inverse cdf method in Yahav and Shmueli (2012) that was proposed in the context of correlated count data generation. Their method utilizes a slightly modified version of the NORTA (Normal to Anything) approach (Nelsen 2006), which involves generation of MVN variates with given univariate marginals and the correlation structure (R_N), and then transforming it into any desired distribution using the inverse cdf. In the Poisson case, NORTA can be implemented by the following steps:

a. Generate a k-dimensional normal vector $\boldsymbol{Z}_N$ from MVN distribution with mean vector $\mathbf{0}$ and a correlation matrix R_N.
b. Transform $\boldsymbol{Z}_N$ to a Poisson vector $\boldsymbol{X}_C$ as follows:
   i. For each element z_i of $\mathbf{Z}_N$, calculate the Normal cdf, $\Phi(z_i)$.
   ii. For each value of $\Phi(z_i)$, calculate the Poisson inverse cdf with a desired corresponding marginal rate λ_i, $\Psi_{\lambda_i}^{-1}(\Phi(z_i))$; where $\Psi_{\lambda_i}(x) = \sum_{i=0}^{x} \frac{e^{-\lambda}\lambda^i}{i!}$.
c. $\boldsymbol{X}_C = \left[\Psi_{\lambda_i}^{-1}(\Phi(z_i)), \ldots, \Psi_{\lambda_k}^{-1}(\Phi(z_k))\right]^T$ is a draw from the desired multivariate count data with correlation matrix R_{POIS}.

An exact theoretical connection between R_N and R_{POIS} has not been established to date. However, it has been shown that a feasible range of correlation between a pair of Poisson variables after the inverse cdf transformation is within $[\underline{\rho} = Cor(\Psi_{\lambda_i}^{-1}(U), \Psi_{\lambda_j}^{-1}(1-U)), \overline{\rho} = Cor(\Psi_{\lambda_i}^{-1}(U), \Psi_{\lambda_j}^{-1}(U))]$, where λ_i and λ_j are the marginal rates, and $U \sim Uniform(0, 1)$. Yahav and Shmueli (2012) proposed a conceptually simple method to approximate the relationship between the two correlations. They have demonstrated that R_{POIS} can be approximated as an exponential function of R_N where the coefficients are the functions of $\underline{\rho}$ and $\overline{\rho}$.

7. For B-N/O-N combinations, find the biserial/polyserial correlation (before discretization of one of the variables) given the point-biserial/point-polyserial correlation (after discretization) by the linearity and constancy arguments proposed by Demirtas and Hedeker (2016). Suppose that X and Y follow a bivariate normal distribution with a correlation of δ_{XY}. Without loss of generality, we may assume that both X and Y are standardized to have a mean of 0 and a variance of 1. Let X_D be the binary variable resulting from a split on X, $X_D = I(X \geq k)$. Thus, $E[X_D] = p$ and $V[X_D] = pq$ where $q = 1 - p$. The correlation between X_D and X, $\delta_{X_D X}$ can be obtained in a simple way, namely, $\delta_{X_D X} = \frac{Cov[X_D, X]}{\sqrt{V[X_D]V[X]}} = E[X_D X]/\sqrt{pq} = E[X|X \geq k]/\sqrt{pq}$. We can also express the relationship between X and Y via the following linear regression model:

$$Y = \delta_{XY} X + \varepsilon \tag{1}$$

where ε is independent of X and Y, and follows $N \sim (0, 1 - \delta_{XY}^2)$. When we generalize this to nonnormal X and/or Y (both centered and scaled), the same relationship can be assumed to hold with the exception that the distribution of ε follows a nonnormal distribution. As long as Eq. 1 is valid,

$$\begin{aligned} Cov[X_D, Y] &= Cov[X_D, \delta_{XY} X + \varepsilon] \\ &= Cov[X_D, \delta_{XY} X] + Cov[X_D, \varepsilon] \\ &= \delta_{XY} Cov[X_D, X] + Cov[X_D, \varepsilon] \,. \end{aligned} \tag{2}$$

Since ε is independent of X, it will also be independent of any deterministic function of X such as X_D, and thus $Cov[X_D, \varepsilon]$ will be 0. As $E[X] = E[Y] =$

0, $V[X] = V[Y] = 1$, $Cov[X_D, Y] = \delta_{X_D Y}\sqrt{pq}$ and $Cov[X, Y] = \delta_{XY}$, Eq. 2 reduces to

$$\delta_{X_D Y} = \delta_{XY}\delta_{X_D X}. \quad (3)$$

In the bivariate normal case, $\delta_{X_D X} = h/\sqrt{pq}$ where h is the ordinate of the normal curve at the point of dichotomization. Equation 3 indicates that the linear association between X_D and Y is assumed to be fully explained by their mutual association with X (Demirtas and Hedeker 2016). The ratio, $\delta_{X_D Y}/\delta_{XY}$ is equal to $\delta_{X_D X} = E[X_D X]/\sqrt{pq} = E[X|X \geq k]/\sqrt{pq}$. It is a constant given p and the distribution of (X, Y). These correlations are invariant to location shifts and scaling, X and Y do not have to be centered and scaled, their means and variances can take any finite values. Once the ratio ($\delta_{X_D X}$) is found, one can compute the biserial correlation when the point-biserial correlation is specified. When X is ordinalized to obtain X_O, the fundamental ideas remain unchanged. If the assumptions of Eqs. 1 and 3 are met, the method is equally applicable to the ordinal case in the context of the relationship between the polyserial (before ordinalization) and point-polyserial (after ordinalization) correlations. The easiest way of computing $\delta_{X_O X}$ is to generate X with a large number of data points, then ordinalize it to obtain X_O, and then compute the sample correlation between X_O and X. X could follow any continuous univariate distribution. However, here X is assumed to be a part of MVN data before discretization.

8. For C-N combinations, use the count version of Eq. 3, which is $\delta_{X_C Y} = \delta_{XY}\delta_{X_C X}$ is valid. The only difference is that we use the inverse cdf method rather than discretization via thresholds as in the binary and ordinal cases.
9. For B-C and O-C combinations, suppose that there are two identical standard normal variables, one underlies the binary/ordinal variable before discretization, the other underlies the count variable before inverse cdf matching. One can find $Cor(O, N)$ by the method of Demirtas and Hedeker (2016). Then, assume $Cor(C, O) = Cor(C, N) * Cor(O, N)$. $Cor(C, O)$ is specified and $Cor(O, N)$ is calculated. Solve for $Cor(C, N)$. Then, find the underlying N-N correlation by Step 8 above (Amatya and Demirtas 2015; Demirtas and Hedeker 2016).
10. Construct an overall, intermediate correlation matrix, Σ^* using the results from Steps 4 through 9, in conjunction with the N-N part that remains untouched when we compute Σ^* from Σ.
11. Check if Σ^* is positive definite. If it is not, find the nearest positive definite correlation matrix by the method of Higham (2002).
12. Generate multivariate normal data with a mean vector of (0, ..., 0) and correlation matrix of Σ^*, which can easily be done by using the Cholesky decomposition of Σ^* and a vector of univariate normal draws. The Cholesky decomposition of Σ^* produces a lower-triangular matrix A for which $AA^T = \Sigma^*$. If $z = (z_1, ..., z_d)$ are d independent standard normal random variables, then $Z = Az$ is a random draw from this distribution.
13. Dichotomize binary, ordinalize ordinal by respective quantiles, go from normal to count by inverse cdf matching.

3 Some Operational Details and an Illustrative Example

The software implementation of the algorithm has been done in **PoisBinOrdNor** package (Demirtas et al. 2016b) within R environment (R Development Core Team 2016). The package has functions for each variable-type pair that are collectively capable of modeling the correlation transitions. More specifically, `corr.nn4bb` function finds the tetrachoric correlation in Step 4, `corr.nn4bn` and `corr.nn4on` functions compute the biserial and polyserial correlations for binary and ordinal variables, respectively, in Step 7, `corr.nn4pbo` function is used to handle B-C and O-C pairs in Step 9, `corr.nn4pn` function is designed for C-N combinations in Step 8, and `corr.nn4pp` function calculates the pre-mapping correlations in Step 6. In addition, polychoric correlations in Step 5 are computed by `ordcont` function in **GenOrd** package (Barbiero and Ferrari 2015). correlation bound check (Step 2) as well as the validation of the specified quantities (Step 3), assembling all the intermediate correlation entries into Σ^* (Step 10), and generating mixed data (Steps 12 and 13) are performed by `validation.specs`, `intermat`, and `genPBONdata` functions, respectively, in **PoisBinOrdNor** package. Positive definiteness checks in Steps 1 and 11 are done by `is.positive.definite` function in **corpcor** package (Schaefer et al. 2015), finding the nearest Σ^* is implemented by `nearPD` function in **Matrix** package (Bates and Maechler 2016), and MVN data are generated by `rmvnorm` function in **mvtnorm** package (Genz et al. 2016).

For illustration, suppose we have two variables in each type. Operationally, **PoisBinOrdNor** package assumes that the variables are specified in a certain order. Let $Y_1 \sim Poisson(3)$, $Y_2 \sim Poisson(5)$, $Y_3 \sim Bernoulli(0.4)$, $Y_4 \sim Bernoulli(0.6)$, Y_5 and Y_6 are ordinal $P(Y_j = i) = p_i$, where $p_i = (0.3, 0.3, 0.4)$ and $(0.5, 0.1, 0.4)$ for $i = 0, 1, 2$ for $j = 5$ and 6, respectively, $Y_7 \sim N(2, 1)$, and $Y_8 \sim N(5, 9)$. The correlation matrix Σ is specified as follows under the assumption that columns (and rows) represent the order above:

$$\Sigma = \begin{bmatrix} 1 & 0.70 & 0.66 & 0.25 & 0.41 & 0.63 & 0.22 & 0.51 \\ 0.70 & 1 & 0.59 & 0.22 & 0.37 & 0.57 & 0.20 & 0.46 \\ 0.66 & 0.59 & 1 & 0.21 & 0.34 & 0.53 & 0.19 & 0.43 \\ 0.25 & 0.22 & 0.21 & 1 & 0.13 & 0.20 & 0.07 & 0.16 \\ 0.41 & 0.37 & 0.34 & 0.13 & 1 & 0.33 & 0.12 & 0.27 \\ 0.63 & 0.57 & 0.53 & 0.20 & 0.33 & 1 & 0.18 & 0.42 \\ 0.22 & 0.20 & 0.19 & 0.07 & 0.12 & 0.18 & 1 & 0.15 \\ 0.51 & 0.46 & 0.43 & 0.16 & 0.27 & 0.42 & 0.15 & 1 \end{bmatrix}$$

The intermediate correlation matrix Σ^*–after validating the feasibility of marginal and correlational specifications and applying all the relevant correlation transition steps– turns out to be (rounded to three digits after the decimal)

$$\Sigma^* = \begin{bmatrix} 1 & 0.720 & 0.857 & 0.325 & 0.477 & 0.776 & 0.226 & 0.523 \\ 0.720 & 1 & 0.757 & 0.282 & 0.424 & 0.693 & 0.202 & 0.466 \\ 0.857 & 0.757 & 1 & 0.336 & 0.470 & 0.741 & 0.241 & 0.545 \\ 0.325 & 0.282 & 0.336 & 1 & 0.186 & 0.299 & 0.089 & 0.203 \\ 0.477 & 0.424 & 0.470 & 0.186 & 1 & 0.438 & 0.135 & 0.305 \\ 0.776 & 0.693 & 0.741 & 0.299 & 0.438 & 1 & 0.216 & 0.504 \\ 0.226 & 0.202 & 0.241 & 0.089 & 0.135 & 0.216 & 1 & 0.150 \\ 0.523 & 0.466 & 0.545 & 0.203 & 0.305 & 0.504 & 0.150 & 1 \end{bmatrix}$$

Generating $N = 10,000$ rows of data based on this eight-variable system yields the following empirical correlation matrix (rounded to five digits after the decimal):

$$\begin{bmatrix} 1 & 0.69823 & 0.67277 & 0.24561 & 0.40985 & 0.63891 & 0.22537 & 0.50361 \\ 0.69823 & 1 & 0.59816 & 0.21041 & 0.36802 & 0.57839 & 0.21367 & 0.45772 \\ 0.67277 & 0.59816 & 1 & 0.20570 & 0.32448 & 0.55564 & 0.20343 & 0.42192 \\ 0.24561 & 0.21041 & 0.20570 & 1 & 0.12467 & 0.20304 & 0.06836 & 0.17047 \\ 0.40985 & 0.36802 & 0.32448 & 0.12467 & 1 & 0.32007 & 0.12397 & 0.26377 \\ 0.63891 & 0.57839 & 0.55564 & 0.20304 & 0.32007 & 1 & 0.17733 & 0.41562 \\ 0.22537 & 0.21367 & 0.20343 & 0.06836 & 0.12397 & 0.17733 & 1 & 0.15319 \\ 0.50361 & 0.45772 & 0.42192 & 0.17047 & 0.26377 & 0.41562 & 0.15319 & 1 \end{bmatrix}$$

The discrepancies between the specified and empirically computed correlations are indiscernibly small and the deviations are within an acceptable range that can be expected in any stochastic process. If we had repeated the experiment many times in a full-blown simulation study, the average differences would be even more negligible. We have observed the similar trends in the behavior of the marginal parameters (not reported for brevity), which lend further support to the presented methodology. The assessment of the algorithm performance in terms of commonly accepted accuracy and precision measures in RNG and imputation settings as well as in other simulated environments can be carried out through the evaluation metric developed in Demirtas (2004a, b, 2005, 2007a, b, 2008, 2009, 2010), Demirtas and Hedeker (2007, 2008a, b, c), Demirtas and Schafer (2003), Demirtas et al. (2007, 2008), and Yucel and Demirtas (2010).

4 Future Directions

The significance of the current study stems from three major reasons: First, data analysts, practitioners, theoreticians, and methodologists across many different disciplines in medical, managerial, social, biobehavioral, and physical sciences will be able to simulate multivariate data of mixed types with relative ease. Second, the proposed work can serve as a milestone for the development of more sophisticated simulation, computation, and data analysis techniques in the digital information, massive data era. Capability of generating many variables of different distributional types, nature, and dependence structures may be a contributing factor for better grasping the operational characteristics of today's intensive data trends (e.g., satellite data, internet traffic data, genetics data, ecological momentary assessment data). Third, these ideas can help to promote higher education and accordingly be instrumental in training graduate students. Overall, it will provide a comprehensive and useful set of computational tools whose generality and flexibility offer promising potential for building enhanced statistical computing infrastructure for research and education.

While this work represents a decent step forward in mixed data generation, it may not be sufficiently complex for real-life applications in the sense that real count and continuous data are typically more complicated than what Poisson and normal distributions accommodate, and it is likely that specification of parameters that control the first two moments and the second order product moment is inadequate. To address these concerns, we plan on building a more inclusive structural umbrella, whose ingredients are as follows: First, the continuous part will be extended to encompass nonnormal continuous variables by the operational utility of the third order power polynomials. This approach is a moment-matching procedure where any given continuous variable in the system is expressed by the sum of linear combinations of powers of a standard normal variate (Fleishman 1978; Vale and Maurelli 1983; Demirtas et al. 2012), which requires the specification of the first four moments. A more elaborate version in the form of the fifth order system will be implemented (Headrick 2010) in an attempt to control for higher order moments to cover a larger area in the skewness-elongation plane and to provide a better approximation to the probability density functions of the continuous variables; and the count data part will be augmented through the generalized Poisson distribution (Demirtas 2017b) that allows under- and over-dispersion, which is usually encountered in most applications, via an additional dispersion parameter. Second, although the Pearson correlation may not be the best association quantity in every situation, all correlations mentioned in this chapter are special cases of the Pearson correlation; it is the most widespread measure of association; and generality of the methods proposed herein with different kinds of variables requires the broadest possible framework. For further broadening the scale, scope, and applicability of the ideas presented in this chapter, the proposed RNG technique will be extended to allow the specification of the Spearman's rho, which is more popular for discrete and heavily skewed continuous distributions, will be incorporated into the algorithm for concurrently generating all four major types of variables. For the continuous-continuous pairs, the connection between the Pearson and Spearman correlations is given in Headrick (2010) through the power coefficients, and these two correlations are known to be equal for the binary-binary pairs. The relationship will be derived for all other variable type combinations. Inclusion of Spearman's rho as an option will allow us to specify nonlinear associations whose monotonic components are reflected in the rank correlation. Third, the expanded fifth order polynomial system will be further augmented to accommodate L-moments and L-correlations (Hosking 1990; Serfling and Xiao 2007) that are based on expectations of certain linear combinations of order statistics. The marginal and product L-moments are known to be more robust to outliers than their conventional counterparts in the sense that they suffer less from the effects of sampling variability, and they enable more secure inferences to be made from small samples about an underlying probability distribution. On a related note, further expansions can be designed to handle more complex associations that involve higher order product moments.

The salient advantages of the proposed algorithm and its augmented versions are as follows: (1) Individual components are well-established. (2) Given their computational simplicity, generality, and flexibility, these methods are likely to be widely used by researchers, methodologists, and practitioners in a wide spectrum of scientific disciplines, especially in the big data era. (3) They could be very useful in graduate-level teaching of statistics courses that involve computation and simulation, and in training graduate students. (4) A specific set of moments for each variable is fairly rare in practice, but a specific distribution that would lead to these moments is very common; so having access to these methods is needed by potentially a large group of people. (5) Simulated variables can be treated as outcomes or predictors in subsequent statistical analyses as the variables are being generated jointly. (6) Required quantities can either be specified or estimated from a real data set. (7) The final product after all these extensions will allow the specification of two prominent types of correlations (Pearson and Spearman correlations) and one emerging type (L-correlations) provided that they are within the limits imposed by marginal distributions. This makes it feasible to generate linear and a broad range of nonlinear associations. (8) The continuous part can include virtually any shape (skewness, low or high peakedness, mode at the boundary, multimodality, etc.) that is spanned by power polynomials; the count data part can be under- or over-dispersed. (9) Ability to jointly generate different types of data may facilitate comparisons among existing data analysis and computation methods in assessing the extent of conditions under which available methods work properly, and foster the development of new tools, especially in contexts where correlations play a significant role (e.g., longitudinal, clustered, and other multilevel settings). (10) The approaches presented here can be regarded as a variant of multivariate Gaussian copula-based methods as (a) the binary and ordinal variables are assumed to have a latent normal distribution before discretization; (b) the count variables go through a correlation mapping procedure via the normal-to-anything approach; and (c) the continuous variables consist of polynomial terms involving normals. To the best of our knowledge, existing multivariate copulas are not designed to have the generality of encompassing all these variable types simultaneously. (11) As the mixed data generation routine is involved with latent variables that are subsequently discretized, it should be possible to see how the correlation structure changes when some variables in a multivariate continuous setting are dichotomized/ordinalized (Demirtas 2016; Demirtas and Hedeker 2016; Demirtas et al. 2016a). An important by-product of this research will be a better understanding of the nature of discretization, which may have significant implications in interpreting the coefficients in regression-type models when some predictors are discretized. On a related note, this could be useful in meta-analysis when some studies discretize variables and some do not. (12) Availability of a general mixed data generation algorithm can markedly facilitate simulated power-sample size calculations for a broad range of statistical models.

References

Amatya, A., & Demirtas, H. (2015). Simultaneous generation of multivariate mixed data with Poisson and normal marginals. *Journal of Statistical Computation and Simulation*, *85*, 3129–3139.

Barbiero, A., & Ferrari, P. A. (2015). Simulation of ordinal and discrete variables with given correlation matrix and marginal distributions. R package GenOrd. https://cran.r-project.org/web/packages/GenOrd

Bates D., & Maechler M. (2016). Sparse and dense matrix classes and methods. R package Matrix. http://www.cran.r-project.org/web/packages/Matrix

Demirtas, H. (2004a). Simulation-driven inferences for multiply imputed longitudinal datasets. *Statistica Neerlandica*, *58*, 466–482.

Demirtas, H. (2004b). Assessment of relative improvement due to weights within generalized estimating equations framework for incomplete clinical trials data. *Journal of Biopharmaceutical Statistics*, *14*, 1085–1098.

Demirtas, H. (2005). Multiple imputation under Bayesianly smoothed pattern-mixture models for non-ignorable drop-out. *Statistics in Medicine*, *24*, 2345–2363.

Demirtas, H. (2006). A method for multivariate ordinal data generation given marginal distributions and correlations. *Journal of Statistical Computation and Simulation*, *76*, 1017–1025.

Demirtas, H. (2007a). Practical advice on how to impute continuous data when the ultimate interest centers on dichotomized outcomes through pre-specified thresholds. *Communications in Statistics-Simulation and Computation*, *36*, 871–889.

Demirtas, H. (2007b). The design of simulation studies in medical statistics. *Statistics in Medicine*, *26*, 3818–3821.

Demirtas, H. (2008). On imputing continuous data when the eventual interest pertains to ordinalized outcomes via threshold concept. *Computational Statistics and Data Analysis*, *52*, 2261–2271.

Demirtas, H. (2009). Rounding strategies for multiply imputed binary data. *Biometrical Journal*, *51*, 677–688.

Demirtas, H. (2010). A distance-based rounding strategy for post-imputation ordinal data. *Journal of Applied Statistics*, *37*, 489–500.

Demirtas, H. (2016). A note on the relationship between the phi coefficient and the tetrachoric correlation under nonnormal underlying distributions. *American Statistician*, *70*, 143–148.

Demirtas, H. (2017a). Concurrent generation of binary and nonnormal continuous data through fifth order power polynomials. *Communications in Statistics–Simulation and Computation, 46,* 489–357.

Demirtas, H. (2017b). On accurate and precise generation of generalized Poisson variates. *Communications in Statistics–Simulation and Computation, 46,* 489–499.

Demirtas, H., Ahmadian, R., Atis, S., Can, F. E., & Ercan, I. (2016a). A nonnormal look at polychoric correlations: Modeling the change in correlations before and after discretization. *Computational Statistics*, *31*, 1385–1401.

Demirtas, H., Arguelles, L. M., Chung, H., & Hedeker, D. (2007). On the performance of bias-reduction techniques for variance estimation in approximate Bayesian bootstrap imputation. *Computational Statistics and Data Analysis*, *51*, 4064–4068.

Demirtas, H., & Doganay, B. (2012). Simultaneous generation of binary and normal data with specified marginal and association structures. *Journal of Biopharmaceutical Statistics*, *22*, 223–236.

Demirtas, H., Freels, S. A., & Yucel, R. M. (2008). Plausibility of multivariate normality assumption when multiply imputing non-Gaussian continuous outcomes: A simulation assessment. *Journal of Statistical Computation and Simulation*, *78*, 69–84.

Demirtas, H., & Hedeker, D. (2007). Gaussianization-based quasi-imputation and expansion strategies for incomplete correlated binary responses. *Statistics in Medicine*, *26*, 782–799.

Demirtas, H., & Hedeker, D. (2008a). Multiple imputation under power polynomials. *Communications in Statistics- Simulation and Computation*, *37*, 1682–1695.

Demirtas, H., & Hedeker, D. (2008b). Imputing continuous data under some non-Gaussian distributions. *Statistica Neerlandica*, *62*, 193–205.

Demirtas, H., & Hedeker, D. (2008c). An imputation strategy for incomplete longitudinal ordinal data. *Statistics in Medicine*, *27*, 4086–4093.

Demirtas, H., & Hedeker, D. (2011). A practical way for computing approximate lower and upper correlation bounds. *The American Statistician*, *65*, 104–109.

Demirtas, H., & Hedeker, D. (2016). Computing the point-biserial correlation under any underlying continuous distribution. *Communications in Statistics- Simulation and Computation*, *45*, 2744–2751.

Demirtas, H., Hedeker, D., & Mermelstein, J. M. (2012). Simulation of massive public health data by power polynomials. *Statistics in Medicine*, *31*, 3337–3346.

Demirtas H., Hu Y., & Allozi R. (2016b). *Data generation with Poisson, binary, ordinal and normal components*, R package PoisBinOrdNor. https://cran.r-project.org/web/packages/PoisBinOrdNor.

Demirtas, H., & Schafer, J. L. (2003). On the performance of random-coefficient pattern-mixture models for non-ignorable drop-out. *Statistics in Medicine*, *22*, 2553–2575.

Demirtas, H., & Yavuz, Y. (2015). Concurrent generation of ordinal and normal data. *Journal of Biopharmaceutical Statistics*, *25*, 635–650.

Emrich, J. L., & Piedmonte, M. R. (1991). A method for generating high-dimensional multivariate binary variates. *The American Statistician*, *45*, 302–304.

Ferrari, P. A., & Barbiero, A. (2012). Simulating ordinal data. *Multivariate Behavioral Research*, *47*, 566–589.

Fleishman, A. I. (1978). A method for simulating non-normal distributions. *Psychometrika*, *43*, 521–532.

Fréchet, M. (1951). Sur les tableaux de corrélation dont les marges sont données. *Annales de l'Université de Lyon Section A*, *14*, 53–77.

Genz, A., Bretz, F., Miwa, T., Mi, X., Leisch, F., Scheipl, F., et al. (2016). Multivariate normal and t distributions. R package mvtnorm. https://cran.r-project.org/web/packages/mvtnorm.

Headrick, T. C. (2010). *Statistical simulation: power method polynomials and other transformations boca raton*. FL: Chapman and Hall/CRC.

Higham, N. J. (2002). Computing the nearest correlation matrix—a problem from finance. *IMA Journal of Numerical Analysis*, *22*, 329–343.

Hoeffding, W. (1994). Scale-invariant correlation theory. In: N.I. Fisher & P.K. Sen (Eds.), T*he collected works of Wassily Hoeffding (the original publication year is 1940)* (pp. 57–107). New York: Springer.

Hosking, J. R. M. (1990). L-moments: Analysis and estimation of distributions using linear combinations of order statistics. *Journal of the Royal Statistical Society, Series B*, *52*, 105–124.

Nelsen, R. B. (2006). *An introduction to copulas*. Berlin, Germany: Springer.

R Development Core Team (2016) *R: A Language and Environment for Statistical Computing*. http://www.cran.r-project.org.

Schaefer, J., Opgen-Rhein, R., Zuber, V., Ahdesmaki, M., Silva, A. D., Strimmer, K. (2015). *Efficient Estimation of Covariance and (Partial) Correlation*. R package corpcor. https://cran.r-project.org/web/packages/BinNonNor.

Serfling, R., & Xiao, P. (2007). A contribution to multivariate L-moments: L-comoment matrices. *Journal of Multivariate Analysis*, *98*, 1765–1781.

Vale, C. D., & Maurelli, V. A. (1983). Simulating multivariate nonnormal distributions. *Psychometrika*, *48*, 465–471.

Yahav, I., & Shmueli, G. (2012). On generating multivariate Poisson data in management science applications. *Applied Stochastic Models in Business and Industry*, *28*, 91–102.

Yucel, R. M., & Demirtas, H. (2010). Impact of non-normal random effects on inference by multiple imputation: A simulation assessment. *Computational Statistics and Data Analysis*, *54*, 790–801.

Improving the Efficiency of the Monte-Carlo Methods Using Ranked Simulated Approach

Hani Michel Samawi

Abstract This chapter explores the concept of using ranked simulated sampling approach (RSIS) to improve the well-known Monte-Carlo methods, introduced by Samawi (1999), and extended to steady-state ranked simulated sampling (SRSIS) by Al-Saleh and Samawi (2000). Both simulation sampling approaches are then extended to multivariate ranked simulated sampling (MVRSIS) and multivariate steady-state ranked simulated sampling approach (MVSRSIS) by Samawi and Al-Saleh (2007) and Samawi and Vogel (2013). These approaches have been demonstrated as providing unbiased estimators and improving the performance of some of the Monte-Carlo methods of single and multiple integrals approximation. Additionally, the MVSRSIS approach has been shown to improve the performance and efficiency of Gibbs sampling (Samawi et al. 2012). Samawi and colleagues showed that their approach resulted in a large savings in cost and time needed to attain a specified level of accuracy.

1 Introduction

The term Monte-Carlo refers to techniques that use random processes to approximate a non-stochastic k-dimensional integral of the form

$$\theta = \int_{R^k} g(\underline{u})d\underline{u}, \tag{1.1}$$

(Hammersley and Handscomb 1964).

The literature presents many approximation techniques, including Monte-Carlo methods. However, as the dimension of the integrals rises, the difficulty of the integration problem increases even for relatively low dimensions (see Evans and Swartz 1995). Given such complications, many researchers are confused about which method

H.M. Samawi (✉)
Department of Biostatistics, Jiann-Ping Hsu College Public Health,
Georgia Southern University, 30460 Statesboro, Georgia
e-mail: hsamawi@georgiasouthern.edu

D.-G. Chen and J.D. Chen (eds.), *Monte-Carlo Simulation-Based Statistical Modeling*, ICSA Book Series in Statistics, DOI 10.1007/978-981-10-3307-0_2

to use; however, the advantages and disadvantages of each method are not the primary concern of this chapter. The focus of this chapter is the use of Monte-Carlo methods in multiple integration approximation.

The motivation for this research is based on the concepts of ranked set sampling (RSS), introduced by McIntyre (1952). The motivation is based on the fact that the ith quantified unit of RSS is simply an observation from $f_{(i)}$, where $f_{(i)}$ is the density function of the ith order statistic of a random sample of size n. When the underlying density is the uniform distribution on (0, 1), $f_{(i)}$ follows a beta distribution with parameters $(i, n - i + 1)$.

Samawi (1999) was the first to explore the idea of RSS (Beta sampler) for integral approximation. He demonstrated that the procedure can improve the simulation efficiency based on the ratio of the variances. Samawi's ranked simulated sampling procedure RSIS generates an independent random sample $U_{(1)}, U_{(2)}, \ldots, U_{(n)}$, which is denoted by RSIS, where $U_{(i)} \sim \beta(i, n - i + 1), \{i = 1, 2, ..., n\}$ and $\beta(., .)$ denotes the beta distribution. The RSIS procedure constitutes an RSS based on random samples from the uniform distribution $U(0, 1)$. The idea is to use this RSIS to compute (1.1) with $k = 1$, instead of using an SRS of size n from $U(0, 1)$, when the range of the integral in (1.1) is (0, 1). In case of arbitrary range (a, b) of the integral in (1.1), Samawi (1999) used the sample: $X_{(1)}, X_{(2)}, ..., X_{(n)}$ and the importance sampling technique to evaluate (1.1), where $X_{(i)} = F_X^{-1}(U_{(i)})$ and $F_X(.)$ is the distribution function of a continuous random variable. He showed theoretically and through simulation studies that using the RSIS sampler for evaluating (1.1) substantially improved the efficiency when compared with the traditional uniform sampler (USS).

Al-Saleh and Zheng (2002) introduced the idea of bivariate ranked set sampling (BVRSS) and showed through theory and simulation that BVRSS outperforms the bivariate simple random sample for estimating the population means. The BVRSS is as follows:
Suppose (X, Y) is a bivariate random vector with the joint probability density function $f_{X,Y}(x, y)$. Then,

1. A random sample of size n^4 is identified from the population and randomly allocated into n^2 pools each of size n^2 so that each pool is a square matrix with n rows and n columns.
2. In the first pool, identify the minimum value by judgment with respect to the first characteristic X, for each of the n rows.
3. For the n minima obtained in Step 2, the actual quantification is done on the pair that corresponds to the minimum value of the second characteristic, Y, identified by judgment. This pair, given the label (1, 1), is the first element of the BVRSS sample.
4. Repeat Steps 2 and 3 for the second pool, but in Step 3, the pair corresponding to the second minimum value with respect to the second characteristic, Y, is chosen for actual quantification. This pair is given the label (1, 2).
5. The process continues until the label (n, n) is ascertained from the n^2th (last) pool.

The procedure described above produces a BVRSS of size n^2. Let $[(X_{[i](j)}, Y_{(i)[j]}), i = 1, 2, \ldots, n \; and \; j = 1, 2, \ldots, n]$ denote the BVRSS sample from $f_{X,Y}(x, y)$ where $f_{X_{[i](j)}, Y_{(i)[j]}}(x, y)$ is the joint probability density function of $(X_{[i](j)} Y_{(i)[j]})$. From Al-Saleh and Zheng (2002),

$$f_{X_{[i](j)}, Y_{(i)[j]}}(x, y) = f_{Y_{(i)[j]}}(y) \frac{f_{X_{(j)}}(x) f_{Y|X}(y|x)}{f_{Y_{[j]}}(y)}, \tag{1.2}$$

where $f_{X_{(j)}}$ is the density of the jth order statistic for an SRS sample of size n from the marginal density of f_X and $f_{Y_{[j]}}(y)$ be the density of the corresponding $Y-$ value given by $f_{Y_{[j]}}(y) = \int_{-\infty}^{\infty} f_{X_{(j)}}(x) f_{Y|X}(y|x)dx$, while $f_{Y_{(i)[j]}}(y)$ is the density of the ith order statistic of an iid sample from $f_{Y_{[j]}}(y)$, i.e.

$$f_{Y_{(i)[j]}}(y) = \text{c.}(\text{F}_{Y_{[j]}}(y))^{i-1}(1 - \text{F}_{Y_{[j]}}(y))^{n-i} f_{Y_{[j]}}(y)$$

where $F_{Y_{[j]}}(y) = \int_{-\infty}^{y} (\int_{-\infty}^{\infty} f_{X_{(j)}}(x) f_{Y|X}(w|x)dx)dw.$

Combining these results, Eq. (1.2) can be written as

$$f_{X_{[i](j)}, Y_{(i)[j]}}(x, y) = c_1 (F_{Y_{[j]}}(y))^{i-1}(1 - F_{Y_{[j]}}(y))^{n-i}(F_X(x))^{j-1}(1 - F_X(x))^{n-j} f(x, y) \tag{1.3}$$

where

$$c_1 = \left(\frac{n!}{(i-1)!(n-i)!}) (\frac{n!}{(j-1)!(n-j)!} \right).$$

Furthermore, Al-Saleh and Zheng (2002) showed that,

$$\frac{1}{n^2} \sum_{j}^{n} \sum_{i}^{n} f_{X_{[i](j)}, Y_{(i)[j]}}(x, y) = f(x, y). \tag{1.4}$$

For a variety of choices of $f(u, v)$, one can have (U, V) bivariate uniform with a probability density function *f(u, v); 0 < u, v < 1*, such that $U \sim U(0, 1)$ and $V \sim U(0, 1)$ (See Johnson 1987). In that case, $[(U_{[i](j)}, V_{(i)[j]}), i = 1, 2, \ldots, n$ and $j = 1, 2, \ldots, n]$ should have a bivariate probability density function given by

$$f_{(j),(i)}(u, v) = \left[\frac{n!}{(i-1)!(n-i)!} \right] \left[\frac{n!}{(j-1)!(n-j)!} \right] [F_{Y_{[j]}}(v)]^{i-1} [1 - F_{Y_{[j]}}(v)]^{n-i} [u]^{j-1} \\ [1-u]^{n-j} f(u, v). \tag{1.5}$$

Samawi and Al-Saleh (2007) extended the work of Samawi (1999) and Al-Saleh and Zheng (2002) for the Monte-Carlo multiple integration approximation of (1.1) when k = 2.

Moreover, to further improve some of the Monte-Carlo methods of integration, Al-Saleh and Samawi (2000) used steady-state ranked set simulated sampling (SRSIS) as introduced by Al-Saleh and Al-Omari (1999). SRSIS has been shown to be simpler and more efficient than Samawi's (1999) method.

In Samawi and Vogel (2013) work, the SRSIS algorithm introduced by Al-Saleh and Samawi (2000) was extended to multivariate case for the approximation of multiple integrals using Monte-Carlo methods. However, to simplify the algorithms, we introduce only the bivariate integration problem; with this foundation, multiple integral problems are a simple extension.

2 Steady-State Ranked Simulated Sampling (SRSIS)

Al-Saleh and Al-Omari (1999) introduced the idea of multistage ranked set sampling (MRSS). To promote the use of MRSS in simulation and Monte-Carlo methods, let $\{X_i^{(s)}; i = 1, 2, \ldots, n, \}$ be an MRSS of size n at stage s. Assume that $X_i^{(s)}$ has probability density function $f_i^{(s)}$ and a cumulative distribution function $F_i^{(s)}$. Al-Saleh and Al-Omeri demonstrated the following properties of MRSS:

1.

$$f(x) = \frac{1}{n}\sum_{i=1}^{n} f_i^{(s)}(x), \tag{2.1}$$

2.

$$\text{If } s \to \infty, \text{ then } \mathrm{F}_i^{(s)}(x) \to \mathrm{F}_i^{(\infty)}(x) = \begin{cases} 0 & if \quad x < Q_{(i-1)/n} \\ nF(x) - (i-1) & if \;\; Q_{(i-1)/n} \le x < Q_{(i)/n} \\ 1 & if \quad x \ge Q_{(i)/n} \end{cases}, \tag{2.2}$$

for $i = 1, 2, \ldots, n$, where Q_α is the 100αth percentile of $F(x)$.

3. If $X \sim U(0, 1)$, then for $i = 1, 2, \ldots, n$, we have

$$F_i^{(\infty)}(x) = \begin{cases} 0 & if \quad x < (i-1)/n \\ nx - (i-1) & if \;\; (i-1)/n \le x < i/n \\ 1 & if \quad x \ge i/n \end{cases}, \tag{2.3}$$

and

$$f_i^{(\infty)}(x) = \begin{cases} n \; if \;\; (i-1)/n \le x < i/n \\ 0 \qquad otherwise. \end{cases} \tag{2.4}$$

These properties imply $\mathrm{X}_i^{(\infty)} \sim U(\frac{i-1}{n}, \frac{i}{n})$, when the underlying distribution function is U(0, 1).

Samawi and Vogel (2013) provided a modification of the Al-Saleh and Samawi (2000) steady-state ranked simulated samples procedure (SRSIS) to bivariate cases (BVSRSIS) as follows:

1. For each (i, j), $j = 1, 2, \ldots, n$ and $i = 1, 2, \ldots, n$ generate independently

   a. ($U_{i(j)}$from U $\left(\frac{j-1}{n}, \frac{j}{n}\right)$ and independent $W_{(i)j}$ from U $\left(\frac{i-1}{n}, \frac{i}{n}\right)$, $i = 1$, 2, ..., n).

2. Generate $Y_{(i)j} = F_Y^{-1}(W_{(i)j})$ and $X_{i(j)} = F_X^{-1}(U_{i(j)})$ from $F_Y(y)$ and $F_X(x)$ respectively.
3. To generate $(X_{[i](j)}, Y_{(i)[j]})$ from $f(x, y)$, generate $U'_{i(j)}$from U $\left(\frac{j-1}{n}, \frac{j}{n}\right)$ and independent $W'_{(i)j}$ from U $\left(\frac{i-1}{n}, \frac{i}{n}\right)$, then

$$X_{[i](j)}|Y_{(i)j} = F^{-1}_{X|Y}(U'_{i(j)}|Y_{(i)j}) \text{ and } Y_{(i)[j]}|X_{i(j)} = F^{-1}_{Y|X}(W'_{(i)j}|X_{i(j)}).$$

The joint density function of $(X_{[i](j)}, Y_{(i)[j]})$ is formed as follows:

$$f^{(\infty)}_{X_{[i](j)}Y_{(i)[j]}}(x, y) = f^{(\infty)}_{X_{[i](j)}}(x) f^{(\infty)}_{Yi|X_{[i](j)}}(y|X_{[i](j)}) = n^2 f_X(x) f_{Y|X_{[i](j)}}(y|X_{[i](j)}),$$
$$Q_{X(j-1)/n} \le x < Q_{X(j)/n}, Q_{Y(i-1)/n} \le y < Q_{Y(i)/n},$$

where $Q_{X(s)}$ and $Q_{Y(v)}$ are the 100 sth percentile of $F_X(x)$ and 100 vth percentile of $F_Y(y)$, respectively. However, for the first stage, both Stokes (1977) and David (1981) showed that $F_{Y|X_{[i](j)}}(y|x) = F_{Y|X}(y|x)$. Al-Saleh and Zheng (2003) demonstrated that joint density is valid for an arbitrary stage, and therefore, valid for a steady state. Therefore,

$$f^{(\infty)}_{X_{[i](j)}Y_{(i)[j]}}(x, y) = f^{(\infty)}_{X_{[i](j)}}(x) f^{(\infty)}_{Yi|X_{[i](j)}}(y|X_{[i](j)}) = n^2 f_X(x) f_{Y|X}(y|x) = n^2 f_{Y,X}(x, y),$$
$$Q_{X(j-1)/n} \le x < Q_{X(j)/n}, Q_{Y(i-1)/n} \le y < Q_{Y(i)/n}. \tag{2.5}$$

Thus, we can write:

$$\frac{1}{n^2}\sum_{i=1}^{n}\sum_{j=1}^{n} f^{(\infty)}_{X_{[i](j)}Y_{(i)[j]}}(x, y) = \sum_{i=1}^{n}\sum_{j=1}^{n} f_{Y,X}(x, y).I[Q_{X(j-1)/n} \le x < Q_{X(j)/n}]I[Q_{Y(i-1)/n} \le y < Q_{Y(i)/n}]$$
$$= f(x, y), \tag{2.6}$$

where I is an indicator variable. Similarly, Eq. (2.5) can be extended by mathematical induction to the multivariate case as follows:

$f^{(\infty)}(x_1, x_2, ..., x_k) = n^k f(x_1, x_2, ..., x_k)$, $Q_{X_i(j-1)/n} \le x_i < Q_{X_i(j)/n}$, $i = 1, \ldots, k$ and $j = 1, 2, .., n$.. In addition, the above algorithm can be extended for k > 2 as follows:

1. For each $(i_l, l = 1, 2, ..., k)$, $i_l = 1, 2, ..., n$ generate independently

$$U_{i_l(i_s)}\text{from U}\left(\frac{i_s - 1}{n}, \frac{i_s}{n}\right), l, s = 1, 2, \ldots, k \text{ and } i_l, i_s = 1, 2, \ldots, n.$$

2. Generate $X_{i_l(i_s)} = F_{X_{i_l}}^{-1}(U_{i_l(i_s)})l, s = 1, 2, \ldots, k$ and $i_l, i_s = 1, 2, \ldots, n,$ from $F_{X_{i_l}}(x), l = 1, 2, \ldots, k$, respectively.
3. Then, generate the multivariate version of the steady-state simulated sample by using any technique for conditional random number generation.

3 Monte-Carlo Methods for Multiple Integration Problems

Very good descriptions of the basics of the various Monte-Carlo methods have been provided by Hammersley and Handscomb (1964), Liu (2001), Morgan (1984), Robert and Casella (2004), and Shreider (1966). The Monte-Carlo methods described include crude, antithetic, importance, control variate, and stratified sampling approaches. However, when variables are related, Monte-Carlo methods cannot be used directly (i.e., similar to the manner that these methods are used in univariate integration problems) because using the bivariate uniform probability density function $f(u, v)$ as a sampler to evaluate Eq. (1.1) with $k = 2$, $f(u, v)$ is not consistent. However, in this context it is reasonable to use the importance sampling method, and therefore, it follows that other Monte-Carlo techniques can be used in conjunction with importance sampling. Thus, our primary concern is importance sampling.

3.1 *Importance Sampling Method*

In general, suppose that f is a density function on R^k such that the closure of the set of points where $g(.)$ is non-zero and the closure set of points where $f(.)$is non-zero. Let $[\underline{U}_i\ i = 1, 2, ..., n]$ be a sample from $f(.)$. Then, because

$$\theta = \int \frac{g(\underline{u})}{f(\underline{u})} f(\underline{u})d\underline{u},$$

Equation (1.1) can be estimated by

$$\hat{\theta} = \frac{1}{n} \sum_{i=1}^{n} \frac{g(\underline{u}_i)}{f(\underline{u}_i)} \quad . \tag{3.1}$$

Equation (3.1) is an unbiased estimator for (1.1), with variance given by

$$Var(\hat{\theta}) = \frac{1}{n}(\int_{R^k} \frac{g(\underline{u})^2}{f(\underline{u})} d\underline{u} - \theta^2).$$

In addition, from the point of view of the strong law of large numbers, it is clear that $\hat{\theta} \to \theta$ almost surely as $n \to \infty$.

A limited number of distributional families exist in a multidimensional context and are commonly used as importance samplers. For example, the multivariate Student's family is used extensively in the literature as an importance sampler. Evans and Swartz (1995) indicated a need for developing families of multivariate distribution that exhibit a wide variety of shapes. In addition, statisticians want distributional families to have efficient algorithms for random variable generation and the capacity to be easily fitted to a specific integrand.

This paper provides a new way of generating a bivariate sample based on the bivariate steady-state sampling (BVSRSIS) that has the potential to extend the existing sampling methods. We also provide a means for introducing new samplers and to substantially improve substantially the efficiency of the integration approximation based on those samplers.

3.2 Using Bivariate Steady-State Sampling (BVSRSIS)

Let

$$\theta = \int g(x, y)dx\, dy. \tag{3.2}$$

To estimate θ, generate a bivariate sample of size n^2 from *f(x, y)*, which mimics $g(x, y)$ and has the same range, such as $[(X_{ij}, Y_{ij}), i = 1, 2, ..., n\, and\, j = 1, 2, ..., n]$. Then

$$\hat{\theta} = \frac{1}{n^2}\sum_{i=1}^{n}\sum_{j=1}^{n}\frac{g(x_{ij}, y_{ij})}{f(x_{ij}, y_{ij})}. \tag{3.3}$$

Equation (3.3) is an unbiased estimate for (3.2) with variance

$$Var(\hat{\theta}) = \frac{1}{n^2}(\int\int\frac{g^2(x, y)}{f(x, y)}dx\, dy - \theta^2). \tag{3.4}$$

To estimate (3.2) using BVSRSIS, generate a bivariate sample of size n^2, as described in above, say $[(X_{[i](j)}, Y_{(i)[j]}), i = 1, 2, ..., n\, and\, j = 1, 2, ..., n]$. Then

$$\hat{\theta}_{BVSRSIS} = \frac{1}{n^2}\sum_{i=1}^{n}\sum_{j=1}^{n}\frac{g(x_{[i](j)}, y_{(i)[j]})}{f(x_{[i](j)}, y_{(i)[j]})}. \tag{3.5}$$

Equation (3.5) is also an unbiased estimate for (3.2) using (2.5). Also, by using (2.5) the variance of (3.5) can be expressed as

$$Var(\hat{\theta}_{BVSRSIS}) = Var(\hat{\theta}) - \frac{1}{n^4} \sum_{i=1}^{n} \sum_{j=1}^{n} (\theta_{g/f}^{(i,j)} - \theta_{g/f})^2, \tag{3.6}$$

where, $\theta_{g/f}^{(i,j)} = E[g(X_{[i](j)}, Y_{(i)[j]})/f(X_{[i](j)}, Y_{(i)[j]})], \theta_{g/f} = E[g(X,Y)/f(X,Y)] = \theta$. The variance of the estimator in (3.6) is less than the variance of the estimator in (3.4).

3.3 Simulation Study

This section presents the results of a simulation study that compares the performance of the importance sampling method described above using BVSRSIS schemes with the performance of the bivariate simple random sample (BVUSS) and BVRSS schemes by Samawi and Al-Saleh (2007) as introduced by Samawi and Vogel (2013).

3.3.1 Illustration for Importance Sampling Method When Integral's Limits Are (0, 1)x(0, 1)

As in Samawi and Al-Saleh (2007), illustration of the impact of BVSRSIS on importance sampling is provided by evaluating the following integral

$$\theta = \int_0^1 \int_0^1 (1+v).\exp(u(1+v))\, du\, dv = 3.671. \tag{3.7}$$

This example uses four bivariate sample sizes: n= 20, 30, 40 and 50. To estimate the variances using the simulation method, we use 2,000 simulated samples from BVUSS and BVSRSIS. Many choices of bivariate and multivariate distributions with uniform marginal on [0, 1] are available (Johnson 1987). However, for this simulation, we chose Plackett's uniform distribution (Plackett 1965), which is given by

$$f(u,v) = \frac{\psi\{(\psi-1)(u+v-2\,uv)+1\}}{\{[1+(u+v)(\psi-1)]^2 - 4\psi(\psi-1)uv\}^{3/2}},\quad 0 < u,\ v < 1,\ \psi > 0. \tag{3.8}$$

The parameter ψ governs the dependence between the components (U, V) distributed according to f. Three cases explicitly indicate the role of ψ (Johnson 1987):

$$\begin{array}{ll} \psi \to 0 & U = 1 - V, \\ \psi = 1 & U \text{ and } V \text{ are independent}, \\ \psi \to \infty & U = V, \end{array}$$

Table 1 Efficiency of estimating (3.7) using BVSRSIS relative to BVUSS and BVRSS

$n\backslash\psi$	1	2
20	289.92 (**8.28**)	273.68 (**9.71**)
30	649.00 (**12.94**)	631.31 (**13.06**)
40	1165.31 (**16.91**)	1086.46 (**18.60**)
50	1725.25 (**21.67**)	1687.72 (**23.03**)

Note Values shown in *bold* were extracted from Samawi and Al-Saleh (2007)

Table 1 presents the relative efficiencies of our estimators using BVRSIS in comparison with using BVUSS and BVSRSIS relative to BVUSS for estimating (3.7).

As illustrated in Table 1, BVSRSIS is clearly more efficient than either BVUSS or BVRSIS when used for estimation.

3.3.2 Illustration When the Integral's Limits Are Arbitrary Subset of R^2

Recent work by Samawi and Al-Saleh (2007) and Samawi and Vogel (2013) used an identical example in which the range of the integral was not (0, 1), and the authors evaluated the bivariate normal distribution (e.g., $g(x, y)$ is the N_2 *(0, 0, 1, 1, ρ)* density.) For integrations with high dimensions and a requirement of low relative error, the evaluation of the multivariate normal distribution function remains one of the unsolved problems in simulation (e.g., Evans and Swartz 1995). To demonstrate how BVSRSIS increases the precision of evaluating the multivariate normal distribution, we illustrate the method by evaluating the bivariate normal distribution as follows:

$$\theta = \int_{-\infty}^{z_1}\int_{-\infty}^{z_2} g(x, y)\, dx\, dy, \tag{3.9}$$

where $g(x, y)$ is the $N_2(0,\ 0,\ 1,\ 1,\ \rho)$ density.

Given the similar shapes of the marginal of the normal and the marginal of the logistic probability density functions, it is natural to attempt to approximate the bivariate normal cumulative distribution function by the bivariate logistic cumulative distribution function. For the multivariate logistic distribution and its properties, see Johnson and Kotz (1972). The density of the bivariate logistic (Johnson and Kotz 1972) is chosen to be

$$f(x, y) = \frac{2!\,\pi^2 e^{-\pi(x+y)/\sqrt{3}}(1 + e^{-\pi z_1/\sqrt{3}} + e^{-\pi z_2/\sqrt{3}})}{3(1 + e^{-\pi x/\sqrt{3}} + e^{-\pi y/\sqrt{3}})^3}, \quad -\infty < x < z_1;\ -\infty < y < z_2. \tag{3.10}$$

It can be shown that the marginal of X is given by

$$f(x) = \frac{\pi e^{-\pi x/\sqrt{3}}(1 + e^{-\pi z_1/\sqrt{3}} + e^{-\pi z_2/\sqrt{3}})}{\sqrt{3}\,(1 + e^{-\pi x/\sqrt{3}} + e^{-\pi z_2/\sqrt{3}})^2}, \quad -\infty < x < z_1. \tag{3.11}$$

Now let $W = Y + \frac{\sqrt{3}}{\pi}\ln(1 + e^{-\pi X/\sqrt{3}} + e^{-\pi z_2/\sqrt{3}})$. Then it can be shown that

$$f(w|x) = \frac{2\,\pi\; e^{-\pi w/\sqrt{3}}}{\sqrt{3}\left(\frac{1+e^{-\pi x/\sqrt{3}}}{1+e^{-\pi x/\sqrt{3}}+e^{-\pi z_2/\sqrt{3}}} + e^{-\pi w/\sqrt{3}}\right)^3}, \tag{3.12}$$

$$-\infty < w < z_2 + \frac{\sqrt{3}}{\pi}\ln\left(1 + e^{-\pi x/\sqrt{3}} + e^{-\pi z_2/\sqrt{3}}\right).$$

To generate from (3.10) proceed as follows:

1. Generate X from (3.11).
2. Generate W independently from (3.12)
3. Let $Y = W - \frac{\sqrt{3}}{\pi}\ln(1 + e^{-\pi X/\sqrt{3}} + e^{-\pi z_2/\sqrt{3}})$.
4. Then the resulting pair (X, Y) has the correct probability density function, as defined in (3.10).

For this illustration, two bivariate sample sizes, $n = 20$ and 40, and different values of ρ and (z_1, z_2) are used. To estimate the variances using simulation, we use 2,000 simulated samples from BVUSS, BVRSIS, and BVSRSIS (Tables 2 and 3).

Notably, when Samawi and Vogel (2013) used identical examples to those used by Samawi and Al-Saleh (2007), a comparison of the simulations showed that Samawi and Vogel (2013) BVSRSIS approach improved the efficiency of estimating the multiple integrals by a factor ranging from 2 to 100.

As expected, the results of the simulation indicated that using BVSRSIS substantially improved the performance of the importance sampling method for integration

Table 2 Efficiency of using BVSRSIS to estimate Eq. (3.9) relative to BVUSS

(z_1, z_2)	$n = 20$				$n = 40$			
	$\rho = \pm 0.20$	$\rho = \pm 0.50$	$\rho = \pm 0.80$	$\rho = \pm 0.95$	$\rho = \pm 0.20$	$\rho = \pm 0.50$	$\rho = \pm 0.80$	$\rho = \pm 0.95$
(0, 0)	5.39 **(6.42)**	9.89 **(21.70)**	5.98 **(144.31)**	49.65 **(171.22)**	9.26 **(12.2)**	22.50 **(63.74)**	15.80 **(526.10)**	158.77 **(612.50)**
(−1, −1)	22.73 **(75.82)**	29.43 **(182.27)**	22.48 **(87.42)**	95.10 **(23.71)**	55.99 **(255.75)**	90.24 **(688.59)**	69.85 **(336.58)**	380.87 **(100.40)**
(−2, −2)	148.30 **(200.46)**	133.21 **(143.62)**	125.37 **(30.45)**	205.99 **(42.75)**	506.63 **(759.81)**	408.66 **(568.77)**	411.19 **(128.94)**	802.73 **(45.55)**
(−1, −2)	173.07 **(91.08)**	281.60 **(42.89)**	216.86 **(08.47)**	24.11[a]	714.92 **(382.01)**	1041.39 **(148.16)**	882.23 **(43.19)**	98.76[a]

Note Values shown in *bold* are results of negative correlations coefficients
[a]Values cannot be obtained due the steep shape of the bivariate distribution for the large negative correlation

Table 3 Relative efficiency of estimating Eq. (3.9) using BVRSIS as compared with using BVUSS

(z_1, z_2)	n = 20				n = 40			
	$\rho = 0.20$	$\rho = 0.50$	$\rho = 0.80$	$\rho = 0.95$	$\rho = 0.20$	$\rho = 0.50$	$\rho = 0.80$	$\rho = 0.95$
(0, 0)	2.39	3.02	2.29	3.21	3.80	4.85	3.73	5.66
(−1, −1)	4.73	4.30	4.04	3.79	7.01	8.28	7.83	7.59
(−2, −2)	8.44	8.47	8.73	5.65	15.69	15.67	16.85	11.02

Source Extracted from Samawi and Al-Saleh (2007)

approximation. BVSRSIS also outperformed the BVRSIS method used by Samawi and Al-Saleh (2007). Moreover, increasing the sample size in both of the above illustrations increases the relative efficiencies of these methods. For instance, in the first illustration, by increasing the sample size from 20 to 50, the relative efficiency of using BVSRSIS as compared with BVUSS to estimate (3.10) is increased from 289.92 to 649.00, with the increase dependent on the dependency between U and V. A similar relationship between sample size and relative efficiencies of these two methods can be demonstrated in the second illustration.

Based on the above conclusions, BVSRSIS can be used in conjunction with other multivariate integration procedures to improve the performance of those methods, and thus providing researchers with a significant reduction in required sample size. With the use of BVSRSIS, researchers can perform integral estimation using substantially fewer simulated numbers. Since using BVSRSIS in simulation does not require any extra effort or programming, we recommend using BVSRSIS to improve the well-known Monte-Carlo method of numerical multiple integration problems. Using BVSRSIS will yield an unbiased and more efficient estimate of those integrals. Moreover, this sampling scheme can be applied successfully to other simulation problems. Last, we recommend using the BVSRSIS method for integrals with a dimension no greater than 3. For higher dimensional integrals, other methods in the literature can be used in conjunction with independent steady ranked simulated sampling.

4 Steady-State Ranked Gibbs Sampler

Many approximation techniques are found in the literature, including Monte-Carlo methods, asymptotic, and Markov chain Monte-Carlo (MCMC) methods such as the Gibbs sampler (Evans and Swartz 1995). Recently, many statisticians have become interested in MCMC methods to simulate complex, nonstandard multivariate distributions. Of the MCMC methods, the Gibbs sampling algorithm is one of the best known and most frequently used MCMC method. The impact of the Gibbs sampler method on Bayesian statistics has been detailed by many authors (e.g., Chib and Greenberg 1994; Tanner 1993) following the work of Tanner and wong (1987) and Gelfand and Smith (1990).

To understand the MCMC process, suppose that we need to evaluate the Monte-Carlo integration *E[f(X)]*, where $f(.)$ is any user-defined function of a random variable X. The MCMC process is as follows: Generate a sequence of random variables, $\{X_0, X_1, X_2, \ldots\}$, such that at each time $t \geq 0$, the next state X_{t+1} is sampled from a distribution $P(X_{t+1}|X_t)$ which depends only on the current state of the chain, X_t. This sequence is called a Markov chain, and $P(.|.)$ is called the transition kernel of the chain. The transition kernel is a conditional distribution function that represents the probability of moving from X_t to the next point X_{t+1} in the support of X. Assume that the chain is time homogenous. Thus, after a sufficiently long burn-in of k iterations, $\{X_t; t = k+1, \ldots, n\}$ will be dependent samples from the stationary distribution. Burn-in samples are usually discarded for this calculation, given an estimator,

$$\bar{f} \approx \frac{1}{n-k} \sum_{t=k+1}^{n} f(\underline{X}_t). \tag{4.1}$$

This average in (4.1) is called an ergodic average. Convergence to the required expectation is ensured by the ergodic theorem. More information and discussions on some of the issues in MCMC can be found in Roberts (1995) and Tierney (1995).

To understand how to construct a Markov chain so that its stationary distribution is precisely the distribution of interest $\pi(.)$, we outline Hastings' (1970) algorithm, which is a generalization of the method first proposed by Metropolis et al. (1953). The method is useful for obtaining a sequence of random samples from a probability distribution for which direct sampling is difficult. The method is as follows: At each time t, the next state X_{t+1} is chosen by first sampling a candidate point Y from a proposal distribution $q(.|X_t)$ (ergodic). Note that the proposal distribution may depend on the current point X_t. The candidate point Y is then accepted with probability $\alpha(X_t, Y)$ where

$$\alpha(X_t, Y) = \min\left(1, \frac{\pi(Y)q(X_t|Y)}{\pi(X_t)q(Y|X_t)}\right). \tag{4.2}$$

If the candidate point is accepted, the next state becomes $X_{t+1} = Y$. If the candidate is rejected, the chain does not move, that is, $X_{t+1} = X_t$. Thus the Metropolis–Hastings algorithm simply requires the following:

Initialize X_0; set t = 0.

Repeat {generate a candidate Y from $q(.|X_t)$

and a value u from a uniform (0, 1), if

$u \leq \alpha(X_t, Y)$ set $X_{t+1} = Y$

Otherwise set $X_{t+1} = X_t$

Increment t}.

A special case of the Metropolis–Hastings algorithm is the Gibbs sampling method proposed by Geman and Geman (1984) and introduced by Gelfand and Smith (1990). To date, most statistical applications of MCMC have used Gibbs sampling. In Gibbs sampling, variables are sampled one at a time from their full conditional distributions.

Gibbs sampling uses an algorithm to generate random variables from a marginal distribution indirectly, without calculating the density. Similar to Casella and George (1992), we demonstrate the usefulness and the validity of the steady-state Gibbs sampling algorithm by exploring simple cases. This example shows that steady-state Gibbs sampling is based only on elementary properties of Markov chains and the properties of BVSRSIS.

4.1 *Traditional (standard) Gibbs Sampling Method*

Suppose that $f(x, y_1, y_2, \ldots, y_g)$ is a joint density function on R^{g+1} and our purpose is to find the characteristics of the marginal density such as the mean and the variance.

$$f_X(x) = \int \cdots \int f(x, y_1, y_2, \ldots, y_g) dy_1, dy_2, \ldots, dy_g \tag{4.3}$$

In cases where (4.3) is extremely difficult or not feasible to perform either analytically or numerically, Gibbs sampling enables the statistician to efficiently generate a sample $X_1, \ldots, X_n \sim f_x(x)$, without requiring $f_x(x)$. If the sample size n is large enough, this method will provide a desirable degree of accuracy for estimating the mean and the variance of $f_x(x)$.

The following discussion of the Gibbs sampling method uses a two-variable case to make the method simpler to follow. A case with more than two variables is illustrated in the simulation study.

Given a pair of random variables (X, Y), Gibbs sampling generates a sample from $f_X(x)$ by sampling from the conditional distribution, $f_{X|Y}(x|y)$ and $f_{Y|X}(y|x)$, which are usually known in statistical models application. The procedure for generating a Gibbs sequence of random variables,

$$X'_0, Y'_0, X'_1, Y'_1, \ldots, X'_k, Y'_k, \tag{4.4}$$

is to start from an initial value $Y'_0 = y'_0$, (which is a known or specified value) and obtaining the rest of the sequence (4.4) iteratively by alternately generating values from

$$\begin{aligned} X'_j &\sim f_{X|Y}(x|Y'_j = y'_j) \\ Y'_{j+1} &\sim f_{Y|X}(y|X'_j = x'_j). \end{aligned} \tag{4.5}$$

For large k and under reasonable conditions (Gelfand and Smith 1990), the final observation in (4.5), namely $X'_j = x'_j$, is effectively a sample point from $f_X(x)$. A natural way to obtain an independent and identically distributed (i.i.d) sample from $f_X(x)$ is to follow the suggestion of Gelfand and Smith (1990) to use Gibbs sampling to find the kth, or final value, from n independent repetitions of the Gibbs sequence in (4.5). Alternatively, we can generate one long Gibbs sequence and use a systematic sampling technique to extract every rth observation. For large enough r, this method will also yield an approximate i.i.d sample from $f_X(x)$. For the advantage and disadvantage of this alternate method see, Gelman and Rubin (1991).

Next, we provide a brief explanation of why Gibbs sampling works under reasonable conditions. Suppose we know the conditional densities $f_{X|Y}(x|y)$ and $f_{Y|X}(y|x)$ of the two random variables X and Y, respectively. Then the marginal density of X, $f_X(x)$ can be determined as follows:

$$f_X(x) = \int f(x, y)dy,$$

where $f(x, y)$ is the unknown joint density of (X, Y). Using the fact that $f_{XY}(x, y) = f_Y(y).f_{Y|X}(x|y)$, then

$$f_X(x) = \int f_Y(y).f_{Y|X}(x|y)dy.$$

Using a similar argument for $f_Y(y)$, then

$$f_X(x) = \int \left[\int f_{X|Y}(x|y) f_{Y|X}(y|t)dy\right] f_X(t)dt = \int g(x, t) f_X(t)dt, \tag{4.6}$$

where $g(x, t) = \int f_{X|Y}(x|y) f_{Y|X}(y|t)dy$ As argued by Gelfand and Smith (1990), Eq. (4.6) defines a fixed-point integral equation for which $f_X(x)$ is the solution and the solution is unique.

4.2 *Steady-State Gibbs Sampling (SSGS): The Proposed Algorithms*

To guarantee an unbiased estimator for the mean, density, and the distribution function of $f_X(x)$, Samawi et al. (2012) introduced two methods for performing steady-state Gibbs sampling. The first method is as follows:

In standard Gibbs sampling, the Gibbs sequence is obtained using the conditional distribution, $f_{X|Y}(x|y)$ and $f_{Y|X}(y|x)$, to generate a sequence of random variables,

$$X'_0, Y'_0, X'_1, Y'_1, \ldots, X'_{k-1}, Y'_{k-1}, \tag{4.7}$$

starting from an initial, specified value $Y'_0 = y'_0$ and iteratively obtaining the rest of the sequence (4.7) by alternately generating values from

$$\begin{aligned} X'_j &\sim f_{X|Y}(x|Y'_j = y'_j) \\ Y'_{j+1} &\sim f_{Y|X}(y|X'_j = x'_j). \end{aligned} \tag{4.8}$$

However, in steady state Gibbs sampling (SSGS), the Gibbs sequence is obtained as follows:

One step before the kth step in the standard Gibbs sampling method, take the last step as

$$\begin{aligned} X'_{i(j)} &\sim F^{-1}_{X|Y}(U_{i(j)}|Y'_{k-1} = y'_{k-1}) \\ Y'_{(i)j} &\sim F^{-1}_{Y|X}(W_{(i)j}|X'_{k-1} = x'_{k-1}), \\ X_{[i](j)} &\sim F^{-1}_{X|Y}(U'_{i(j)}|Y'_{(i)j} = y'_{(i)j}) \\ Y_{(i)[j]} &\sim F^{-1}_{Y|X}(W'_{(i)j}|X'_{i(j)} = x'_{i(j)}) \end{aligned} \tag{4.9}$$

where $\{U_{i(j)}, U'_{i(j)}\}$ from $U\left(\frac{j-1}{n}, \frac{j}{n}\right)$ and $\{W_{(i)j}, W'_{(i)j}\}$ from $U\left(\frac{i-1}{n}, \frac{i}{n}\right)$ as described above. Clearly, this step does not require extra computer time since we generate the Gibbs sequences from uniform distributions only. Repeat this step independently for $i = 1, 2, \ldots, n$ and $j = 1, 2, \ldots, n$ to get an independent sample of size n^2, namely $[(X_{[i](j)}, Y_{(i)[j]}), i = 1, 2, \ldots, n \text{ and } j = 1, 2, \ldots, n]$. For large k and under reasonable conditions (Gelfand and Smith 1990), the final observation in Eq. (4.9), namely $(X_{[i](j)} = x_{[i](j)}, Y_{(i)[j]} = y_{(i)[j]})$ is effectively a sample point from (2.5). Using the properties of SRSIS, $[(X_{[i](j)}, Y_{(i)[j]}), i = 1, 2, \ldots, n \text{ and } j = 1, 2, \ldots, n]$, will produce unbiased estimators for the marginal means and distribution functions. Alternatively, we can generate one long standard Gibbs sequence and use a systematic sampling technique to extract every rth observation using a similar method as described above. Again, a SSGS sample will be obtained as

$$\begin{aligned} X'_{i(j)} &\sim F^{-1}_{X|Y}(U_{i(j)}|Y'_{r-1} = y'_{r-1}), \\ Y'_{(i)j} &\sim F^{-1}_{Y|X}(W_{(i)j}|X'_{r-1} = x'_{r-1}), \\ X_{[i](j)} &\sim F^{-1}_{X|Y}(U'_{i(j)}|Y'_{(i)j} = y'_{(i)j}), \\ Y_{(i)[j]} &\sim F^{-1}_{Y|X}(W'_{(i)j}|X'_{i(j)} = x'_{i(j)}), \end{aligned} \tag{4.10}$$

where $\{U_{i(j)}, U'_{i(j)}\}$ from $U\left(\frac{j-1}{n}, \frac{j}{n}\right)$ and $\{W_{(i)j}, W'_{(i)j}\}$ from $U\left(\frac{i-1}{n}, \frac{i}{n}\right)$ to obtain an independent sample of size n^2, that is, $[(X_{[i](j)}, Y_{(i)[j]}), i = 1, 2, \ldots, n$ and $j = 1, 2, \ldots, n]$.

Using the same arguments as in (4.1), suppose we know the conditional densities $f_{X|Y}(x|y)$ and $f_{Y|X}(y|x)$ of the two random variables X and Y, respectively. Equation (4.6) is the limiting form of the Gibbs iteration scheme, showing how sampling from conditionals produces a marginal distribution. As in Gelfand and Smith (1990) for $k \to \infty$, $X'_{k-1} \sim f_X(x)$ and $Y'_{k-1} \sim f_Y(y)$ and hence $F^{-1}_{Y|X}(W_{(i)j}|x'_{k-1}) = Y'_{(i)j} \sim f^{\infty}_{Y_{(i)}}(y)$, $F^{-1}_{Y|X}(U_{i(j)}|x'_{k-1}) = X'_{(i)j} \sim f^{\infty}_{X_{(j)}}(x)$ where $W_{(i)j} \sim U\left(\frac{i-1}{n}, \frac{i}{n}\right)$ and $U_{i(j)} \sim U\left(\frac{j-1}{n}, \frac{j}{n}\right)$. Therefore,

$$F^{-1}_{X|Y'_{(i)j}}(U'_{i(j)}|Y'_{(i)j}) = X_{[i](j)}|Y'_{(i)j} \sim f^{\infty}_{X_{[i](j)}|Y'_{(i)j}}(x|Y'_{(i)j}) \text{ and}$$
$$F^{-1}_{Y|X'_{i(j)}}(W'_{(i)j}|X'_{i(j)}) = Y_{(i)[j]}|X'_{i(j)} \sim f^{\infty}_{Y_{(i)[j]}|X'_{i(j)}}(y|X'_{i(j)}).$$

This step produces an independent bivariate steady-state sample, $[(X_{[i](j)}, Y_{(i)[j]}), i = 1, 2, \ldots, n$ and $j = 1, 2, \ldots, n]$, where some characteristic of the marginal distributions are to be investigated. To see how to apply this bivariate steady-state sample Gibbs sampling, using (2.5) we get

$$\begin{aligned} f_{X[i](j)}(x) &= \int_{Q_{\frac{(i-1)}{n}}}^{Q_{\frac{(i)}{n}}} n^2 f_Y(y).f_{X|Y}(x|y)dy = \int_{Q_{\frac{(i-1)}{n}}}^{Q_{\frac{(i)}{n}}} nf_{X|Y}(x|y)n \int_{Q_{\frac{(j-1)}{n}}}^{Q_{\frac{(j)}{n}}} f_X(t)f_{Y|X}(y|t)dtdy \\ &= \int_{Q_{\frac{(j-1)}{n}}}^{Q_{\frac{(j)}{n}}} \left[\int_{Q_{\frac{(i-1)}{n}}}^{Q_{\frac{(i)}{n}}} nf_{Y|X}(y|t)f_{X|Y}(x|y)dy \right] nf_X(t)dt \\ &= \int_{Q_{\frac{(j-1)}{n}}}^{Q_{\frac{(j)}{n}}} \left[\int_{Q_{\frac{(i-1)}{n}}}^{Q_{\frac{(i)}{n}}} nf_{Y|X}(y|t)f_{X|Y}(x|y)dy \right] f_{X[i](j)}(t)dt. \end{aligned} \tag{4.11}$$

As argued by Gelfand and Smith (1990), Eq. (4.11) defines a fixed-point integral equation for which $f_{X[i](j)}(x)$ is the solution and the solution is unique.

We next show how SSGS can improve the efficiency of estimating the sample means of a probability density function $f(x)$.

Theorem 4.1 (Samawi et al. 2012). *Under the same conditions of the standard Gibbs sampling, the bivariate SSGS sample above* $[(X_{[i](j)}, Y_{(i)[j]}), i = 1, 2, \ldots, n$ *and* $j = 1, 2, \ldots, n]$ *from* $f(x, y)$ *provides the following:*

1. *Unbiased estimator of the marginal means of X and/or Y. Hence* $E(\bar{X}_{SSGS}) = \mu_x$, *where* $\mu_x = E(X)$.
2. $Var(\bar{X}_{SSGS}) \leq Var(\bar{X})$, *where* $\bar{X} = \frac{\sum_{i=1}^{n^2} X_i}{n^2}$.

Proof Using (2.5),

$$E(\bar{X}_{SSGS}) = \frac{1}{n^2}\sum_{i=1}^{n}\sum_{j=1}^{n} E(X_{[i](j)}) = \frac{1}{n^2}\sum_{i=1}^{n}\sum_{j=1}^{n}\int_{Q_{Y(i-1)/n}}^{Q_{Y(i)/n}}\int_{Q_{X(j-1)/n}}^{Q_{X(j)/n}} x\, f^{(\infty)}_{X_{[i](j)Y(i)[j]}}(x, y)dxdy$$

$$= \frac{1}{n^2}\sum_{i=1}^{n}\sum_{j=1}^{n}\int_{Q_{Y(i-1)/n}}^{Q_{Y(i)/n}}\int_{Q_{X(j-1)/n}}^{Q_{X(j)/n}} xn^2 f(x, y)dxdy$$

$$= \sum_{j=1}^{n}\int_{Q_{X(j-1)/n}}^{Q_{X(j)/n}} x\, f_x(x)dx \sum_{i=1}^{n}\int_{Q_{Y(i-1)/n}}^{Q_{Y(i)/n}} f_{Y|X}(y|x)dy = E(X) = \mu_x.$$

Similarly,

$$\mathrm{var}(\bar{X}_{SSGS}) = \frac{1}{n^4}\sum_{i=1}^{n}\sum_{j=1}^{n} Var(X_{[i](j)}) = \frac{1}{n^4}\sum_{i=1}^{n}\sum_{j=1}^{n}\int_{Q_{Y(i-1)/n}}^{Q_{Y(i)/n}}\int_{Q_{X(j-1)/n}}^{Q_{X(j)/n}} (x-\mu_{x_{[i](j)}})^2 f^{(\infty)}X_{[i](j)Y(i)[j]}(x, y)dxdy$$

$$= \frac{1}{n^4}\sum_{i=1}^{n}\sum_{j=1}^{n}\int_{Q_{Y(i-1)/n}}^{Q_{Y(i)/n}}\int_{Q_{X(j-1)/n}}^{Q_{X(j)/n}} (x-\mu_{x_{[i](j)}}\pm\mu_x)^2 n^2 f(x, y)dxdy$$

$$= \frac{1}{n^2}\sum_{i=1}^{n}\sum_{j=1}^{n}\int_{Q_{Y(i-1)/n}}^{Q_{Y(i)/n}}\int_{Q_{X(j-1)/n}}^{Q_{X(j)/n}} [(x-\mu_x)-(\mu_{[i](j)}-\mu_x]^2 f(x, y)dxdy$$

and,

$$\mathrm{var}(\bar{X}_{SSGS}) = \frac{1}{n^2}\sum_{i=1}^{n}\sum_{j=1}^{n}\int_{Q_{Y(i-1)/n}}^{Q_{Y(i)/n}}\int_{Q_{X(j-1)/n}}^{Q_{X(j)/n}} \{(x-\mu_x)^2 - 2(x-\mu_x)(\mu_{x_{[i](j)}}-\mu_x)$$

$$+(\mu_{x_{[i](j)}}-\mu_x)^2\} f(x, y)dxdy$$

$$= \frac{1}{n^2}\sum_{i=1}^{n}\sum_{j=1}^{n}\int_{Q_{Y(i-1)/n}}^{Q_{Y(i)/n}}\int_{Q_{X(j-1)/n}}^{Q_{X(j)/n}} (x-\mu_x)^2 f(x, y)dxdy - \frac{1}{n^2}\sum_{i=1}^{n}\sum_{j=1}^{n}$$

$$\int_{Q_{Y(i-1)/n}}^{Q_{Y(i)/n}}\int_{Q_{X(j-1)/n}}^{Q_{X(j)/n}} 2(x-\mu_x)(\mu_{x_{[i](j)}}-\mu_x) f(x, y)dxdy$$

$$+\frac{1}{n^2}\sum_{i=1}^{n}\sum_{j=1}^{n}\int_{Q_{Y(i-1)/n}}^{Q_{Y(i)/n}}\int_{Q_{X(j-1)/n}}^{Q_{X(j)/n}} (\mu_{x_{[i](j)}}-\mu_x)^2 f(x, y)dxdy$$

$$= \frac{1}{n^2}\sum_{i=1}^{n}\sum_{j=1}^{n}\int_{Q_{Y(i-1)/n}}^{Q_{Y(i)/n}}\int_{Q_{X(j-1)/n}}^{Q_{X(j)/n}} (x-\mu_x)^2 f(x, y)dxdy - \frac{1}{n^2}\sum_{i=1}^{n}\sum_{j=1}^{n}$$

$$\int_{Q_{Y(i-1)/n}}^{Q_{Y(i)/n}} \int_{Q_{X(j-1)/n}}^{Q_{X(j)/n}} (\mu_{x_{[i](j)}} - \mu_x)^2 f(x, y) dx dy$$

$$= \frac{\sigma_X^2}{n^2} - \frac{1}{n^2} \sum_{i=1}^{n} \sum_{j=1}^{n} \int_{Q_{Y(i-1)/n}}^{Q_{Y(i)/n}} \int_{Q_{X(j-1)/n}}^{Q_{X(j)/n}} (\mu_{x_{[i](j)}} - \mu_x)^2 f(x, y) dx dy \le V(\bar{X}) = \frac{\sigma_X^2}{n^2},$$

where $\sigma_X^2 = Var(X)$. Similar results can be obtained for the marginal mean of Y and the marginal distributions of X and Y. Note that as compared with using standard Gibbs sampling, using SSGS not only provides a gain in efficiency by but also reduces the sample size required to achieve a certain accuracy in estimating the marginal means and distributions. For very complex situations, the smaller required sample size can substantially reduce computation time. To provide insight to the gain in efficiency by using SSGS, we next conduct a simulation study.

4.3 Simulation Study and Illustrations

This section presents the results of a simulation study comparing the performance of the SSGS with the standard Gibbs sampling methods. To compare the performance of our proposed algorithm, we used the same illustrations as Casella and George (1992). For these examples, four bivariate samples of sizes, $n = 10, 20$, and 50 and Gibbs sequence length $k = 20, 50$ and 100 and $r = 20, 50$, and 100 in the long sequence Gibbs sampler. To estimate the variances of the estimators using the simulation method, we completed 5,000 replications. Using the 5,000 replications, we estimate the efficiency of our procedure relative to the traditional (i.e., standard) Gibbs sampling method by $\text{eff}(\hat{\theta}, \hat{\theta}_{SSGS}) = \frac{Var(\hat{\theta})}{Var(\hat{\theta}_{SSGS})}$, where θ is the parameter of interest.

Example 1 Casella and George (1992).
X and Y have the following joint distribution, $f(x, y) \propto \binom{m}{x} y^{x+\alpha-1}(1-y)^{m-x+\beta-1}$, $x = 0, 1, \ldots, m, 0 \le y \le 1$. Assume our purpose is to determine certain characteristics of the marginal distribution $f(x)$ of X. In Gibbs sampling method, we use the conditional distributions $f(x|y) \sim Binomial(m, y)$ and $f(y|x) \sim Beta(x + \alpha, m - x + \beta)$.

Tables 4 and 5 show that, relative to the standard Gibbs sampling method, SSGS improves the efficiency of estimating the marginal means. The amount of improvement depends on two factors: (1) which parameters we intend to estimate, and (2) the conditional distributions used in the process. Moreover, using the short or long Gibbs sampling sequence has only a slight effect on the relative efficiency.

Table 4 Standard Gibbs sampling method compared With the Steady-State Gibbs Sampling (SSGS) method (Beta-Binomial distribution)

n^2	k	Sample mean Gibbs sampling of X	Sample mean SSGS of X	Relative efficiency	Sample mean Gibbs sampling of Y	Sample mean SSGS of Y	Relative efficiency
$m = 5, \alpha = 2,$ and $\beta = 4$							
100	20	1.672	1.668	**3.443**	0.340	0.334	**3.787**
	50	1.667	1.666	**3.404**	0.333	0.333	**3.750**
	100	1.667	1.666	**3.328**	0.333	0.333	**3.679**
400	20	1.666	1.666	**3.642**	0.333	0.333	**3.861**
	50	1.669	1.667	**3.495**	0.333	0.333	**3.955**
	100	1.668	1.667	**3.605**	0.333	0.333	**4.002**
2500	20	1.666	1.666	**3.760**	0.333	0.333	**4.063**
	50	1.668	1.667	**3.786**	0.333	0.333	**3.991**
	100	1.667	1.667	**3.774**	0.333	0.333	**4.007**
$m = 16, \alpha = 2,$ and $\beta = 4$							
100	20	5.321	5.324	**1.776**	0.333	0.333	**1.766**
	50	5.334	5.334	**1.771**	0.333	0.333	**1.766**
	100	5.340	5.337	**1.771**	0.334	0.334	**1.769**
400	20	5.324	5.327	**1.805**	0.333	0.333	**1.811**
	50	5.333	5.333	**1.816**	0.333	0.333	**1.809**
	100	5.330	5.331	**1.803**	0.333	0.333	**1.806**
2500	20	5.322	5.325	**1.820**	0.333	0.333	**1.828**
	50	5.334	5.334	**1.812**	0.333	0.333	**1.827**
	100	5.334	5.333	**1.798**	0.333	0.333	**1.820**

Note The exact mean of x is equal to 5/3 and the exact mean of y is equal to 1/3 for the first case

Example 2 Casella and George (1992).
Let X and Y has the following conditional distributions that are exponential distributions, restricted to the interval $(0, B)$, that is $f(x|y) \propto ye^{-yx}, 0 < x < B < \infty$ and $f(y|x) \propto xe^{-yx}, 0 < y < B < \infty$.

Similarly, Table 6 shows that SSGS improves the efficiency of marginal means estimation relative to standard Gibbs sampling. Again, using short or long Gibbs sampler sequence has only a slight effect on the relative efficiency.

Example 3 Casella and George (1992).
In this example, a generalization of the joint distribution is $f(x, y, m) \propto \binom{m}{x} y^{x+\alpha-1}$ $(1-y)^{nx+\beta-1}, e^{-\lambda} \frac{\lambda^m}{m!}, \lambda > 0, x = 0, 1, \ldots, m, 0 \leq y \leq 1, m = 1, 2, \ldots.$

Again, suppose we are interested in calculating some characteristics of the marginal distribution $f(x)$ of X. In Gibbs sampling method, we use the conditional distributions $f(x|y, m) \sim Binomial(m, y)$, $f(y|x, m) \sim Beta(x + \alpha, m - x + \beta)$ and

Table 5 Comparison of the Long Gibbs sampling method and the Steady-State Gibbs Sampling (SSGS) method (Beta-Binomial distribution)

n^2	r	Sample mean Gibbs sampling of X	Sample mean SSGS of X	Relative efficiency	Sample mean Gibbs sampling of Y	Sample mean SSGS of Y	Relative efficiency
$m = 5, \alpha = 2,$ and $\beta = 4$							
100	20	1.667	1.666	**3.404**	0.333	0.333	**3.655**
	50	1.665	1.665	**3.506**	0.333	0.333	**3.670**
	100	1.668	1.667	**3.432**	0.334	0.333	**3.705**
400	20	1.667	1.667	**3.623**	0.333	0.333	**4.014**
	50	1.666	1.666	**3.606**	0.333	0.333	**3.945**
	100	1.667	1.667	**3.677**	0.333	0.333	**3.997**
2500	20	1.667	1.666	**3.814**	0.333	0.333	**4.011**
	50	1.667	1.667	**3.760**	0.333	0.333	**4.125**
	100	1.667	1.667	**3.786**	0.333	0.333	**4.114**
$m = 16, \alpha = 2,$ and $\beta = 4$							
100	20	5.338	5.338	**1.770**	0.334	0.334	**1.785**
	50	5.335	5.334	**1.767**	0.334	0.334	**1.791**
	100	5.335	5.334	**1.744**	0.334	0.333	**1.763**
400	20	5.332	5.332	**1.788**	0.333	0.333	**1.820**
	50	5.337	5.336	**1.798**	0.333	0.333	**1.815**
	100	5.332	5.333	**1.821**	0.333	0.333	**1.820**
2500	20	5.332	5.332	**1.809**	0.333	0.333	**1.821**
	50	5.335	5.335	**1.832**	0.333	0.333	**1.827**
	100	5.333	5.333	**1.825**	0.333	0.333	**1.806**

Note The exact mean of x is equal to 5/3 and the exact mean of y is equal to 1/3

$f(m|x, y) \propto e^{-(1-y)\lambda} \frac{[(1-y)\lambda]^{m-x}}{(m-x)!}$, $m = x, x + 1, \ldots$. For this example, we used the following parameters: $m = 5, \alpha = 2,$ and $\beta = 4$.

Similarly, Table 7 illustrates the improved efficiency of using SSGS for marginal means estimation, relative to standard Gibbs sampling. Again using a short or long Gibbs sampling sequence has only a slight effect on the relative efficiency. Note that this example is a three-dimensional problem, which shows the improved efficiency depends on the parameters under consideration.

We show that SSGS converges in the same manner as in the standard Gibbs sampling method. However, Sects. 3 and 4 indicate that SSGS is more efficient than standard Gibbs sampling for estimating the means of the marginal distributions using the same sample size. In the examples provided above, the SSGS efficiency (versus standard Gibbs) ranged from 1.77 to 6.6, depending on whether Gibbs sampling used the long or short sequence method and the type of conditional distributions used in the process. Using SSGS yielded a reduced sample size, and thus, reduces

Table 6 Relative efficiency of Gibbs sampling method and Steady-State Gibbs Sampling (SSGS) method (Exponential Distribution)

Standard Gibbs Algorithm $B = 5$							
n^2	k	Sample mean Gibbs sampling of X	Sample mean SSGS of X	Relative efficiency	Sample mean Gibbs sampling of Y	Sample mean SSGS of Y	Relative efficiency
100	20	1.265	1.264	**4.255**	1.264	1.263	**4.132**
	50	1.267	1.267	**4.200**	1.265	1.264	**4.203**
	100	1.267	1.267	**4.100**	1.263	1.265	**4.241**
400	20	1.263	1.264	**4.510**	1.266	1.265	**4.651**
	50	1.265	1.265	**4.341**	1.262	1.263	**4.504**
	100	1.263	1.264	**4.436**	1.265	1.265	**4.345**
2500	20	1.264	1.264	**4.461**	1.264	1.264	**4.639**
	50	1.264	1.264	**4.466**	1.265	1.265	**4.409**
	100	1.265	1.264	**4.524**	1.265	1.264	**4.525**
Long Gibbs Algorithm							
n^2	r	$B = 5$					
100	20	1.265	1.265	**4.305**	1.264	1.265	**4.349**
	50	1.267	1.264	**4.254**	1.261	1.264	**4.129**
	100	1.264	1.265	**4.340**	1.265	1.265	**4.272**
400	20	1.265	1.264	**4.342**	1.263	1.264	**4.543**
	50	1.265	1.265	**4.434**	1.265	1.265	**4.446**
	100	1.266	1.265	**4.387**	1.264	1.264	**4.375**
2500	20	1.264	1.264	**4.403**	1.264	1.264	**4.660**
	50	1.265	1.265	**4.665**	1.264	1.264	**4.414**
	100	1.265	1.265	**4.494**	1.264	1.264	**4.659**

computing time. For example, if the efficiency of using SSGS is 4, then the sample size needed for estimating the simulation's distribution mean, or other distribution characteristics, when using the ordinary Gibbs sampling method is 4 times greater than when using SSGS to achieve the same accuracy and convergence rate. Additionally, our SSGS sample produces unbiased estimators, as shown by theorem 4.1. Moreover, the bivariate steady-state simulation depends on n^2 simulated sample size to produce an unbiased estimate. However, in k dimensional problem, multivariate steady-state simulation depends on n^k simulated sample size to produce an unbiased estimate. Clearly, this sample size is not practical and will increase the simulated sample size required. To overcome this problem in high dimensional cases, we can use the independent simulation method described by Samawi (1999) that needs only a simulated sample of size n regardless of the number of dimensions. This approach slightly reduces the efficiency of using steady-state simulation. In conclusion, SSGS

Table 7 Gibbs Sampling Method and Steady-State Gibbs Sampling (SSGS) method (Beta-Binomial Distribution and Poisson Distribution)

Standard Gibbs Algorithm $\lambda = 5, \alpha = 2,$ and $\beta = 4$

n^3	k	Gibbs mean of X	SSGS mean X	Relative efficiency	Gibbs mean of Y	SSGS mean Y	Relative efficiency	Gibbs mean of Z	SSGS mean Z	Relative efficiency
1000	20	1.667	1.757	**1.883**	0.333	0.343	**6.468**	5.001	5.043	**5.333**
	50	1.665	1.756	**1.891**	0.333	0.343	**6.298**	4.999	5.043	**5.341**
	100	1.666	1.756	**1.881**	0.333	0.343	**6.352**	5.000	5.043	**5.331**
8000	20	1.667	1.758	**1.899**	0.333	0.343	**6.510**	5.000	5.042	**5.338**
	50	1.666	1.757	**1.897**	0.333	0.343	**6.477**	5.001	5.039	**5.321**
	100	1.666	1.755	**1.894**	0.333	0.343	**6.480**	5.000	5.039	**5.318**
27000	20	1.668	1.756	**1.909**	0.333	0.343	**6.591**	4.999	5.039	**5.340**
	50	1.667	1.755	**1.899**	0.333	0.343	**6.584**	5.001	5.039	**5.338**
	100	1.667	1.757	**1.892**	0.333	0.343	**6.583**	5.000	5.039	**5.339**

Long Gibbs Algorithm

n^3	r	$\lambda = 5, \alpha = 2,$ and $\beta = 4$								
1000	20	1.670	1.757	**1.883**	0.333	0.343	**6.468**	5.001	5.043	**5.333**
	50	1.665	1.756	**1.891**	0.333	0.343	**6.298**	4.999	5.043	**5.341**
	100	1.666	1.756	**1.881**	0.333	0.343	**6.352**	5.000	5.043	**5.331**
8000	20	1.665	1.758	**1.899**	0.333	0.343	**6.510**	5.000	5.042	**5.338**
	50	1.666	1.757	**1.897**	0.333	0.343	**6.477**	5.001	5.039	**5.321**
	100	1.667	1.755	**1.894**	0.333	0.343	**6.480**	5.000	5.039	**5.318**
27000	20	1.668	1.756	**1.909**	0.333	0.343	**6.591**	4.999	5.039	**5.340**
	50	1.667	1.755	**1.899**	0.333	0.343	**6.584**	5.001	5.039	**5.338**
	100	1.667	1.757	**1.892**	0.333	0.343	**6.583**	5.000	5.039	**5.339**

The exact mean of x is equal to 5/3 and the exact mean of y is equal to 1/3

performs at least as well as standard Gibbs sampling and SSGS offers greater accuracy. Thus, we recommend using SSGS whenever a Gibbs sampling procedure is needed. Further investigation is needed to explore additional applications and more options for using the SSGS approach.

References

Al-Saleh, M. F., & Al-Omari, A. I., (1999). Multistage ranked set sampling. *Journal of Statistical Planning and Inference*, *102*(2), 273–286.

Al-Saleh, M. F., & Samawi, H. M. (2000). On the efficiency of Monte-Carlo methods using steady state ranked simulated samples. *Communication in Statistics- Simulation and Computation*, *29*(3), 941–954. doi:10.1080/03610910008813647.

Al-Saleh, M. F., & Zheng, G. (2002). Estimation of multiple characteristics using ranked set sampling. *Australian & New Zealand Journal of Statistics*, *44*, 221–232. doi:10.1111/1467-842X.00224.

Al-Saleh, M. F., & Zheng, G. (2003). Controlled sampling using ranked set sampling. *Journal of Nonparametric Statistics*, *15*, 505–516. doi:10.1080/10485250310001604640.

Casella, G., & George, I. E. (1992). Explaining the Gibbs sampler. *American Statistician*, *46*(3), 167–174.

Chib, S., & Greenberg, E. (1994). Bayes inference for regression models with ARMA (p, q) errors. *Journal of Econometrics*, *64*, 183–206. doi:10.1016/0304-4076(94)90063-9.

David, H. A. (1981). *Order statistics*(2nd ed.). New York, NY: Wiley.

Evans, M., & Swartz, T. (1995). Methods for approximating integrals in statistics with special emphasis on Bayesian integration problems. *Statistical Science*, *10*(2), 254–272. doi:10.1214/ss/1177009938.

Gelfand, A. E., & Smith, A. F. M. (1990). Sampling-based approaches to calculating marginal densities. *Journal of the American Statistical Association*, *85*, 398–409. doi:10.1080/01621459.1990.10476213.

Gelman, A., & Rubin, D. (1991). *An overview and approach to inference from iterative simulation [Technical report]*. Berkeley, CA: University of California-Berkeley, Department of Statistics.

Geman, S., & Geman, D. (1984). Stochastic relaxation, Gibbs distributions and the Bayesian restoration of images. *IEEE Transactions on Pattern Analysis and Machine Intelligence*, *6*, 721–741. doi:10.1109/TPAMI.1984.4767596.

Hammersley, J. M., & Handscomb, D. C. (1964). Monte-Carlo methods. London, UK: Chapman & Hall. doi:10.1007/978-94-009-5819-7.

Hastings, W. K. (1970). Monte-Carlo sampling methods using Markov chains and their applications. *Biometrika*, *57*, 97–109. doi:10.1093/biomet/57.1.97.

Johnson, M. E. (1987). *Multivariate statistical simulation*. New York, NY: Wiley. doi:10.1002/9781118150740.

Johnson, N. L., & Kotz, S. (1972). *Distribution in statistics: Continuous multivariate distributions*. New York, NY: Wiley.

Liu, J. S. (2001). *Monte-Carlo strategies in scientific computing*. New York, NY: Springer.

McIntyre, G. A. (1952). A method of unbiased selective sampling, using ranked sets. *Australian Journal of Agricultural Research*, *3*, 385–390. doi:10.1071/AR9520385.

Metropolis, N., Rosenbluth, A. W., Rosenbluth, M. N., Teller, A. H., & Teller, E. (1953). Equations of state calculations by fast computing machine. *Journal of Chemical Physics*, *21*, 1087–1091. doi:10.1063/1.1699114.

Morgan, B. J. T. (1984). *Elements of simulation*. London, UK: Chapman & Hall. doi:10.1007/978-1-4899-3282-2.

Plackett, R. L. (1965). A class of bivariate distributions. *Journal of the American Statistical Association, 60*, 516–522. doi:10.1080/01621459.1965.10480807.

Robert, C., & Casella, G. (2004). *Monte-Carlo statistical methods* (2nd ed.). New York, NY: Springer. doi:10.1007/978-1-4757-4145-2.

Roberts, G. O. (1995). Markov chain concepts related to sampling algorithms. In W. R. Gilks, S. Richardson, & D. J. Spiegelhalter (Eds.), *Markov Chain Monte-Carlo in practice* (pp. 45–57). London, UK: Chapman & Hall.

Samawi, H. M. (1999). More efficient Monte-Carlo methods obtained by using ranked set simulated samples. *Communication in Statistics-Simulation and Computation, 28*, 699–713. doi:10.1080/03610919908813573.

Samawi H. M., & Al-Saleh, M. F. (2007). On the approximation of multiple integrals using multivariate ranked simulated sampling. *Applied Mathematics and Computation, 188*, 345–352. doi:10.1016/j.amc.2006.09.121.

Samawi, H. M., Dunbar, M., & Chen, D. (2012). Steady state ranked Gibbs sampler. *Journal of Statistical Simulation and Computation, 82*, 1223–1238. doi:10.1080/00949655.2011.575378.

Samawi, H. M., & Vogel, R. (2013). More efficient approximation of multiple integration using steady state ranked simulated sampling. *Communications in Statistics-Simulation and Computation, 42*, 370–381. doi:10.1080/03610918.2011.636856.

Shreider, Y. A. (1966). *The Monte-Carlo method.* Oxford, UK: Pergamom Press.

Stokes, S. L. (1977). Ranked set sampling with concomitant variables. *Communications in Statistics—Theory and Methods, 6*, 1207–1211. doi:10.1080/03610927708827563.

Tanner, M. A. (1993). *Tools for statistical inference* (2nd ed.). New York, NY: Wiley. doi:10.1007/978-1-4684-0192-9.

Tanner, M. A., & Wong, W. (1987). The calculation of posterior distribution by data augmentation (with discussion). *Journal of the American Statistical Association, 82*, 528–550. http://dx.doi.org/10.1080/01621459.1987.10478458.

Tierney, L. (1995). Introduction to general state-space Markov chain theory. In W. R. Gilks, S. Richardson, & D. J. Spiegelhalter (Eds.), *Markov Chain Monte-Carlo in practice* (pp. 59–74). London, UK: Chapman & Hall.

Normal and Non-normal Data Simulations for the Evaluation of Two-Sample Location Tests

Jessica R. Hoag and Chia-Ling Kuo

Abstract Two-sample location tests refer to the family of statistical tests that compare two independent distributions via measures of central tendency, most commonly means or medians. The t-test is the most recognized parametric option for two-sample mean comparisons. The pooled t-test assumes the two population variances are equal. Under circumstances where the two population variances are unequal, Welch's t-test is a more appropriate test. Both of these t-tests require data to be normally distributed. If the normality assumption is violated, a non-parametric alternative such as the Wilcoxon rank-sum test has potential to maintain adequate type I error and appreciable power. While sometimes considered controversial, pretesting for normality followed by the F-test for equality of variances may be applied before selecting a two-sample location test. This option results in multi-stage tests as another alternative for two-sample location comparisons, starting with a normality test, followed by either Welch's t-test or the Wilcoxon rank-sum test. Less commonly utilized alternatives for two-sample location comparisons include permutation tests, which evaluate statistical significance based on empirical distributions of test statistics. Overall, a variety of statistical tests are available for two-sample location comparisons. Which tests demonstrate the best performance in terms of type I error and power depends on variations in data distribution, population variance, and sample size. One way to evaluate these tests is to simulate data that mimic what might be encountered in practice. In this chapter, the use of Monte Carlo techniques are demonstrated to simulate normal and non-normal data for the evaluation of two-sample location tests.

J.R. Hoag · C.-L. Kuo (✉)
Department of Community Medicine and Health Care, Connecticut Institute for Clinical and Translational Science, University of Connecticut Health Center, Farmington, USA
e-mail: kuo@uchc.edu

J.R. Hoag
e-mail: hoag@uchc.edu

D.-G. Chen and J.D. Chen (eds.), *Monte-Carlo Simulation-Based Statistical Modeling*, ICSA Book Series in Statistics, DOI 10.1007/978-981-10-3307-0_3

1 Introduction

Statistical tests for two-sample location comparison include t-tests, Wilcoxon rank-sum test, permutation tests, and multi-stage tests. The t-test is the most recognized parametric test for two-sample mean comparison. It assumes independent identically distributed (i.i.d) samples from two normally distributed populations. The pooled t-test (also called Student's t-test) additionally assumes the equality of variances while the Welch's t-test does not. Other tests such as Wilcoxon rank-sum test (also called Mann-Whitney test) and Fisher-Pitman permutation tests make no such distributional assumptions; thus, are theoretically robust against non-normal distributions. Multi-stage tests that comprise preliminary tests for normality and/or equality of variance before a two-sample location comparison test is attempted may also be used as alternatives.

When the normal assumption is met, t-tests are optimal. The pooled t-test is also robust to the equal variance assumption, but only when the sample sizes are equal (Zimmerman 2004). Welch's t-test does not alter the type I error rate regardless of unequal sample sizes and performs as well as the pooled t-test when the equality of variances is held (Zimmerman 2004; Kohr and Games 1974), but it can result in substantial type II error under heavily non-normal distributions and for small sample sizes, e.g. Beasley et al. (2009). Under circumstances of equal variance with either equal or unequal sample sizes less than or equal to 5, De Winter recommended the pooled t-test over Welch's t-test, provided an adequately large population effect size, due to a significant loss in statistical power associated with Welch's t-test (de Winter 2013). The power loss associated with Welch's t-test in this scenario is attributed to its lower degrees of freedom compared to the pooled t-test (de Winter 2013).

In samples with known non-normal distributions, the Wilcoxon rank-sum test is far superior than parametric tests, with power advantages actually increasing with increasing sample size (Sawilowsky 2005). The Wilcoxon rank-sum test, however, is not a solution to heterogeneous variance. Simply, heterogeneous variance is mitigated but does not disappear when actual data is converted to ranks (Zimmerman 1996). In addition to the Wilcoxon rank-sum test, permutation tests have been recommended as a supplement to t-tests across a range of conditions, with emphasis on samples with non-normal distributions. Boik (1987), however, found that the Fisher-Pitman permutation test (Ernst et al. 2004) is no more robust than the t-test in terms of type I error rate under circumstances of heterogeneous variance.

A pretest for equal variance followed by the pooled t-test or Welch's t-test, however appealing, fails to protect type I error rate (Zimmerman 2004; Rasch et al. 2011; Schucany and Tony Ng 2006). With the application of a pre-test for normality (e.g. Shapiro-Wilk Royston 1982), however, it remains difficult to fully assess the normality assumption—acceptance of the null hypothesis is predicated only on insufficient evidence to reject it (Schucany and Tony Ng 2006). A three-stage procedure which also includes a test for equal variance (e.g. F-test or Levenes test) following a normality test is also commonly applied in practice, but as Rash and colleagues have pointed out, pretesting biases type I and type II conditional error rates, and may be altogether unnecessary (Rochon et al. 2012; Rasch et al. 2011). Current recommendations suggest that pretesting is largely unnecessary and that the t-test should be replaced with Welch's t-test in general practice because it is robust against heterogeneous variance (Rasch et al. 2011; Welch 1938).

In summary, the choice of a test for two-sample location comparison depends on variations in data distribution, population variance, and sample size. Previous studies suggested Welch's t-test for normal data and Wilcoxon rank-sum test for non-normal data without heterogeneous variance (Ruxton 2006). Some recommended Welch's t-test for general use (Rasch et al. 2011; Zimmerman 1998). However, Welch's t-test is not a powerful test when the data is extremely non-normal or the sample size is small (Beasley et al. 2009). Previous Monte Carlo simulations tuned the parameters ad hoc and compared a limited selection of two-sample location tests. The simulation settings were simplified by studying either non-normal data only, unequal variances only, small sample sizes only, or unequal sample sizes only.

In this chapter, simulation experiments are designed to mimic what is often encountered in practice. Sample sizes are calculated for a broad range of power based on the pooled t-test, i.e. assuming normally distributed data with equal variance. A medium effect size is assumed (Cohen 2013) as well as multiple sample size ratios. Although the sample sizes are determined assuming normal data with equal variance, the simulations consider normal and moderately non-normal data and allow for heterogeneous variance. The simulated data are used to compare two-sample location tests including parametric tests, non-parametric tests, and permutation-based tests. The goal of this chapter is to provide insight into these tests and how Monte Carlo simulation techniques can be applied to demonstrate their evaluation.

2 Statistical Tests

Capital letters are used to represent random variables and lowercase letters represent realized values. Additionally, vectors are presented in bold. Assume $\mathbf{x} = \left[x_1, x_2, .., x_{n_1}\right]$ and $\mathbf{y} = \left[y_1, y_2, ..., y_{n_2}\right]$ are two i.i.d. samples from two independent populations with means μ_1 and μ_2 and variances σ_1^2 and σ_2^2. Denote by $\bar{x}$ and $\bar{y}$ the sample means and s_1^2 ands_2^2 the sample variances. Let $\mathbf{z} = \left[z_1, ..., z_{n_1}, z_{n_1+1}, ..., z_{(n_1+n_2)}\right]$ represent a vector of group labels with $z_i = 1$ associated with x_i for $i = 1, 2, ..., n_1$ and $z_{(n_1+j)} = 2$ associated with y_j for $j = 1, 2, ..., n_2$. The combined x's and y's are ranked from largest to smallest. For tied observations, the average rank is assigned and each is still treated uniquely. Denote by r_i the rank associated with x_i and s_i the rank associated with y_i. The sum of ranks in x's is given by $R = \sum_{i=1}^{n_1} r_i$ and by $S = \sum_{i=1}^{n_2} s_i$ for the sum of ranks in y's.

Next,the two-sample location tests considered in the simulations are reviewed. The performance of these tests are evaluated by type I error and power.

2.1 *t*-Test

The pooled t-test and Welch's t-test assume that each sample is randomly sampled from a population that is approximately normally distributed. The pooled t-test further assumes that the two variances are equal ($\sigma_1^2 = \sigma_2^2$). The test statistic for the null hypothesis that the two population means are equal ($H_0 : \mu_1 = \mu_2$) is given by

$$t_{\text{pooled}} = \frac{\bar{x} - \bar{y}}{s_p\sqrt{\frac{1}{n_1} + \frac{1}{n_2}}}, \tag{1}$$

where $s_p^2 = \frac{(n_1-1)s_1^2+(n_2-1)s_2^2}{n_1+n_2-2}$. When the null hypothesis is true and n_1 and n_2 are sufficiently large, t_{pooled} follows a t-distribution with degrees of freedom $n_1 + n_2 - 2$. Welch's t-test allows for unequal variances and tests against the null hypothesis by

$$t_{\text{welch}} = \frac{\bar{x} - \bar{y}}{\sqrt{\frac{s_1^2}{n_1} + \frac{s_2^2}{n_2}}}. \tag{2}$$

The asymptotic distribution of t_{welch} is approximated by the t-distribution with degrees of freedom,

$$v = \frac{\left(\frac{s_1^2}{n_1} + \frac{s_2^2}{n_2}\right)^2}{\frac{s_1^4}{n_1^2(n_1-1)} + \frac{s_2^4}{n_2^2(n_2-1)}}. \tag{3}$$

Alternatively, the F-test based on the test statistic,

$$F = \frac{s_1^2}{s_2^2}, \tag{4}$$

can be applied to test the equality of variances. If the null hypothesis is rejected, Welch's t-test is preferred; otherwise, the pooled t-test is used.

2.2 Wilcoxon Rank-Sum Test

The Wilcoxon rank-sum test is essentially a rank-based test. It uses the rank data to compare two distributions. The test statistic to test against the null hypothesis that the two distributions are same is given by

$$U = n_1 n_2 + [n_2(n_2 + 1)]/2 - R\,. \tag{5}$$

Under the null hypothesis, U is asymptotically normally distributed with the mean $(n_1 n_2)/2$ and the variance $n_1 n_2 (n_1 + n_2 + 1)/12$.

2.3 Two-Stage Test

The choice between t-test and Wilcoxon rank-sum test can be based on whether the normality assumption is met for both samples. Pior to performing the mean comparison test, the Shapiro-Wilk test (Royston 1982) can be used to evaluate normality. If both p-values are greater than the significance level α, t-test is used; otherwise, Wilcoxon rank-sum test is used. The chosen t-test here is Welch's t-test for its robustness against heterogeneous variance. Since the two normality tests for each sample are independent, the overall type I error rate (i.e. the family-wise error rate) that at least one hypothesis is incorrectly rejected is thus controlled at 2α. When $\alpha = 2.5\%$, it results in the typical value 5% for the family-wise error rate.

2.4 Permutation Test

Let the group labels in $\mathbf{z}$ be shuffled for B times. Each produces a new vector of group labels, $\mathbf{z}_k = \left(z_{k1}, \ldots, z_{kn_1}, z_{k(n_1+1)}, \ldots, z_{k(n_1+n_2)}\right)$, for $k = 1, 2, \ldots, B$ where z_{kj} is the new group label for x_j if $j \le n_1$ and for $y_{(j-n_1)}$ if $j > n_1$. Given the k-th permuted data set, $\mathbf{x}_k$ and $\mathbf{y}_k$, where $\mathbf{x}_k = \{\{x_j,\ j : z_{kj} = 1,\ j = 1, 2, \ldots, n_1\}, \{y_j,\ j : z_{k(n_1+j)} = 1,\ j = 1, 2, \ldots, n_2\}\}$ and $\mathbf{y}_k = \{\{x_j,\ j : z_{kj} = 2, j = 1, 2, \ldots, n_1\}, \{y_j, j : z_{k(n_1+j)} = 2,\ j = 1, 2, \ldots, n_2\}\}$, Welch's t-test and Wilcoxon rank-sum test are applied and the test statistics and p-values are calculated. Denoted by s_k^t and s_k^w the test statistics and q_k^t and q_k^w the p-values at the k-th permutation associated with Welch's t-test and Wilcoxon rank-sum test, respectively. Similarly, denote by s_o^t and s_o^w the observed test statistics and p_o^t and p_o^w the observed p-values, calculated using the observed data, $\mathbf{x}$ and $\mathbf{y}$. Define $m_k(p_k^t, p_k^w)$ as the minimum of p_k^t and p_k^w. The permutation p-values of Welch's t-test and Wilcoxon rank-sum test are given by

$$p_{\text{p.welch}} = \frac{2}{B} \min \left\{ \sum_{k=1}^{B} I(s_k^t \le s_o^t),\ \sum_{k=1}^{B} I(s_k^t > s_o^t) \right\}, \tag{6}$$

and

$$p_{\text{p.wilcox}} = \frac{2}{B} \min \left\{ \sum_{k=1}^{B} I(s_k^w \le s_o^w),\ \sum_{k=1}^{B} I(s_k^w > s_o^w) \right\}, \tag{7}$$

where $I(\cdot) = 1$ if the condition in the parentheses is true; otherwise, $I(\cdot) = 0$. Similarly, the p-value associated with the minimum p-value is given by

$$p_{\text{minp}} = \frac{1}{B} \sum_{k=1}^{B} I\left(m_k(p_k^t, p_k^w) \le m_o(p_o^t, p_o^w)\right). \tag{8}$$

Overall, 10 two-sample location tests are compared including t-tests, Wilcoxon rank-sum test, two-stage test, and permutation tests. For convenience, when presenting results, each test is referenced by an abbreviated notation, shown below in parentheses.

- t-tests: pooled t-test (**pooled**), Welch's t-test (**welch**), permutation Welch's t-test (**p.welch**), robust t-test (**robust.t**, F-test for the equality of variances followed by pooled t-test or Welch's t-test)
- Wilcoxon rank-sum tests: Wilcoxon rank-sum test (**wilcox**), permutation Wilcoxon rank-sum test (**p.wilcox**)
- two-stage tests: two-stage test with the first-stage α level at 0.5%, 2.5%, and 5% (**2stage0.5**, **2stage2.5**, **2stage5**)
- minimum p-value: minimum p-value of permutation Welch's t-test and Wilcoxon rank-sum test (**minp**)

3 Simulations

Simulated data were used to test the null hypothesis $H_0 : \mu_1 = \mu_2$ versus the alternative hypothesis $H_1 : \mu_1 \neq \mu_2$. Assume $\mathbf{x} = \{x_1, ..., x_{n_1}\}$ was simulated from $N(\mu_1, \sigma_1^2)$ and $\mathbf{y} = \{y_1, ..., y_{n_2}\}$ was simulated from $N(\mu_2, \sigma_2^2)$. Without losing generality, let σ_1 and σ_2 be set equal to 1. Let μ_1 be zero, μ_2 be 0 under the null hypothesis and 0.5 under the alternative hypothesis, the conventional value suggested by Cohen (2013) for a medium effect size. n_1 is set equal to or double the size of n_2. n_1 and n_2 were chosen to detect the difference between μ_1 and μ_2 for 20, 40, 60, or 80% power at 5% significance level. When $n_1 = n_2$, the sample size required per group was 11, 25, 41, and 64, respectively. When $n_1 = 2n_2$, n_1 =18, 38, 62, 96 and n_2 was half of n_1 in each setting. The same sample sizes were used for null simulations, non-normal, and heteroscedastic settings. Power analysis was conducted using G*Power software (Faul et al. 2007).

Fleishman's power method to simulate normal and non-normal data is based on a polynomial function given by

$$x = f(\omega) = a + b\omega + c\omega^2 + d\omega^3, \tag{9}$$

where ω is a random value from the standard normal with mean 0 and standard deviation 1. The coefficients a, b, c, and d are determined by the first four moments of X with the first two moments set to 0 and 1. For distributions with mean and standard deviation different from 0 and 1, the data can be shifted and/or rescaled after being simulated. Let γ_3 denote the skewness and γ_4 denote the kurtosis. γ_3 and γ_4 are both set to 0 if X is normally distributed. The distribution is left-skewed if $\gamma_3 < 0$ and right-skewed if $\gamma_3 > 0$. γ_4 is smaller than 0 for a platokurtotic distribution and greater than 0 for a leptokurtotic distribution. By the 12 moments of the standard normal distribution, we can derive the equations below and solve a, b, c, and d via the Newton-Raphson method or any other non-linear root-finding method,

$$a = -c \tag{10}$$

$$b^2 + 6bd + 2c^2 + 15d^2 - 1 = 0 \tag{11}$$

$$2c\left(b^2 + 24bd + 105d^2 + 2\right) - \gamma_3 = 0 \tag{12}$$

$$24\left[bd + c^2\left(1 + b^2 + 28bd\right) + d^2\left(12 + 48bd + 141c^2 + 225d^2\right)\right] - \gamma_4 = 0 \tag{13}$$

One limitation of Fleishman's power method is that it does not cover the entire domain of skewness and kurtosis. Given γ_3, the relationship between γ_3 and γ_4 is described by the inequation, $\gamma_4 \geq \gamma_3^2 - 2$ (Devroye 1986). Precisely, the empirical lower bound of γ_4 given $\gamma_3 = 0$ is -1.151320 (Headrick and Sawilowsky 2000).

By Fleishman's power method, three conditions were investigated: (1) heterogeneous variance, (2) skewness, and (3) kurtosis. Each was investigated using equal and unequal sample sizes to achieve a broad range of power. $\mu_1 = \mu_2 = 0$ under the null hypothesis. $\mu_1 = 0$ and $\mu_2 = 0.5$ when the alternative hypothesis is true. Other parameters were manipulated differently. For (1), normal data was simulated with equal and unequal variances by letting $\sigma_1 = 1$ and σ_2= 0.5, 1, 1.5. For (2), skewed data was simulated assuming equal variance at 1, equal kurtosis at 0, and equal skewness at 0, 0.4, 0.8. Similarly, for (3), kurtotic data was simulated assuming equal variance at 1, equal skewness at 0, and equal kurtosis at 0, 5, 10. To visualize the distributions from which the data were simulated, 10^6 data points were simulated from the null distributions and used to create density plots (see Fig. 1). To best visualize the distributions, the data range was truncated at 5 and −5. The left panel

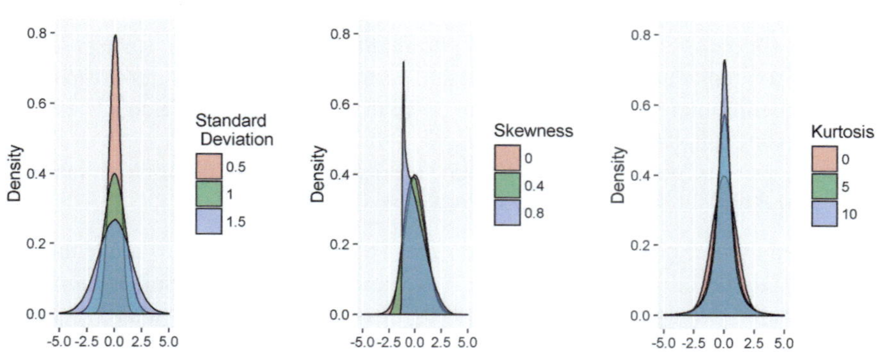

Fig. 1 Distributions to simulate heteroscedastic, skewed, and kurtotic data when the null hypothesis of equal population means is true (*left*: normal distributions with means at 0 and standard deviations at 0.5, 1, and 1.5; *middle*: distributions with means at 0, standard deviations at 1, skewness at 0, 0.4, and 0.8, and kurtosis at 0; *right*: distributions with means at 0, standard deviations at 1, skewness at 0, and kurtosis at 0, 5, and 10)

demonstrates the distributions for the investigation of heterogeneous variance. One sample is simulated from the standard normal (green). The other sample is simulated from the normal with mean 0 and standard deviation 0.5 (red) or 1.5 (blue). For skewness and kurtosis, the two distributions were assumed to be exactly the same under the null hypothesis. Although these distributions are for null simulations, the distributions for power simulations are the same for one sample and shifted from 0 to 0.5 for the second sample.

The number of simulation replicates was 10,000 and 1,000 for null simulations and power simulations, respectively. The number of permutation replicates for the permutation tests at each simulation replicate was 2,000, i.e. $B = 2000$. The significance level for two-location comparison was set to 0.05. All the simulations were carried out in R 3.1.2 (Team 2014). The Fleishman coefficients (a, b, c, and d) given the first four moments (the first two moments 0 and 1) were derived via the R function "Fleishman.coef.NN" (Demirtas et al. 2012).

4 Results

The simulation results are presented in figures. In each figure, the type I error or power is presented on the y-axis and the effective sample size per group to achieve 20%, 40%, 60%, and 80% power is presented on the x-axis assuming the pooled t-test is appropriate. The effective sample size per group for two-sample mean comparison is defined as

$$n_e = \frac{2n_1 n_2}{n_1 + n_2}. \tag{14}$$

Each test is represented by a colored symbol. Given an effective sample size, the results of

1. t-tests (pooled t-test, theoretical and permutation Welch's t-tests, robust t-test)
2. Wilcoxon rank-sum tests (theoretical and permutation Wilcoxon rank-sum tests) and the minimum p-value (minimum p-value of permutation Welch's t-test and Wilcoxon rank-sum test)
3. two-stage tests (normality test with the α level at 0.5, 2.5, and 5% for both samples followed by Welch's t-test or Wilcoxon rank-sum test)

are aligned in three columns from left to right. The results for $n_1 = n_2$ are in the left panels and those for $n_1 = 2n_2$ are in the right panels. In the figures that present type I error results, two horizontal lines $y = 0.0457$ and $y = 0.0543$ are added to judge whether the type I error is correct. The type I error is considered correct if it falls within the 95% confidence interval of the 5% significance level (0.0457, 0.0543). We use *valid* to describe tests that maintain a correct type I error and *liberal* and *conservative* for tests that result in a type I error above and under the nominal level, respectively.

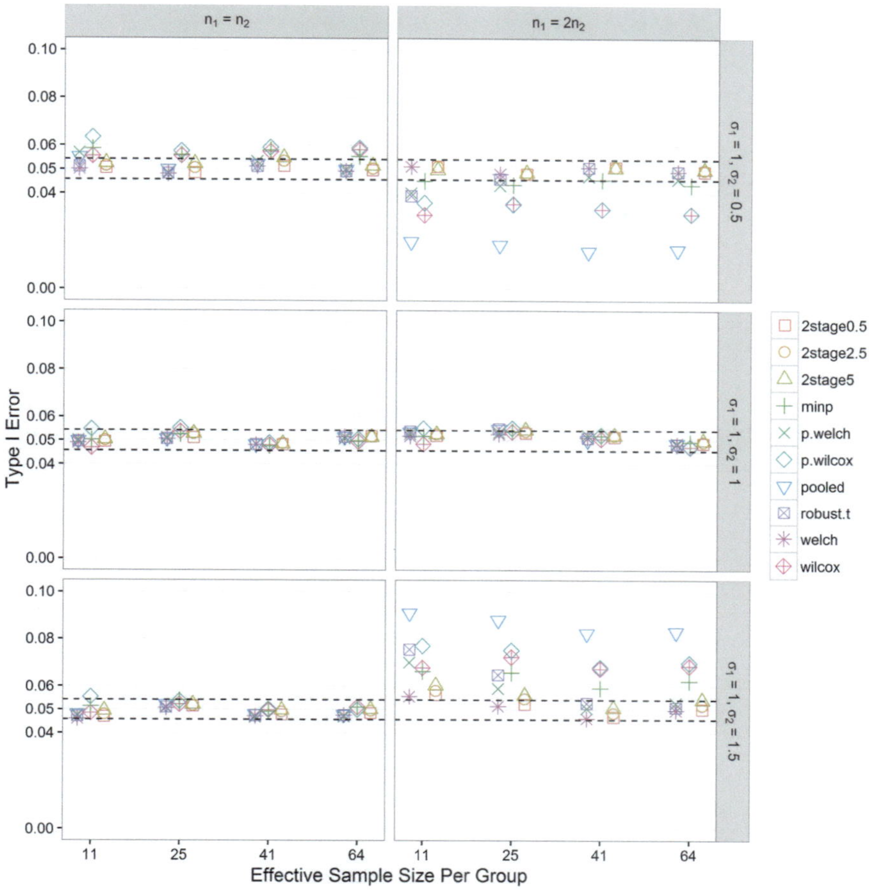

Fig. 2 Null simulation results for normal data with equal or unequal variances. $y = 0.0457$ and $y = 0.0543$ are added to judge whether the type I error is correct. The type I error is considered correct if it falls within the 95% confidence interval of the 5% significance level (0.0457, 0.0543). The tests are grouped into three groups from *left* to *right* for result presentation, (1) t-tests: pooled t-test (**pooled**), theoretical and permutation Welch's t-tests (**welch** and **p.welch**), robust t-test (**robust.t**), (2) Wilcoxon rank-sum test (**wilcox**), permutation Wilcoxon rank-sum test (**p.wilcox**) and the minimum p-value (**minp**), (3) two-stage tests: two-stage tests with the first stage α level at 0.5, 2.5, 5% (**2stage0.5**, **2stage2.5**, **2stage5**)

4.1 Heterogeneous Variance

The null simulation results are presented in Fig. 2. All tests maintain a correct type I error when the sample sizes are equal. Wilcoxon rank-sum tests and the minimum p-value have a slightly inflated type I error when the variances are unequal and relatively small ($\sigma_1 = 1$ and $\sigma_2 = 0.5$). When the sample sizes are unequal, all tests are valid as long as the variances are equal. When the variances are unequal, however, neither the pooled t-test nor the Wilcoxon rank-sum test maintain a correct type I

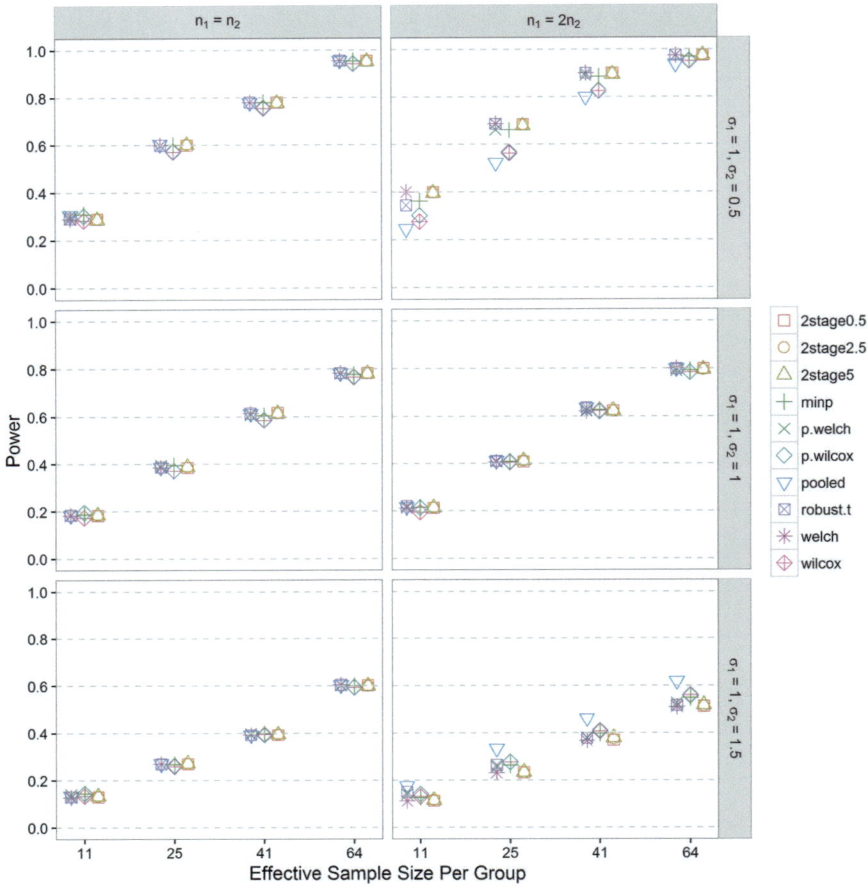

Fig. 3 Power simulation results for normal data with equal or unequal variances. The tests are grouped into three groups from *left* to *right* for result presentation, (1) t-tests: pooled t-test (**pooled**), theoretical and permutation Welch's t-tests (**welch** and **p.welch**), robust t-test (**robust.t**), (2) Wilcoxon rank-sum test (**wilcox**), permutation Wilcoxon rank-sum test (**p.wilcox**) and the minimum p-value (**minp**), (3) two-stage tests: two-stage tests with the first stage α level at 0.5, 2.5, 5% (**2stage0.5**, **2stage2.5**, **2stage5**)

error. Both exhibit a conservative type I error when $\sigma_1 = 1$ and $\sigma_2 = 0.5$ and a liberal type I error when $\sigma_1 = 1$ and $\sigma_2 = 1.5$. The pooled t-test in particular is more strongly influenced than the Wilcoxon rank-sum test. The minimum p-value involves Wilcoxon rank-sum test and is thus similarly affected but not as severely as the Wilcoxon rank-sum test alone. The robust t-test and permutation Welch's t-test behave similarly when the sample size is small. Welch's t-test and two-stage tests are the only tests with type I error protected for heterogeneous variance regardless of sample sizes.

The power simulation results are presented in Fig. 3. Unsurprisingly, the settings of $\sigma_1 = 1, \sigma_2 = 0.5$ result in higher power than $\sigma_1 = 1, \sigma_2 = 1$, and than $\sigma_1 = 1, \sigma_2 =$

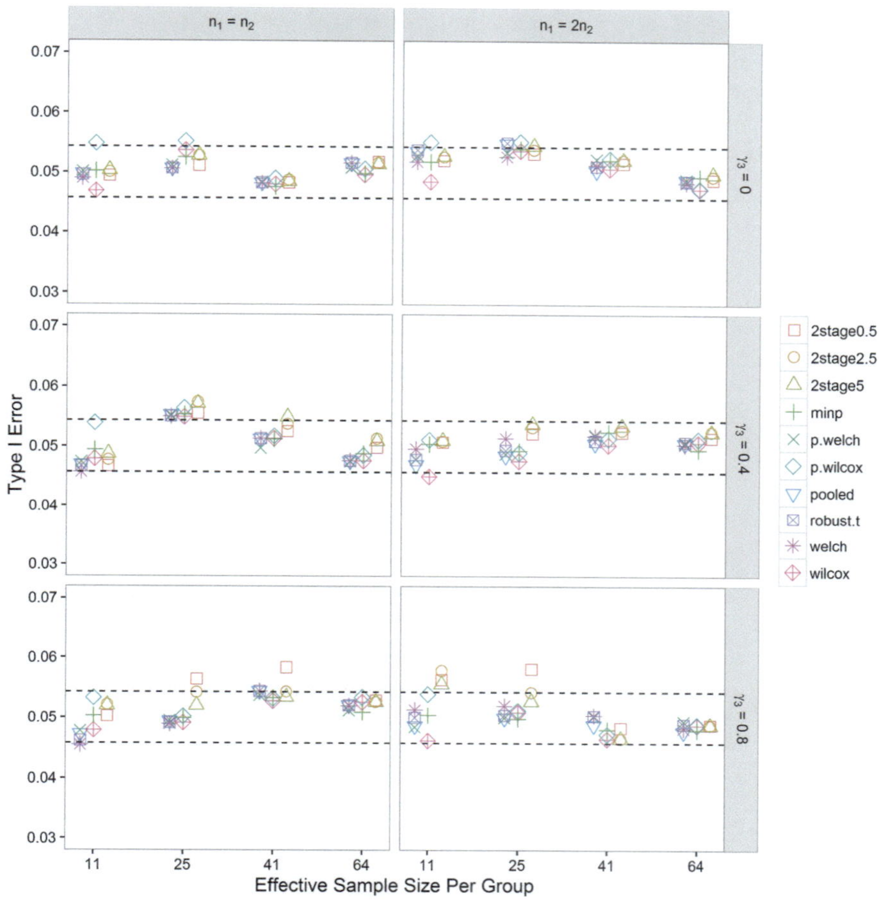

Fig. 4 Null simulation results for skewed data. $y = 0.0457$ and $y = 0.0543$ are added to judge whether the type I error is correct. The type I error is considered correct if it falls within the 95% confidence interval of the 5% significance level (0.0457, 0.0543). The tests are grouped into three groups from *left* to *right* for result presentation, (1) t-tests: pooled t-test (**pooled**), theoretical and permutation Welch's t-tests (**welch** and **p.welch**), robust t-test (**robust.t**), (2) Wilcoxon rank-sum test (**wilcox**), permutation Wilcoxon rank-sum test (**p.wilcox**) and the minimum p-value (**minp**), (3) two-stage tests: two-stage tests with the first stage α level at 0.5, 2.5, 5% (**2stage0.5**, **2stage2.5**, **2stage5**)

1.5. All the valid tests have similar power. Tests with an inflated type I error tend to be slightly more powerful. In contrast, tests that have a conservative type I error tend to be slightly less powerful.

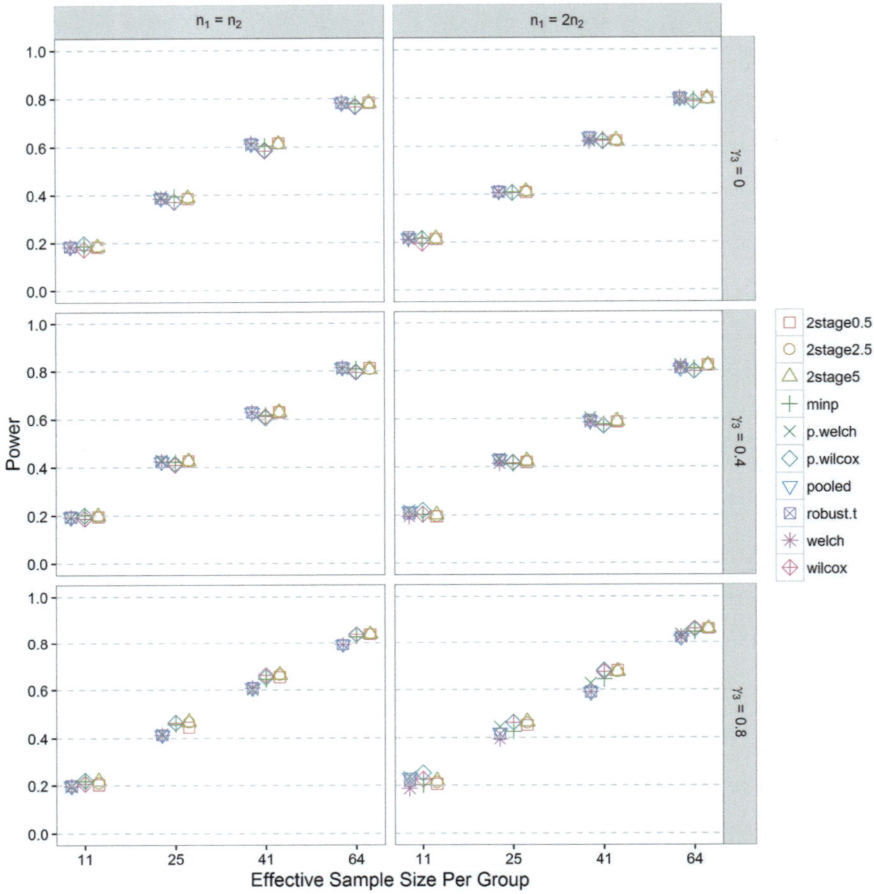

Fig. 5 Power simulation results for skewed data. The tests are grouped into three groups from *left* to *right* for result presentation, (1) t-tests: pooled t-test (**pooled**), theoretical and permutation Welch's t-tests (**welch** and **p.welch**), robust t-test (**robust.t**), (2) Wilcoxon rank-sum test (**wilcox**), permutation Wilcoxon rank-sum test (**p.wilcox**) and the minimum p-value (**minp**), (3) two-stage tests: two-stage tests with the first stage α level at 0.5, 2.5, 5% (**2stage0.5**, **2stage2.5**, **2stage5**)

4.2 Skewness

The two sample size ratios result in similar patterns of type I error and power. All tests maintain a correct type I error across the spectrum of sample size and skewness except 2stage0.5, which has an inflated type I error when the sample size is insufficiently large and the skewness is 0.8 (Fig. 4). A similar pattern for power is observed within groups (three groups in three columns). With an increase in skewness, the power of t-tests (pooled t-test, theoretical and permutation Welch's t-test, robust t-test) remains at the level as planned and is slightly lower than other tests when the skewness reaches 0.8 (Fig. 5).

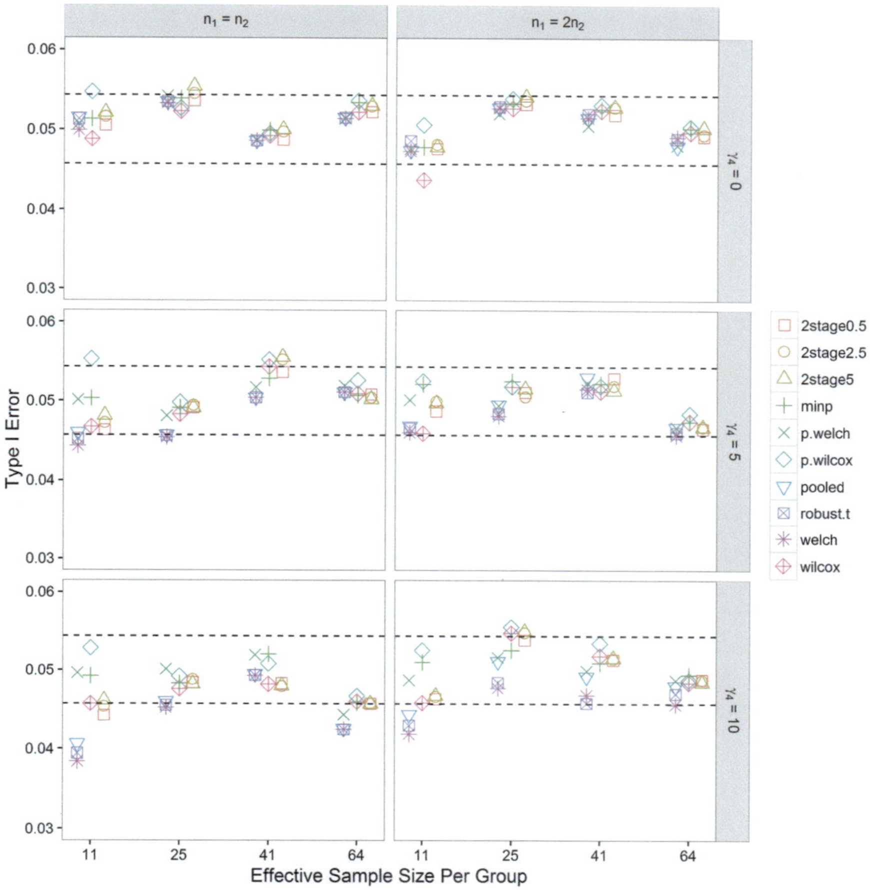

Fig. 6 Null simulation results for kurtotic data. $y = 0.0457$ and $y = 0.0543$ are added to judge whether the type I error is correct. The type I error is considered correct if it falls within the 95% confidence interval of the 5% significance level (0.0457, 0.0543). The tests are grouped into three groups from *left* to *right* for result presentation, (1) t-tests: pooled t-test (**pooled**), theoretical and permutation Welch's t-tests (**welch** and **p.welch**), robust t-test (**robust.t**), (2) Wilcoxon rank-sum test (**wilcox**), permutation Wilcoxon rank-sum test (**p.wilcox**) and the minimum p-value (**minp**), (3) two-stage tests: two-stage tests with the first stage α level at 0.5, 2.5, 5% (**2stage0.5**, **2stage2.5**, **2stage5**)

4.3 *Kurtosis*

Based on Fig. 6, all tests maintain a correct type I error when the effective sample size is equal or greater than 25. When the effective sample size is small, e.g. $n_e = 11$, and the kurtosis is 10, t-tests (with the exception of permutation Welch's t-test) exhibit a conservative type I error—the type I error of permutation Welch's t-test is protected with consistently similar power to other t-tests. Figure 7 displays tests within groups

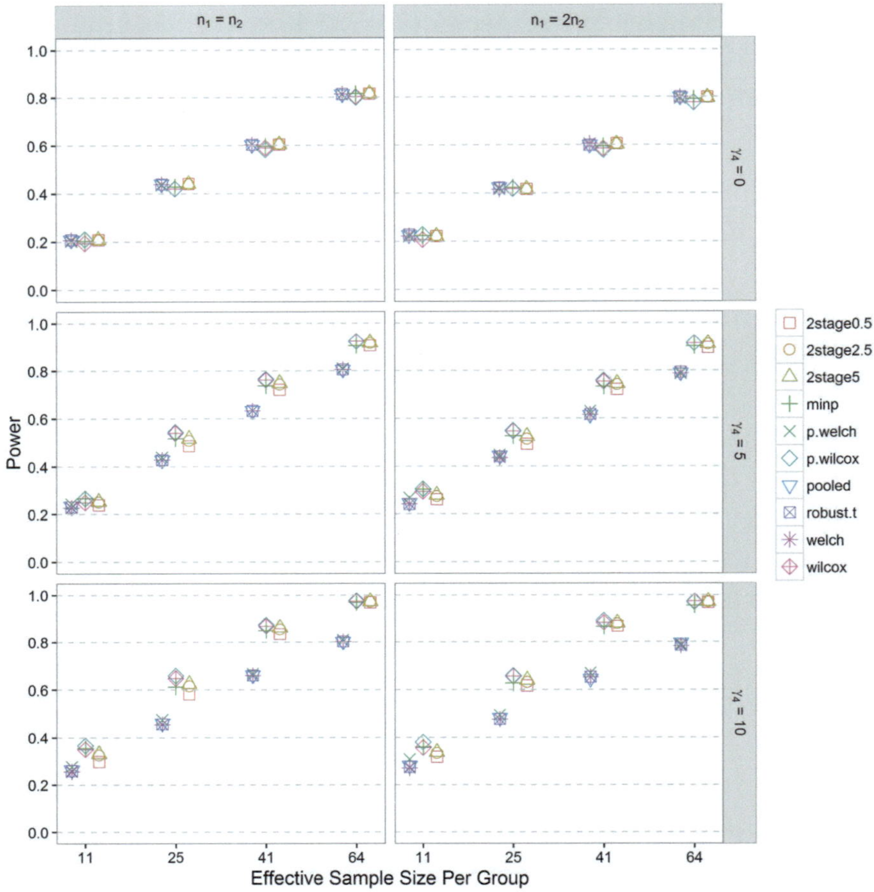

Fig. 7 Power simulation results for kurtotic data. The tests are grouped into three groups from *left* to *right* for result presentation, (1) t-tests: pooled t-test (**pooled**), theoretical and permutation Welch's t-tests (**welch** and **p.welch**), robust t-test (**robust.t**), (2) Wilcoxon rank-sum test (**wilcox**), permutation Wilcoxon rank-sum test (**p.wilcox**) and the minimum p-value (**minp**), (3) two-stage tests: two-stage tests with the first stage α level at 0.5, 2.5, 5% (**2stage0.5**, **2stage2.5**, **2stage5**)

(three groups in three columns) with similar power. While not apparent when the sample size is small, t-tests are not as compelling as other tests when the skewness is 5 or greater. Wilcoxon rank-sum tests are the best performing tests in this setting, but the power gain is not significant when compared to the minimum p-value and two-stage tests.

5 Discussion

The goal of this chapter was to conduct an evaluation of the variety of tests for two-sample location comparison using Monte Carlo simulation techniques. In conclusion, heterogeneous variance is not a problem for any of the tests when the sample sizes are equal. For unequal sample sizes, Welch's t-test maintains robustness against heterogeneous variance. The interaction between heterogeneous variance and sample size is consistent with the statement of (Kohr and Games 1974) for general two-sample location comparison that a test is conservative when the larger sample has the larger population variance and is liberal when the smaller sample has the larger population variance. When the normality assumption is violated by moderately skewed or kurtotic data, Welch's t-test and other t-tests maintain a correct type I error so long as the sample size is sufficiently large. The t-tests, however, are not as powerful as the Wilcoxon rank-sum test and others.

As long as the effective sample size is 11 or greater, distributions of Welch's t-test and the Wilcoxon rank sum test statistics are well approximated by their theoretical distributions. There is no need to apply a permutation test for a small sample size of that level if a theoretical test is appropriately applied. A permutation test is not a solution to heterogeneous variance but may protect type I error when the normality is violated and the sample size is small.

An extra test to test for equality of variance or for normality may fail to protect type I error, e.g. robust.t for heterogeneous variance and 2stage0.5 for skewed data when the sample size is insufficiently large. While t-tests are not sensitive to non-normal data for the protection of type I error, a conservative normality test with α as low as 0.5% can lead to a biased type I error. In simulations, the two-stage tests perform well against heterogeneous variance, but this is only the case because the simulation settings assume both distributions are normal and the two-stage tests in fact are Welch's t-test most of the time. The two-stage tests are meant to optimize the test based on whether the normality assumption is met. These tests do not handle heterogeneous variance. Under circumstances of non-normal distributions and heterogeneous variance, it has been shown by Rasch et al. that two-stage tests may lead to an incorrect type I error and lose power (Rasch et al. 2011).

The literature and these simulation results reveal that the optimal test procedure is a sensible switch between Welch's t-test and the Wilcoxon rank sum test. Multi-stage tests such as robust t-test and two-stage tests are criticized for their biased type I error. The minimum p-value simply takes the best result of Welch's t and Wilcoxon rank-sum tests and is not designed specifically for heterogeneous variance or non-normal data distributions. While it shares the flaw of multi-stage tests with no protection of type I error and secondary power, the minimum p-value is potentially robust to heterogeneous variance and non-normal distributions.

A simple way to protect type I error and power against heterogeneous variance is to plan a study with two samples of equal size. In this setting, Wilcoxon rank-sum test maintains robustness to non-normal data and competitive with Welch's t-test when the normality assumption is met. If the sample sizes are unequal, the minimum p-value is a robust option but computationally more expensive. Otherwise, heterogeneous variance and deviation from a normal distribution must be weighed in selecting between Welch's t-test or the Wilcoxon rank-sum test. Alternatively, Zimmerman and Zumbo recommended that Welchs t-test performs better than the Wilcoxon rank sum test under conditions of both equal and unequal variance on data that has been pre-ranked (Zimmerman and Zumbo 1993).

Although moderately non-normal data has been the focus of this chapter, the conclusions are still useful as a general guideline. When data is extremely non-normal, none of the tests will be appropriate. Data transformation may be applied to meet the normality assumption Osborne (2005, 2010). Not every distribution, however, can be transformed to normality (e.g. L-shaped distributions). In this scenario, data dichotomization may be applied to simplify statistical analysis if information loss is not a concern (Altman and Royston 2006). Additional concerns include multimodal distributions which may be a result of subgroups or data contamination (Marrero 1985)—a few outliers can distort the distribution completely. All of these considerations serve as a reminder that prior to selecting a two-sample location test, the first key to success is full practical and/or clinical understanding of the data being assessed.

References

Altman, D. G., & Royston, P. (2006). The cost of dichotomising continuous variables. *Bmj*, *332*(7549), 1080.

Beasley, T. M., Erickson, S., & Allison, D. B. (2009). Rank-based inverse normal transformations are increasingly used, but are they merited? *Behavior Genetics*, *39*(5), 580–595.

Boik, R. J. (1987). The fisher-pitman permutation test: A non-robust alternative to the normal theory f test when variances are heterogeneous. *British Journal of Mathematical and Statistical Psychology*, *40*(1), 26–42.

Cohen, J. (2013). *Statistical power analysis for the behavioral sciences*. Academic press.

de Winter, J. C. (2013). Using the students t-test with extremely small sample sizes. *Practical Assessment, Research & Evaluation*, *18*(10), 1–12.

Demirtas, H., Hedeker, D., & Mermelstein, R. J. (2012). Simulation of massive public health data by power polynomials. *Statistics in Medicine*, *31*(27), 3337–3346.

Devroye, L. (1986). Sample-based non-uniform random variate generation. In *Proceedings of the 18th conference on Winter simulation*, pp. 260–265. ACM.

Ernst, M. D., et al. (2004). Permutation methods: a basis for exact inference. *Statistical Science*, *19*(4), 676–685.

Faul, F., Erdfelder, E., Lang, A.-G., & Buchner, A. (2007). G* power 3: A flexible statistical power ana lysis program for the social, behavioral, and biomedical sciences. *Behavior Research Methods*, *39*(2), 175–191.

Headrick, T. C., & Sawilowsky, S. S. (2000). Weighted simplex procedures for determining boundary points and constants for the univariate and multivariate power methods. *Journal of Educational and Behavioral Statistics*, *25*(4), 417–436.

Kohr, R. L., & Games, P. A. (1974). Robustness of the analysis of variance, the welch procedure and a box procedure to heterogeneous variances. *The Journal of Experimental Education*, *43*(1), 61–69.

Marrero, O. (1985). Robustness of statistical tests in the two-sample location problem. *Biometrical Journal, 27*(3), 299–316.
Osborne, J. (2005). Notes on the use of data transformations. *Practical Assessment, Research and Evaluation, 9*(1), 42–50.
Osborne, J. W. (2010). Improving your data transformations: Applying the box-cox transformation. *Practical Assessment, Research & Evaluation, 15*(12), 1–9.
Rasch, D., Kubinger, K. D., & Moder, K. (2011). The two-sample t-test: pre-testing its assumptions does not pay off. *Statistical Papers, 52*(1), 219–231.
Rochon, J., Gondan, M., & Kieser, M. (2012). To test or not to test: Preliminary assessment of normality when comparing two independent samples. *BMC Medical Research Methodology, 12*(1), 81.
Royston, J. (1982). An extension of shapiro and wilk's w test for normality to large samples. *Applied Statistics*, 115–124.
Ruxton, G. D. (2006). The unequal variance t-test is an underused alternative to student's t-test and the mann-whitney u test. *Behavioral Ecology, 17*(4), 688–690.
Sawilowsky, S. S. (2005). Misconceptions leading to choosing the t-test over the wilcoxon mann-whitney test for shift in location parameter.
Schucany, W. R., & Tony Ng H. (2006). Preliminary goodness-of-fit tests for normality do not validate the one-sample student t. *Communications in Statistics Theory and Methods, 35*(12), 2275–2286.
Team, R. C. (2014). R: A language and environment for statistical computing. R foundation for statistical computing, vienna, austria, 2012.
Welch, B. L. (1938). The significance of the difference between two means when the population variances are unequal. *Biometrika, 29*(3–4), 350–362.
Zimmerman, D. W. (1996). A note on homogeneity of variance of scores and ranks. *The Journal of Experimental Education, 64*(4), 351–362.
Zimmerman, D. W. (1998). Invalidation of parametric and nonparametric statistical tests by concurrent violation of two assumptions. *The Journal of Experimental Education, 67*(1), 55–68.
Zimmerman, D. W. (2004). A note on preliminary tests of equality of variances. *British Journal of Mathematical and Statistical Psychology, 57*(1), 173–181.
Zimmerman, D. W., & Zumbo, B. D. (1993). Rank transformations and the power of the student t-test and welch t'test for non-normal populations with unequal variances. *Canadian Journal of Experimental Psychology/Revue canadienne de psychologie expérimentale, 47*(3), 523.

Anatomy of Correlational Magnitude Transformations in Latency and Discretization Contexts in Monte-Carlo Studies

Hakan Demirtas and Ceren Vardar-Acar

Abstract This chapter is concerned with the assessment of correlational magnitude changes when a subset of the continuous variables that may marginally or jointly follow nearly any distribution in a multivariate setting is dichotomized or ordinalized. Statisticians generally regard discretization as a bad idea on the grounds of power, information, and effect size loss. Despite this undeniable disadvantage and legitimate criticism, its widespread use in social, behavioral, and medical sciences stems from the fact that discretization could yield simpler, more interpretable, and understandable conclusions, especially when large audiences are targeted for the dissemination of the research outcomes. We do not intend to attach any negative or positive connotations to discretization, nor do we take a position of advocacy for or against it. The purpose of the current chapter is providing a conceptual framework and computational algorithms for modeling the correlation transitions under specified distributional assumptions within the realm of discretization in the context of the latency and threshold concepts. Both directions (identification of the pre-discretization correlation value in order to attain a specified post-discretization magnitude, and the other way around) are discussed. The ideas are developed for bivariate settings; a natural extension to the multivariate case is straightforward by assembling the individual correlation entries. The paradigm under consideration has important implications and broad applicability in the stochastic simulation and random number generation worlds. The proposed algorithms are illustrated by several examples; feasibility and performance of the methods are demonstrated by a simulation study.

H. Demirtas (✉)
Division of Epidemiology and Biostatistics (MC923), University of Illinois at Chicago, 1603 West Taylor Street, Chicago, IL 60612, USA
e-mail: demirtas@uic.edu

C. Vardar-Acar
Department of Statistics, Middle East Technical University, Ankara, Turkey
e-mail: cvardar@metu.edu.tr

D.-G. Chen and J.D. Chen (eds.), *Monte-Carlo Simulation-Based Statistical Modeling*, ICSA Book Series in Statistics, DOI 10.1007/978-981-10-3307-0_4

1 Introduction

Unlike natural (true) dichotomies such as male versus female, conductor versus insulator, vertebrate versus invertebrate, and in-patient versus out-patient, some binary variables are derived through dichotomization of underlying continuous measurements. Such artificial dichotomies often arise across many scientific disciplines. Examples include obesity status (obese versus non-obese) based on body mass index, preterm versus term babies given the gestation period, high versus low need of social interaction, small versus large tumor size, early versus late response time in surveys, young versus old age, among many others. In the ordinal case, discretization is equally commonly encountered in practice. Derived polytomous variables such as young-middle-old age, low-medium-high income, cold-cool-average-hot temperature, no-mild-moderate-severe depression are obtained based on nominal age, income, temperature, and depression score, respectively. While binary is a special case of ordinal, for the purpose of illustration, integrity, and clarity, separate arguments are presented throughout the chapter. On a terminological note, we use the words binary/dichotomous and ordinal/polytomous interchangeably to simultaneously reflect the preferences of statisticians/psychometricians. Obviously, polytomous variables can normally be ordered (ordinal) or unordered (nominal). For the remainder of the chapter, the term "polytomous" is assumed to correspond ordered variables.

Discretization is typically shunned by statisticians for valid reasons, the most prominent of which is the power and information loss. In most cases, it leads to a diminished effect size as well as reduced reliability and strength of association. However, simplicity, better interpretability and comprehension of the effects of interest, and superiority of some categorical data measures such as odds ratio have been argued by proponents of discretization. Those who are against it assert that the regression paradigm is general enough to account for interactive effects, outliers, skewed distributions, and nonlinear relationships. In practice, especially substantive researchers and practitioners employ discretization in their works. For conflicting views on relative perils and merits of discretization, see MacCallum et al. (2002) and Farrington and Loeber (2000), respectively. We take a neutral position; although it is not a recommended approach from the statistical theory standpoint, it frequently occurs in practice, mostly driven by improved understandability-based arguments. Instead of engaging in fruitless philosophical discussions, we feel that a more productive effort can be directed towards finding answers when the discretization is performed, which motivates the formation of this chapter's major goals: (1) The determination of correlational magnitude changes when some of the continuous variables that may marginally or jointly follow almost any distribution in a multivariate setting are dichotomized or ordinalized. (2) The presentation of a conceptual and computational framework for modeling the correlational transformations before and after discretization.

A correlation between two continuous variables is usually computed as the common Pearson correlation. If one or both variables is/are dichotomized/ordinalized by a threshold concept of underlying continuous variables, different naming conventions are assigned to the correlations. A correlation between a continuous and a dichotomized/ordinalized variable is a biserial/polyserial and point-biserial/point-polyserial correlation before and after discretization, respectively. When both variables are dichotomized/ordinalized, the correlation between the two latent continuous variables is known as the tetrachoric/polychoric correlation. The phi coefficient is the correlation between two discretized variables; in fact, the term phi coefficient is reserved for dichotomized variables, but for lack of a better term we call it the "ordinal phi coefficient" for ordinalized variables. All of these correlations are special cases of the Pearson correlation.

Correlations are naturally altered in magnitude by discretization. In the binary case, there is a closed form, double numerical integration formula that connects the correlations before and after dichotomization under the normality assumption when both variables are dichotomized (Emrich and Piedmonte 1991). Demirtas and Doganay (2012) enhanced this Gaussian copula approach, along with the algebraic relationship between the biserial and point-biserial correlations, in the context of joint generation of binary and normal variables. Demirtas and Hedeker (2016) further extended this correlational connection to nonnormal variables via linearity and constancy arguments when only one variable in a bivariate setting is dichotomized. Going back to the scenario where the dichotomization is performed on both variables, Demirtas (2016) proposed algorithms that find the phi coefficient when the tetrachoric correlation is specified (and the other way around), under any distributional assumption for continuous variables through added operational utility of the power polynomials (Fleishman 1978; Vale and Maurelli 1983). In the ordinal case, modeling the correlation transition can only be performed iteratively under the normality assumption (Ferrari and Barbiero 2012). Demirtas et al. (2016a) augmented the idea of computing the correlation before or after discretization when the other one is specified and vice versa, to an ordinal setting. The primary purpose of this chapter is providing several algorithms that are designed to connect pre- and post-discretization correlations under specified distributional assumptions in simulated environments. More specifically, the following relationships are established: (a) tetrachoric correlation/phi coefficient, (b) biserial/point-biserial correlations, (c) polychoric correlation/ordinal phi coefficient, and (d) polyserial/point-polyserial correlations, where (a)–(b) and (c)–(d) are relevant to binary and ordinal data, respectively; (b)–(d) and (a)–(c) pertain to situations where only one or both variables is/are discretized, respectively. In all of these cases, the marginal distributions that are needed for finding skewness (symmetry) and elongation (peakedness) values for the underlying continuous variables, proportions for binary and ordinal variables, and associational quantities in the form of the Pearson correlation are assumed to be specified.

This work is important and of interest for the following reasons: (1) The link between these types of correlations has been studied only under the normality assumption; however, the presented level of generality that encompasses a comprehensive range of distributional setups is a necessary progress in computational statistics. (2) As simulation studies are typically based on replication of real data characteristics and/or specified hypothetical data trends, having access to the latent data as well as the eventual binary/ordinal data may be consequential for exploring a richer spectrum of feasible models that are applicable for a given data-analytic problem involving correlated binary/ordinal data. (3) The set of techniques sheds light on how correlations are related before and after discretization; and has potential to broaden our horizon on its relative advantages and drawbacks. (4) The algorithms work for a very broad class of underlying bivariate latent densities and binary/ordinal data distributions, allowing skip patterns for the latter, without requiring the identical distribution assumption on either type of variables. (5) The required software tools for the implementation are minimal, one only needs a numerical double integration solver for the binary-binary case, a computational platform with univariate random number generation (RNG) capabilities for the binary/ordinal-continuous case, an iterative scheme that connects the polychoric correlations and the ordinal phi coefficients under the normality assumption for the ordinal-ordinal case, a polynomial root-finder and a nonlinear equations set solver to handle nonnormal continuous variables. (6) The algorithmic steps are formulated for the bivariate case by the nature of the problem, but handling the multivariate case is straightforward by assembling the correlation matrix entries. (7) The collection of the algorithms can be regarded as an operational machinery for developing viable RNG mechanisms to generate multivariate latent variables as well as subsequent binary/ordinal variables given their marginal shape characteristics and associational structure in simulated settings, potentially expediting the evolution of the mixed data generation routines. (8) The methods could be useful in advancing research in meta-analysis domains where variables are discretized in some studies and remained continuous in some others.

The organization of the chapter is as follows. In Sect. 2, necessary background information is given for the development of the proposed algorithms. In particular, how the correlation transformation works for discretized data through numerical integration and an iterative scheme for the binary and ordinal cases, respectively, under the normality assumption, is outlined; a general nonnormal continuous data generation technique that forms a basis for the proposed approaches is described (when both variables are discretized), and an identity that connects correlations before and after discretization via their mutual associations is elaborated upon (when only one variable is discretized). In Sect. 3, several algorithms for finding one quantity given the other are provided under various combinations of cases (binary versus ordinal, directionality in terms of specified versus computed correlation with respect to pre- versus post-discretization, and whether discretization is applied on one versus both variables) and some illustrative examples, representing a broad range of distributional shapes that can be encountered in real applications, are presented for the purpose of exposition. In Sect. 4, a simulation study for evaluating the method's performance in a multivariate setup by commonly accepted accuracy (unbiasedness) measures in both directions is discussed. Section 5 includes concluding remarks, future research directions, limitations, and extensions.

2 Building Blocks

This section gives necessary background information for the development of the proposed algorithms in modeling correlation transitions. In what follows, correlation type and related notation depend on the three factors: (a) before or after discretization, (b) only one or both variables is/are discretized, and (c) discretized variable is dichotomous or polytomous. To establish the notational convention, for the remainder of the chapter, let Y_1 and Y_2 be the continuous variables where either Y_1 only or both are discretized to yield X_1 and X_2 depending on the correlation type under consideration (When Y's are normal, they are denoted as Z, which is relevant for the normal-based results and for the nonnormal extension via power polynomials). To distinguish between binary and ordinal variables, the symbols B and O appear in the subscripts. Furthermore, for avoiding any confusion, the symbols BS, TET, PS, and $POLY$ are made a part of $\delta_{Y_1Y_2}$ and $\delta_{Z_1Z_2}$ to differentiate among the biserial, tetrachoric, polyserial, and polychoric correlations, respectively. For easier readability, we include Table 1 that shows the specific correlation types and associated notational symbols based on the three above-mentioned factors.

2.1 *Dichotomous Case: Normality*

Borrowing ideas from the RNG literature, if the underlying distribution before dichotomization is bivariate normal, the relationship between the phi coefficient and the tetrachoric correlation is known (Emrich and Piedmonte 1991; Demirtas and Doganay 2012; Demirtas 2016). Let X_{1B} and X_{2B} represent binary variables such that $E[X_{jB}] = p_j = 1 - q_j$ for $j = 1, 2$, and $Cor[X_{1B}, X_{2B}] = \delta_{X_{1B}X_{2B}}$, where p_1, p_2, and $\delta_{X_{1B}X_{2B}}$ are given. Let $Z_1 = Y_1$ and $Z_2 = Y_2$ be the corresponding standard normal variables, and let Φ be the cumulative distribution func-

Table 1 Terminological and notational convention for different correlation types depending on the three self-explanatory factors

When	Discrete data type	Discretized	Name	Symbol
Before	Dichotomous	Y_1 only	Biserial correlation	$\delta_{Y_1Y_2^{BS}}$
After			Point-biserial correlation	$\delta_{X_{1B}Y_2}$
Before		Both Y_1 and Y_2	Tetrachoric correlation	$\delta_{Y_1Y_2^{TET}}$
After			Phi coefficient	$\delta_{X_{1B}X_{2B}}$
Before	Polytomous	Y_1 only	Polyserial correlation	$\delta_{Y_1Y_2^{PS}}$
After			Point-polyserial correlation	$\delta_{X_{1O}Y_2}$
Before		Both Y_1 and Y_2	Polychoric correlation	$\delta_{Y_1Y_2^{POLY}}$
After			Ordinal phi coefficient	$\delta_{X_{1O}X_{2O}}$

tion for a standard bivariate normal random variable with correlation coefficient $\delta_{Z_1 Z_2^{TET}}$. Obviously, $\Phi[z_1, z_2, \delta_{Z_1 Z_2^{TET}}] = \int_{-\infty}^{z_1} \int_{-\infty}^{z_2} f(z_1, z_2, \delta_{Z_1 Z_2^{TET}}) dz_1 dz_2$, where $f(z_1, z_2, \delta_{Z_1 Z_2^{TET}}) = [2\pi(1 - \delta^2_{Z_1 Z_2^{TET}})^{1/2}]^{-1} \times exp\Big[- (z_1^2 - 2\delta_{Z_1 Z_2^{TET}} z_1 z_2 + z_2^2)/ (2(1 - \delta^2_{Z_1 Z_2^{TET}}))\Big]$. The phi coefficient ($\delta_{X_{1B} X_{2B}}$) and the tetrachoric correlation ($\delta_{Z_1 Z_2^{TET}}$) are linked via the equation

$$\Phi[z(p_1), z(p_2), \delta_{Z_1 Z_2^{TET}}] = \delta_{X_{1B} X_{2B}} (p_1 q_1 p_2 q_2)^{1/2} + p_1 p_2 \tag{1}$$

where $z(p_j)$ denotes the p_j^{th} quantile of the standard normal distribution for $j = 1, 2$. As long as $\delta_{X_{1B} X_{2B}}$ is within the feasible correlation range (Hoeffding 1940, Fréchet 1951; Demirtas and Hedeker 2011), the solution is unique. Once Z_1 and Z_2 are generated, the binary variables are derived by setting $X_{jB} = 1$ if $Z_j \geq z(1 - p_j)$ and 0 otherwise for $j = 1, 2$. While RNG is not our ultimate interest in this work, we can use Eq. 1 for bridging the phi coefficient and the tetrachoric correlation.

When only one of the normal variables (Z_1) is dichotomized, i.e., $X_{1B} = I(Z_1 \geq z(1 - p_1)$, it is relatively easy to show that $\delta_{X_{1B} Z_2} / \delta_{Z_1 Z_2^{BS}} = \delta_{X_{1B} Z_1} = h/\sqrt{p_1 q_1}$ where h is the ordinate of the normal curve at the point of dichotomization (Demirtas and Hedeker 2016).

Real data often do not conform to the assumption of normality; hence most simulation studies should take nonnormality into consideration. The next section investigates the situation where one or both continuous variables is/are nonnormal.

2.2 *Dichotomous Case: Beyond Normality*

Extending the limited and restrictive normality-based results to a broad range of distributional setups requires the employment of the two frameworks (an RNG routine for multivariate continuous data and a derived linear relationship in the presence of discretization), which we outline below.

We first tackle the relationship between the tetrachoric correlation $\delta_{Y_1 Y_2^{TET}}$ and the phi coefficient $\delta_{X_{1B} X_{2B}}$ under nonnormality via the use of the power polynomials (Fleishman 1978), which is a moment-matching procedure that simulates nonnormal distributions often used in Monte-Carlo studies, based on the premise that real-life distributions of variables are typically characterized by their first four moments. It hinges upon the polynomial transformation, $Y = a + bZ + cZ^2 + dZ^3$, where Z follows a standard normal distribution, and Y is standardized (zero mean and unit variance).[1] The distribution of Y depends on the constants a, b, c, and d, that can be computed for specified or estimated values of skewness ($\nu_1 = E[Y^3]$) and excess kurtosis ($\nu_2 = E[Y^4] - 3$). The procedure of expressing any given variable by the sum of linear combinations of powers of a standard normal variate is capable of

[1] We drop the subscript in Y as we start with the univariate case.

covering a wide area in the skewness-elongation plane whose bounds are given by the general expression $\nu_2 \geq \nu_1^2 - 2$.[2]

Assuming that $E[Y] = 0$, and $E[Y^2] = 1$, by utilizing the moments of the standard normal distribution, the following set of equations can be derived:

$$a = -c \tag{2}$$

$$b^2 + 6bd + 2c^2 + 15d^2 - 1 = 0 \tag{3}$$

$$2c(b^2 + 24bd + 105d^2 + 2) - \nu_1 = 0 \tag{4}$$

$$24[bd + c^2(1 + b^2 + 28bd) + d^2(12 + 48bd + 141c^2 + 225d^2)] - \nu_2 = 0 \tag{5}$$

These equations can be solved by the Newton-Raphson method, or any other plausible root-finding or nonlinear optimization routine. More details for the Newton-Raphson algorithm for this particular setting is given by Demirtas et al. (2012). The polynomial coefficients are estimated for centered and scaled variables; the resulting data set should be back-transformed to the original scale by multiplying every data point by the standard deviation and adding the mean. Centering-scaling and the reverse of this operation are linear transformations, so it does not change the values of skewness, kurtosis, and correlations. Of note, we use the words symmetry and skewness interchangeably. Similarly, kurtosis, elongation, and peakedness are meant to convey the same meaning.

The multivariate extension of Fleishman's power method (Vale and Maurelli 1983) plays a central role for the remainder of this chapter. The procedure for generating multivariate continuous data begins with computation of the constants given in Eqs. 2–5, independently for each variable. The bivariate case can be formulated in matrix notation as shown below. First, let Z_1 and Z_2 be variables drawn from standard normal populations; let $\mathbf{z}'$ be the vector of normal powers 0 through 3, $\mathbf{z_j}' = [1, Z_j, Z_j^2, Z_j^3]$; and let $\mathbf{w}'$ be the weight vector that contains the power function weights a, b, c, and d, $\mathbf{w_j}' = [a_j, b_j, c_j, d_j]$ for $j = 1, 2$. The nonnormal variable Y_j is then defined as the product of these two vectors, $Y_j = \mathbf{w_j}'\mathbf{z_j}$. Let $\delta_{Y_1 Y_2}$ be the correlation between two nonnormal variables Y_1 and Y_2 that correspond to the normal variables Z_1 and Z_2, respectively.[3] As the variables are standardized, meaning $E(Y_1) = E(Y_2) = 0$, $\delta_{Y_1 Y_2} = E(Y_1 Y_2) = E(\mathbf{w_1}'\mathbf{z_1}\mathbf{z_2}'\mathbf{w_2}) = \mathbf{w_1}'\mathscr{R}\mathbf{w_2}$, where $\mathscr{R}$ is the expected matrix product of $\mathbf{z_1}$ and $\mathbf{z_2}'$:

$$\mathscr{R} = E(\mathbf{z_1}\mathbf{z_2}') = \begin{bmatrix} 1 & 0 & 1 & 0 \\ 0 & \delta_{Z_1 Z_2} & 0 & 3\delta_{Z_1 Z_2} \\ 1 & 0 & 2\delta^2_{Z_1 Z_2} + 1 & 0 \\ 0 & 3\delta_{Z_1 Z_2} & 0 & 6\delta^3_{Z_1 Z_2} + 9\delta_{Z_1 Z_2} \end{bmatrix},$$

---

[2] In fact, equality is not possible for continuous distributions.

[3] $\delta_{Y_1 Y_2}$ is the same as $\delta_{Y_1 Y_2^{TET}}$ or $\delta_{Y_1 Y_2^{POLY}}$ depending on if discretized variables are binary or ordinal, respectively. For the general presentation of the power polynomials, we do not make that distinction.

where $\delta_{Z_1Z_2}$ is the correlation between Z_1 and Z_2. After algebraic operations, the following relationship between $\delta_{Y_1Y_2}$ and $\delta_{Z_1Z_2}$ in terms of polynomial coefficients ensues:

$$\delta_{Y_1Y_2} = \delta_{Z_1Z_2}(b_1b_2 + 3b_1d_2 + 3d_1b_2 + 9d_1d_2) + \delta^2_{Z_1Z_2}(2c_1c_2) + \delta^3_{Z_1Z_2}(6d_1d_2) \quad (6)$$

Solving this cubic equation for $\delta_{Z_1Z_2}$ gives the intermediate correlation between the two standard normal variables that is required for the desired post-transformation correlation $\delta_{Y_1Y_2}$. Clearly, correlations for each pair of variables should be assembled into a matrix of intercorrelations in the multivariate case. For a comprehensive source and detailed account on the power polynomials, see Headrick (2010).

In the dichotomization context, the connection between the underlying nonnormal ($\delta_{Y_1Y_2}$) and normal correlations ($\delta_{Z_1Z_2}$) in Eq. 6, along with the relationship between the tetrachoric correlation ($\delta_{Z_1Z_2^{TET}}$) and the phi coefficient ($\delta_{X_{1B}X_{2B}}$) conveyed in Eq. 1, is instrumental in *Algorithms-1a* and *-1b* in Sect. 3.

To address the situation where only one variable (Y_1) is dichotomized, we now move to the relationship of biserial ($\delta_{Y_1Y_2^{BS}}$) and point-biserial ($\delta_{X_{1B}Y_2}$) correlations in the absence of the normality assumption, which merely functions as a starting point below. Suppose that Y_1 and Y_2 jointly follow a bivariate normal distribution with a correlation of $\delta_{Y_1Y_2^{BS}}$. Without loss of generality, we may assume that both Y_1 and Y_2 are standardized to have a mean of 0 and a variance of 1. Let X_{1B} be the binary variable resulting from a split on Y_1, $X_{1B} = I(Y_1 \geq k)$, where k is the point of dichotomization. Thus, $E[X_{1B}] = p_1$ and $V[X_{1B}] = p_1q_1$ where $q_1 = 1 - p_1$. The correlation between X_{1B} and Y_1, $\delta_{X_{1B}Y_1}$ can be obtained in a simple way, namely, $\delta_{X_{1B}Y_1} = \frac{Cov[X_{1B}, Y_1]}{\sqrt{V[X_{1B}]V[Y_1]}} = E[X_{1B}Y_1]/\sqrt{p_1q_1} = E[Y_1|Y_1 \geq k]/\sqrt{p_1q_1}$. We can also express the relationship between Y_1 and Y_2 via the following linear regression model:

$$Y_2 = \delta_{Y_1Y_2^{BS}} Y_1 + \varepsilon \quad (7)$$

where ε is independent of Y_1 and Y_2, and follows $N \sim (0, 1 - \delta^2_{Y_1Y_2^{BS}})$. When we generalize this to nonnormal Y_1 and/or Y_2 (both centered and scaled), the same relationship can be assumed to hold with the exception that the distribution of ε follows a nonnormal distribution. As long as Eq. 7 is valid,

$$\begin{aligned} Cov[X_{1B}, Y_2] &= Cov[X_{1B}, \delta_{Y_1Y_2^{BS}} Y_1 + \varepsilon] \\ &= Cov[X_{1B}, \delta_{Y_1Y_2^{BS}} Y_1] + Cov[X_{1B}, \varepsilon] \\ &= \delta_{Y_1Y_2^{BS}} Cov[X_{1B}, Y_1] + Cov[X_{1B}, \varepsilon]\ . \end{aligned} \quad (8)$$

Since ε is independent of Y_1, it will also be independent of any deterministic function of Y_1 such as X_{1B}, and thus $Cov[X_{1B}, \varepsilon]$ will be 0. As $E[Y_1] = E[Y_2] = 0$, $V[Y_1] = V[Y_2] = 1$, $Cov[X_{1B}, Y_1] = \delta_{X_{1B}Y}\sqrt{p_1q_1}$ and $Cov[Y_1, Y_2] = \delta_{Y_1Y_2^{BS}}$, Eq. 8 reduces to

$$\delta_{X_{1B}Y_2} = \delta_{Y_1Y_2^{BS}} \delta_{X_{1B}Y_1}\ . \quad (9)$$

In the bivariate normal case, $\delta_{X_{1B}Y_1} = h/\sqrt{p_1q_1}$ where h is the ordinate of the normal curve at the point of dichotomization. Equation 9 indicates that the linear association between X_{1B} and Y_2 is assumed to be fully explained by their mutual association with Y_1 (Demirtas and Hedeker 2016). The ratio, $\delta_{X_{1B}Y_2}/\delta_{Y_1Y_2^{BS}}$ is equal to $\delta_{X_{1B}Y_1} = E[X_{1B}Y_1]/\sqrt{p_1q_1} = E[Y_1|Y_1 \geq k]/\sqrt{p_1q_1}$, which is a constant given p_1 and the distribution of (Y_1, Y_2). These correlations are invariant to location shifts and scaling, Y_1 and Y_2 do not have to be centered and scaled, their means and variances can take any finite values. Once the ratio ($\delta_{X_{1B}Y_1}$) is found (it could simply be done by generating Y_1 and dichotomizing it to yield X_{1B}), one can compute the point-biserial ($\delta_{X_{1B}Y_2}$) or biserial $\delta_{Y_1Y_2^{BS}}$ correlation when the other one is specified. This linearity-constancy argument that jointly emanates from Eqs. 7 and 9, will be the crux of *Algorithm-2* given in Sect. 3.

2.3 Polytomous Case: Normality

In the ordinal case, although the relationship between the polychoric correlation ($\delta_{Y_1Y_2^{POLY}}$) and the ordinal phi correlation ($\delta_{X_{1O}X_{2O}}$) can be written in closed form, as explained below, the solution needs to be obtained iteratively even under the normality assumption since no nice recipe such as Eq. 1 is available. In the context of correlated ordinal data generation, Ferrari and Barbiero (2012) proposed an iterative procedure based on a Gaussian copula, in which point-scale ordinal data are generated when the marginal proportions and correlations are specified. For the purposes of this chapter, one can utilize their method to find the corresponding polychoric correlation or the ordinal phi coefficient when one of them is given under normality. The algorithm in Ferrari and Barbiero (2012) serves as an intermediate step in formulating the connection between the two correlations under any distributional assumption on the underlying continuous variables.

Concentrating on the bivariate case, suppose $\mathbf{Z} = (Z_1, Z_2) \sim N(0, \Delta_{Z_1Z_2})$, where $\mathbf{Z}$ denotes the bivariate standard normal distribution with correlation matrix $\Delta_{Z_1Z_2}$ whose off-diagonal entry is $\delta_{Z_1Z_2^{POLY}}$. Let $\mathbf{X} = (X_{1O}, X_{2O})$ be the bivariate ordinal data where underlying $\mathbf{Z}$ is discretized based on corresponding normal quantiles given the marginal proportions, with a correlation matrix $\Delta_{Z_1Z_2}$. If we need to sample from a random vector (X_{1O}, X_{2O}) whose marginal cumulative distribution functions (cdfs) are F_1, F_2 tied together via a Gaussian copula, we generate a sample (z_1, z_2) from $\mathbf{Z} \sim N(0, \Delta_{Z_1Z_2})$, then set $\mathbf{x} = (x_{1O}, x_{2O}) = (F_1^{-1}(u_1), F_2^{-1}(u_2))$ when $\mathbf{u} = (u_1, u_2) = (\Phi(z_1), \Phi(z_2))$, where Φ is the cdf of the standard normal distribution. The correlation matrix of $\mathbf{X}$, denoted by $\Delta_{X_{1O}X_{2O}}$ (with an off-diagonal entry $\delta_{X_{1O}X_{2O}}$) obviously differs from $\Delta_{Z_1Z_2}$ due to discretization. More specifically, $|\delta_{X_{1O}X_{2O}}| < |\delta_{Z_1Z_2^{POLY}}|$ in large samples. The relationship between $\Delta_{X_{1O}X_{2O}}$ and $\Delta_{Z_1Z_2}$ is established resorting to the following formula (Cario and Nelson 1997):

$$E[X_{1O}X_{2O}] = E[F_1^{-1}(\Phi(Z_1))F_2^{-1}(\Phi(Z_1))] = \int_{-\infty}^{\infty}\int_{-\infty}^{\infty} F_1^{-1}(\Phi(Z_1))F_2^{-1}(\Phi(Z_1))f(z_1, z_2)dz_1dz_2 \quad (10)$$

where $f(z_1, z_2)$ is the bivariate standard normal probability function (pdf) with correlation matrix $\Delta_{Z_1Z_2}$, which implies that $\Delta_{X_{1O}X_{2O}}$ is a function of $\Delta_{Z_1Z_2}$. If X_1 and X_2 are ordinal random variables with cardinality k_1 and k_2, respectively, Eq. 10 reduces to a sum of $k_1 \times k_2$ integrals of $f(z_1, z_2)$ over a rectangle, i.e., $k_1 \times k_2$ differences of the bivariate cdf computed at two distinct points in $\Re^2$, as articulated by Ferrari and Barbiero (2012).

The relevant part of the algorithm is as follows:

1. Generate standard bivariate normal data with the correlation $\delta^0_{Z_1Z_2^{POLY}}$ where $\delta^0_{Z_1Z_2^{POLY}} = \delta_{X_{1O}X_{2O}}$ (Here, $\delta^0_{Z_1Z_2^{POLY}}$ is the initial polychoric correlation).
2. Discretize Z_1 and Z_2, based on the cumulative probabilities of the marginal distribution F_1 and F_2, to obtain X_{1O} and X_{2O}, respectively.
3. Compute $\delta^1_{X_{1O}X_{2O}}$ through X_{1O} and X_{2O} (Here, $\delta^1_{X_{1O}X_{2O}}$ is the ordinal phi coefficient after the first iteration).
4. Execute the following loop as long as $|\delta^v_{X_{1O}X_{2O}} - \delta_{X_{1O}X_{2O}}| > \varepsilon$ and $1 \leq v \leq v_{max}$ (v_{max} and ε are the maximum number of iterations and the maximum tolerated absolute error, respectively, both quantities are set by users):
   (a) Update $\delta^v_{Z_1Z_2^{POLY}}$ by $\delta^v_{Z_1Z_2^{POLY}} = \delta^{v-1}_{Z_1Z_2^{POLY}} g(v)$, where $g(v) = \delta_{X_{1O}X_{2O}}/\delta^v_{X_{1O}X_{2O}}$. Here, $g(v)$ serves as a correction coefficient, which ultimately converges to 1.
   (b) Generate bivariate normal data with $\delta^v_{Z_1Z_2^{POLY}}$ and compute $\delta^{v+1}_{X_{1O}X_{2O}}$ after discretization.

Again, our focus in the current work is not RNG per se, but the core idea in Ferrari and Barbiero (2012) is a helpful tool that links $\delta_{Y_1Y_2^{POLY}}$ and $\delta_{X_{1O}X_{2O}}$ for ordinalized data through the intermediary role of normal data between ordinal and nonnormal continuous data (Sect. 2.4).

When only one of the normal variables (Z_1) is ordinalized, no nice formulas such as the one in Sect. 2.1 given in the binary data context are available. The good news is that a much more general procedure that accommodates any distributional assumption on underlying continuous variables is available. The process that relates the polyserial ($\delta_{Z_1Z_2^{PS}}$) and point-polyserial ($\delta_{X_{1O}Z_2}$) correlations is available by extending the arguments substantiated in Sect. 2.2 to the ordinal data case.

2.4 Polytomous Case: Beyond Normality

When both variables are ordinalized, the connection between the polychoric correlation ($\delta_{Y_1Y_2^{POLY}}$) and the ordinal phi coefficient ($\delta_{X_{1O}X_{2O}}$) can be established by a two-stage scheme, in which we compute the normal, intermediate correlation ($\delta_{Z_1Z_2^{POLY}}$) from the ordinal phi coefficient by the method in Ferrari and Barbiero (2012) (presented in Sect. 2.3) before we find the nonnormal polychoric correlation via the power polynomials (Eq. 6). The other direction (computing $\delta_{X_{1O}X_{2O}}$ from $\delta_{Y_1Y_2^{POLY}}$) can be implemented by executing the same steps in the reverse order. The associated computational routines are presented in Sect. 3 (*Algorithms-3a* and *-3b*).

The correlational identity given in Eq. 9 holds for ordinalized data as well when only one variable is ordinalized (Demirtas and Hedeker 2016); the ordinal version of the equation can be written as $\delta_{X_{1O}Y_2} = \delta_{Y_1 Y_2^{PS}} \delta_{X_{1O}Y_1}$. The same linearity and constancy of ratio arguments equally apply in terms of the connection between the polyserial ($\delta_{Y_1 Y_2^{PS}}$) and point-polyserial ($\delta_{X_{1O}Y_1}$) correlations; the fundamental utility and operational characteristics are parallel to the binary case. Once the ratio ($\delta_{X_{1O}Y_1}$) is found by generating Y_1 and discretizing it to obtain X_{1O}, one can easily compute either of these quantities given the other. This will be pertinent in *Algorithm-4* below.

The next section puts all these concepts together from an algorithmic point of view with numerical illustrations.

3 Algorithms and Illustrative Examples

We work with eight distributions to reflect some common shapes that can be encountered in real-life applications. The illustrative examples come from bivariate data with *Weibull* and *Normal mixture* marginals. In what follows, *W* and *NM* stand for *Weibull* and *Normal mixture*, respectively. The *W* density is $f(y|\gamma, \delta) = \frac{\delta}{\gamma^\delta} y^{\delta-1} exp(-(\frac{y}{\gamma})^\delta)$ for $y > 0$, and $\gamma > 0$ and $\delta > 0$ are the scale and shape parameters, respectively. The *NM* density is $f(y|\pi, \mu_1, \sigma_1, \mu_2, \sigma_2) = \frac{\pi}{\sigma_1\sqrt{2\pi}} exp\big(-\frac{1}{2}(\frac{y-\mu_1}{\sigma_1})^2\big) + \frac{(1-\pi)}{\sigma_2\sqrt{2\pi}} exp\big(-\frac{1}{2}(\frac{y-\mu_2}{\sigma_2})^2\big)$, where $0 < \pi < 1$ is the mixing parameter. Since it is a mixture, it can be unimodal or bimodal. Depending on the choice of parameters, both distributions can take a variety of shapes. We use four sets of parameter specifications for each of these distributions: For *W* distribution, (γ, δ) pairs are chosen to be $(1, 1)$, $(1, 1.2)$, $(1, 3.6)$, and $(1, 25)$, corresponding to mode at the boundary, positively skewed, nearly symmetric, and negatively skewed shapes, respectively. For *NM* distribution, the parameter set $(\pi, \mu_1, \sigma_1, \mu_2, \sigma_2)$ is set to $(0.5, 0, 1, 3, 1)$, $(0.6, 0, 1, 3, 1)$, $(0.3, 0, 1, 2, 1)$, and $(0.5, 0, 1, 2, 1)$, whose shapes are bimodal-symmetric, bimodal-asymmetric, unimodal-negatively skewed, and unimodal-symmetric, respectively. These four variations of the *W* and *NM* densities are plotted in Fig. 1 (W/NM: the first/second columns) in the above order of parameter values, moving from top to bottom. Finally, as before, p_1 and p_2 represent the binary/ordinal proportions. In the binary case, they are single numbers. In the ordinal case, the marginal proportions are denoted as $P(X_i = j) = p_{ij}$ for $i = 1, 2$ and $j = 1, 2, ...k_i$, and $p_i = (p_{i1}, p_{i2}, ..., p_{ik_i})$, in which skip patterns are allowed. Furthermore, if the users wish to start the ordinal categories from 0 or any integer other than 1, the associational implications remain unchanged as correlations are invariant to the location shifts. Of note, the number of significant digits reported throughout the chapter varies by the computational sensitivity of the quantities.

Algorithm-1a: Computing the tetrachoric correlation ($\delta_{Y_1 Y_2^{TET}}$) from the phi coefficient ($\delta_{X_{1B}X_{2B}}$): The algorithm for computing $\delta_{Y_1 Y_2^{TET}}$ when $\delta_{X_{1B}X_{2B}}$, p_1, p_2, and the key distributional characteristics of Y_1 and Y_2 (ν_1 and ν_2) are specified, is as follows:

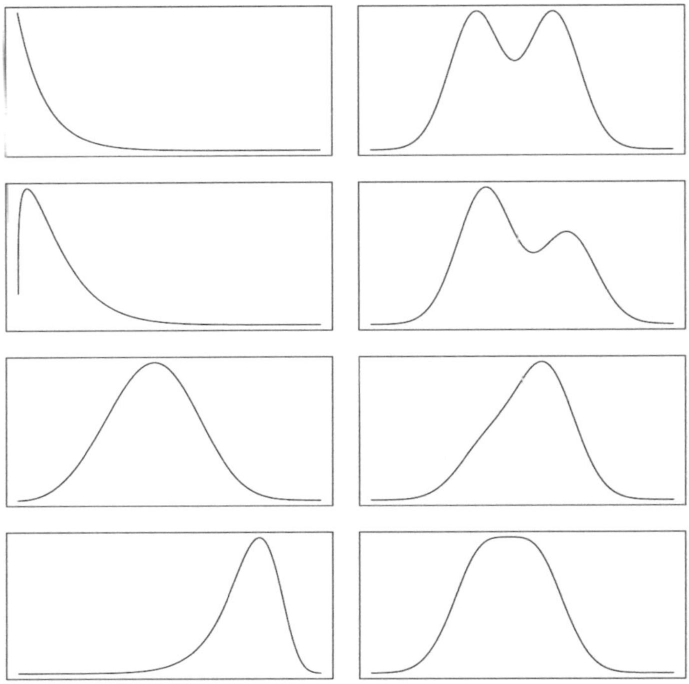

Fig. 1 Density functions of *Weibull* (*first column*) and *Normal Mixture* (*second column*) distributions for chosen parameter values that appear in the text

1. Solve Eq. 1 for $\delta_{Z_1 Z_2^{TET}}$.
2. Compute the power coefficients (a, b, c, d) for Y_1 and Y_2 by Eqs. 2–5.
3. Plug all quantities obtained in Steps 1–2 into Eq. 6, and solve for $\delta_{Y_1 Y_2^{TET}}$.

Suppose $Y_1 \sim W(1, 1)$, $Y_2 \sim NM(0.5, 0, 1, 3, 1)$, $(p_1, p_2) = (0.85, 0.15)$, and $\delta_{X_{1B} X_{2B}} = 0.1$. Solving for $\delta_{Z_1 Z_2^{TET}}$ in Eq. 1 (Step 1) yields 0.277. The power coefficients (a, b, c, d) in Eqs. 2–5 (Step 2) turn out to be $(-0.31375, 0.82632, 0.31375, 0.02271)$ and $(0.00004, 1.20301, -0.00004, -0.07305)$ for Y_1 and Y_2, respectively. Substituting these into Eq. 6 (Step 3) gives $\delta_{Y_1 Y_2^{TET}} = 0.243$. Similarly, for $(p_1, p_2) = (0.10, 0.30)$ and $\delta_{X_{1B} X_{2B}} = 0.5$, $\delta_{Z_1 Z_2^{TET}} = 0.919$ and $\delta_{Y_1 Y_2^{TET}} = 0.801$. The upper half of Table 2 includes a few more combinations.

Algorithm-1b: Computing the phi coefficient ($\delta_{X_{1B} X_{2B}}$) from the tetrachoric correlation ($\delta_{Y_1 Y_2^{TET}}$): The quantities that need to be specified are the same as in *Algorithm-1a* , and the steps are as follows:

1. Compute the power coefficients (a, b, c, d) for Y_1 and Y_2 by Eqs. 2–5.
2. Solve Eq. 6 for $\delta_{Z_1 Z_2^{TET}}$.
3. Plug $\delta_{Z_1 Z_2^{TET}}$ into Eq. 1, and solve for $\delta_{X_{1B} X_{2B}}$.

With the same pair of distributions, where $Y_1 \sim W(1, 1)$ and $Y_2 \sim NM(0.5, 0, 1, 3, 1)$, suppose $(p_1, p_2) = (0.85, 0.15)$ and $\delta_{Y_1 Y_2^{TET}} = -0.4$. After solving for the

Table 2 Computed values of the tetrachoric correlation ($\delta_{Y_1 Y_2^{TET}}$) or the phi coefficient ($\delta_{X_{1B} X_{2B}}$) when one of them is specified, with two sets of proportions for $Y_1 \sim W(1, 1)$ and $Y_2 \sim NM(0.5, 0, 1, 3, 1)$

p_1	p_2	$\delta_{X_{1B} X_{2B}}$	$\delta_{Z_1 Z_2^{TET}}$	$\delta_{Y_1 Y_2^{TET}}$
0.85	0.15	−0.6	−0.849	−0.755
		−0.3	−0.533	−0.472
		0.1	0.277	0.243
0.10	0.30	−0.2	−0.616	−0.540
		0.3	0.572	0.502
		0.5	0.919	0.801
p_1	p_2	$\delta_{Y_1 Y_2^{TET}}$	$\delta_{Z_1 Z_2^{TET}}$	$\delta_{X_{1B} X_{2B}}$
0.85	0.15	−0.4	−0.456	−0.246
		0.2	0.227	0.085
		0.6	0.685	0.173
0.10	0.30	−0.5	−0.570	−0.192
		0.1	0.114	0.052
		0.7	0.801	0.441

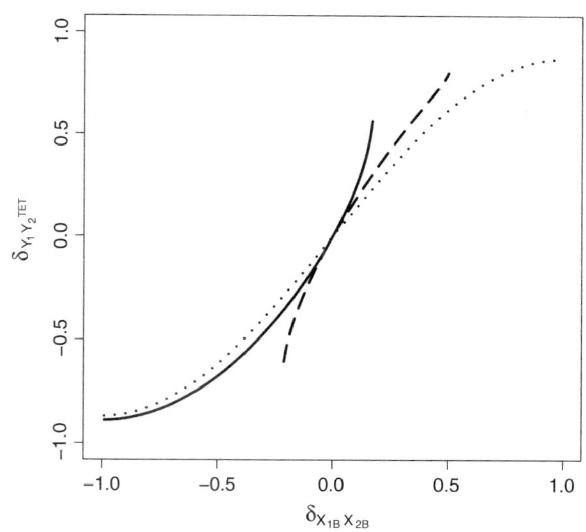

Fig. 2 $\delta_{X_{1B} X_{2B}}$ versus $\delta_{Y_1 Y_2^{TET}}$ for $Y_1 \sim W(1, 1)$ and $Y_2 \sim NM(0.5, 0, 1, 3, 1)$, where solid, dashed, and dotted curves represent $(p_1, p_2) = (0.85, 0.15)$, $(0.10, 0.30)$, and $(0.50, 0.50)$, respectively; the range differences are due to the Fréchet-Hoeffding bounds

power coefficients (Step 1), Steps 2 and 3 yield $\delta_{Z_1 Z_2^{TET}} = -0.456$ and $\delta_{X_{1B} X_{2B}} = -0.246$, respectively. Similarly, when $(p_1, p_2) = (0.10, 0.30)$ and $\delta_{Y_1 Y_2^{TET}} = 0.7$, $\delta_{Z_1 Z_2^{TET}} = 0.801$ and $\delta_{X_{1B} X_{2B}} = 0.441$. The lower half of Table 2 includes a few more combinations. More comprehensively, Fig. 2 shows the comparative behavior of $\delta_{X_{1B} X_{2B}}$ and $\delta_{Y_1 Y_2^{TET}}$ when the proportion pairs take three different values, with the addition of $(p_1, p_2) = (0.50, 0.50)$ to the two pairs above, for this particular distributional setup.

Table 3 Computed values of $\hat{c}_1 = \delta_{X_{1B}Y_1}$ and $\hat{c}_2 = \delta_{X_{2B}Y_2}$ that connect the biserial ($\delta_{Y_1Y_2^{BS}}$) and point-biserial correlations ($\delta_{X_{1B}Y_2}$ or $\delta_{X_{2B}Y_1}$), where $Y_1 \sim W(1, 1.2)$ and $Y_2 \sim NM(0.6, 0, 1, 3, 1)$

p_1 or p_2	$\hat{c}_1$	$\hat{c}_2$
0.05	0.646	0.447
0.15	0.785	0.665
0.25	0.809	0.782
0.35	0.795	0.848
0.45	0.760	0.858
0.55	0.710	0.829
0.65	0.640	0.772
0.75	0.554	0.692
0.85	0.436	0.585
0.95	0.261	0.392

Algorithm-2: Computing the biserial correlation ($\delta_{Y_1Y_2^{BS}}$) from the point-biserial correlation ($\delta_{X_{1B}Y_2}$) and the other way around: One only needs to specify the distributional form of Y_1 (the variables that is to be dichotomized) and the proportion (p_1) for this algorithm (See Sect. 2.2). The steps are as follows:

1. Generate Y_1 with a large number of data points (e.g., $N = 100, 000$).
2. Dichotomize Y_1 to obtain X_{1B} through the specified value of p_1, and compute the sample correlation, $\delta_{X_{1B}Y_1} = \hat{c}_1$.
3. Find $\delta_{X_{1B}Y_2}$ or $\delta_{Y_1Y_2^{BS}}$ by $\delta_{X_{1B}Y_2}/\delta_{Y_1Y_2^{BS}} = \hat{c}_1$ by Eq. 9.

In this illustration, we assume that $Y_1 \sim W(1, 1.2)$, $Y_2 \sim NM(0.6, 0, 1, 3, 1)$, and $\delta_{Y_1Y_2^{BS}} = 0.60$. Y_1 is dichotomized to obtain X_{1B} where $E(X_{1B}) = p_1 = 0.55$. After following Steps 1 and 2, $\hat{c}_1$ turns out to be 0.710, and accordingly $\delta_{X_{1B}Y_2} = \hat{c}_1\delta_{Y_1Y_2^{BS}} = 0.426$. Similarly, if the specified value of δ_{X_{1B},Y_2} is 0.25, then $\delta_{Y_1Y_2^{BS}} = \delta_{X_{1B}Y_2}/\hat{c}_1 = 0.352$. The fundamental ideas remain the same if Y_2 is dichotomized (with a proportion p_2) rather than Y_1. In that case, with a slight notational difference, the new equations would be $\delta_{X_{2B}Y_2} = \hat{c}_2$ and $\delta_{X_{2B}Y_1}/\delta_{Y_1Y_2^{BS}} = \hat{c}_2$. Table 3 shows $\hat{c}_1$ and $\hat{c}_2$ values when p_1 or p_2 ranges between 0.05 and 0.95 with an increment of 0.10. We further generated bivariate continuous data with the above marginals and the biserial correlations between −0.85 and 0.90 with an increment of 0.05. We then dichotomized Y_1 where p_1 is 0.15 and 0.95, and computed the empirical point-biserial correlation. The lower- and upper-right graphs in Fig. 3 the plot of the algorithmic value of $\hat{c}_1$ in Step 2 and $\delta_{Y_1Y_2^{BS}}$ versus $\delta_{X_{1B}Y_2}$, respectively, where the former is a theoretical and $\delta_{X_{1B}Y_2}$ in the latter is an empirical quantity. As expected, the two $\hat{c}_1$ values are the same as the slopes of the linear lines of $\delta_{Y_1Y_2^{BS}}$ versus $\delta_{X_{1B}Y_2}$, lending support for how plausibly *Algorithm-2* is working. The procedure is repeated under the assumption that Y_2 is dichotomized rather than Y_1 (lower graphs in Fig. 3).

Algorithm-3a: Computing the polychoric correlation ($\delta_{Y_1Y_2^{POLY}}$) from the ordinal phi coefficient ($\delta_{X_{1O}X_{2O}}$): The algorithm for computing $\delta_{Y_1Y_2^{POLY}}$ when $\delta_{X_{1O}X_{2O}}$, p_1,

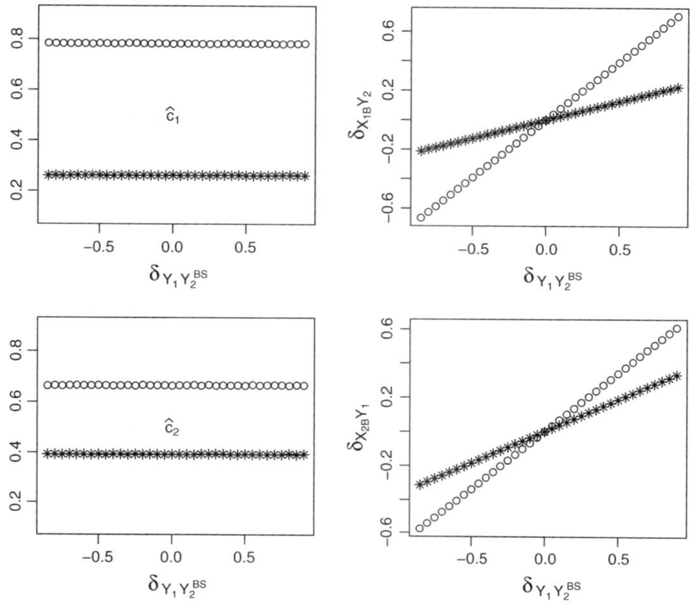

Fig. 3 Plots of $\hat{c}_1$ (*upper-left*), $\hat{c}_2$ (*lower-left*), empirical $\delta_{X_{1B}Y_2}$ versus $\delta_{Y_1 Y_2^{BS}}$ (*upper-right*), and $\delta_{X_{2B}Y_1}$ versus $\delta_{Y_1 Y_2^{BS}}$ for p_1 or p_2 is 0.15 (shown by o) or 0.95 (shown by *), where $Y_1 \sim W(1, 1.2)$ and $Y_2 \sim NM(0.6, 0, 1, 3, 1)$

p_2 and the key distributional characteristics of Y_1 and Y_2 (ν_1 and ν_2) are specified, is as follows:

1. Use the method in Ferrari and Barbiero (2012), outlined in Sect. 2.3 for finding $\delta_{Z_1 Z_2^{POLY}}$.
2. Compute the power coefficients (a, b, c, d) for Y_1 and Y_2 by Eqs. 2–5.
3. Plug all quantities obtained in Steps 1–2 into Eq. 6, and solve for $\delta_{Y_1 Y_2^{POLY}}$.

Suppose $Y_1 \sim W(1, 3.6)$, $Y_2 \sim NM(0.3, 0, 1, 2, 1)$, $(p_1, p_2) = ((0.4, 0.3, 0.2, 0.1), (0.2, 0.2, 0.6))$, and $\delta_{X_{1O}X_{2O}} = -0.7$. Solving for $\delta_{Z_1 Z_2^{POLY}}$ in Step 1 yields -0.816. The power coefficients (a, b, c, d) in Eqs. 2–5 (Step 2) turn out to be $(-0.00010, 1.03934, 0.00010, -0.01268)$ for Y_1 and $(0.05069, 1.04806, -0.05069, -0.02626)$ for Y_2. Substituting these into Eq. 6 (Step 3) gives $\delta_{Y_1 Y_2^{POLY}} = -0.813$. Similarly, for $(p_1, p_2) = ((0.1, 0.1, 0.1, 0.7), (0.8, 0.1, 0.1))$ and $\delta_{X_{1O}X_{2O}} = 0.2$, $\delta_{Z_1 Z_2^{POLY}} = 0.441$ and $\delta_{Y_1 Y_2^{POLY}} = 0.439$. The upper half of Table 4 includes a few more combinations.

Algorithm-3b: Computing the ordinal phi coefficient ($\delta_{X_{1O}X_{2O}}$) from the polychoric correlation ($\delta_{Y_1 Y_2^{POLY}}$): The required quantities that need specification are the same as in *Algorithm-3a*, and the steps are as follows:

Table 4 Computed values of the polychoric correlation ($\delta_{Y_1 Y_2^{POLY}}$) or the ordinal phi coefficient ($\delta_{X_{1O} X_{2O}}$) given the other, with two sets of proportions for $Y_1 \sim W(1, 3.6)$ and $Y_2 \sim NM(0.3, 0, 1, 2, 1)$

p_1	p_2	$\delta_{X_{1O} X_{2O}}$	$\delta_{Z_1 Z_2^{POLY}}$	$\delta_{Y_1 Y_2^{POLY}}$
(0.4, 0.3, 0.2, 0.1)	(0.2, 0.2, 0.6)	−0.7	−0.816	−0.813
		−0.3	−0.380	−0.378
		0.4	0.546	0.544
		0.6	0.828	0.826
(0.1, 0.1, 0.1, 0.7)	(0.8, 0.1, 0.1)	−0.6	−0.770	−0.767
		−0.4	−0.568	−0.566
		−0.1	−0.167	−0.166
		0.2	0.441	0.439
p_1	p_2	$\delta_{Y_1 Y_2^{POLY}}$	$\delta_{Z_1 Z_2^{POLY}}$	$\delta_{X_{1O} X_{2O}}$
(0.4, 0.3, 0.2, 0.1)	(0.2, 0.2, 0.6)	−0.8	−0.802	−0.686
		−0.2	−0.201	−0.155
		0.5	0.502	0.368
		0.7	0.702	0.511
(0.1, 0.1, 0.1, 0.7)	(0.8, 0.1, 0.1)	−0.8	−0.802	−0.638
		−0.2	−0.201	−0.122
		0.5	0.502	0.219
		0.7	0.702	0.263

1. Compute the power coefficients (a, b, c, d) for Y_1 and Y_2 by Eqs. 2–5.
2. Solve Eq. 6 for $\delta_{Z_1 Z_2^{TET}}$.
3. Solve for $\delta_{X_{1O} X_{2O}}$ given $\delta_{Z_1 Z_2^{TET}}$ by the method in Ferrari and Barbiero (2012).

With the same set of specifications, namely, $Y_1 \sim W(1, 3.6)$, $Y_2 \sim NM(0.3, 0, 1, 2, 1)$, and $(p_1, p_2) = ((0.4, 0.3, 0.2, 0.1), (0.2, 0.2, 0.6))$, suppose $\delta_{Y_1 Y_2^{POLY}} = 0.5$. After solving for the power coefficients (Step 1), Steps 2 and 3 yield $\delta_{Z_1 Z_2^{POLY}} = 0.502$ and $\delta_{X_{1O} X_{2O}} = 0.368$. Similarly, when $(p_1, p_2) = ((0.1, 0.1, 0.1, 0.7), (0.8, 0.1, 0.1))$ and $\delta_{Y_1 Y_2^{POLY}} = 0.7$, $\delta_{Z_1 Z_2^{POLY}} = 0.702$ and $\delta_{X_{1O} X_{2O}} = 0.263$. The lower half of Table 4 includes a few more combinations. A more inclusive set of results is given in Fig. 4, which shows the relative trajectories of $\delta_{X_{1O} X_{2O}}$ and $\delta_{Y_1 Y_2^{POLY}}$ when the proportion sets take three different values, with the addition of $(p_1, p_2) = ((0.25, 0.25, 0.25, 0.25), (0.05, 0.05, 0.9))$ to the two sets above.

Algorithm-4: Computing the polyserial correlation ($\delta_{Y_1 Y_2^{PS}}$) from the point-polyserial correlation ($\delta_{X_{1O} Y_2}$) and the other way around: The following steps enable us to calculate either one of these correlations when the distribution of Y_1 (the variable that is subsequently ordinalized) and the ordinal proportions (p_1) are specified (See Sect. 2.4):

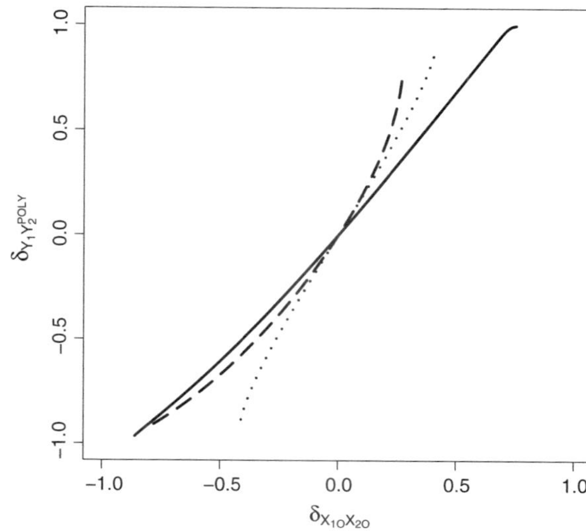

Fig. 4 $\delta_{X_{1O}X_{2O}}$ versus $\delta_{Y_1Y_2^{POLY}}$ for $Y_1 \sim W(1, 3.6)$ and $Y_2 \sim NM(0.3, 0, 1, 2, 1)$, where *solid, dashed,* and *dotted curves* represent $(p_1, p_2) = ((0.4, 0.3, 0.2, 0.1), (0.2, 0.2, 0.6))$, $((0.1, 0.1, 0.1, 0.7), (0.8, 0.8, 0.1))$, and $((0.25, 0.25, 0.25, 0.25), (0.05, 0.05, 0.9))$, respectively; the range differences are due to the Fréchet-Hoeffding bounds

1. Generate Y_1 with a large number of data points (e.g., $N = 100,000$).
2. Ordinalize Y_1 to obtain X_{1O} through the specified value of p_1, and compute the sample correlation, $\delta_{X_{1O}Y_1} = \hat{c}_1$.
3. Find $\delta_{X_{1O}Y_2}$ or $\delta_{Y_1Y_2^{PS}}$ by $\delta_{X_{1O}Y_2}/\delta_{Y_1Y_2^{PS}} = \hat{c}_1$ by Eq. 9.

For illustrative purposes, we assume that $Y_1 \sim W(1, 25)$, $Y_2 \sim NM(0.5, 0, 1, 2, 1)$, and $\delta_{Y_1, Y_2^{PS}} = 0.6$. Y_1 is ordinalized to obtain X_{1O}, where $p_1 = (0.4, 0.3, 0.2, 0.1)$. After following Steps 1 and 2, $\hat{c}_1$ turns out to be 0.837, and accordingly $\delta_{X_{1O}Y_2} = \hat{c}_1\delta_{Y_1Y_2^{PS}} = 0.502$. Similarly, if the specified value of $\delta_{X_{1O}Y_2}$ is 0.3, then $\delta_{Y_1Y_2^{PS}} = \delta_{X_{1O}Y_2}/\hat{c}_1 = 0.358$. The core ideas remain unchanged if Y_2 is dichotomized (with a proportion p_2) instead of Y_1, in which the new equations become $\delta_{X_{2O}Y_2} = \hat{c}_2$ and $\delta_{X_{2O}Y_1}/\delta_{Y_1Y_2^{PS}} = \hat{c}_2$. Several $\hat{c}_1$ and $\hat{c}_2$ values are tabulated for corresponding p_1 or p_2 specifications in Table 5. Figure 5 provides a comparison between the theoretical (suggested by *Algorithm-4*) and empirical point-polyserial correlations ($\delta_{X_{1O}Y_2}$) for specified polyserial correlation ($\delta_{Y_1Y_2^{PS}}$) values in the range of −0.95 and 0.95 (upper graph) and the scatter plot of the differences between the two quantities (lower graph) are given. We first generated bivariate continuous data using the above distributional assumptions, then ordinalized Y_1, computed the empirical post-discretization correlations, and made a comparison that is shown in the two graphs in Fig. 5, which collectively suggest that the procedure is working properly.

Some operational remarks: All computing work has been done in $\mathscr{R}$ software (R Development Core Team, 2016). In the algorithms that involve the power polynomials, a stand-alone computer code in Demirtas and Hedeker (2008a) was used to solve the system of equations. More sophisticated programming implementations such as `fleishman.coef` function in **BinNonNor** package (Inan and Demirtas 2016), `Param.fleishman` function in **PoisNonNor** package (Demirtas et al. 2016b), and `Fleishman.coef.NN` function in **BinOrdNonNor** package (Demirtas et al.

Table 5 Computed values of $\hat{c}_1 = \delta_{X_{1O}Y_1}$ and $\hat{c}_2 = \delta_{X_{2O}Y_2}$ that connect the polyserial ($\delta_{Y_1 Y_2^{BS}}$) and point-polyserial correlations ($\delta_{X_{1O}Y_2}$ or $\delta_{X_{2O}Y_1}$), where $Y_1 \sim W(1, 25)$ and $Y_2 \sim NM(0.5, 0, 1, 2, 1)$

p_1	p_2	$\hat{c}_1$	$\hat{c}_2$
(0.4, 0.3, 0.2, 0.1)	–	0.837	–
(0.1, 0.2, 0.3, 0.4)	–	0.927	–
(0.1, 0.4, 0.4, 0.1)	–	0.907	–
(0.4, 0.1, 0.1, 0.4)	–	0.828	–
(0.7, 0.1, 0.1, 0.1)	–	0.678	–
–	(0.3, 0.4, 0.3)	–	0.914
–	(0.6, 0.2, 0.2)	–	0.847
–	(0.1, 0.8, 0.1)	–	0.759
–	(0.1, 0.1, 0.8)	–	0.700
–	(0.4, 0.4, 0.2)	–	0.906

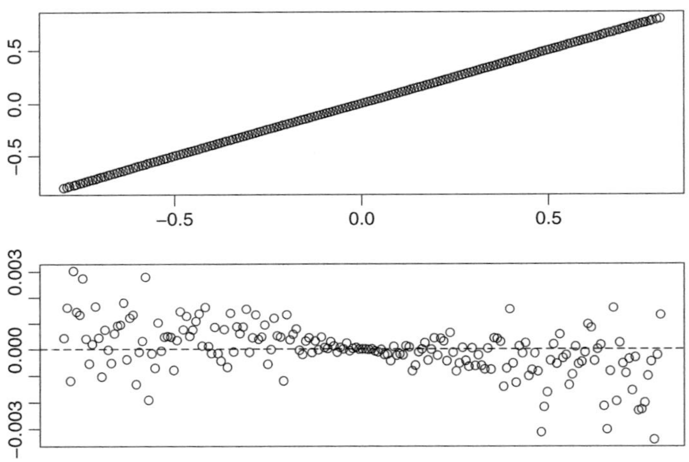

Fig. 5 The plot of the theoretical (x axis) versus empirical (y axis) point-polyserial correlations ($\delta_{X_{1O}Y_2}$) given the specified polyserial correlation ($\delta_{Y_1 Y_2^{PS}}$) values (*upper graph*) and the scatter plot of the differences between the two quantities (*lower graph*), where $Y_1 \sim W(1, 25)$ and $Y_2 \sim NM(0.5, 0, 1, 2, 1)$

2016c) can also be employed. The root of the third order polynomials in Eq. 6 was found by `polyroot` function in the base package. The tetrachoric correlation and the phi coefficient in Eq. 1 was computed by `phi2tetra` function in **psych** package (Revelle 2016) and `pmvnorm` function in **mvtnorm** package (Genz et al. 2016), respectively. Finding the polychoric correlation given the ordinal phi coefficient and the opposite direction were performed by `ordcont` and `contord` functions in **GenOrd** package (Barbiero and Ferrari 2015), respectively.

4 Simulations in a Multivariate Setting

By the problem definition and design, all development has been presented in bivariate settings. For assessing how the algorithms work in a broader multivariate context and for highlighting the generality of our approach, we present two simulation studies that involve the specification of either pre- or post-discretization correlations.

Simulation work is devised around five continuous variables, and four of these are subsequently dichotomized or ordinalized. Referring to the *Weibull* and *Normal mixture* densities in the illustrative examples, the distributional forms are as follows: $Y_1 \sim W(1, 3.6)$, $Y_2 \sim W(1, 1.2)$, $Y_3 \sim NM(0.3, 0, 1, 2, 1)$, $Y_4 \sim NM(0.5, 0, 1, 2, 1)$, and $Y_5 \sim NM(0.5, 0, 1, 3, 1)$. $Y_1, ..., Y_4$ are to be discretized with proportions $p_1 = 0.6$, $p_2 = 0.3$, $p_3 = (0.4, 0.2, 0.2, 0.2)$, and $p_4 = (0.1, 0.6, 0.3)$, respectively. Two dichotomized (Y_1 and Y_2), two ordinalized (Y_3 and Y_4), and one continuous (Y_5) variables form a sufficient environment, in which all types of correlations mentioned in this work are covered. For simplicity, we only indicate if the correlations are pre- or post-discretization quantities without distinguishing between different types in terms of naming and notation in this section. We investigate both directions: (1) The pre-discretization correlation matrix is specified; the theoretical (algorithmic) post-discretization quantities were computed; data were generated, discretized with the prescription guided by the proportions, and empirical correlations were found via $n = 1000$ simulation replicates to see how closely the algorithmic and empirical values are aligned on average. (2) The post-discretization matrix is specified; correlation among latent variables were computed via the algorithms; data were generated with this correlation matrix; then the data were dichotomized or ordinalized to gauge if we obtain the specified post-discretization correlations on average. In Simulation 1, the pre-discretization correlation matrix (Σ_{pre}) representing the correlation structure among continuous variables, is defined as

$$\Sigma_{pre} = \begin{bmatrix} 1.00 & 0.14 & -0.32 & 0.56 & 0.54 \\ 0.14 & 1.00 & -0.10 & 0.17 & 0.17 \\ -0.32 & -0.10 & 1.00 & -0.40 & -0.38 \\ 0.56 & 0.17 & -0.40 & 1.00 & 0.67 \\ 0.54 & 0.17 & -0.38 & 0.67 & 1.00 \end{bmatrix},$$

where the variables follow the order of $(Y_1, ..., Y_5)$. Let $\Sigma[i, j]$ denote the correlation between variables i and j, where $i, j = 1, ..., 5$. The theoretical post-discretization values (under the assumption that the algorithms function properly and yield the true values) were computed. More specifically, $\Sigma_{post}[1, 2]$ was found by *Algorithm-1b*, $\Sigma_{post}[1, 5]$ and $\Sigma_{post}[2, 5]$ by *Algorithm-2*, $\Sigma_{post}[1, 3]$, $\Sigma_{post}[1, 4]$, $\Sigma_{post}[2, 3]$, $\Sigma_{post}[2, 4]$, and $\Sigma_{post}[3, 4]$ by *Algorithm-3b*, $\Sigma_{post}[3, 5]$ and $\Sigma_{post}[4, 5]$ by *Algorithm-4*. These values collectively form a post-discretization correlation matrix (Σ_{post}), which serves as the *True Value (TV)*. The empirical post-discretization correlation estimates were calculated after generating $N = 1,000$ rows

Table 6 Results of Simulation 1 (the reported quantities are defined in the text)

Parameter	Σ_{pre}	$TV(\Sigma_{post})$	AE	RB	PB	SB
$\Sigma[1,2]$	0.14	0.08942	0.08934	0.00008	0.09	0.03
$\Sigma[1,3]$	−0.32	−0.22936	−0.23385	0.00449	1.96	1.40
$\Sigma[1,4]$	0.56	0.38946	0.39128	0.00182	0.47	0.72
$\Sigma[1,5]$	0.54	0.43215	0.43425	0.00210	0.49	0.84
$\Sigma[2,3]$	−0.10	−0.07420	−0.07319	0.00101	1.36	0.32
$\Sigma[2,4]$	0.17	0.12175	0.12073	0.00102	0.84	0.33
$\Sigma[2,5]$	0.17	0.13681	0.13883	0.00202	1.48	0.65
$\Sigma[3,4]$	−0.40	−0.31908	−0.31922	0.00014	0.04	0.05
$\Sigma[3,5]$	−0.38	−0.34082	−0.34671	0.00589	1.73	2.15
$\Sigma[4,5]$	0.67	0.58528	0.58773	0.00245	0.42	1.28

of multivariate latent, continuous data (Y_1, ..., Y_5) by the specified Σ_{pre}, followed by discretization of (Y_1, ..., Y_4). The whole process was repeated for $n = 1,000$ times.

We evaluated the quality of estimates by three commonly accepted accuracy measures: (a) *Raw Bias (RB)*, (b) *Percentage Bias (PB)*, and (c) *Standardized Bias (SB)* (Demirtas 2004a, 2007a,b, 2008). They all are functions of the average estimate (*AE*), and their definitions are well-established: When the parameter of interest is δ, $RB = |E[\hat{\delta} - \delta]|$ (absolute average deviation), $PB = 100 * |E[\hat{\delta} - \delta]/\delta|$ (absolute average deviation as a percentage of the true value), and $SB = 100 * |E[\hat{\delta} - \delta]|/V[\hat{\delta}]^{1/2}$ (absolute average deviation with respect to the overall uncertainty in the system). A procedure is typically regarded as working properly if $RB < 5$ and $SB < 50$ (Demirtas et al. 2007). In Table 6, we tabulate Σ_{pre}, $TV(\Sigma_{post})$, AE, RB, PB, and SB. All the three accuracy quantities demonstrate negligibly small deviations from the true values; they are within acceptable limits, suggesting that the set of algorithms provides unbiased estimates.

In Simulation 2, we take the reverse route by specifying the post-discretization correlation matrix (Σ_{post}), which serves as the *True Value (TV)*, in the following way:

$$\Sigma_{post} = \begin{bmatrix} 1.00 & 0.24 & 0.18 & 0.10 & 0.38 \\ 0.24 & 1.00 & 0.20 & -0.11 & 0.42 \\ 0.18 & 0.20 & 1.00 & -0.07 & 0.29 \\ 0.10 & -0.11 & -0.07 & 1.00 & -0.16 \\ 0.38 & 0.42 & 0.29 & -0.16 & 1.00 \end{bmatrix}$$

The corresponding pre-discretization matrix was found via the algorithms. The theoretical $\Sigma_{pre}[1,2]$ was computed by *Algorithm-1a*, $\Sigma_{pre}[1,5]$ and $\Sigma_{pre}[2,5]$ by *Algorithm-2*, $\Sigma_{pre}[1,3]$, $\Sigma_{pre}[1,4]$, $\Sigma_{pre}[2,3]$, $\Sigma_{pre}[2,4]$, and $\Sigma_{pre}[3,4]$ by *Algorithm-3a*, $\Sigma_{pre}[3,5]$ and $\Sigma_{pre}[4,5]$ by *Algorithm-4*. These values jointly form a pre-discretization correlation matrix (Σ_{pre}). The empirical post-discretization

Table 7 Results of Simulation 2 (the reported quantities are defined in the text)

Parameter	TV (Σ_{post})	Σ_{pre}	AE	RB	PB	SB
$\Sigma[1,2]$	0.24	0.40551	0.24000	0.00000	0.00	0.00
$\Sigma[1,3]$	0.18	0.24916	0.17727	0.00273	1.52	0.90
$\Sigma[1,4]$	0.10	0.14551	0.10041	0.00041	0.41	0.13
$\Sigma[1,5]$	0.38	0.47484	0.38022	0.00022	0.06	0.08
$\Sigma[2,3]$	0.20	0.29496	0.20729	0.00729	3.64	2.44
$\Sigma[2,4]$	−0.11	−0.15954	−0.10601	0.00399	3.63	1.27
$\Sigma[2,5]$	0.42	0.52188	0.43347	0.01347	3.21	5.50
$\Sigma[3,4]$	−0.07	−0.08884	−0.07001	0.00001	0.01	0.01
$\Sigma[3,5]$	0.29	0.32334	0.29443	0.00443	1.53	1.58
$\Sigma[4,5]$	−0.16	−0.18316	−0.16004	0.00004	0.02	0.01

correlation estimates were calculated after generating $N = 1,000$ rows of multivariate latent, continuous data (Y_1, ..., Y_5) by the computed Σ_{pre} before discretization of (Y_1, ..., Y_4). As before, this process is repeated for $n = 1,000$ times. In Table 7, we tabulate $TV(\Sigma_{post})$, Σ_{pre}, AE, RB, PB, and SB. Again, the discrepancies between the expected and empirical quantities are minimal by the three accuracy criteria, providing substantial support for the proposed method.

These results indicate compelling and promising evidence in favor of the algorithms herein. Our evaluation is based on accuracy (unbiasedness) measures. Precision is another important criterion in terms of the quality and performance of the estimates (Demirtas et al. 2008; Demirtas and Hedeker 2008b; Yucel and Demirtas 2010). We address the precision issues by plotting the correlation estimates across all simulation replicates in both scenarios (Fig. 6). The estimates closely match the true values shown in Tables 6 and 7, with a healthy amount of variation that is within the limits of Monte-Carlo simulation error. On a cautious note, however, there seems to be slightly more variation in Simulation 2, which is natural since there are two layers of randomness (additional source of variability).

5 Discussion

If the discretization thresholds and underlying continuous measurements are available, one can easily compute all types of correlations that appear herein. The scale of the work in this chapter is far broader, as it is motivated by and oriented towards computing the different correlational magnitudes before and after discretization when one of these quantities is specified in the context of simulation and RNG, in both directions.

The set of proposed techniques is driven by the idea of augmenting the normal-based results concerning different types of correlations to any bivariate continuous

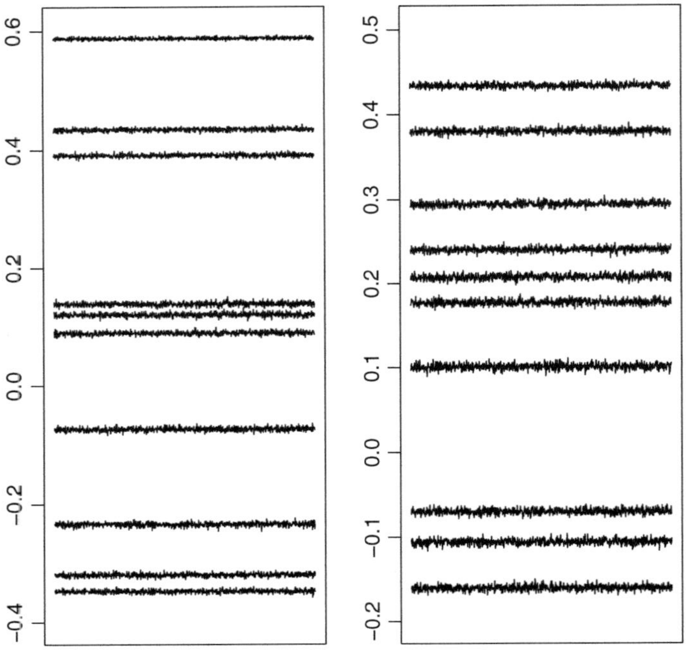

Fig. 6 The trace plot of correlation estimates for Simulation 1 (*left graph*) and Simulation 2 (*right graph*) across $n = 1,000$ replicates; they closely match the true values shown in Tables 6 and 7

setting. Nonnormality is handled by the power polynomials that map the normal and nonnormal correlations. The approach works as long as the marginal characteristics (skewness and elongation parameters for continuous data and proportion values for binary/ordinal data) and the degree of linear association between the two variables are legitimately defined, regardless of the shape of the underlying bivariate continuous density. When the above-mentioned quantities are specified, one can connect correlations before and after discretization in a relatively simple manner.

One potential limitation is that power polynomials cover most of the feasible symmetry-peakedness plane ($\nu_2 \geq \nu_1^2 - 2$), but not entirely. In an attempt to span a larger space, one can utilize the fifth order polynomial systems (Demirtas 2017; Headrick 2002), although it may not constitute an ultimate solution. In addition, a minor concern could be that unlike binary data, the marginal proportions and the second order product moment (correlation) do not fully define the joint distribution for ordinal data. In other words, odds ratios and correlations do not uniquely determine each other. However, in overwhelming majority of applications, the specification of the first and second order moments suffices for practical purposes; and given the scope of this work, which is modeling the transition between different pairs of correlations, this complication is largely irrelevant. Finally, the reasons we base our algorithms on the Pearson correlation (rather than the Spearman correlation) are that it is much more common in RNG context and in practice; and the differences between the two are negligibly small in most cases. Extending this method for encompassing the Spearman correlation will be taken up in future work resorting to a variation of the sorting idea that appeared in Demirtas and Hedeker (2011), allowing us to capture any monotonic relationship in addition to the linear relationships. On a related note, further expansions can be imagined to accommodate more complex associations that involve higher order moments.

The positive characteristics and salient advantages of these algorithms are as follows:

- They work for an extensive class of underlying bivariate latent distributions whose components are allowed to be non-identically distributed. Nearly all continuous shapes and skip patterns for ordinal variables are permissible.
- The required software tools for the implementation are rather basic, users merely need a computational platform with numerical double integration solver for the binary-binary case, univariate RNG capabilities for the binary/ordinal-continuous case, an iterative scheme that connects the polychoric correlations and the ordinal phi coefficients under the normality assumption for the ordinal-ordinal case, a polynomial root-finder and a nonlinear equations set solver to handle nonnormal continuous variables.
- The description of the connection between the two correlations is naturally given for the bivariate case. The multivariate extension is easily manageable by assembling the individual correlation entries. The way the techniques work is independent of the number of variables; the curse of dimensionality is not an issue.
- The algorithms could be conveniently used in meta-analysis domains where some studies discretize variables and some others do not.
- Assessing the magnitude of change in correlations before and after ordinalization is likely to be contributory in simulation studies where we replicate the specified trends especially when simultaneous access to the latent data and the eventual binary/ordinal data is desirable.
- One can more rigorously fathom the nature of discretization in the sense of knowing how the correlation structure is transformed after dichotomization or ordinalization.
- The proposed procedures can be regarded as a part of sensible RNG mechanisms to generate multivariate latent variables as well as subsequent binary/ordinal variables given their marginal shape characteristics and associational structure in simulated environments, potentially expediting the development of novel mixed data generation routines, especially when an RNG routine is structurally involved with generating multivariate continuous data as an intermediate step. In conjunction with the published works on joint binary/normal (Demirtas and Doganay 2012), binary/nonnormal continuous (Demirtas et al. 2012), ordinal/normal (Demirtas and Yavuz 2015), count/normal (Amatya and Demirtas 2015), and multivariate ordinal data generation (Demirtas 2006), the ideas presented in this chapter might serve as a milestone for concurrent mixed data generation schemes that span binary, ordinal, count, and nonnormal continuous data.
- These algorithms may be instrumental in developing multiple imputation strategies for mixed longitudinal or clustered data as a generalization of the incomplete

data methods published in Demirtas and Schafer (2003), Demirtas (2004b, 2005), and Demirtas and Hedeker (2007, 2008c). Concomitantly, they can be helpful in improving rounding techniques in multiple imputation (Demirtas 2009, 2010).

Wrapping it up, this work is inspired by the development of algorithms that are designed to model the magnitude of change in correlations when discretization is employed. In this regard, the proposed algorithms could be of working functionality in identifying the relationships between different types of correlations before and after discretization, and have noteworthy advantages for simulation purposes. As a final note, a software implementation of the algorithms can be accessed through the recent $\mathscr{R}$ package **CorrToolBox** (Allozi and Demirtas 2016).

References

Allozi, R., & Demirtas, H. (2016). Modeling Correlational Magnitude Transformations in Discretization Contexts, R package **CorrToolBox**. https://cran.r-project.org/web/packages/CorrToolBox.

Amatya, A., & Demirtas, H. (2015). Simultaneous generation of multivariate mixed data with Poisson and normal marginals. *Journal of Statistical Computation and Simulation*, *85*, 3129–3139.

Barbiero, A., & Ferrari, P.A. (2015). Simulation of Ordinal and Discrete Variables with Given Correlation Matrix and Marginal Distributions, R package **GenOrd**. https://cran.r-project.org/web/packages/GenOrd.

Cario, M. C., & Nelson, B. R. (1997). *Modeling and generating random vectors with arbitrary marginal distributions and correlation matrix (Technical Report)*. Department of Industrial Engineering and Management Services: Northwestern University, Evanston, IL, USA.

Demirtas, H. (2004a). Simulation-driven inferences for multiply imputed longitudinal datasets. *Statistica Neerlandica*, *58*, 466–482.

Demirtas, H. (2004b). Assessment of relative improvement due to weights within generalized estimating equations framework for incomplete clinical trials data. *Journal of Biopharmaceutical Statistics*, *14*, 1085–1098.

Demirtas, H. (2005). Multiple imputation under Bayesianly smoothed pattern-mixture models for non-ignorable drop-out. *Statistics in Medicine*, *24*, 2345–2363.

Demirtas, H. (2006). A method for multivariate ordinal data generation given marginal distributions and correlations. *Journal of Statistical Computation and Simulation*, *76*, 1017–1025.

Demirtas, H. (2007a). Practical advice on how to impute continuous data when the ultimate interest centers on dichotomized outcomes through pre-specified thresholds. *Communications in Statistics-Simulation and Computation*, *36*, 871–889.

Demirtas, H. (2007b). The design of simulation studies in medical statistics. *Statistics in Medicine*, *26*, 3818–3821.

Demirtas, H. (2008). On imputing continuous data when the eventual interest pertains to ordinalized outcomes via threshold concept. *Computational Statistics and Data Analysis*, *52*, 2261–2271.

Demirtas, H. (2009). Rounding strategies for multiply imputed binary data. *Biometrical Journal*, *51*, 677–688.

Demirtas, H. (2010). A distance-based rounding strategy for post-imputation ordinal data. *Journal of Applied Statistics*, *37*, 489–500.

Demirtas, H. (2016). A note on the relationship between the phi coefficient and the tetrachoric correlation under nonnormal underlying distributions. *American Statistician*, *70*, 143–148.

Demirtas, H. (2017). Concurrent generation of binary and nonnormal continuous data through fifth order power polynomials, *Communications in Statistics- Simulation and Computation. 46*, 344–357.

Demirtas, H., Ahmadian, R., Atis, S., Can, F. E., & Ercan, I. (2016a). A nonnormal look at polychoric correlations: Modeling the change in correlations before and after discretization. *Computational Statistics, 31*, 1385–1401.

Demirtas, H., Arguelles, L. M., Chung, H., & Hedeker, D. (2007). On the performance of bias-reduction techniques for variance estimation in approximate Bayesian bootstrap imputation. *Computational Statistics and Data Analysis, 51*, 4064–4068.

Demirtas, H., & Doganay, B. (2012). Simultaneous generation of binary and normal data with specified marginal and association structures. *Journal of Biopharmaceutical Statistics, 22*, 223–236.

Demirtas, H., Freels, S. A., & Yucel, R. M. (2008). Plausibility of multivariate normality assumption when multiply imputing non-Gaussian continuous outcomes: A simulation assessment. *Journal of Statistical Computation and Simulation, 78*, 69–84.

Demirtas, H., & Hedeker, D. (2007). Gaussianization-based quasi-imputation and expansion strategies for incomplete correlated binary responses. *Statistics in Medicine, 26*, 782–799.

Demirtas, H., & Hedeker, D. (2008a). Multiple imputation under power polynomials. *Communications in Statistics- Simulation and Computation, 37*, 1682–1695.

Demirtas, H., & Hedeker, D. (2008b). Imputing continuous data under some non-Gaussian distributions. *Statistica Neerlandica, 62*, 193–205.

Demirtas, H., & Hedeker, D. (2008c). An imputation strategy for incomplete longitudinal ordinal data. *Statistics in Medicine, 27*, 4086–4093.

Demirtas, H., & Hedeker, D. (2011). A practical way for computing approximate lower and upper correlation bounds. *The American Statistician, 65*, 104–109.

Demirtas, H., & Hedeker, D. (2016). Computing the point-biserial correlation under any underlying continuous distribution. *Communications in Statistics- Simulation and Computation, 45*, 2744–2751.

Demirtas, H., Hedeker, D., & Mermelstein, J. M. (2012). Simulation of massive public health data by power polynomials. *Statistics in Medicine, 31*, 3337–3346.

Demirtas, H., & Schafer, J. L. (2003). On the performance of random-coefficient pattern-mixture models for non-ignorable drop-out. *Statistics in Medicine, 22*, 2553–2575.

Demirtas, H., Shi, Y., & Allozi, R. (2016b). Simultaneous generation of count and continuous data, R package PoisNonNor. https://cran.r-project.org/web/packages/PoisNonNor.

Demirtas, H., Wang, Y., & Allozi, R. (2016c) Concurrent generation of binary, ordinal and continuous data, R package BinOrdNonNor. https://cran.r-project.org/web/packages/BinOrdNonNor.

Demirtas, H., & Yavuz, Y. (2015). Concurrent generation of ordinal and normal data. *Journal of Biopharmaceutical Statistics, 25*, 635–650.

Emrich, J. L., & Piedmonte, M. R. (1991). A method for generating high-dimensional multivariate binary variates. *The American Statistician, 45*, 302–304.

Farrington, D. P., & Loeber, R. (2000). Some benefits of dichotomization in psychiatric and criminological research. *Criminal Behaviour and Mental Health, 10*, 100–122.

Ferrari, P. A., & Barbiero, A. (2012). Simulating ordinal data. *Multivariate Behavioral Research, 47*, 566–589.

Fleishman, A. I. (1978). A method for simulating non-normal distributions. *Psychometrika, 43*, 521–532.

Fréchet, M. (1951). Sur les tableaux de corrélation dont les marges sont données. *Annales de l'Université de Lyon Section A, 14*, 53–77.

Genz, A., Bretz, F., Miwa, T., Mi, X., Leisch, F., Scheipl, F., Bornkamp, B., Maechler, M., & Hothorn, T. (2016). Multivariate normal and t distributions, R package mvtnorm. https://cran.r-project.org/web/packages/mvtnorm.

Headrick, T. C. (2002). Fast fifth-order polynomial transforms for generating univariate and multivariate nonnormal distributions. *Computational Statistics and Data Analysis, 40*, 685–711.

Headrick, T. C. (2010). *Statistical Simulation: Power Method Polynomials and Other Transformations Boca Raton*. FL: Chapman and Hall/CRC.

Hoeffding, W. (1994). Scale-invariant correlation theory. In N. I. Fisher & P. K. Sen (Eds.), *The Collected Works of Wassily Hoeffding* (the original publication year is 1940) (pp. 57–107). New York: Springer.

Inan, G., & Demirtas, H. (2016). Data generation with binary and continuous non-normal components, R package BinNonNor. https://cran.r-project.org/web/packages/BinNonNor.

MacCallum, R. C., Zhang, S., Preacher, K. J., & Rucker, D. D. (2002). On the practice of dichotomization of quantitative variables. *Psychological Methods*, *7*, 19–40.

R Development Core Team. (2016). R: A Language and Environment for Statistical Computing. http://www.cran.r-project.org.

Revelle, W. (2016). Procedures for psychological, psychometric, and personality researchmultivariate normal and t distributions, R package psych. https://cran.r-project.org/web/packages/psych.

Vale, C. D., & Maurelli, V. A. (1983). Simulating multivariate nonnormal distributions. *Psychometrika*, *48*, 465–471.

Yucel, R. M., & Demirtas, H. (2010). Impact of non-normal random effects on inference by multiple imputation: A simulation assessment. *Computational Statistics and Data Analysis*, *54*, 790–801.

Monte-Carlo Simulation of Correlated Binary Responses

Trent L. Lalonde

Abstract Simulation studies can provide powerful conclusions for correlated or longitudinal response data, particularly for relatively small samples for which asymptotic theory does not apply. For the case of logistic modeling, it is necessary to have appropriate methods for simulating correlated binary data along with associated predictors. This chapter presents a discussion of existing methods for simulating correlated binary response data, including comparisons of various methods for different data types, such as longitudinal versus clustered binary data generation. The purposes and issues associated with generating binary responses are discussed. Simulation methods are divided into four main approaches: using a marginally specified joint probability distribution, using mixture distributions, dichotomizing non-binary random variables, and using a conditionally specified distribution. Approaches using a completely specified joint probability distribution tend to be more computationally intensive and require determination of distributional properties. Mixture methods can involve mixtures of discrete variables only, mixtures of continuous variables only, and mixtures involving both continuous and discrete variables. Methods that involve discretizing non-binary variables most commonly use normal or uniform variables, but some use count variables such as Poisson random variables. Approaches using a conditional specification of the response distribution are the most general, and allow for the greatest range of autocorrelation to be simulated. The chapter concludes with a discussion of implementations available using R software.

1 Introduction

Correlated binary data occur frequently in practice, across disciplines such as health policy analysis, clinical biostatistics, econometric analyses, and education research. For example, health policy researchers may record whether or not members of a household have health insurance; econometricians may be interested in whether small

T.L. Lalonde (✉)
Department of Applied Statistics and Research Methods, University of Northern Colorado, Greeley, CO, USA
e-mail: trent.lalonde@unco.edu

D.-G. Chen and J.D. Chen (eds.), *Monte-Carlo Simulation-Based Statistical Modeling*, ICSA Book Series in Statistics, DOI 10.1007/978-981-10-3307-0_5

businesses within various urban districts have applied for financial assistance; higher education researchers might study the probabilities of college attendance for students from a number of high schools. In all of these cases the response of interest can be represented as a binary outcome, with a reasonable expectation of autocorrelation among those responses. Correspondingly, analysis of correlated binary outcomes has received considerable and long-lasting attention in the literature (Stiratelli et al. 1984; Zeger and Liang 1986; Prentice 1988; Lee and Nelder 1996; Molenberghs and Verbeke 2006). The most common models fall under the class of correlated binary logistic regression modeling.

While appropriately developed logistic regression models include asymptotic estimator properties, Monte-Carlo simulation can be used to augment the theoretical results of such large-sample distributional properties. Monte-Carlo simulation can be used to confirm such properties, and perhaps more importantly, simulation methods can be used to complement large-sample distributional properties with small-sample results. Therefore it is important to be able to simulate binary responses with specified autocorrelation so that correlated binary data models can benefit from simulation studies.

Throughout the chapter, the interest will be in simulating correlated binary outcomes, Y_{ij}, where i indicates a cluster of correlated responses and j enumerates the responses. It will be assumed that the simulated outcomes have specified marginal probabilities, π_{ij}, and pairwise autocorrelation, $\rho_{ij,ik}$. The term "cluster" will be used to refer to a homogenous group of responses known to have autocorrelation, such as an individual in a longitudinal study or a group in a correlated study. For many of the algorithms presented in this discussion, a single cluster will be considered for simplicity, in which case the first subscript of the Y_{ij} will be omitted and Y_i, π_i, and ρ_{ij} will be used instead. Some methods will additionally require the specification of joint probabilities, higher-order correlations, or predictors.

1.1 Binary Data Issues

There are a number of important factors to consider when developing or evaluating a correlated data simulation technique. Common issues include computational feasibility or simplicity, the incorporation of predictors or covariates, variation of parameters between and within clusters, and the ability to effectively control expectations, variation, and autocorrelation. For binary data in particular, there are a number of additional concerns that are not relevant to the more traditional normal, continuous data generation. Simulation of binary responses is typically based on a probability of interest, or the expectation of the Bernoulli distribution. Many authors have noted that the pairwise joint probabilities for binary data, $\pi_{i,j} = P(Y_i = 1, Y_j = 1)$, are restricted by the marginal probabilities, according to,

$$max(0, \pi_i + \pi_j - 1) \leq \pi_{i,j} \leq min(\pi_i, \pi_j),$$

which imposes restrictions on joint distributions according to the desired marginal probabilities. In addition, it is necessary to properly account for the inherent mean-variance relationship associated with Bernoulli data, $\text{Var}(Y_i) = \pi_i(1 - \pi_i)$, imposing further restrictions on higher-order moments based on the desired marginal probabilities.

Many methods of simulating correlated binary data suffer from restricted ranges of produced autocorrelation, which should typically span $(-1, 1)$. For correlated binary outcomes $Y_1, \ldots, Y_N$ with marginal expectations $\pi_1, \ldots, \pi_N$, Prentice (1988) argued that the pairwise correlation between any two responses Y_i and Y_j must lie within the range (l, u), where

$$l = max\left(-\sqrt{(\pi_i\pi_j)/(1-\pi_i)(1-\pi_j)}, -\sqrt{(1-\pi_i)(1-\pi_j)/(\pi_i\pi_j)}\right),$$
$$u = min\left(\sqrt{\pi_i(1-\pi_j)/\pi_j(1-\pi_i)}, \sqrt{\pi_j(1-\pi_i)/\pi_i(1-\pi_j)}\right), \tag{1}$$

to satisfy the requirements for the joint distribution. This implies that, depending on the desired marginal probabilities, using a fully specified joint probability distribution can lead to simulated values with restricted ranges of pairwise correlation. The restrictions of Eq. 1 can be counterintuitive to researchers who are used to the unconstrained correlation values of normal variables.

Some methods of simulating correlated binary outcomes struggle to control changes in probabilities across clusters of correlated data. Due to the typically non-linear nature of the relationships between response probabilities and predictors, many methods also fail to incorporate predictors into data simulation. These issues, among others, must be considered when developing and selecting a method for simulating correlated binary data.

This chapter presents a thorough discussion of existing methods for simulating correlated binary data. The methods are broadly categorized into four groups: correlated binary outcomes produced directly from a fully specified joint probability distribution, from mixtures of discrete or continuous variables, from dichotomized continuous or count variables, and from conditional probability distributions. The oldest literature is available for fully specified joint probability distributions, but often these methods are computationally intensive and require specification of higher-order probabilities or correlations at the outset. Mixture approaches have been used to combine products of binary variables to induce the desired autocorrelation, while other mixtures involving continuous variables require dichotomization of the resulting values. Dichotomizing normal or uniform variables is a widely implemented approach to producing binary data, although less well-known approaches have also been pursued, such as dichotomizing counts. Conditional specification of a binary distribution typically makes use of "prior" binary outcomes or predictors of interest, and tends to lead to the greatest range of simulated autocorrelation. Each of these methods for generating correlated binary data will be discussed through a chronological perspective, with some detail provided and some detail left to the original publications. The chapter concludes with general recommendations for the most effective binary data simulation methods.

2 Fully Specified Joint Probability Distributions

The method for simulating correlated binary outcomes with the longest-tenured presence in the literature, is full specification of a joint probability distribution for correlated binary variates. The joint pdf can either be written explicitly in closed form, or derived by exhaustive listing of possible outcomes and associated probabilities. In all cases the generation of data relies on the method of Devroye (1986).

2.1 Simulating Binary Data with a Joint PDF

Given a joint probability density function for any type of binary data, correlated or independent, Devroye (1986) described an effective method for generating appropriate sequences of binary outcomes. Assume a joint pdf has been fully specified for the binary random vector $\mathbf{Y}^T = (Y_1, \ldots, Y_N)$, such that marginal probabilities can be calculated for any combination of binary outcomes. For finite N, probabilities can be calculated for all 2^N possible outcome vectors, denoted $p_0, \ldots, p_{2^N-1}$. Realizations for the random vector $\mathbf{Y}$ can be generated according to the following algorithm.

Generating Binary Values with a Joint PDF

1. Order the probabilities from smallest to largest, $p_{(0)}, \ldots, p_{(2^N-1)}$.
2. Define cumulative values z_j according to:

$$\begin{aligned} z_0 &= 0, \\ z_j &= z_{j-1} + p_{(j)}, \\ z_{2^N-1} &= 1. \end{aligned}$$

3. Generate a standard uniform variate U on (0, 1).
4. Select j such that $z_j \leq U < z_{j+1}$.
5. The random sequence is the binary representation of the integer j.

The method is relatively simple, relying on calculating probabilities associated with all possible N-dimensional binary sequences, and generating a single random uniform variable to produce an entire vector of binary outcomes. Devroye (1986) has argued that this method produces appropriate binary sequences according to the pdf provided, and the method is relied on extensively in situations in which a joint pdf can be constructed. In fact, relying on this algorithm, many authors have pursued methods of constructing a joint pdf as a means to generate vectors of binary responses.

2.2 Explicit Specification of the Joint PDF

Simulation of correlated binary outcomes is generally thought to begin with Bahadur (1961). Bahadur (1961) presented a joint pdf for N correlated binary random variables as follows. Let y_i indicate realizations of binary variables, $\pi = P(Y_i = 1)$ represent the constant expectation of the binary variables, all equivalent, and ρ_{ij} be the autocorrelation between two variables Y_i and Y_j. Then the joint pdf can be written,

$$f(y_1, \ldots, y_N) = 1 + \frac{\sum_{i \neq j} \rho_{ij} (-1)^{(y_i + y_j)} \pi^{(2 - y_i - y_j)} (1 - \pi)^{(y_i + y_j)}}{\pi(1 - \pi)}.$$

While Bahadur (1961) expressed the joint pdf in terms of lagged correlations for an autoregressive time series, the idea expands generally to clustered binary data. In practice it is necessary to estimate range restrictions for the autocorrelation to ensure $f(y_1, \ldots, y_N) \in (0, 1)$. These restrictions depend on values of π, and are typically determined empirically (Farrell and Sutradhar 2006). Using the algorithm of Devroye (1986), the model of Bahadur (1961) can be used to simulate values for a single cluster or longitudinal subject, then repeated for additional clusters. This allows the probability π to vary across clusters, however, π is assumed constant within clusters, reducing the ability to incorporate effects of covariates. Most crucially, the pdf given by Bahadur (1961) will become computationally burdensome for high-dimensional data simulations (Lunn and Davies 1998; Farrell and Sutradhar 2006).

2.3 Derivation of the Joint PDF

Instead of relying on the joint pdf of Bahadur (1961), authors have pursued the construction of joint probability distributions not according to a single pdf formula, but instead by using desired properties of the simulated data to directly calculate all possible probabilities associated with N-dimensional sequences of binary outcomes. Often these methods involve iterative processes, solutions to linear or nonlinear systems of equations, and matrix and function inversions. They are computationally complex but provide complete information about the distribution of probabilities without using a closed-form pdf.

Lee (1993) introduced a method that relies on a specific copula distribution, with binary variables defined according to a relationship between the copula parameters and the desired probabilities and correlation for the simulated binary responses. A copula (Genest and MacKay 1986a, b) is a multivariate distribution for random variables $(X_1, \ldots, X_N)$ on the N-dimensional unit space, such that each marginal distribution for X_i is uniform on the the domain (0, 1). The variables $X_1, \ldots, X_N$ can be used to generate binary variables $Y_1, \ldots, Y_N$ by dichotomizing according to $Y_i = I(X_i > \pi_i)$, where π_i is the desired expectation of the binary variable Y_i and $I()$ takes the value of 1 for a true argument and 0 otherwise. However, Lee (1993) did not use a simple dichotomization of continuous variables.

In order to extend this idea to allow for autocorrelation, Lee (1993) proposed using the copula with exchangeable correlation, the Archimidian copula proposed by Genest and MacKay (1986a). Use of such a copula will induce Pearson correlation between any two random binary variables Y_i and Y_j given by,

$$\rho_{ij} = \frac{\pi_{0,0} - \pi_i \pi_j}{\sqrt{\pi_i \pi_j (1 - \pi_i)(1 - \pi_j)}}, \tag{2}$$

where $\pi_{0,0}$ is the joint probability that both variables Y_i and Y_j take the value 0. Because of the restriction that $\pi_{0,0} \leq min(\pi_i, \pi_j)$, it follows that $\pi_{0,0} > \pi_i \pi_j$ and therefore the induced Pearson correlation will always be positive when this method is applied. The method of Lee (1993) requires the correlation of Eq. 2 to be constant within clusters.

Constructing the Joint Distribution by Archimidian Copula

1. For a single cluster, determine marginal probabilities $\pi_1, \ldots, \pi_N$ for correlated binary variables $Y_1, \ldots, Y_N$ with desired constant autocorrelation ρ.
2. A value for the Archimidian copula distribution parameter Ψ can be calculated based on the values of π_i and ρ. The parameter Ψ takes values on (0, 1], where a value of 1 indicates independence, while as $\Psi \to 0$ the Kendall correlation converges to 1.
3. Using the Archimidian copula distribution, solve linearly for the joint probabilities of $Y_1, \ldots, Y_N$ with respect to the binary representation of each integer j, where $0 \leq j \leq 2^N - 1$,

$$P_j = P(Y_1 = j_1, \ldots, Y_N = j_N), \tag{3}$$

   where $j_1 \ldots j_N$ is the binary representation of j. Calculation of this probability using the Archimidian copula has closed form but requires computation of the inverse of a CDF-like function.
4. Given the probabilities P_0, P_{2^N-1}, simulate a binary sequence according to the algorithm of Devroye (1986).
5. Repeat for additional clusters.

The method of Lee (1993) allows inclusion of predictors when determining the marginal π_i, which clearly allows these probabilities to vary within clusters. However, the method is restricted to positive, exchangeable autocorrelation that cannot vary within groups, and requires the solution to numerous systems of equations and the inverse of a CDF-like function.

Kang and Jung (2001) discussed an alternative to this method such that the probabilities P_j can be obtained by solving a nonlinear system of equations relating P_j to the first two desired moments, π_i and the correlation $\text{Corr}(Y_i, Y_j) = \rho_{ij}$. However, the necessity of solving a system of nonlinear equations does not decrease the computational complexity of the algorithm as compared to the copula approach.

The method of Gange (1995) simplifies the method of Lee (1993) but can add computational complexity. Suppose that, in addition to specifying the desired marginal probabilities π_i associated with the correlated binary variables, Y_i, pairwise and higher-order joint probabilities are also specified. That is, joint probabilities $\pi_{i,j} = P(Y_i = y_i, Y_j = y_j)$ and higher-order joint probabilities $\pi_{i_1,\ldots,i_k} = P(Y_{i_1} = y_{i_1}, \ldots, Y_{i_k} = y_{i_k})$ can be specified, up to order k. Given such probabilities, a full joint probability density function is constructed through the Iterative Proportional Fitting algorithm.

The main idea of the Iterative Proportional Fitting algorithm is to equate derivation of the joint pdf to fitting a log-linear model to a contingency table of order 2^N, including interactions up to order k. Each pairwise or higher-order joint probability is treated as a constraint on the implicit contingency table, and has an associated nonlinear equation. Solving this system of nonlinear equations corresponding to the higher-order joint probabilities is equated to the standard likelihood-based solution to log-linear model fitting. Predicted values from the log-linear model give the probabilities associated with all possible outcome combinations.

Constructing the Joint Distribution using the Iterative Proportional Fitting Algorithm

1. For a single cluster, specify the desired marginal probabilities π_i, pairwise probabilities $\pi_{i,j}$, and up to k-order joint probabilities $\pi_{i_1,\ldots,i_k}$.
2. Construct a log-linear model with interactions up to order k corresponding to the constraints specified by the probabilities up to order k.
3. Fit the log-linear model to obtain estimated probabilities corresponding to the joint marginal pdf.
4. Simulate binary values according to the algorithm of Devroye (1986).
5. Repeat for additional clusters.

The method of Gange (1995) allows for covariates to be included in determining initial probabilities and pairwise or higher-order joint probabilities, and Gange (1995) describes how the pairwise probabilities can be connected to the working correlation structure of the Generalized Estimating Equations, allowing for specific correlation structures to be derived. However, specific correlation structures are not explicitly exemplified by Gange (1995). Further, this method requires an iterative procedure to solve a system of nonlinear equations, which can become computationally burdensome with increased dimension of each cluster.

3 Specification by Mixture Distributions

Many approaches to simulating correlated binary variables rely on the properties of random variables defined as mixtures of other random variables with various distributions. Authors have pursued mixtures of discrete distributions, continuous distributions, and combinations of discrete and continuous distributions. Methods relying solely on mixtures of continuous distributions tend to struggle to represent correlated binary responses.

3.1 Mixtures Involving Discrete Distributions

Many attempts to circumvent the computational burden of the methods of Bahadur (1961), Lee (1993), and Gange (1995) have turned to generating binary responses through mixtures of discrete random variables. Kanter (1975) presented a method that directly uses autoregression to simulate properly correlated binary outcomes. Suppose it is of interest to simulate a series of binary values Y_i with constant expectation π. Kanter (1975) proposed to use the model,

$$Y_i = U_i\left[Y_{i-1} \oplus W_i\right] + (1 - U_i)W_i,$$

where $\oplus$ indicates addition modulo 2, U_i can be taken to be Bernoulli with probability π_U and W_i can be taken to be Bernoulli with probability $\pi_W = (\pi(1 - \pi_U))/(1 - 2\pi\pi_U)$. Assuming Y_{i-1}, U_i, and W_i to be independent, it can be shown that all Y_i have expectation π and that the autocorrelation between any two outcomes is

$$\mathrm{Corr}(Y_i, Y_j) = \left(\frac{\pi_U(1 - 2\pi)}{1 - 2\pi\pi_U}\right)^{|i-j|}.$$

The method of Kanter (1975) requires $\pi_U \in (0, min((1 - \pi)/\pi, 1))$, and includes restrictions on the autocorrelation in the simulated data based on the probabilities used. In particular, Farrell and Sutradhar (2006) showed that no negative autocorrelation can be generated with any probability π chosen to be less than or equal to 0.50. In addition, this method does not allow for easy variation of probabilities within clusters or series.

Use of a mixture of binary random variables was updated by Lunn and Davies (1998), who proposed a simple method for generating binary data for multiple clusters simultaneously, and for various types of autocorrelation structure within groups.

Suppose the intention is to simulate random binary variables Y_{ij} such that the expectation of each is cluster-dependent, π_i, and the autocorrelation can be specified. First assume a positive, constant correlation ρ_i is desired within clusters. Then simulate binary values according to the equation,

$$Y_{ij} = (1 - U_{ij})W_{ij} + U_{ij}Z_i, \tag{4}$$

where U_{ij} can be taken to be Bernoulli with probability $\pi_U = \sqrt{\rho_i}$, and both W_{ij} and Z_i can be Bernoulli with success probability π_i, all independent. Then it can be shown that $\mathrm{E}[Y_{ij}] = \pi_i$ and $\mathrm{Corr}(Y_{ij}, Y_{ik}) = \rho_i$. Lunn and Davies (1998) explain how to adjust their method to allow for additional correlation structures within clusters, by changing the particular mixture from Eq. 4,

$$\begin{aligned} Y_{ij} &= (1 - U_{ij})W_{ij} + U_{ij}Z_i && \text{(Exchangeable)},\\ Y_{ij} &= (1 - U_{ij})W_{ij} + U_{ij}Y_{i,j-1} && \text{(Autoregressive)},\\ Y_{ij} &= (1 - U_{ij})W_{ij} + U_{ij}W_{i,j-1} && \text{(M-Dependent)}. \end{aligned}$$

This algorithm requires constant outcome probabilities within clusters. In order to accommodate varying probabilities within clusters, Lunn and Davies (1998) proposed a simple transformation of generated binary responses,

$$\tilde{Y}_{ij} = A_{ij}Y_{ij},$$

where A_{ij} is taken to be Bernoulli with success probability α_{ij}, independent of all other simulated variables. Then the $\tilde{Y}_{ij}$ satisfy the previous distributional requirements with $\mathrm{E}[\tilde{Y}_{ij}] = \alpha_{ij} max(\pi_i)$. Lunn and Davies (1998) acknowledge that this multiplicative transformation imposes an additional multiplicative correction to the correlation between generated binary values. While this transformation allows the probabilities to vary within clusters, it does not easily incorporate predictors into this variation, instead accommodating known success probabilities that may differ across responses within clusters.

Generating Binary Values using a Binary Distribution Mixture

1. For a single cluster i, determine the desired probability π_i and autocorrelation structure, along with cluster-dependent correlation ρ_i.
2. Depending on the autocorrelation structure, generate U_{ij}, W_{ij}, and possibly Z_i.
3. Calculate Y_{ij} using the appropriate mixture.
4. If necessary, transform Y_{ij} to $\tilde{Y}_{ij}$ to accommodate varying probabilities within-cluster.
5. Repeat for other clusters.

The issue of allowing within-cluster variation of success probabilities was addressed by Oman and Zucker (2001) in a method that is truly a combination of mixture variable simulation and dichotomization of continuous values. Oman and Zucker (2001) argued that the cause of further restricted ranges of autocorrelation in simulated binary data, beyond the limits from Eq. 1 given by Prentice (1988), is the combination of varying probabilities within clusters and the inherent mean-variance relationship of binary data. Assume the interest is in simulating binary variables Y_{ij} with probabilities π_{ij}, varying both between and within clusters. To define the joint probability distribution of any two binary outcomes Y_{ij} and Y_{ik}, define

$$P(Y_{ij} \times Y_{ik} = 1) = (1 - \nu_{jk})\pi_{ij}\pi_{ik} + \nu_{jk} min(\pi_{ij}, \pi_{ik}), \tag{5}$$

where ν_{jk} is chosen to reflect the desired correlation structure within groups, as follows. The joint distribution specified by Eq. 5 allows the correlation between any two responses within a cluster, denoted ρ_{jk}, to be written $\rho_{jk} = \nu_{jk} \times max(\rho_{st})$, where the maximum is taken over all values of correlation within cluster i. The process of generating binary values is constructed to accommodate this joint distribution as follows. Define responses according to

$$Y_{ij} = I(Z_{ij} \leq \theta_{ij}),$$

where $\theta_{ij} = F^{-1}(\pi_{ij})$ for any continuous CDF F, and Z_{ij} is defined as an appropriate mixture. Similarly to the method of Lunn and Davies (1998), Oman and Zucker (2001) provide mixtures for common correlation structures,

$$\begin{aligned}
Z_{ij} &= U_{ij} X_{i0} + (1 - U_{ij}) X_{ij} && \text{(Exchangeable)},\\
Z_{ij} &= U_{ij} X_{i,j-1} + (1 - U_{ij}) X_{ij} && \text{(Moving Average)},\\
Z_{ij} &= U_{ij} Z_{i,j-1} + (1 - U_{ij}) X_{ij} && \text{(Autoregressive)},
\end{aligned}$$

where U_{ij} can be taken to be Bernoulli with probability $\pi_U = \sqrt{\nu_{jk}}$, and all X_{ij} are independently distributed according to the continuous distribution F. The different correlation structures are induced by both adjusting the mixture as given above and also by defining ν_{jk} accordingly. For example, constant ν across all i, j with the first mixture will produce exchangeable correlation among the binary outcomes, while choosing $\nu_{ij} = \gamma^{|j_1 - j_2|}$ with the third mixture will produce autoregressive correlation.

Generating Binary Values using Binary and Continuous Distribution Mixture

1. For a single cluster i, determine the marginal probabilities π_{ij}.
2. Decide the target correlation structure.
3. Based on the target correlation structure, select ν_{jk} and the mixture function accordingly.

4. Generate U_{ij} as Bernoulli, X_{ij} according to continuous F, and calculate Z_{ij} and θ_{ij}.
5. Define each outcome as $Y_{ij} = I(Z_{ij} \leq \theta_{ij})$.
6. Repeat for additional clusters.

Oman and Zucker (2001) noted that covariates can be incorporated by defining $\theta_{ij} = \mathbf{x}_{ij}^T \boldsymbol{\beta}$ as the systematic component of a generalized linear model and taking F^{-1} to be the associated link function. It is an interesting idea to use the inverse link function from a generalized linear model to help connect predictors to the determination of whether the binary realization will be 0 or 1. However, the method described continues to suffer from restricted ranges of autocorrelation, most notably that the correlations between binary responses must all be positive.

3.2 *Mixtures Involving Continuous Distributions*

Instead of using proper binary distributions, non-normal values can be simulated using linear combinations of standard normal variables to represent the known moment-based properties of the desired data distribution. Many such methods are based on the work of Fleishman (1978), and later extended (Headrick 2002a, 2010, 2011). In general, the idea is to simulate *any* non-normal distribution as a polynomial mixture of normal variables,

$$Y = \sum_{i=1}^{m} c_i Z^{(i-1)},$$

where Y indicates the desired random variable, Z is a standard normal random variable, and the c_i represent coefficients chosen according to the desired distribution of Y. Fleishman (1978) derived a system of nonlinear equations that, given the target distribution mean, variance, skewness, and kurtosis, could be solved for coefficients c_1, c_2, c_3, and c_4 to produce the third-order polynomial approximation to the desired distribution. The intention is to use an expected probability π to estimate those first four moments for the Bernoulli distribution, and approximate accordingly using standard normals. This process has been extended numerous times, including to multivariate data (Vale and Maurelli 1983), to higher-order polynomials (Headrick 2002a), to using percentiles in place of standard moments to reflect instead the median, the inter-decile range, the left-right tail-weight ratio, and the tail-weight factor (Karian and Dudewicz 1999), and to control autocorrelation in multivariate data (Koran et al. 2015).

The Fleishman process and similar methods derived from it suffer from some consistent issues. First, the systems of nonlinear equations have a limited range of solutions for the necessary coefficients, and consequently can only be used to represent limited ranges of values for moments or percentile statistics. Therefore the ranges of probabilities and correlation values in the simulated data will be limited.

Secondly, the resulting mixture random variable is a power function of standard normal variables, which generally will not reflect the true mean-variance relationship necessary for binary data. While the values of mean and variance using the Fleishman method can in some cases reflect those of an appropriate sample, the dynamic relationship between changing mean and variation will not be captured in general. Thus as the mean changes, the variance generally will not show a corresponding change. Finally, the method does not readily account for the effects of predictors in simulating responses. In short, such methods are poorly equipped to handle independent binary data, let alone correlated binary outcomes.

4 Simulation by Dichotomizing Variates

Perhaps the most commonly implemented methods for simulating correlated binary outcomes are those that involve dichotomization of other types of variables. The most frequent choice is to dichotomize normal variables, although defining thresholds for uniform variables is also prevalent.

4.1 Dichotomizing Normal Variables

Many methods of dichotomizing normal variables have been proposed. The method of Emrich and Piedmonte (1991), one of the most popular, controls the probabilities and pairwise correlations of resulting binary variates. Assume it is of interest to simulate binary variables Y_i with associated probabilities π_i and pairwise correlations given by $Corr(Y_i, Y_j) = \rho_{ij}$. Begin by solving the following equation for the normal pairwise correlation, δ_{ij}, using the bivariate normal CDF Φ,

$$\Phi(z(\pi_i), z(\pi_j), \delta_{ij}) = \rho_{ij}\sqrt{\pi_i(1-\pi_i)\pi_j(1-\pi_j)} + \pi_i\pi_j,$$

where $z()$ indicates the standard normal quantile function. Next generate one N-dimensional multivariate normal variable $\mathbf{Z}$ with mean $\mathbf{0}$ and correlation matrix with components δ_{ij}. Define the correlated binary realizations using

$$Y_i = I(Z_i \le z(\pi_i)).$$

Emrich and Piedmonte (1991) showed that the sequence $Y_1, \ldots, Y_N$ has the appropriate desired probabilities π_i and correlation values ρ_{ij}.

Generating Binary Values by Dichotomizing Normals

1. For a single cluster, determine the marginal probabilities π_i and autocorrelation values ρ_{ij}.
2. Using the bivariate normal CDF, solve for normal pairwise correlation values δ_{ij}.
3. Generate one N-dimensional normal variable $\mathbf{Z}$ with correlation components given by δ_{ij}.
4. Define the binary values as $Y_i = I(Z_i \leq z(\pi_i))$, where $z()$ is the standard normal quantile function.
5. Repeat for additional clusters.

This method is straightforward and allows probabilities to vary both within and between clusters. A notable disadvantage of this method is the necessity of solving a system of nonlinear equations involving the normal CDF, which increases computational burden with large-dimensional data generation.

4.2 *Iterated Dichotomization*

Headrick (2002b) proposed a method of simulating multiple clusters of correlated binary data using JMASM3, an iterative method of dichotomizing two sets of binary variables in two steps. This method is unique in that it allows for autocorrelation both within and between clusters of binary data.

Assume it is of interest to simulate N clusters of binary data with correlation within clusters denoted by $\rho_{Y_{ij},Y_{ik}}$, and correlation between clusters by $\rho_{Y_{ij},Y_{kl}}$. Given probabilities $\pi_1, \ldots, \pi_N$, correlated binary variables $X_1, \ldots, X_N$ can be defined using random uniform variables $U_1, \ldots, U_N$ as follows. Define $X_1 = I(U_1 < \pi_1)$. Define successive X_i as

$$X_i = \begin{cases} X_1, & U_i < \pi_i \\ X_1 + 1, & U_i > \pi_i \text{ and } X_1 = 0 \\ 1 - X_1, & U_i > \pi_i \text{ and } X_1 = 1. \end{cases}$$

This generates a cluster of binary values, each correlated with X_1. Next simulate binary values Y_{ij}, where i indicates the cluster and j indicates individual outcomes, as follows, where the U_{ij} are independent uniform variables on (0, 1),

$$Y_{ij} = \begin{cases} X_i, & U_{ij} < \pi_{ij} \\ X_i + 1, & U_{ij} > \pi_{ij} \text{ and } X_i = 0 \\ 1 - X_i, & U_{ij} > \pi_{ij} \text{ and } X_i = 1. \end{cases}$$

Headrick (2002b) shows that the threshold values π_{ij} can be obtained by solving a nonlinear system in terms of the specified correlation values $\rho_{Y_{ij},Y_{ik}}$ and $\rho_{Y_{ij},Y_{kl}}$. The order of the nonlinear system corresponds to the number of correlation values specified at the outset. Headrick (2002b) also provides expressions for the marginal probabilities $P(Y_{ij} = 1)$ in terms of the first-stage probabilities π_i and the second-stage thresholds π_{ij}; however, it is not clear that these marginal probabilities can be controlled from the start.

Generating Binary Values by Iterated Dichotomization

1. Determine the autocorrelation desired within and between N clusters of binary responses, and select first-stage probabilities $\pi_1, \ldots, \pi_N$.
2. Simulate X_1 as $I(U_1 < \pi_1)$, where U_1 is a random uniform realization on $(0, 1)$.
3. Generate the remaining first-stage binary outcomes X_i according to X_1 and corresponding random uniform variables U_i.
4. Solve for the second-stage thresholds, π_{ij}, given the desired within and between correlation values, $\rho_{Y_{ij},Y_{ik}}$ and $\rho_{Y_{ij},Y_{kl}}$, respectively.
5. Generate the second-stage binary outcomes Y_{ij} according to X_i and corresponding random uniform variables U_{ij}.

The iterated dichotomization algorithm allows for control of autocorrelation both within and between clusters, and does not require complete specification of the joint probability distribution. However, it does not clearly accommodate predictors or reduce in complexity for common correlation structures; it does not allow easy specification of the marginal binary outcome probabilities; and it requires the solution of a potentially high-dimensional system of nonlinear equations.

4.3 Dichotomizing Non-normal Variables

The method of Park et al. (1996) simulates correlated binary values using a dichotomization of counts, and in the process avoids the necessity of solving any system of nonlinear equations. Assume an interest in generating N correlated binary variables $Y_1, \ldots, Y_N$, with probabilities $\pi_1, \ldots, \pi_N$ and associated pairwise correlations ρ_{ij}. Begin by generating N counts $Z_1, \ldots, Z_N$ using a collection of M Poisson random variables $X_1(\lambda_1), \ldots, X_M(\lambda_M)$, as linear combinations,

$$Z_1 = \sum_{i \in S_1} X_i(\lambda_i),$$

$$Z_2 = \sum_{i \in S_2} X_i(\lambda_i),$$

$$\vdots$$

$$Z_N = \sum_{i \in S_N} X_i(\lambda_i).$$

Notice that each count Z_i is a combination of a specific set of the Poisson random variables, denoted by S_i. The number of Poisson variables, M, the associated means, λ_i, and the sets used in the sums, S_i, are all determined algorithmically based on the desired probabilities and correlations. Each binary value is then defined by dichotomizing, $Y_i = I(Z_i = 0)$.

Park et al. (1996) describe the determination of M, λ_i, and S_i as follows. The Poisson means λ_i can be constructed as linear combinations of parameters α_{ij}, $1 \leq i, j \leq N$. The α_{ij} can be calculated based on the desired probabilities and pairwise correlations,

$$\alpha_{ij} = \ln\left(1 + \rho_{ij}\sqrt{(1-\pi_i)(1-\pi_j)/(\pi_i \pi_j)}\right).$$

Given all of the α_{ij}, define λ_k to be the smallest positive α_{ij}, $i, j \geq k$, until the first mean λ_L matches the magnitude of the largest positive α_{ij}. Then set $M = L$, let each mean λ_i remain as determined by the α_{ij}, and define each summation set S_i as those means composed of positive α_{ij} (Park et al. (1996).

Generating Binary Values by Dichotomizing Linear Poisson Mixtures

1. For a single cluster, determine the individual probabilities π_i and pairwise correlations ρ_{ij}.
2. Using the probabilities and correlation values, calculate the parameters α_{ij}.
3. Using the parameters α_{ij}, determine the number of Poisson variables, M, the Poisson means, λ_i, and the summation sets, S_i, for each count variable Z_i.
4. Calculate $Z_1, \ldots, Z_N$, and define the binary responses as $Y_i = I(Z_i = 0)$.
5. Repeat for additional clusters.

The method of Park et al. (1996) is computationally efficient and does not require solving systems of nonlinear equations. It allows for varying probabilities and correlation values within and between clusters, and can be adjusted to incorporate covariates into the probabilities. However, there is still a restriction on the range of correlations available through this algorithm.

5 Conditionally Specified Distributions

Recent attention has been paid to conditionally specifying the distribution of correlated binary variables for the purposes of simulation. While the mixture distributions can be viewed as conditional specifications, in such cases discussed Sect. 3 the mixtures were defined so that the marginal distributions of the resulting binary variables were completely specified. In this section the discussion focuses on situations without full specification of the marginal outcome distribution. Instead, the distributions are defined using predictor values or prior outcome values.

5.1 The Linear Conditional Probability Model

Qaqish (2003) introduced a method of simulating binary variates using autoregressive-type relationships to simulate autocorrelation. Each outcome value is conditioned on prior outcomes, a relationship referred to as the conditional linear family. The conditional linear family is defined by parameter values that are so-called reproducible in the following algorithm, or those that result in conditional means within the allowable range (0, 1).

Suppose the interest is in simulating correlated binary variables Y_i with associated probabilities π_i, and variance-covariance structure defined for each response by its covariation with all previous responses, $\mathbf{s}_i = \text{Cov}([Y_1, \ldots, Y_{i-1}]^T, Y_i)$. Qaqish (2003) argued that the expectation of the conditional distribution of any response Y_i, given all previous responses, can be expressed in the form,

$$\begin{aligned} \text{E}[Y_i | [Y_1, \ldots, Y_{i-1}]^T] &= \pi_i + \boldsymbol{\kappa}_i^T \left([Y_1, \ldots, Y_{i-1}]^T - [\pi_1, \ldots, \pi_{i-1}]^T \right) \\ &= \pi_i + \sum_{j=1}^{i-1} \kappa_{ij} (Y_j - \pi_j), \end{aligned} \tag{6}$$

where the components of $\boldsymbol{\kappa}_i$ are selected corresponding to the desired variance-covariance structure according to

$$\boldsymbol{\kappa}_i = [\text{Cov}([Y_1, \ldots, Y_{i-1}]^T)]^{-1} \mathbf{s}_i.$$

The correlated binary variables are then generated such that Y_1 is Bernoulli with probability π_1, and all subsequent variables are random Bernoulli with probability given by the conditional mean in Eq. 6. It is straightforward to show that such a sequence will have the desired expectation $[\pi_1, \ldots, \pi_N]^T$ and autocorrelation defined by the variance-covariance $\mathbf{s}_i$. Qaqish (2003) provides simple expressions for κ_{ij} to produce exchangeable, auto-regressive, and moving average correlation structures, as follows,

$$\kappa_{ij} = \frac{\rho}{1+(i-1)\rho}\left(\frac{V_{ii}}{V_{jj}}\right)^{1/2} \quad \text{(Exchangeable)},$$

$$\lambda_i = \pi_i + \rho(y_{i-1} - \pi_{i-1})\left(\frac{V_{ii}}{V_{i-1,i-1}}\right)^{1/2} \quad \text{(Autoregressive)},$$

$$\kappa_{ij} = \frac{\beta^j - \beta^{-j}}{\beta^{-i} - \beta^i}\left(\frac{V_{ii}}{V_{jj}}\right)^{1/2} \quad \text{(Moving Average)},$$

where V_{ii} represents diagonal elements of the response variance-covariance, $\lambda_i = \mathrm{E}[Y_i|[Y_1, \ldots, Y_{i-1}]^T]$ represents the conditional expectation, and $\beta = [(1 - 4\rho^2)^{1/2} - 1]/2\rho$ with ρ the decaying correlation for autoregressive models and the single time-lag correlation for moving average models.

Generating Binary Values by the Linear Conditional Probability Model

1. For a single cluster, determine the individual probabilities π_i.
2. Select the desired correlation structure, and the corresponding constants κ_i and conditional means λ_i.
3. Generate the first response, Y_1, as a random Bernoulli with success probability π_1.
4. Generate subsequent responses according to the appropriate conditional probability.
5. Repeat for additional clusters.

An interesting property of the method presented by Qaqish (2003) is the nature of including prior binary outcomes. The terms $\kappa_{ij}(Y_j - \pi_j)$ show that a binary response variable is included relative to its expectation and transformed according to a constant related to the desired autocorrelation. While this method does not explicitly include predictors in the simulation algorithm, predictors could be included as part of each expected value π_i. The method clearly allows for both positive and negative values of autocorrelation, unlike many other proposed methods, but restrictions on the values of the autocorrelation remain as discussed by Qaqish (2003).

5.2 *Non-linear Dynamic Conditional Probability Model*

The most general method in this discussion is based on the work of Farrell and Sutradhar (2006), in which a nonlinear version of the linear conditional probability model proposed by Qaqish (2003) is constructed. The model of Farrell and Sutradhar (2006) is conditioned not only on prior binary outcomes in an autoregressive-type of sequence, but also on possible predictors to be considered in data generation. This approach allows for the inclusion of covariates in the conditional mean, allows for the probabilities to vary both between and within clusters, and allows for the greatest range of both positive and negative values of autocorrelation. However, the nonlinear conditional probability approach of Farrell and Sutradhar (2006) does not explicitly provide methods for controlling the probabilities and correlation structure at the outset of data simulation.

Assume an interest in simulating correlated binary variables, Y_i, where each outcome is to be associated with a vector of predictors, $\mathbf{x}_i$, through a vector of parameters, $\boldsymbol{\beta}$. Farrell and Sutradhar (2006) proposed using the non-linear conditional model,

$$\mathrm{E}[Y_i|[Y_1, \ldots, Y_{i-1}], \mathbf{x}_i] = \frac{exp(\mathbf{x}_i^T \boldsymbol{\beta} + \sum_{k=1}^{i-1} \gamma_k Y_k)}{1 + exp(\mathbf{x}_i^T \boldsymbol{\beta} + \sum_{k=1}^{i-1} \gamma_k Y_k)}. \quad (7)$$

Instead of beginning with desired probabilities and pairwise correlations or an associated correlation structure, the method of Farrell and Sutradhar (2006) focuses on the model relating predictors to the conditional response probability. This allows the greatest flexibility in producing values of correlation at the expense of control of the correlation structure. Farrell and Sutradhar (2006) show that, for a simple auto-regression including only the immediately previous response Y_{i-1}, the marginal expectation and correlation can be calculated based on the nonlinear dynamic model,

$$\mu_i = \mathrm{E}[Y_i] = P(Y_i = 1|Y_0 = 1) + \mathrm{E}[Y_{i-1}]\left[P(Y_i = 1|Y_1 = 1) - P(Y_i = 1|Y_0 = 1)\right],$$

$$\mathrm{Corr}(Y_i, Y_j) = \sqrt{\frac{\mu_i(1-\mu_i)}{\mu_j(1-\mu_j)}} \prod_{k \in (i,j]} \left[P(Y_k = 1|Y_1 = 1) - P(Y_k = 1|Y_0 = 1)\right].$$

Because the conditional probabilities $P(Y_k = 1|Y_1 = 1)$ and $P(Y_k = 1|Y_0 = 1)$ can vary within (0, 1), Farrell and Sutradhar (2006) argue that the marginal correlation can vary unrestricted between -1 and 1.

Generating Binary Values by the Nonlinear Dynamic Conditional Probability Model

1. For a single cluster of correlated binary data, select predictors $\mathbf{x}_i$, coefficients $\boldsymbol{\beta}$, and autoregressive coefficients γ_k.
2. Simulate Y_1 as Bernoulli with probability $\pi_1 = (exp(\mathbf{x}_i^T \boldsymbol{\beta}))/(1 + exp(\mathbf{x}_i^T \boldsymbol{\beta}))$.
3. Simulate subsequent Y_i according to the conditional probability $\mathrm{E}[Y_i|[Y_1, \ldots, Y_{i-1}], \mathbf{x}_i]$.
4. Repeat for additional clusters.

The intuition behind such an approach is that the predictors $\mathbf{x}_i$ and also the previous outcome variables $Y_1, \ldots, Y_{i-1}$ are combined linearly but related to the conditional mean through the inverse logit function, as in Eq. 7. The inverse logit function will map any real values to the range (0, 1), thus avoiding the concern of reproducibility discussed by Qaqish (2003).

6 Software Discussion

Few of the methods discussed are readily available in software. The R packages *bindata* and *BinNor* utilize discretizations of normal random variables, but not according to Emrich and Piedmonte (1991). The method of Emrich and Piedmonte (1991) is implemented in the *generate.binary()* function within the *MultiOrd* package, and also within the functions of the *mvtBinaryEP* package.

The *rbin()* function in the *SimCorMultRes* package explicitly uses threshold values to transform continuous values into binary values (Touloumis 2016). Touloumis (2016) proposed defining the marginal distribution of each binary value according to a CDF applied to the systematic component associated with a model,

$$P(Y_{ij} = 1) = F(\mathbf{x}_{ij}^T \boldsymbol{\beta}),$$

where F is a cumulative distribution function of a continuous random variable, and $\mathbf{x}_{ij}^T \boldsymbol{\beta}$ is a linear combination of predictors and parameter values. Each binary outcome is then defined according to $Y_{ij} = I(e_{ij} \leq \mathbf{x}_{ij}^T \boldsymbol{\beta})$, where $e_{ij} \sim F$ and is independent across clusters. While the method of Touloumis (2016) does clearly accommodate predictors and allow probabilities to vary within clusters, it is not clear the limitations on the range of autocorrelation, nor can the autocorrelation be easily controlled, as with other methods.

The package *binarySimCLF* implements the method of Qaqish (2003). Authors have individually implemented the methods of Park et al. (1996), Fleishman (1978), and publications include code for various software, such as with Lee (1993) and Headrick (2002b). However, there is a clear bias in available software: users prefer dichotomization of normal or uniform random variables to produce binary data.

Table 1 provides a brief summary of the simulation methods considered, along with some common advantages and disadvantages of each class of algorithms. Methods requiring the full specification of the joint probability distribution, while allowing complete control of the simulated data properties, tend to be complicated and computationally expensive. Alternatively, the mixture methods tend to have simpler algorithms and computational burden, and generally allow the user to specify correlation structures at the outset. The method of Oman and Zucker (2001) would seem to be the ideal such approach. Dichotomizing normal and uniform variables remain the most commonly implemented methods for simulating correlated binary outcomes, but most require computations involving systems of nonlinear equations. The approach of Emrich and Piedmonte (1991) remains prevalent and, accepting the computational burden, is an understandable method that allows easy control of correlation structures. Methods involving conditionally defined distributions are more

Table 1 Advantages and disadvantages of methods of correlated binary outcome simulation

Simulation type	Prominent example
Fully specified joint distribution	Lee (1993), using the Archimidian copula
	Advantages
	Control of probabilities, correlation, and higher-order moments
	Disadvantages
	Computationally expensive, nonlinear systems, function/matrix inversions
Mixture distributions	Oman and Zucker (2001), using a mixture of binary and continuous variables
	Advantages
	Simple algorithms, controlled correlation structures
	Disadvantages
	Constant probabilities within clusters, no predictors
Dichotomizing variables	Emrich and Piedmonte (1991), using dichotomized multivariate normals
	Advantages
	Short algorithms, probabilities vary within clusters, Correlation between clusters
	Disadvantages
	Nonlinear systems, computationally expensive
Conditional distributions	Qaqish (2003), using the linear conditional probability model
	Advantages
	Widest range of correlations, controlled correlation structures, predictors
	Disadvantages
	Complicated algorithms, requiring conditional means and covariances

recent, and allow for the greatest range of correlations to be simulated. The algorithm of Qaqish (2003) is an ideal example, with few disadvantages other than a slightly limited range of correlation values, but allowing the inclusion of predictors, prior outcomes, and easily specified common correlation structures.

References

Bahadur, R. R. (1961). A representation of the joint distribution of responses to n dichotomous items. *Stanford Mathematical Studies in the Social Sciences*, *6*, 158–168.

Devroye, L. (1986). *Non-uniform random variate generation* (1st ed.). Springer, New York.

Emrich, L. J., & Piedmonte, M. R. (1991). A method for generating high-dimensional multivariate binary variates. *The American Statistician: Statistical Computing*, *45*(4), 302–304.

Farrell, P. J., & Sutradhar, B. C. (2006). A non-linear conditional probability model for generating correlated binary data. *Statistics & Probability Letters*, *76*, 353–361.

Fleishman, A. I. (1978). A method for simulating non-normal distributions. *Psychometrika*, *43*, 521–532.

Gange, S. J. (1995). Generating multivariate categorical variates using the iterative proportional fitting algorithm. *The American Statistician*, *49*(2), 134–138.

Genest, C., & MacKay, R. J. (1986a). Copules archimediennes et familles de lois bidimenionnelles dont les marges sont donnees. *Canadian Journal of Statistics*, *14*, 280–283.

Genest, C., & MacKay, R. J. (1986b). The joy of copulas: Bivariate distributions with uniform marginals. *The American Statistician*, *40*, 549–556.

Headrick, T. C. (2002a). Fast fifth-order polynomial transforms for generating univariate and multivariate non normal distributions. *Computational Statistics & Data Analysis*, *40*, 685–711.

Headrick, T. C. (2002b). Jmasm3: A method for simulating systems of correlated binary data. *Journal of Modern Applied Statistical Methods*, *1*, 195–201.

Headrick, T. C. (2010). *Statistical simulation: Power method polynomials and other transformations* (1st ed.). Chapman & Hall/CRC, New York.

Headrick, T. C. (2011). A characterization of power method transformations through l-moments. *Journal of Probability and Statistics*, *2011*.

Kang, S. H., & Jung, S. H. (2001). Generating correlated binary variables with complete specification of the joint distribution. *Biometrical Journal*, *43*(3), 263–269.

Kanter, M. (1975). Autoregression for discrete processes mod 2. *Journal of Applied Probability*, *12*, 371–375.

Karian, Z. A., & Dudewicz, E. J. (1999). Fitting the generalized lambda distribution to data: A method based on percentiles. *Communications in Statistics: Simulation and Computation*, *28*, 793–819.

Koran, J., Headrick, T. C., & Kuo, T. C. (2015). Simulating univariate and multivariate no normal distributions through the method of percentiles. *Multivariate Behavioral Research*, *50*, 216–232.

Lee, A. J. (1993). Generating random binary deviates having fixed marginal distributions and specified degrees of association. *The American Statistician: Statistical Computing*, *47*(3), 209–215.

Lee, Y., & Nelder, J. A. (1996). Hierarchical generalized linear models. *Journal of the Royal Statistical Society, Series B (Methodological)*, *58*(4), 619–678.

Lunn, A. D., & Davies, S. J. (1998). A note on generating correlated binary variables. *Biometrika*, *85*(2), 487–490.

Molenberghs, G., & Verbeke, G. (2006). *Models for discrete longitudinal data* (1st ed.). Springer.

Oman, S. D., & Zucker, D. M. (2001). Modelling and generating correlated binary variables. *Biometrika*, *88*(1), 287–290.

Park, C. G., Park, T., & Shin, D. W. (1996). A simple method for generating correlated binary variates. *The American Statistician*, *50*(4), 306–310.

Prentice, R. L. (1988). Correlated binary regression with covariates specific to each binary observation. *Biometrics*, *44*, 1033–1048.

Qaqish, B. F. (2003). A family of multivariate binary distributions for simulating correlated binary variables with specified marginal means and correlations. *Biometrika*, *90*(2), 455–463.

Stiratelli, R., Laird, N., & Ware, J. H. (1984). Random-effects models for serial observations with binary response. *Biometrics*, *40*, 961–971.

Touloumis, A. (2016). Simulating correlated binary and multinomial responses with simcormultres. *The Comprehensive R Archive Network* 1–5.

Vale, C. D., & Maurelli, V. A. (1983). Simulating multivariate no normal distributions. *Psychometrika*, *48*, 465–471.

Zeger, S. L., & Liang, K. Y. (1986). Longitudinal data analysis for discrete and continuous outcomes. *Biometrics*, *42*, 121–130.

Quantifying the Uncertainty in Optimal Experiment Schemes via Monte-Carlo Simulations

H.K.T. Ng, Y.-J. Lin, T.-R. Tsai, Y.L. Lio and N. Jiang

Abstract In the process of designing life-testing experiments, experimenters always establish the optimal experiment scheme based on a particular parametric lifetime model. In most applications, the true lifetime model is unknown and need to be specified for the determination of optimal experiment schemes. Misspecification of the lifetime model may lead to a substantial loss of efficiency in the statistical analysis. Moreover, the determination of the optimal experiment scheme is always relying on asymptotic statistical theory. Therefore, the optimal experiment scheme may not be optimal for finite sample cases. This chapter aims to provide a general framework to quantify the sensitivity and uncertainty of the optimal experiment scheme due to misspecification of the lifetime model. For the illustration of the methodology developed here, analytical and Monte-Carlo methods are employed to evaluate the robustness of the optimal experiment scheme for progressive Type-II censored experiment under the location-scale family of distributions.

H.K.T. Ng (✉)
Department of Statistical Science, Southern Methodist University,
Dallas, TX 75275, USA
e-mail: ngh@mail.smu.edu

Y.-J. Lin
Department of Applied Mathematics, Chung Yuan Christian University, Chung-Li District,
Taoyuan city 32023, Taiwan
e-mail: yujaulin@cycu.edu.tw

T.-R. Tsai
Department of Statistics, Tamkang University, Tamsui District, New Taipei City, Taiwan
e-mail: trtsai@stat.tku.edu.tw

Y.L. Lio · N. Jiang
Department of Mathematical Sciences, University of South Dakota,
Vermillion, SD 57069, USA
e-mail: Yuhlong.Lio@usd.edu

N. Jiang
e-mail: Nan.Jiang@usd.edu

D.-G. Chen and J.D. Chen (eds.), *Monte-Carlo Simulation-Based Statistical Modeling*, ICSA Book Series in Statistics, DOI 10.1007/978-981-10-3307-0_6

1 Introduction

In designing life-testing experiments for industrial and medical settings, experimenters always assume a parametric model for the lifetime distribution and then determine the optimal experiment scheme by optimizing a specific objective function based on the assumed model. In most applications, the optimal experimental design is model dependent because the optimality criterion is usually a function of the information measures. Therefore, prior information about the unknown model derived from physical/chemical theory, engineering pre-test results, or past experience with similar experimental units is needed to determine the optimal experiment scheme in practice. However, this prior information may not be accurate and hence the optimal experiment scheme may not perform as well as one expected. In other words, the objective function may not be optimized when the optimal experiment scheme based on inaccurate prior information is used. In addition, to determine the optimal experiment scheme, we rely on the asymptotic statistical theory in most cases. For instance, A-optimality is developed to minimize the variances of the estimators of the model parameters while these variances are always replaced by the asymptotic ones during the process of optimization. There is no guarantee that the optimal experiment scheme obtained based on asymptotic theory can be more efficient than a non-optimal experiment scheme in finite sample situations.

For these reasons, it is important to have a systematic procedure to quantify the sensitivity and uncertainty of the optimal experiment scheme due to model misspecification. It will be useful to see whether a proposed experiment scheme is robust to model misspecification. If a design is indeed robust, it would then assure the practitioners that misspecification in the model would not result in an unacceptable change in the precision of the estimates of model parameters. In this chapter, we discuss the analytical and Monte-Carlo methods for quantifying the sensitivity and uncertainty of the optimal experiment scheme and evaluate the robustness of the optimal experiment scheme.

Let $\boldsymbol{\theta}$ be the parameter vector of lifetime distribution of test items. The commonly used procedures for the determination of the optimal experiment scheme are described as follows: A-optimality, that minimizes the trace of the variance-covariance matrix of the maximum likelihood estimators (MLEs) of elements of $\boldsymbol{\theta}$, provides an overall measure of variability from the marginal variabilities. It is particularly useful when the correlation between the MLEs of the parameters is low. It is also pertinent for the construction of marginal confidence intervals for the parameters in $\boldsymbol{\theta}$. D-optimality, that minimizes the determinant of the variance-covariance matrix of the MLEs of components of $\boldsymbol{\theta}$, provides an overall measure of variability by taking into account the correlation between the estimates. It is particularly useful when the estimates are highly correlated. It is also pertinent for the construction of joint confidence regions for the parameters in $\boldsymbol{\theta}$. V-optimality, that minimizes the variance of the estimator of lifetime distribution percentile.

In Sect. 2, we present the notation and general methods for quantifying the uncertainty in the optimal experiment scheme with respect to changes in model.

Then, in Sect. 3, we focus on progressive Type-II censoring with location-scale family of distributions. The procedure for determining the optimal scheme with progressive censoring and some commonly used optimal criteria are presented in Sects. 3.1 and 3.2, respectively. In Sect. 3.3, some numerical illustrations via analytical and simulation approaches based on extreme value (Gumbel), logistic, and normal distributions are discussed. Discussions based on the numerical results are provided in Sect. 3.4. A numerical example is presented in Sect. 4. Finally, some concluding remarks are given in Sect. 5.

2 Quantifying the Uncertainty in the Optimal Experiment Scheme

In a life-testing experiment, let the lifetimes of test items follow a family of statistical model $\mathscr{M}$. We are interested in determining the optimal experiment scheme that optimizes the objective function $Q(S, \mathscr{M}_0)$, where S denotes an experiment scheme and $\mathscr{M}_0$ denotes the true model. In many situations, the determination of the optimal experiment scheme requires a specification of the unknown statistical model $\mathscr{M}$ and hence the optimal experiment scheme depends on the specified model $\mathscr{M}$. For instance, in experimental design of multi-level stress testing, (Ka et al, 2011) and Chan et al. (2016) considered the extreme value regression model and derived the expected Fisher information matrix. Consequently the optimal experiment schemes obtained in Ka et al (2011) and Chan et al. (2016) are specifically for the extreme value regression model, which may not be optimal for other regression models. Here, we denote the optimal experiment scheme based on a specified model $\mathscr{M}$ as $S^*(\mathscr{M})$. In the ideal situation, the model specified for the optimal experimental scheme is the true model, i.e., $Q(S^*(\mathscr{M}_0), \mathscr{M}_0) = \inf_S Q(S(\mathscr{M}_0), \mathscr{M}_0)$ and $S^*(\mathscr{M}_0) = \arg\inf_S Q(S(\mathscr{M}_0), \mathscr{M}_0)$.

On the other hand, in determining the optimal experiment scheme, experimenter always relies on the asymptotic results which are derived based on the sample size goes to infinity. Nevertheless, in practice, the number of experimental units can be used in an experiment is finite and thus the use of the asymptotic theory may not be appropriate. For instance, for A-optimality, the aim is to minimize the variances of the estimated model parameters. This is always attained through minimizing the trace of the inverse of the Fisher information matrix or equivalently, the trace of the asymptotic variance-covariance matrix of MLEs. However, the asymptotic variance-covariance matrix may not correctly reflect the true variations of the estimators when the sample size is finite, and hence the optimal experiment scheme may not be optimal or as efficient as expected in finite sample situations. Therefore, large-scale Monte-Carlo simulations can be used to estimate the objective functions and evaluate the performance of the optimal experiment scheme. For quantifying the sensitivity and uncertainty of the optimal experiment scheme $S^*(\mathscr{M})$, we describe two possible approaches by comparing experimental schemes and objective functions in the following subsections.

2.1 Comparing Experimental Schemes

Let the specified model for obtaining the optimal experiment scheme be $\mathscr{M}^*$, then the optimal experiment scheme is

$$S^*(\mathscr{M}^*) = \arg\inf_{S} Q(S(\mathscr{M}^*), \mathscr{M}^*).$$

To quantify the sensitivity of the optimal experiment scheme $S^*(\mathscr{M}^*)$, we consider a different model $\mathscr{M}$ and establish the optimal experiment scheme based on $\mathscr{M}$ as

$$S^*(\mathscr{M}) = \arg\inf_{S} Q(S(\mathscr{M}), \mathscr{M}).$$

Comparing the experimental scheme $S^*(\mathscr{M}^*)$ and $S^*(\mathscr{M})$ will provide us some insights on the sensitivity of the optimal experiment scheme $S^*(\mathscr{M}^*)$. If $S^*(\mathscr{M}^*)$ is insensitivity to the change in the model $\mathscr{M}$, then $S^*(\mathscr{M}^*)$ and $S^*(\mathscr{M})$ will be similar to each other. Depending on the nature of the life-testing experiments, different measures of similarity of two experimental schemes can be considered to quantify the sensitivity of the optimal experiment scheme $S^*(\mathscr{M}^*)$. When apply this approach, evaluation of the optimal experiment scheme for different models is needed.

2.2 Comparing Values of Objective Functions

To quantify the sensitivity of the optimal experiment scheme $S^*(\mathscr{M}^*)$, another approach is to compare the objective function of the optimal experiment scheme $S^*(\mathscr{M}^*)$ under the model $(\mathscr{M})$ which is believed to be the true model. Specifically, we can compute the objective function when the experiment scheme $S^*(\mathscr{M}^*)$ is adopted but the model is $\mathscr{M}$, i.e., to compute $Q(S^*(\mathscr{M}^*), \mathscr{M})$. If the optimal experiment scheme $S^*(\mathscr{M}^*)$ is insensitivity to the change in the model $\mathscr{M}$, then $Q(S^*(\mathscr{M}^*), \mathscr{M}^*)$ will be similar to $Q(S^*(\mathscr{M}^*), \mathscr{M})$ or $Q(S^*(\mathscr{M}), \mathscr{M})$. When apply this approach, evaluation of the objective function $Q(S^*(\mathscr{M}^*), \mathscr{M})$ is needed.

3 Progressive Censoring with Location-Scale Family of Distributions

In this section, we illustrate the proposed methodology through the optimal progressive Type-II censoring schemes (see, for example, Balakrishnan and Aggarwala 2000; Balakrishnan 2007; Balakrishnan and Cramer 2014). We consider that the underline statistical model, $\mathscr{M}$, used for this purpose is a member of the log-location-scale family of distributions. Specifically, the log-lifetimes of the units on test have a

Table 1 Examples of functional forms of $g(\cdot)$ and $G(\cdot)$

	Extreme value (EV)	Logistic (LOGIS)	Normal (NORM)
$g(z)$	$\exp[z - \exp(z)]$	$\exp(-z)/[1+\exp(-z)]^2$	$\frac{1}{\sqrt{2\pi}}\exp(-z^2/2)$
$G(z)$	$1 - \exp[-\exp(z)]$	$1/[1+\exp(-z)]$	$\int_{-\infty}^{z}\frac{1}{\sqrt{2\pi}}\exp(-x^2/2)dx$

location-scale distribution with probability density function (p.d.f.)

$$f_X(x;\mu,\sigma) = \frac{1}{\sigma}g\left(\frac{x-\mu}{\sigma}\right), \tag{1}$$

and cumulative distribution function (c.d.f.)

$$F_X(x;\mu,\sigma) = G\left(\frac{x-\mu}{\sigma}\right), \tag{2}$$

where $g(\cdot)$ is the standard form of the p.d.f. $f_X(x;\mu,\sigma)$ and $G(\cdot)$ is the standard form of the c.d.f. $F_X(x;\mu,\sigma)$ when $\mu = 0$ and $\sigma = 1$. The functional forms g and G are completely specified and they are parameter-free, but the location and scale parameters, $-\infty < \mu < \infty$ and $\sigma > 0$ of $f_X(x;\mu,\sigma)$ and $F_X(x;\mu,\sigma)$, are unknown. Many well-known properties for location-scale family of distributions had been established in the literature (e.g., Johnson et al. 1994). This is a rich family of distributions that include the normal, extreme value and logistic models as special cases. The functional forms of $g(\cdot)$ and $G(\cdot)$ for extreme value, logistic and normal, distributions are summarized in Table 1.

A progressively Type-II censored life-testing experiment is described in order. Let n independent units be placed on a life-test with corresponding lifetimes T_1, $T_2, \ldots, T_n$ that are independent and identically distributed (i.i.d.) with p.d.f. $f_T(t;\boldsymbol{\theta})$ and c.d.f. $F_T(t;\boldsymbol{\theta})$, where $\boldsymbol{\theta}$ denotes the vector of unknown parameters. Prior to the experiment, the number of complete observed failures $m < n$ and the censoring scheme $(R_1, R_2, \ldots, R_m)$, where $R_j \geq 0$ and $\sum_{j=1}^{m} R_j + m = n$ are pre-fixed. During the experiment, R_j functioning items are removed (or censored) randomly from the test when the j-th failure is observed. Note that in the analysis of lifetime data, instead of working with the parametric model for T_i, it is often more convenient to work with the equivalent model for the log-lifetimes $X_i = \log T_i$, for $i = 1, 2, \ldots, n$. The random variables X_i, $i = 1, 2, \ldots, n$, are i.i.d. with p.d.f. $f_X(x;\mu,\sigma)$ and c.d.f. $F_X(x;\mu,\sigma)$.

3.1 Maximum Likelihood Estimation

Let the m completely observed (ordered) log-lifetimes from a progressively Type-II censored experiment be $X_{i:m:n}$, $i = 1, 2 \ldots, m$ and their observed values be $x_{i:m:n}$, $i = 1, 2, \ldots, m$. The likelihood function based on $x_{i:m:n} (1 \leq i \leq m)$ is

$$L(\mu, \sigma) = c' \prod_{i=1}^{m} f_X(x_{i:m:n}; \mu, \sigma) \left[1 - F_X(x_{i:m:n}; \mu, \sigma)\right]^{R_i}, \quad x_{1:m:n} < x_{2:m:n} < \cdots < x_{m:m:n}, \tag{3}$$

where c' is the normalizing constant given by

$$c' = n(n - R_1 - 1) \cdots (n - R_1 - R_2 - \cdots - R_{m-1} - m + 1).$$

The MLEs of μ and σ are the values of μ and σ which maximizes (3). For location-scale family of distributions described in Eqs. (1) and (2), the log-likelihood function can be expressed as

$$\begin{aligned} \ell(\mu, \sigma) &= \ln L(\mu, \sigma) \\ &= \ln c' - m \ln \sigma + \sum_{i=1}^{m} g\left(\frac{x_{i:m:n} - \mu}{\sigma}\right) \\ &\quad + \sum_{i=1}^{m} R_i \ln \left[1 - G\left(\frac{x_{i:m:n} - \mu}{\sigma}\right)\right]. \end{aligned}$$

We denote the MLEs of the parameters μ and σ by $\hat{\mu}$ and $\hat{\sigma}$, respectively. Computational algorithms for obtaining the MLEs of the parameters of some commonly used location-scale distributions are available in many statistical software packages such as R (R Core Team 2016), SAS and JMP.

The expected Fisher information matrix of the MLEs can be obtained as

$$\mathbf{I}(\mu, \sigma) = -\begin{bmatrix} E\left(\frac{\partial^2 \ell(\mu,\sigma)}{\partial \mu \partial \mu}\right) & E\left(\frac{\partial^2 \ell(\mu,\sigma)}{\partial \mu \partial \sigma}\right) \\ E\left(\frac{\partial^2 \ell(\mu,\sigma)}{\partial \mu \partial \sigma}\right) & E\left(\frac{\partial^2 \ell(\mu,\sigma)}{\partial \sigma \partial \sigma}\right) \end{bmatrix} = \begin{bmatrix} I_{\mu\mu} & I_{\mu\sigma} \\ I_{\mu\sigma} & I_{\sigma\sigma} \end{bmatrix}. \tag{4}$$

Then, the asymptotic variance-covariance matrix of the MLEs can be obtained by inverting the expected Fisher information matrix in Eq. (4) as

$$\mathbf{V}(\mu, \sigma) = \mathbf{I}^{-1}(\mu, \sigma) = \begin{bmatrix} Var(\hat{\mu}) & Cov(\hat{\mu}, \hat{\sigma}) \\ Cov(\hat{\mu}, \hat{\sigma}) & Var(\hat{\sigma}) \end{bmatrix} = \sigma^2 \begin{bmatrix} V_{11} & V_{12} \\ V_{12} & V_{22} \end{bmatrix}. \tag{5}$$

The computational formulas of the elements in the Fisher information matrix of (4) can be found in Balakrishnan et al. (2003), Ng et al. (2004) and Dahmen et al. (2012). For fixed values of n and m and a specific progressive censoring scheme $(R_1, R_2, \ldots, R_m)$, we can compute the expected Fisher information matrix and the asymptotic variance-covariance matrix of the MLEs from Eqs. (4) and (5).

3.2 Optimal Criteria

To determine the optimal scheme under progressive Type-II censoring , we consider the following optimal criteria:

[1] *D*-optimality

For *D*-optimality,we search for the censoring scheme that maximizes the determinant of the Fisher information matrix, $\det(\mathbf{I}(\mu, \sigma))$. For a given censoring scheme $S = (R_1, R_2, \ldots, R_m)$ with a specific model $\mathcal{M}$, the objective function is

$$Q_D(S, \mathcal{M}) = I_{\mu\mu} I_{\sigma\sigma} - I^2_{\mu\sigma}. \tag{6}$$

We denote the optimal experiment scheme for *D*-optimality with a specific model $\mathcal{M}$ as $S^*_D(\mathcal{M})$.

[2] *A*-optimality

For *A*-optimality, we aim to minimize the variances of the estimators of the model parameters. This can be achieved by designing an experiment that minimizes the trace of the asymptotic variance-covariance matrix, $tr[\mathbf{V}(\mu, \sigma)]$. For a given experiment scheme $S = (R_1, R_2, \ldots, R_m)$ with a specific model $\mathcal{M}$, the objective function is

$$Q_A(S, \mathcal{M}) = V_{11} + V_{22}. \tag{7}$$

We denote the optimal experiment scheme for *A*-optimality with a specified model $\mathcal{M}$ as $S^*_A(\mathcal{M})$.

[3] *V*-optimality

For *V*-optimality, we aim to minimize the variance of the estimator of 100δ-th percentile of the log-lifetime distribution, $0 < \delta < 1$, i.e.,

$$\hat{q}_\delta = \hat{\mu} + \hat{\sigma} G^{-1}(\delta).$$

For a given censoring scheme $S = (R_1, R_2, \ldots, R_m)$ with a specific model $\mathcal{M}$, the objective function is

$$\begin{aligned} Q_{V_\delta}(S, \mathcal{M}) &= Var(\hat{q}_\delta) = Var(\hat{\mu} + \hat{\sigma} G^{-1}(\delta)) \\ &= V_{11} + [G^{-1}(\delta)]^2 V_{22} + 2G^{-1}(\delta) V_{12}, \end{aligned} \tag{8}$$

where $G^{-1}(\cdot)$ is the inverse c.d.f. of the standard location-scale distribution. We denote the optimal experimental scheme for V-optimality with a specified model $\mathscr{M}$ as $S^*_{V_\delta}(\mathscr{M})$.

When the values of n and m are chosen in advance, that depend on the availability of units, experimental facilities and cost considerations, we can determine the optimal censoring scheme $(R_1, R_2, \ldots, R_m)$. In the finite sample situation, we can list all possible censoring schemes and compute the corresponding objective functions, and then determine the optimal censoring schemes, respectively, through an extensive search.

3.3 *Numerical Illustrations*

For illustrative purpose, we consider the true underline lifetime model of the test units to be Weibull (i.e., the log-lifetimes follow an extreme value distribution, $\mathscr{M}_0 = EV$) and we are interested in investigating the effect of misspecification of the underline lifetime model as log-logistic (i.e., the log-lifetimes follow a logistic distribution, $\mathscr{M}^* = LOGIS$). We also consider the case that the true underline lifetime model for the test units to be lognormal (i.e., the log-lifetimes follow a normal distribution, $\mathscr{M}_0 = NOR$) and we are interested in investigating the effect of misspecification of the underline lifetime model as Weibull ($\mathscr{M}^* = EV$).

3.3.1 Analytical Approach

In this subsection, we evaluate the sensitivities of the optimal progressive Type-II censoring scheme analytically based on the expected Fisher information matrix and the asymptotic variance-covariance matrix of the MLEs. For the specific model $\mathscr{M}^*$, we determine the optimal progressive Type-II censoring schemes under different optimal criteria, $S^*_D(\mathscr{M}^*)$, $S^*_A(\mathscr{M}^*)$, $S^*_{V_{.95}}(\mathscr{M}^*)$ and $S^*_{V_{.05}}(\mathscr{M}^*)$ from Eqs. (4) to (8). For the true model $\mathscr{M}_0$, we also determine the optimal progressive Type-II censoring schemes under different optimal criteria, $S^*_D(\mathscr{M}_0)$, $S^*_A(\mathscr{M}_0)$ and $S^*_{V_{.95}}(\mathscr{M}_0)$ and $S^*_{V_{.05}}(\mathscr{M}_0)$. Then, we can compare these experimental schemes $S^*(\mathscr{M}^*)$ and $S^*(\mathscr{M}_0)$. In addition, we compute the objective functions based on the optimal censoring scheme under the specified model $\mathscr{M}^*$ while the true underline model is $\mathscr{M}_0$, i.e., we compute $Q_D(S^*_D(\mathscr{M}^*), \mathscr{M}_0)$, $Q_A(S^*_A(\mathscr{M}^*), \mathscr{M}_0)$, and $Q_{V_\delta}(S^*_{V_\delta}(\mathscr{M}^*), \mathscr{M}_0)$ for $\delta = 0.05$ and 0.95. The results for $n = 10$, $m = 5(1)9$ with ($\mathscr{M}^* = LOGIS$, $\mathscr{M}_0 = EV$) and ($\mathscr{M}^* = EV$, $\mathscr{M}_0 = NOR$) are presented in Tables 2 and 3, respectively.

3.3.2 Simulation Approach

In this subsection, we use Monte-Carlo simulation to evaluate the performance of the optimal censoring scheme by comparing the objective functions based on the asymptotic variance-covariance matrix and the simulated values. Moreover, we can also evaluate the sensitivities of the optimal censoring schemes when the model is misspecified based on Monte-Carlo simulation. In our simulation study, 20,000 sets of progressively Type-II censored data are generated from the true model $\mathcal{M}_0 = EV$ and the MLEs of the parameters μ and σ are obtained. The simulated values of the objective functions for the progressive censoring schemes ($n = 10, m = 5(1)9$) in Tables 2 and 3 are presented in Table 4.

3.4 Discussions

3.4.1 Comparing Experimental Schemes

From Tables 2 and 3, we can compare the optimal censoring schemes under models $\mathcal{M}^* = LOGIS$ and $\mathcal{M}_0 = EV$. While comparing the optimal censoring schemes of the same optimal criterion under these two models, we can see that these two optimal censoring schemes are different except the case when $n = 10, m = 9$ for V-optimality with $\delta = 0.95$. In some cases, the two optimal censoring schemes can be very different from each other. For example, when $n = 10, m = 7$ for A-optimality, the optimal censoring scheme under model $\mathcal{M}^* = LOGIS$ is (0, 0, 0, 0, 0, 0, 3) while the optimal censoring scheme under model $\mathcal{M}_0 = EV$ is (0, 0, 3, 0, 0, 0, 0). These results indicate that when the model is misspecified, the optimal censoring scheme based on logistic distribution may not be optimal under the true model (i.e., extreme value distribution).

Nevertheless, the analytical approach proposed here will be useful for practitioners to choose an appropriate censoring scheme which is robust with respect to model misspecification. For instance, from Table 2 with $n = 10, m = 5$ for D-optimality, if one believes that the model is $\mathcal{M}^* = LOGIS$ but also suspects that the model might be $\mathcal{M}_0 = EV$, then from Table 2, it may not be the best option to use the optimal censoring scheme (0, 0, 0, 0, 5) with optimal objective function $\det(\mathbf{I}) = 45.05$ because there is a substantial loss in efficient if the underline model is $\mathcal{M}_0 = EV$ ($\det(\mathbf{I}) = 32.89$ compared to the optimal value 54.52 under $\mathcal{M}_0 = EV$ in Table 3). In this situation, it may better to use a non-optimal censoring scheme such as (0, 3, 0, 0, 2) which gives objective function $\det(\mathbf{I}) = 44.64$ under $\mathcal{M}^* = LOGIS$ and $\det(\mathbf{I}) = 47.09$ under $\mathcal{M}_0 = EV$.

Table 2 Optimal progressive censoring schemes for $n = 10$, $m = 5(1)9$ with $\mathscr{M}^* = LOGIS$ and $\mathscr{M}_0 = EV$

m	$1 - m/n$	Optimality criterion	Optimal censoring scheme ($S^*(LOGIS)$)	$Q(S^*(LOGIS), LOGIS)$				$Q(S^*(LOGIS), EV)$			
				$tr[\mathbf{V}]$	$det[\mathbf{I}]$	$Var(\hat{q}_{0.95})$	$Var(\hat{q}_{0.05})$	$tr[\mathbf{V}]$	$det[\mathbf{I}]$	$Var(\hat{q}_{0.95})$	$Var(\hat{q}_{0.05})$
5	50%	[1]	$R_5 = 5$	0.3269	45.05	0.8276	0.5002	0.3866	32.89	0.2585	2.0065
		[2]	$R_2 = 3, R_5 = 2$	0.3160	44.64	0.7568	0.4834	0.3071	47.09	0.3070	1.1967
		[3, $\delta = 0.95$]	$R_1 = 4, R_5 = 1$	0.3271	39.22	0.7233	0.5295	0.3099	45.61	0.3211	1.1975
		[3, $\delta = 0.05$]	$R_3 = 4, R_5 = 1$	0.3253	43.48	0.8038	0.4786	0.3304	41.89	0.3304	0.2751
6	40%	[1]	$R_6 = 4$	0.2699	60.24	0.6249	0.4560	0.3009	48.04	0.2514	1.4809
		[2]	$R_2 = 1, R_6 = 3$	0.2677	59.88	0.6074	0.4508	0.2890	51.24	0.2585	1.3522
		[3, $\delta = 0.95$]	$R_1 = 2, R_6 = 2$	0.2712	56.82	0.5850	0.4673	0.2752	55.62	0.2791	1.1679
		[3, $\delta = 0.05$]	$R_3 = 3, R_6 = 1$	0.2735	57.47	0.6242	0.4393	0.2679	59.91	0.2655	1.1131
7	30%	[1], [2]	$R_7 = 3$	0.2341	76.34	0.4984	0.4196	0.2480	67.13	0.2486	1.1194
		[3, $\delta = 0.95$]	$R_1 = 1, R_7 = 2$	0.2370	72.99	0.4882	0.4250	0.2395	72.44	0.2618	0.9882
		[3, $\delta = 0.05$]	$R_3 = 2, R_7 = 1$	0.2394	71.94	0.5085	0.4097	0.2312	79.35	0.2578	0.8980
8	20%	[1], [2], [3, $\delta = 0.95$]	$R_8 = 2$	0.2111	90.91	0.4183	0.3899	0.2134	91.14	0.2470	0.8574
		[3, $\delta = 0.05$]	$R_3 = 1, R_8 = 1$	0.2150	87.72	0.4283	0.3862	0.2070	100.05	0.2511	0.7572
9	10%	[1], [2], [3, $\delta = 0.95$]									
		[3, $\delta = 0.05$]	$R_9 = 1$	0.1966	104.17	0.3696	0.3668	0.1895	121.94	0.2448	0.6578

Table 3 Optimal progressive censoring schemes for $n = 10$, $m = 5(1)9$ with $\mathscr{M}^* = EV$ and $\mathscr{M}_0 = NOR$

m	$1 - m/n$	Optimality criterion	Optimal censoring scheme $(S^*(EV))$	$Q(S^*(EV), EV)$				$Q(S^*(EV), NOR)$			
				$tr[\mathbf{V}]$	$det[\mathbf{I}]$	$Var(\hat{q}_{0.95})$	$Var(\hat{q}_{0.05})$	$tr[\mathbf{V}]$	$det[\mathbf{I}]$	$Var(\hat{q}_{0.95})$	$Var(\hat{q}_{0.05})$
5	50%	[1], [2]	$R_2 = 5$	0.2918	54.52	0.3062	1.0207	0.2369	83.23	0.4651	0.2896
		[3, $\delta = 0.95$]	$R_5 = 5$	0.3866	32.89	0.2585	2.0065	0.2595	71.10	0.6150	0.2879
		[3, $\delta = 0.05$]	$R_1 = 5$	0.2948	54.21	0.3474	0.9247	0.2433	78.79	0.4293	0.3252
6	40%	[1], [2]	$R_2 = 4$	0.2494	73.65	0.2887	0.8253	0.2098	103.78	0.3891	0.2765
		[3, $\delta = 0.95$]	$R_5 = 4$	0.2737	57.78	0.2546	1.1951	0.2166	95.83	0.4633	0.2691
		[3, $\delta = 0.05$]	$R_1 = 4$	0.2549	72.31	0.3178	0.7797	0.2155	99.00	0.3684	0.3021
7	30%	[1]	$R_2 = 3$	0.2212	94.22	0.2742	0.6997	0.1895	125.73	0.3344	0.2647
		[2]	$R_3 = 3$	0.2205	92.57	0.2630	0.7387	0.1890	125.44	0.3492	0.2570
		[3, $\delta = 0.95$]	$R_6 = 3$	0.2347	76.28	0.2484	0.9661	0.1905	119.04	0.3735	0.2643
		[3, $\delta = 0.05$]	$R_1 = 3$	0.2261	92.38	0.2936	0.6778	0.1939	121.23	0.3227	0.2821
8	20%	[1]	$R_2 = 2$	0.2005	116.21	0.2617	0.6115	0.1736	149.08	0.2932	0.2541
		[2]	$R_3 = 2$	0.1994	115.44	0.2548	0.6298	0.1728	149.31	0.3007	0.2490
		[3, $\delta = 0.95$]	$R_7 = 2$	0.2070	99.10	0.2453	0.7791	0.1623	178.78	0.3726	0.1840
		[3, $\delta = 0.05$]	$R_1 = 2$	0.2039	114.44	0.2732	0.6017	0.1765	145.47	0.2871	0.2645
9	10%	[1]	$R_2 = 1$	0.1846	139.63	0.2505	0.5457	0.1607	173.84	0.2611	0.2443
		[2]	$R_4 = 1$	0.1836	138.40	0.2452	0.5614	0.1600	173.81	0.2664	0.2408
		[3, $\delta = 0.95$]	$R_7 = 1$	0.1853	131.56	0.2427	0.6026	0.1597	171.68	0.2652	0.2470
		[3, $\delta = 0.05$]	$R_1 = 1$	0.1862	138.48	0.2557	0.5423	0.1621	171.73	0.2586	0.2491

Table 4 Simulated values of the objective functions under $\mathscr{M}_0 = EV$ for the progressive censoring schemes ($n = 10, m = 5(1)9$) presented in Tables 2 and 3 with $\mathscr{M}^* = LOGIS$

m	$1 - m/n$	Optimality	Censoring scheme	Simulated values			
				$tr[\mathbf{V}]$	$det[\mathbf{I}]$	$Var(\hat{q}_{0.95})$	$Var(\hat{q}_{0.05})$
5	50%	$S^*_{V_{.05}}(\mathscr{M}_0)$	$R_1 = 5$	0.3221	47.06	0.3192	1.1115
		$S^*_D(\mathscr{M}_0), S^*_A(\mathscr{M}_0)$	$R_2 = 5$	0.3247	46.15	0.3636	1.0105
		$S^*_{V_{.95}}(\mathscr{M}_0), S^*_D(\mathscr{M}^*)$	$R_5 = 5$	0.4086	31.55	0.6121	1.0571
		$S^*_A(\mathscr{M}^*)$	$R_2 = 3, R_5 = 2$	0.3580	37.25	0.4780	1.0504
		$S^*_{V_{.95}}(\mathscr{M}^*)$	$R_1 = 4, R_5 = 1$	0.3443	38.94	0.3935	1.1363
		$S^*_{V_{.05}}(\mathscr{M}^*)$	$R_3 = 4, R_5 = 1$	0.3575	38.91	0.4827	0.9793
6	40%	$S^*_D(\mathscr{M}^*), S^*_{V_{.95}}(\mathscr{M}_0)$	$R_6 = 4$	0.3165	45.78	0.4329	1.0088
		$S^*_A(\mathscr{M}^*)$	$R_2 = 1, R_6 = 3$	0.3047	48.30	0.3978	1.0050
		$S^*_{V_{.95}}(\mathscr{M}^*)$	$R_1 = 2, R_6 = 2$	0.2860	53.08	0.3413	1.0234
		$S^*_{V_{.05}}(\mathscr{M}^*)$	$R_3 = 3, R_6 = 1$	0.2873	54.81	0.3456	0.9496
		$S^*_D(\mathscr{M}_0), S^*_A(\mathscr{M}_0)$	$R_2 = 4$	0.2720	64.31	0.2762	0.9475
		$S^*_{V_{.05}}(\mathscr{M}_0)$	$R_1 = 4$	0.2719	64.37	0.2556	1.0180
7	30%	$S^*_D(\mathscr{M}^*), S^*_A(\mathscr{M}^*)$	$R_7 = 3$	0.2582	63.95	0.3118	0.9573
		$S^*_{V_{.95}}(\mathscr{M}^*)$	$R_1 = 1, R_7 = 2$	0.2463	70.16	0.2777	0.9437
		$S^*_{V_{.05}}(\mathscr{M}^*)$	$R_3 = 2, R_7 = 1$	0.2461	72.50	0.2651	0.9174
		$S^*_D(\mathscr{M}_0)$	$R_2 = 3$	0.2315	87.17	0.2176	0.8879
		$S^*_A(\mathscr{M}_0)$	$R_3 = 3$	0.2321	85.05	0.2320	0.8628
		$S^*_{V_{.95}}(\mathscr{M}_0)$	$R_6 = 3$	0.2452	71.88	0.2841	0.8976
		$S^*_{V_{.05}}(\mathscr{M}_0)$	$R_1 = 3$	0.2395	81.81	0.2129	0.9645

(continued)

Table 4 (continued)

m	$1-m/n$	Optimality	Censoring scheme	Simulated values			
				$tr[\mathbf{V}]$	$det[\mathbf{I}]$	$Var(\hat{q}_{0.95})$	$Var(\hat{q}_{0.05})$
8	20%	$S_D^*(\mathscr{M}^*), S_A^*(\mathscr{M}^*), S_{V_{.95}}^*(\mathscr{M}^*)$	$R_8=2$	0.2222	85.70	0.2319	0.9101
		$S_{V_{.05}}^*(\mathscr{M}^*)$	$R_3=1, R_8=1$	0.2150	93.98	0.2120	0.8807
		$S_D^*(\mathscr{M}_0)$	$R_2=2$	0.2117	103.89	0.1866	0.8696
		$S_A^*(\mathscr{M}_0)$	$R_5=2$	0.2104	100.24	0.2046	0.8418
		$S_{V_{.95}}^*(\mathscr{M}_0)$	$R_7=2$	0.2145	93.09	0.2183	0.8762
		$S_{V_{.05}}^*(\mathscr{M}_0)$	$R_1=2$	0.2114	106.82	0.1767	0.8870
9	10%	$S_{V_{.05}}^*(\mathscr{M}_0)$	$R_1=1$	0.1939	128.64	0.1533	0.8477
		$S_D^*(\mathscr{M}_0)$	$R_2=1$	0.1909	129.84	0.1563	0.8280
		$S_A^*(\mathscr{M}_0)$	$R_8=1$	0.1921	120.37	0.1717	0.8349
		$S_D^*(\mathscr{M}^*), S_A^*(\mathscr{M}^*), S_V^*(\mathscr{M}^*), S_{V_{.95}}^*(\mathscr{M}_0)$	$R_9=1$	0.1950	115.70	0.1734	0.8673

3.4.2 Comparing Values of Objective Functions

By comparing the values of the objective functions $Q(S^*(\mathscr{M}^*), \mathscr{M}_0)$ with $Q(S^*(\mathscr{M}^*), \mathscr{M}^*)$ and $Q(S^*(\mathscr{M}_0), \mathscr{M}_0)$, we can observe that the model misspecification has a more substantial effect for V-optimality and a relatively minor effect for A-optimality. For instance, we compare $Q(S^*(\mathscr{M}^*), \mathscr{M}_0)$ and $Q(S^*(\mathscr{M}_0), \mathscr{M}_0)$ by considering $n = 10$ and $m = 5$, the optimal censoring scheme for V-optimality with $\delta = 0.95$ under $\mathscr{M}^* = LOGIS$ is $(4, 0, 0, 0, 1)$ with $Q_{V_{.95}}(S^*(LOGIS), EV) = 0.3211$ (Table 2), while the optimal censoring scheme under $\mathscr{M}_0 = EV$ is (0, 0, 0, 0, 5) with $Q_{V_{.95}}(S^*(EV), EV) = 0.2585$ which gives 19.50% loss of efficient. In contrast, consider $n = 10$ and $m = 5$, the optimal censoring scheme for A-optimality under $\mathscr{M}^* = LOGIS$ is $(0, 3, 0, 0, 2)$ with $Q_A(S^*(LOGIS), EV) = 0.3071$ (Table 2), while the optimal censoring scheme under $\mathscr{M}_0 = EV$ is (0, 5, 0, 0, 0) with $Q_A(S^*(EV), EV) = 0.2918$ (Table 3) which gives 4.98% loss of efficient. We have a similar observation when we compare the objective functions $Q(S^*(\mathscr{M}^*), \mathscr{M}^*)$ and $Q(S^*(\mathscr{M}^*), \mathscr{M}_0)$. For example, in Table 3, when $n = 10$, $m = 6$, the optimal censoring scheme for V-optimality with $\delta = 0.95$ under $\mathscr{M}^* = EV$ is $(0, 0, 0, 0, 4)$ with $Q_{V_{.95}}(S^*(EV), EV) = 0.2546$. If the censoring scheme (0, 0, 0, 0, 4) is applied when the true model is normal ($\mathscr{M}_0 = NOR$), the asymptotic variance of the estimator of 95-th percentile is $Var(\hat{q}_{0.95}) = 0.4633$, which is clearly not the minimum variance that can be obtained because the censoring scheme (4, 0, 0, 0, 0) yields $Var(\hat{q}_{0.95}) = 0.3684$. Based on the results from our simulation studies, one should be cautious when the quantity of interest is one of those extreme percentiles (e.g., 1-st, 5-th, 95-th, 99-th percentiles) because the optimal censoring schemes could be sensitive to the change of the model.

Based on the simulation approach, we observe that the optimal censoring schemes determined based on asymptotic theory of the MLEs may not be optimal even when the underline model is correctly specified. Since the Monte-Carlo simulation is a numerically mimic of the real data analysis procedure in practice, the results are showing that when the analytical value of the objective function of the optimal censoring scheme and the values of the objective functions of other censoring schemes are closed, it is likely that those non-optimal censoring schemes will perform better than the optimal censoring scheme. We would suggest the practitioners to use Monte-Carlo simulation in comparing with other progressive censoring schemes and choose the optimal one. However, since the number of possible censoring schemes can be numerous when n and m are large, it will not be feasible to use Monte-Carlo simulation to compare all the possible censoring schemes. Therefore, in practice, we can use the analytical approach to identify the optimal censoring scheme and some near optimal censoring schemes, then Monte-Carlo simulation can be used to choose the best censoring schemes among those candidates. This approach will be illustrated in the example which will be presented in the next section.

4 Illustrative Example

R Core Team (2016) presented a progressively Type-II censored sample based on the breakdown data on insulating fluids tested at 34 kV from Nelson (1982). The progressively censored data presented in R Core Team (2016) has $n = 19$ and $m = 8$ with censoring scheme (0, 0, 3, 0, 3, 0, 0, 5). Suppose that we want to re-run the same experiment with $n = 19$ and $m = 8$ and we are interested in using the optimal censoring scheme that minimizing the variances of the parameter estimators (i.e., A-optimality) or minimizing the variance of the estimator of the 95-th percentile of the lifetime distribution (i.e., V-optimality with $\delta = 0.95$). We can first identify the top k optimal censoring schemes based on the asymptotic variances and then use Monte-Carlo simulation to evaluate the performances of those censoring schemes.

Since R Core Team (2016) discussed the linear inference under progressive Type-II censoring when the lifetime distribution is Weibull and used the breakdown data on insulating fluids as a numerical example, we assume here the underline lifetime distribution to be Weibull and determine the top ten censoring schemes subject to the A-optimality and V-optimality with $\delta = 0.05$. To study the effect of model misspecification to the optimal censoring schemes, we compute the objective functions for these censoring schemes when the true underline lifetime distribution is lognormal. Then, we also use Monte-Carlo simulation to evaluate the performances of these censoring schemes. To reduce the effect of Monte-Carlo simulation errors, we used 100,000 simulations to obtain the simulated variances. These results are presented in Tables 5 and 6 for A-optimality and V-optimality with $\delta = 0.95$, respectively.

As we illustrated in the previous section, the optimal censoring scheme under a particular model may not be optimal if the model is misspecified. The same observation is obtained in this numerical example. For A-optimality, from Table 5, instead of choosing the optimal censoring scheme (0, 11, 0, 0, 0, 0, 0, 0) based on asymptotic variances, the censoring scheme (1, 9, 1, 0, 0, 0, 0, 0) can be a better option based on the simulation results. For V-optimality with $\delta = 0.95$, from Table 6, instead of choosing the optimal censoring scheme (0, 0, 0, 0, 0, 0, 11, 0) based on asymptotic variances, one may adopt (0, 0, 0, 0, 2, 0, 9, 0) as the censoring scheme because it gives a smaller simulated $Var(\hat{q}_{0.95})$ and the performance of this censoring under $\mathscr{M}_0 = NOR$ is better than (0, 0, 0, 0, 0, 0, 11, 0).

5 Concluding Remarks

In this chapter, we propose analytical and simulation approaches to quantify the uncertainty in optimal experiment schemes systematically. The robustness of the optimal progressive Type-II censoring scheme with respect to changes of model is studied. We have shown that the optimal censoring schemes are sensitive to misspecification of models, especially when the V-optimal criterion is under consideration. In practice, we would recommend the use of Monte-Carlo simulation to verify if the

Table 5 Top ten optimal censoring scheme for A-optimalitywith $n = 19$, $m = 8$ based on asymptotic variances and 100,000 simulations

Censoring scheme	$M^* = EV$ (Asymptotic)		$M^* = EV$ (Simulated)		$M_0 = NORM$ (Asymptotic)		$M_0 = NORM$ (Simulated)	
	Q_A $(S(EV), EV)$	$Q_{V.95}$ $(S^*(EV), EV)$	Q_A $(S(EV), EV)$	$Q_{V.95}$ $(S^*(EV), EV)$	Q_A $(S(EV), NOR)$	$Q_{V.95}$ $(S^*(EV), NOR)$	Q_A $(S(EV), NOR)$	$Q_{V.95}$ $(S^*(EV), NOR)$
(0, 11, 0, 0, 0, 0, 0, 0)	0.1770	0.1909	0.1903	0.2022	0.1397	0.2579	0.1530	0.3163
(0, 10, 1, 0, 0, 0, 0, 0)	0.1772	0.1881	0.1919	0.2078	0.1395	0.2614	0.1535	0.3200
(1, 10, 0, 0, 0, 0, 0, 0)	0.1772	0.1922	0.1904	0.2026	0.1403	0.2581	0.1535	0.3154
(0, 9, 2, 0, 0, 0, 0, 0)	0.1773	0.1858	0.1905	0.2095	0.1393	0.2642	0.1524	0.3218
(1, 9, 1, 0, 0, 0, 0, 0)	0.1774	0.1893	0.1892	0.2057	0.1400	0.2615	0.1523	0.3185
(0, 8, 3, 0, 0, 0, 0, 0)	0.1775	0.1839	0.1901	0.2115	0.1392	0.2665	0.1534	0.3256
(2, 9, 0, 0, 0, 0, 0, 0)	0.1775	0.1935	0.1916	0.2017	0.1408	0.2582	0.1530	0.3139
(1, 8, 2, 0, 0, 0, 0, 0)	0.1776	0.1870	0.1897	0.2071	0.1399	0.2643	0.1543	0.3239
(0, 7, 4, 0, 0, 0, 0, 0)	0.1776	0.1823	0.1907	0.2148	0.1391	0.2684	0.1525	0.3250
(0, 6, 5, 0, 0, 0, 0, 0)	0.1777	0.1809	0.1918	0.2162	0.1390	0.2700	0.1533	0.3296

Table 6 Top ten optimal censoring scheme for V-optimality with $\delta = 0.95$, $n = 19$, $m = 8$ based on asymptotic variances and 100,000 simulations

Censoring scheme	$M^* = EV$ (Asymptotic)		$M^* = EV$ (Simulated)		$M_0 = NORM$ (Asymptotic)		$M_0 = NORM$ (Simulated)	
	Q_A $(S(EV), EV)$	$Q_{V.95}$ $(S^*(EV), EV)$	Q_A $(S(EV), EV)$	$Q_{V.95}$ $(S^*(EV), EV)$	Q_A $(S(EV), NOR)$	$Q_{V.95}$ $(S^*(EV), NOR)$	Q_A $(S(EV), NOR)$	$Q_{V.95}$ $(S^*(EV), NOR)$
(0, 0, 0, 0, 0, 0, 11, 0)	0.2327	0.1456	0.2531	0.3895	0.1582	0.3911	0.1649	0.4154
(0, 0, 0, 0, 0, 1, 10, 0)	0.2321	0.1457	0.2514	0.3861	0.1580	0.3899	0.1664	0.4198
(0, 0, 0, 0, 0, 2, 9, 0)	0.2314	0.1459	0.2499	0.3816	0.1577	0.3885	0.1644	0.4131
(0, 0, 0, 0, 1, 0, 10, 0)	0.2313	0.1460	0.2487	0.3816	0.1575	0.3879	0.1666	0.4203
(0, 0, 0, 0, 0, 3, 8, 0)	0.2307	0.1461	0.2484	0.3800	0.1573	0.3869	0.1649	0.4148
(0, 0, 0, 0, 1, 1, 9, 0)	0.2307	0.1461	0.2477	0.3780	0.1572	0.3866	0.1650	0.4158
(0, 0, 0, 0, 0, 4, 7, 0)	0.2298	0.1463	0.2472	0.3768	0.1569	0.3850	0.1649	0.4162
(0, 0, 0, 0, 1, 2, 8, 0)	0.2299	0.1463	0.2476	0.3773	0.1569	0.3850	0.1645	0.4152
(0, 0, 0, 1, 0, 0, 10, 0)	0.2304	0.1463	0.2497	0.3820	0.1570	0.3854	0.1651	0.4150
(0, 0, 0, 0, 2, 0, 9, 0)	0.2298	0.1464	0.2470	0.3763	0.1567	0.3844	0.1646	0.4136

optimal censoring schemes are delivering significant superior results compared to other censoring schemes.

The current study is limited to the progressive Type-II censoring; it will be of interest to apply the proposed approaches to other optimal experiment design problems and to study the effect of model misspecification. Moreover, the methodologies and illustrations presented in this chapter are mainly focused on misspecification of the underlying statistical model. In the case that the determination of the optimal experiment scheme on the specified value of parameters, the methodologies developed here can be applied as well. For instance, in experiment design of multi-level stress testing with extreme value regression under censoring (Ka et al. 2011 and Chan et al. 2016), the expected Fisher information matrix depends on the proportions of observed failures which are functions of the unknown parameters, and consequently the optimal experimental scheme also depends on the unknown parameters. Quantifying the uncertainty of the optimal experimental scheme due to parameter misspecification can be proceed in a similar manner as presented in this chapter.

R function for compute the objective functions for a specific progressive Type-II censoring scheme for extreme value distribution

```
##############################################
## Function to compute the objective        ##
## functions for a specific progressive     ##
## Type-II censoring scheme                 ##
##############################################

#######################################
# Input values:                       #
# nn: Sample size                     #
# mm: Effective sample size           #
# ir: Censoring scheme (length = mm)  #
#######################################

###################################################
# Output values:                                  #
# dfi: Determinant of Fisher information matrix   #
# tvar: Trace of variance-covariance matrix       #
# vq95: Variance of the MLE of 95-th percentile   #
# vq05: Variance of the MLE of 5-th percentile    #
###################################################

objpcs <- function(mm, nn, ir)
{
rr <- numeric(mm)
cc <- numeric(mm)
aa <- matrix(0, mm, mm)
epcos <- numeric(mm)
epcossq <- numeric(mm)
rpcs <- numeric(mm)
gg <- numeric(mm)

##Compute rr##
for (jj in 1:mm)
{rr[jj] <- mm - jj + 1 + sum(ir[jj:mm])}
```

```
##Compute cc##
cc (?) <- rr (?)
for (ii in 2:mm)
{cc[ii] <- cc[ii-1]*rr[ii] }

##Compute aa##
for (jj in 1:mm)
{for (ii in 1:jj)
{aa[ii,jj] = 1
for (kk in 1:jj)
{if (kk != ii) {aa[ii,jj] <- aa[ii,jj]/(rr[kk] - rr[ii])}
} }}

##Compute E(Z_i:m:n) and E(Z_i:m:n∧2) ##

for (ii in 1:mm)
{psum <- 0
psumsq <- 0
for (ll in 1:ii) {psum <- psum + aa[ll,ii]*(digamma(1) -
log(rr[ll]))/(rr[ll])}
for (ll in 1:ii) {psumsq <- psumsq + aa[ll,ii]*(
(digamma(1)∧2) - 2*digamma(1)*log(rr[ll])
+ log(rr[ll])*log(rr[ll]) + pi*pi/6)/rr[ll]}
epcos[ii] <- cc[ii]*psum
epcossq[ii] <- cc[ii]*psumsq }

##Elements of Fisher Information Matrix##

i22 <- sum(1 + 2*epcos + epcossq)
i12 <- -sum(1 + epcos)
i11 <- mm

dfi <- det(matrix(c(i11, i12, i12, i22), 2, 2))
vcov <- solve(matrix(c(i11, i12, i12, i22), 2, 2))
inv05 <- log(-log(0.95))
inv95 <- log(-log(0.05))

tvar <- vcov[1,1] + vcov[2,2]
vq95 <- vcov[1,1] + inv95*inv95*vcov[2,2] + 2*inv95*vcov[1,2]
vq05 <- vcov[1,1] + inv05*inv05*vcov[2,2] + 2*inv05*vcov[1,2]
out <- c(dfi, tvar, vq95, vq05)
names(out) <- c("dfi", "tvar", "vq95", "vq05")
return(out)}

## Example: n = 10, m = 5, censoring scheme = (5,0,0,0,0)

objpcs(5, 10, c(5,0,0,0,0))
dfi          tvar        vq95        vq05
54.2144581   0.2947966   0.3473800   0.9247358
```

References

Balakrishnan, N., & Aggarwala, R. (2000). *Progressive censoring: Theory, methods and applications*. Boston: Birkhäuser.

Balakrishnan, N. (2007). Progressive censoring methodology: An appraisal (with discussion). *Test, 16*, 211–296.

Balakrishnan, N., & Cramer, E. (2014). *The art of progressive censoring: Applications to reliability and quality*. Boston, MA: Birkhäuser.

Balakrishnan, N., Kannan, N., Lin, C. T., & Ng, H. K. T. (2003). Point and interval estimation for gaussian distribution, based on progressively Type-II censored samples. *IEEE Transactions on Reliability*, *52*, 90–95.

Chan, P. S., Balakrishnan, N., So, H. Y., & Ng, H. K. T. (2016). Optimal sample size allocation for multi-level stress testing with exponential regression under Type-I censoring. *Communications in Statistics—Theory and Methods, 45*, 1831–1852.

Dahmen, K., Burkschat, M., & Cramer, E. (2012). A- and D-optimal progressive Type-II censoring designs based on Fisher information. *Journal of Statistical Computation and Simulation*, *82*, 879–905.

Johnson, N. L., Kotz, S., & Balakrishnan, N. (1994) *Continuous univariate distribution* (Vol. 1, 2nd ed.). New York: Wiley

Ka, C. Y., Chan, P. S., Ng, H. K. T., & Balakrishnan, N. (2011). Optimal sample size allocation for multi-level stress testing with extreme value regression under Type-II Censoring, *Statistics, 45*, 257–279.

Ng, H. K. T., Chan, P. S., & Balakrishnan, N. (2004). Optimal progressive censoring plan for the Weibull distribution. *Technometrics*, *46*, 470–481.

R Core Team. (2016). *A language and environment for statistical computing*. Vienna, Austria: R Foundation for Statistical Computing

Viveros, R., & Balakrishnan, N. (1994). Interval estimation of parameters of life from progressively censored data. *Technometrics*, *36*, 84–91.

Part II
Monte-Carlo Methods in Missing Data

Markov Chain Monte-Carlo Methods for Missing Data Under Ignorability Assumptions

Haresh Rochani and Daniel F. Linder

Abstract Missing observations are a common occurrence in public health, clinical studies and social science research. Consequences of discarding missing observations, sometimes called complete case analysis, are low statistical power and potentially biased estimates. Fully Bayesian methods using Markov Chain Monte-Carlo (MCMC) provide an alternative model-based solution to complete case analysis by treating missing values as unknown parameters. Fully Bayesian paradigms are naturally equipped to handle this situation by augmenting MCMC routines with additional layers and sampling from the full conditional distributions of the missing data, in the case of Gibbs sampling. Here we detail ideas behind the Bayesian treatment of missing data and conduct simulations to illustrate the methodology. We consider specifically Bayesian multivariate regression with missing responses and the missing covariate setting under an ignorability assumption. Applications to real datasets are provided.

1 Introduction

Complete data are rarely available in epidemiological, clinical and social research, especially when a requirement of the study is to collect information on a large number of individuals or on a large number of variables. Analyses that improperly treat missing data can lead to more bias and loss of efficiency, which may limit generalizability of results to a wider population and can diminish our ability to understand true underlying phenomena. In applied research, linear regression models are an important tool to characterize relationships among variables. The four common approaches for inference in regression models with missing data are: Maximum like-

H. Rochani (✉)
Department of Biostatistics, Jiann-Ping Hsu College of Public Health,
Georgia Southern University, Statesboro, GA, Georgia
e-mail: hrochani@georgiasouthern.edu

D.F. Linder
Department of Biostatistics and Epidemiology, Medical College of Georgia,
Augusta University, Augusta, GA, Georgia

D.-G. Chen and J.D. Chen (eds.), *Monte-Carlo Simulation-Based Statistical Modeling*, ICSA Book Series in Statistics, DOI 10.1007/978-981-10-3307-0_7

lihood (ML), Multiple imputation (MI), Weighted Estimating Equations (WEE) and Fully Bayesian (FB) (Little and Rubin 2014). This chapter focuses on FB methods for regression models with missing multivariate responses and models with missing covariates. The Bayesian approach provides a natural framework for making inferences about regression coefficients with incomplete data, where certain other methods may be viewed as special cases or related. For instance, the *maximum a posteriori* (MAP) estimate from a FB approach under uniform improper priors leads to ML estimates; therefore ML can be viewed as a special case of Bayesian inference. Moreover, in MI the "imputation" step is based on the sampling from a posterior predictive distribution. Overall, FB methods are general and can be powerful tools for dealing with incomplete data, since they easily accommodate missing data without having extra modeling assumptions.

2 Missing Data Mechanisms

For researchers, it is crucial to have some understanding of the underlying missing mechanism for the variables under investigation so that parameter estimates are accurate and precise. Rubin (1976) defined the taxonomy of missing data mechanisms based on how the probability of a missing value relates to the data itself. This taxonomy has been widely adopted in the statistical literature. There are mainly three types of missing data mechanisms:

Missing Completely At Random (MCAR):-

MCAR mechanisms assume that the probability of missingness in the variable of interest does not depend on the values of that variable that are either missing or observed. We begin by denoting the data as D and M as the missing indicator matrix, which has values 1 if the variable is observed and 0 if the variable is not observed. For MCAR, missingness in D is independent of the data being observed or missing, or equivalently $p(M|D, \theta) = p(M|\theta)$, where θ are the unknown parameters.

Missing At Random (MAR):-

A MAR mechanism assumes that the probability of missingness in the variable of interest is associated only with components of observed variables and not on the components that are missing. In mathematical terms, it can be written as $p(M|D, \theta) = p(M|D_{obs}, \theta)$.

Missing Not At Random (MNAR):-

Finally, an MNAR assumption allows missingness in the variable of interest to depend on the unobserved values in the data set. In notation, it can be written as $p(M|D, \theta) = p(M|D_{miss}, \theta)$.

3 Data Augmentation

Data augmentation (DA) within MCMC algorithms is a widely used routine that handles incomplete data by treating missing values as unknown parameters, which are sampled during iterations and then marginalized away in the final computation of various functionals of interest, like for instance when computing posterior means (Tanner and Wong 1987). To illustrate how inferences from data with missing variables may be improved via data augmentation, we first distinguish between *complete* data, *observed* data and *complete-cases. Complete* data refers to the data that are observed or intended to be observed, which includes response variables, covariates of interest and missing data indicators. Conversely, the *observed* data comprise the observed subset of complete data. Furthermore, *complete-cases* comprise of the data after excluding all observations for which the outcome or any of the inputs are missing. In a Bayesian context, for most missing data problems, the observed data posterior $p(\theta|D_{obs})$ is intractable and cannot easily be sampled from. However, when D_{obs} is 'augmented' by assumed values of D_{miss}, the resulting complete data posterior, $p(\theta|D_{obs}, D_{miss})$ becomes much easier to handle. For DA, at each iteration t, we can sample $\left(D_{obs}^{(t)}, \theta^{(t)}\right)$ by

1. $D_{miss}^{(t)} \sim p\left(D_{miss}|D_{obs}, \theta^{(t-1)}\right)$
2. $\theta^{(t)} \sim p\left(\theta|D_{obs}, D_{miss}^{(t)}\right)$

where θ may be sampled by using one of the various MCMC algorithms discussed in the previous chapters of this book. A stationary distribution of the above transition kernel is $p(\theta, D_{miss}|D_{obs})$ where upon marginalization over the imputed missing values one arrives at the desired target density $p(\theta|D_{obs})$. In the following section, we detail how these ideas may be implemented with missing multivariate responses in the context of Gibbs sampling.

4 Missing Response

It is quite common to have missing responses in multivariate regression models, particularly when repeated measures are taken and missing values occur at later follow up times. In this section, we focus on the multivariate model with missing responses, in which the missingness depends only on the covariates that are fully observed. For some literature dealing with the missing response scenario, see (Little and Rubin 2014; Daniels and Hogan 2008; Schafer 1997)

4.1 Method: Multivariate Normal Model

In this section, we consider regression models where we have n subjects measured at (d) occasions with the same set of (p) predictors. For instance, this would be the case for subjects with baseline levels of a predictor and responses measured over time, or clusters in which predictors are measured at the cluster level. Suppose $y_1, y_2, \ldots y_d$ are d possibly correlated random variables with respective means $\mu_1, \mu_2, \ldots \mu_d$ and variance-covariance matrix $\boldsymbol{\Sigma}$. In terms of matrix notation, we denote the $n \times d$ response matrix as $(\boldsymbol{Y})$ and the $n \times p$ design matrix as $(\boldsymbol{X})$, which can be expressed as

$$\boldsymbol{Y} = \begin{bmatrix} Y_{11} & Y_{12} & Y_{13} & \ldots & Y_{1d} \\ Y_{21} & Y_{22} & Y_{23} & \ldots & Y_{2d} \\ \vdots & \vdots & \vdots & \ddots & \vdots \\ Y_{n1} & Y_{n2} & Y_{n3} & \ldots & Y_{nd} \end{bmatrix} \qquad \boldsymbol{X} = \begin{bmatrix} X_{11} & X_{12} & X_{13} & \ldots & X_{1p} \\ X_{21} & X_{22} & X_{23} & \ldots & X_{2p} \\ \vdots & \vdots & \vdots & \ddots & \vdots \\ X_{n1} & X_{n2} & X_{n3} & \ldots & X_{np} \end{bmatrix}$$

Our multivariate regression model for response vector $\boldsymbol{Y}_i = (Y_{i1}, Y_{i2}, \ldots, Y_{id})^\top$ with covariate vector $X_i = \left(X_{i1}, X_{i2}, \ldots, X_{ip}\right)$ can be written as $\boldsymbol{Y}_i | X_i \overset{ind}{\sim} \mathcal{N}\left(\boldsymbol{\mu}_i, \boldsymbol{\Sigma}_i\right)$, where $\boldsymbol{\mu}_i$ is a $d \times 1$ vector and $\boldsymbol{\Sigma}$ is a $d \times d$ matrix. Furthermore, $\mu_{ij} = E\left(\boldsymbol{Y}_i | X_i\right) = X_i \boldsymbol{\beta}$ where $\boldsymbol{\beta}$ is a $p \times 1$ vector of regression coefficients. The key assumption for this particular model is that the measured covariate terms X_{ik} are the same for each component of the observations Y_{ij} where $1 \le j \le d$. Additionally, since we are assuming that $\boldsymbol{\beta} \in \mathbb{R}^p$, the mean response for observation i is the same for each $1 \le j \le d$. Both the assumption of common baseline covariates and a vector $\boldsymbol{\beta}$ can easily be extended to time varying covariates and a matrix of coefficients $\boldsymbol{\beta} \in \mathbb{R}^{d \times p}$ with only minor changes to notation; however, we use the current scenario for illustration purposes. Since we are only focusing on the ignorable missing mechanism, we will consider the scenario where missing in $\boldsymbol{Y}$ depends on non-missing predictors.

We define, $\boldsymbol{Y}_{i(obs)} = (Y_{i1}, Y_{i2}, \ldots, Y_{id^*})^\top$ and $\boldsymbol{Y}_{i(miss)} = \left(Y_{i(d^*+1)}, Y_{i2}, \ldots, Y_{id}\right)^\top$, where $d^* < d$. The variance-covariance matrix, $\boldsymbol{\Sigma}$, can be partitioned as

$$\boldsymbol{\Sigma} = \begin{pmatrix} \boldsymbol{\Sigma}_{(obs)} & \boldsymbol{\Sigma}_{(obs,miss)} \\ \boldsymbol{\Sigma}^\top_{(obs,miss)} & \boldsymbol{\Sigma}_{(miss)} \end{pmatrix}$$

For the Bayesian solution to the missing data problem, after we specify the complete data model and noninformative independence Jeffreys' prior for parameters, $p(\boldsymbol{\beta}, \boldsymbol{\Sigma}) \propto |\boldsymbol{\Sigma}|^{-\frac{d+1}{2}}$, we would then like to draw inferences based on the observed data posterior $p\left(\boldsymbol{\beta}, \boldsymbol{\Sigma} | \boldsymbol{Y}_{obs}, \boldsymbol{X}\right)$. However, as discussed previously, the complete data posterior $p\left(\boldsymbol{\beta}, \boldsymbol{\Sigma} | \boldsymbol{Y}, \boldsymbol{X}\right)$ is easier to sample from than the observed data posterior. In this particular situation, full conditional distributions for complete data and parameters of interest are easy to derive. For posterior sampling using data augmentation, at each iteration t, samples from $\left(\boldsymbol{Y}^{(t)}_{(miss)}, \boldsymbol{\beta}^{(t)}, \boldsymbol{\Sigma}^{(t)} | \boldsymbol{Y}_{obs}, \boldsymbol{X}\right)$ can be obtained by first sampling $\boldsymbol{Y}^{(t)}_{i(miss)} | \cdots \sim p\left(\boldsymbol{Y}_{i(miss)} | \boldsymbol{Y}_{i(obs)}, \boldsymbol{\beta}^{(t-1)}, \boldsymbol{\Sigma}^{t-1)}, \boldsymbol{X}\right)$ for $1 \le i \le n$ and then sampling $\left(\boldsymbol{\beta}^{(t)}, \boldsymbol{\Sigma}^{(t)}\right) | \cdots \sim p\left(\boldsymbol{\beta}, \boldsymbol{\Sigma} | \boldsymbol{Y}^{(t)}_{(miss)}, \boldsymbol{Y}_{(obs)}, \boldsymbol{X}\right)$. In the data augmentation step, $p\left(\boldsymbol{Y}_{i(miss)} | \boldsymbol{Y}_{i(obs)}, \boldsymbol{\beta}^{(t-1)}, \boldsymbol{\Sigma}^{(t-1)}, \boldsymbol{X}\right)$ is a normal density with mean $\boldsymbol{\mu}_i^*$ and variance-covariance matrix $\boldsymbol{\Sigma}_i^*$, where

$$\begin{aligned}
\boldsymbol{\mu}_i^* &= \text{vec}(X_i\boldsymbol{\beta}^{(t-1)})_{d-d^*} + \left(\boldsymbol{\Sigma}^{\top(t-1)}_{(obs,miss)} \boldsymbol{\Sigma}^{-1(t-1)}_{(obs)}\right) \text{vec}\left(\boldsymbol{Y}_{i(obs)} - X_i\boldsymbol{\beta}^{(t-1)}\right)_{d^*} \\
\boldsymbol{\Sigma}_i^* &= \boldsymbol{\Sigma}^{(t-1)}_{(miss)} - \boldsymbol{\Sigma}^{\top(t-1)}_{(obs,miss)} \boldsymbol{\Sigma}^{-1(t-1)}_{(obs)} \boldsymbol{\Sigma}^{(t-1)}_{(obs,miss)}
\end{aligned}$$

and $\text{vec}(x)_d$ denotes the scalar value x stacked in a vector of length d. Samples from $p\left(\boldsymbol{\beta}, \boldsymbol{\Sigma} | \boldsymbol{Y}^{(t)}_{(miss)}, \boldsymbol{Y}_{(obs)}, \boldsymbol{X}\right)$ can be obtained by Gibbs sampling since the full conditional posteriors for each parameter are analytic and can be written as $p(\boldsymbol{\beta}|\ldots) \sim \mathcal{N}\left(\boldsymbol{\mu}_\beta, \boldsymbol{V}_\beta\right)$ and $p(\boldsymbol{\Sigma}|\ldots) \sim \mathcal{W}^{-1}(nd + d, \boldsymbol{\psi})$, where

$$\begin{aligned}
\boldsymbol{\mu}_\beta &= \left(\sum_i (\boldsymbol{X}_i^\top \boldsymbol{\Sigma}^{-1} \boldsymbol{X}_i)\right)^{-1} \left[\sum_i (\boldsymbol{X}_i^\top \boldsymbol{\Sigma}^{-1} \boldsymbol{Y}_i)\right] \\
\boldsymbol{V}_\beta &= \left(\sum_i (\boldsymbol{X}_i^\top \boldsymbol{\Sigma}^{-1} \boldsymbol{X}_i)\right)^{-1} \\
\boldsymbol{\psi} &= \sum_i (\boldsymbol{Y}_i - \boldsymbol{X}_i\boldsymbol{\beta})(\boldsymbol{Y}_i - \boldsymbol{X}_i\boldsymbol{\beta})^\top
\end{aligned}$$

In the above notation, the bold symbol $\boldsymbol{X}_i$ represents the $d \times p$ matrix with rows X_i, $\mathcal{N}$ denotes a multivariate normal density and $\mathcal{W}^{-1}$ the inverse Wishart density.

4.2 *Simulation*

A simulation study was conducted to compare the bias and root mean squared error (RMSE) of regression coefficients ($\boldsymbol{\beta}$) for complete case analysis and the FB approach under a MAR assumption for our multivariate normal model. Data augmentation using Gibbs sampling was performed to compare the performance of the estimators by using various proportions of missing values in the multivariate response variable as shown in Tables 1 and 2. In the simulation, at each iteration, three covariates (X_1, X_2, X_3) were generated, in which X_1 was binary, and X_2, X_3 were continuous. X_1 was sampled from the binomial distribution with success probability of 0.4, while X_2 and X_3 were sampled from the normal distribution with mean $= 0$ variance $= 1$. Furthermore, the multivariate response variable, Y, was generated from $\mathcal{N}(\beta_0 + \beta_1 x_1 + \beta_2 x_2 + \beta_3 x_3, \boldsymbol{\Sigma})$ where

$$\boldsymbol{\Sigma} = \begin{pmatrix} 4 & 2 & 4 \\ 2 & 4 & 2 \\ 2 & 2 & 4 \end{pmatrix}$$

Table 1 Bias comparison of regression coefficients for complete-cases and data augmentation

Missing %			Complete cases			Data augmentation		
y_1	y_2	y_3	β_1	β_2	β_3	β_1	β_2	β_3
5	5	5	0.0194	0.0045	0.0001	0.0208	0.0002	0.0008
10	5	5	0.0009	0.0022	0.0033	0.0058	0.0047	0.0038
20	5	5	0.0006	0.0036	0.0015	0.0020	0.0007	0.0026
20	5	5	0.0006	0.0036	0.0015	0.0020	0.0007	0.0026
5	10	5	0.0053	0.0084	0.0018	0.0090	0.0130	0.0002
5	10	5	0.0068	0.0190	0.0086	0.0147	0.0151	0.0067
10	10	5	0.0253	0.0033	0.0053	0.0266	0.0061	0.0081
10	10	5	0.0069	0.0047	0.0001	0.0005	0.0088	0.0002
20	10	5	0.0047	0.0244	0.0039	0.0017	0.0067	0.0025
5	20	5	0.0099	0.0057	0.0001	0.0235	0.0067	0.0032
5	20	5	0.0270	0.0085	0.0057	0.0198	0.0044	0.0021
10	20	5	0.0043	0.0155	0.0045	0.0029	0.0129	0.0029
10	20	5	0.0148	0.0119	0.0055	0.0132	0.0010	0.0019
20	20	5	0.0105	0.0052	0.0012	0.0234	0.0071	0.0091
5	5	10	0.0052	0.0055	0.0042	0.0068	0.0027	0.0039
5	10	10	0.0250	0.0186	0.0082	0.0175	0.0139	0.0065
10	10	10	0.0032	0.0132	0.0028	0.0066	0.0111	0.0013
20	10	10	0.0042	0.0156	0.0004	0.0017	0.0075	0.0058
5	20	10	0.0227	0.0056	0.0042	0.0073	0.0047	0.0010
10	20	10	0.0268	0.0009	0.0046	0.0110	0.0037	0.0014
20	20	10	0.0264	0.0036	0.0144	0.0176	0.0095	0.0072
20	5	20	0.0347	0.0064	0.0006	0.0164	0.0092	0.0015
5	10	20	0.0240	0.0081	0.0045	0.0069	0.0021	0.0037
10	10	20	0.0016	0.0039	0.0083	0.0028	0.0038	0.0006
20	10	20	0.0097	0.0046	0.0033	0.0079	0.0001	0.0043
5	20	20	0.0236	0.0093	0.0052	0.0223	0.0054	0.0041
10	20	20	0.0158	0.0080	0.0016	0.0093	0.0047	0.0010
20	20	20	0.0146	0.0034	0.0003	0.0164	0.0034	0.0050

$\beta_0 = 0$ and $\beta_1 = \beta_2 = \beta_3 = 1$. The sample size for each iteration was 100. Various proportions of missing values were created in Y_1, Y_2 and Y_3 that depend only on non-missing $\boldsymbol{X} = (X_1, X_2, X_3)$ in order to simulate a MAR missing mechanism. To model the missing probability for the variable Y_1; $Pr\ (Y_{i1} = \text{missing}|\boldsymbol{X})$, we consider the logistic model as follows;

$$Pr\ (Y_{i1} = \text{missing}|\boldsymbol{X}) = \frac{exp\ (\gamma_0 + X_{i1} + X_{i2} + X_{i3})}{1 + exp\ (\gamma_0 + X_{i1} + X_{i2} + X_{i3})} = p_i.$$

Table 2 RMSE comparison of regression coefficients for complete-cases and data augmentation

Missing %			Complete cases			Data augmentation		
y_1	y_2	y_3	β_1	β_2	β_3	β_1	β_2	β_3
5	5	5	0.1396	0.0910	0.0440	0.1424	0.0797	0.0419
10	5	5	0.1514	0.0962	0.0461	0.1332	0.0824	0.0441
20	5	5	0.1613	0.1065	0.0521	0.1428	0.0833	0.0368
20	5	5	0.1613	0.1065	0.0461	0.1428	0.0833	0.0368
5	10	5	0.1441	0.0922	0.0516	0.1421	0.0831	0.0438
5	10	5	0.1515	0.0849	0.0461	0.1420	0.0797	0.0399
10	10	5	0.1740	0.0899	0.0491	0.1513	0.0778	0.0400
10	10	5	0.1688	0.0859	0.0477	0.1530	0.0788	0.0438
20	10	5	0.1816	0.1032	0.0555	0.1704	0.0802	0.0441
5	20	5	0.1604	0.0882	0.0456	0.1519	0.0847	0.0428
5	20	5	0.1700	0.0963	0.0515	0.1396	0.0711	0.0434
10	20	5	0.1666	0.1026	0.0562	0.1603	0.0889	0.0465
10	20	5	0.1578	0.0982	0.0539	0.1389	0.0748	0.0424
20	20	5	0.1850	0.0976	0.0507	0.1631	0.0775	0.0473
5	5	10	0.1624	0.0887	0.0428	0.1514	0.0741	0.0362
5	10	10	0.1561	0.0876	0.0428	0.1515	0.0713	0.0368
10	10	10	0.1634	0.1086	0.0457	0.1470	0.0887	0.0403
20	10	10	0.1714	0.0848	0.0477	0.1266	0.0771	0.0368
5	20	10	0.1756	0.0957	0.0539	0.1498	0.0737	0.0420
10	20	10	0.1579	0.0957	0.0504	0.1307	0.0727	0.0394
20	20	10	0.2012	0.1043	0.0589	0.1502	0.0815	0.0433
20	5	20	0.1857	0.1082	0.0533	0.1355	0.0782	0.0444
5	10	20	0.1561	0.0998	0.0546	0.1391	0.0848	0.0458
10	10	20	0.1745	0.0985	0.0528	0.1432	0.0864	0.0385
20	10	20	0.1794	0.1049	0.0495	0.1532	0.0725	0.0464
5	20	20	0.2022	0.1054	0.0570	0.1628	0.0809	0.0457
10	20	20	0.1853	0.1156	0.0556	0.1493	0.0856	0.0459
20	20	20	0.2067	0.1159	0.0593	0.1652	0.0924	0.0414

Based on this probability, we generated a binary variable to indicating whether Y_{i1} was missing: $I_i \sim$ Bernoulli $(1, p_i)$. If $I_i = 1$ then we deleted the corresponding value of Y_{i1}. This process ensures that missingness in Y_1 depends on the $\boldsymbol{X}$ which are fully observed. We generated missing values in Y_2 in a similar manner. Inference for $\boldsymbol{\beta} = (\beta_0, \beta_1, \beta_2, \beta_3)$ in the complete case analysis was performed by using the complete data posterior mean. The complete case posterior, $p(\boldsymbol{\beta}, \boldsymbol{\Sigma}|\boldsymbol{Y}, \boldsymbol{X})$, was sampled via the Gibbs sampling algorithm (Geman and Geman 1984) as discussed in the previous section. Table 3 shows that the bias for regression coefficients are negligible under both complete case analysis and data augmentation for various

Table 3 Bias comparison of regression coefficients for complete-cases and data augmentation

Missing %		Complete cases			Data augmentation		
x_1	x_2	β_1	β_2	β_3	β_1	β_2	β_3
5	5	0.0029	0.0003	0.0026	0.0122	0.0263	0.0042
5	10	0.0112	0.0086	0.0010	0.0231	0.0273	0.0017
5	15	0.0008	0.0048	0.0105	0.0110	0.0393	0.0143
5	20	0.0032	0.0025	0.005	0.0094	0.0489	0.0088
10	5	0.0045	0.0088	0.0038	0.0206	0.0327	0.0145
10	10	0.0007	0.0021	0.0041	0.0218	0.0398	0.0164
10	15	0.0062	0.0055	0.0022	0.0282	0.0377	0.0052
10	20	0.0114	0.0008	0.0015	0.0172	0.0511	0.0133
15	5	0.0075	0.0035	0.0012	0.0376	0.0232	0.0237
15	10	0.0102	0.0045	0.0016	0.0222	0.0380	0.0298
15	15	0.0025	0.0042	0.0045	0.0264	0.0396	0.0289
15	20	0.0078	0.0034	0.0045	0.0240	0.0478	0.0221
20	5	0.0146	0.0005	0.0008	0.0529	0.0290	0.0325
20	10	0.0137	0.0026	0.0023	0.0426	0.0296	0.0381
20	15	0.0103	0.0042	0.0035	0.0423	0.0430	0.0456
20	20	0.0011	0.0028	0.0033	0.0257	0.0573	0.0406

proportions of missing values of the response variable. However, the RMSEs are smaller for DA as compared to the complete case analysis under different proportions of missingness in the response variable (Table 2).

4.3 *Prostate Specific Antigen (PSA) Data*

This section focuses on a real data application of the fully Bayesian approach for analyzing the multivariate normal model as discussed in the previous section. We will illustrate the application by using the prostate specific antigen (PSA) data which was published by Etzioni et al. (1999). This was a sub study of the beta-carotene and retinol trial (CARET) with 71 cases of prostate cancer patients and 70 controls (i.e. subjects not diagnosed with prostate cancer by the time of analysis, matched to cases on date of birth). In the PSA dataset, in addition to baseline age, there were two biomarkers measured over time (9 occasions): free PSA (fpsa) and total PSA (tpsa). For illustrative purposes, we investigate the effect of baseline age on the first three fpsa measurement in patients. There were missing values in the fpsa variable at occasion 2 and 3 for 14 patients in the study. Under the assumption of missingness being dependent only on the fully observed baseline age, we can obtain estimates of the regression coefficient for age $\left(\beta_{age} = 0.0029\right)$ with $SE = 0.00026$ using the fully Bayesian modeling approach. A similar model fit with complete-case analysis gives estimates for baseline age as 0.0061 with $SE = 0.00033$.

5 Missing Covariates

In addition to missing outcomes being a common occurrence in experimental studies, missing covariates are frequently encountered as well. In this section, we focus on missing covariates in which the missingness depends on the fully observed response. We specialize our analysis to the normal regression model. We direct the reader to the multiple methods proposed in the literature for handling missing covariates (Ibrahim et al. 1999a, 2002; Lipsitz and Ibrahim 1996; Satten and Carroll 2000; Xie and Paik 1997).

5.1 *Method*

Let Y_i be the ith response and X_{ij} be the jth covariate for ith subject. A missing value in the jth covariate for the ith subject can be represented as a parameter, M_{ij}. Let R_i be the set that indexes covariates which are observed for the ith subject. The normal regression model with missing covariates can be represented as

$$\mu_i = \sum_{j \in R_i} X_{ij}\beta_j + \sum_{j \in R_i^c} M_{ij}\beta_j \tag{1}$$

where $Y_i \sim \mathcal{N}\left(\mu_i, \sigma^2\right)$. Our goal is to estimate regression coefficients in the presence of missing data. For notation purposes, we can write the collection of missing data as the parameter, $\boldsymbol{M} = \left(M_{i1}, M_{i2} \dots M_{ip}\right)$, with corresponding prior $p(\boldsymbol{M})$. Each of the missing parameters, M_{ij}, will be assigned a prior distribution $p\left(M_{ip}\right)$. The complete data posterior $p\left(\boldsymbol{\beta}, \sigma^2, \boldsymbol{M}|\boldsymbol{Y}, \boldsymbol{X}\right)$ can be determined by Bayes rule as follows:

$$p\left(\boldsymbol{\beta}, \sigma^2, \boldsymbol{M}|\boldsymbol{Y}, \boldsymbol{X}\right) = \frac{p\left(\boldsymbol{\beta}, \sigma^2, \boldsymbol{M}\right) L\left(\boldsymbol{Y}|\boldsymbol{\beta}, \sigma^2, \boldsymbol{M}, \boldsymbol{X}\right)}{\int p\left(\boldsymbol{\beta}, \sigma^2, \boldsymbol{M}\right) L\left(\boldsymbol{Y}|\boldsymbol{\beta}, \sigma^2, \boldsymbol{M}, \boldsymbol{X}\right) d\boldsymbol{\beta} d\sigma^2 d\boldsymbol{M}} \tag{2}$$

The posterior in Eq. 2 depends on the missing covariate parameters $\boldsymbol{M}$. However, our main interest is in posterior inference about $\boldsymbol{\beta}$ and σ^2, and the desired posterior distribution $p\left(\boldsymbol{\beta}, \sigma^2|\boldsymbol{Y}, \boldsymbol{X}\right)$ can be obtained by

$$p\left(\boldsymbol{\beta}, \sigma^2|\boldsymbol{Y}, \boldsymbol{X}\right) = \int p\left(\boldsymbol{\beta}, \sigma^2, \boldsymbol{M}|\boldsymbol{Y}, \boldsymbol{X}\right) d\boldsymbol{M} \tag{3}$$

In general, Eq. 3 involves multi-dimensional integrals which do not have closed forms and will be high-dimensional even for low fractions of missing covariate values. In fact, the dimension of the integration problem is the same as the number of missing values. Posterior sampling can be performed based on various MCMC methods, such as via the Gibbs sampler after specifying full conditionals or using a random walk Metropolis algorithm discussed in Chap. 1 of this book.

An important issue in our current setting is the appropriate prior specification for missing covariates. There are many ways to assign missing covariate distributions; however, some modeling strategies, especially in presence of large fractions of missing data, can lead to a vast number of nuisance parameters. Hence, MCMC methods to sample posteriors can then become computationally intensive and inefficient even if the parameters are identifiable. Thus, it is essential to reduce the number of the nuisance parameters by specifying effective joint covariate distributions. Ibrahim et al. (1999b); Lipsitz and Ibrahim (1996) proposed the strategy of modeling the joint distribution of missing covariates as a sequence of one-dimensional distributions, which can be given by

$$\begin{aligned} p\left(M_{i1}, \ldots M_{ip} | \boldsymbol{\alpha}\right) = & p\left(M_{ip} | M_{i1} \ldots M_{ip-1}, \alpha_p\right) \\ & \times p\left(M_{ip-1} | M_{i1} \ldots M_{ip-2}, \alpha_{p-1}\right) \ldots\ldots \\ & \times p\left(M_{i2} | M_{i1}, \alpha_2\right) p\left(M_{i1} | \alpha_1\right) \end{aligned} \tag{4}$$

where $\boldsymbol{\alpha} = \left(\alpha_1, \alpha_2, \ldots \alpha_p\right)^T$. In specification of one dimensional conditional distributions in Eq. 4, suppose we know that $\boldsymbol{X} = \left(X_{i1}, X_{i2}, \ldots, X_{ip}\right)$ contains all continuous predictors, then a sequence of univariate normal distributions can be assigned to $p\left(M_{ip} | M_{i1} \ldots M_{ip-1}, \alpha_p\right)$. If $\boldsymbol{X}$ contains all categorical variables, say binary, then the guideline is to specify a sequence of logistic regressions for each $p\left(M_{ip} | M_{i1} \ldots M_{ip-1}, \alpha_p\right)$. One can consider the sequence of multinomial distributions in the case where all variables in $\boldsymbol{X}$ have more than two levels. Similarly, for all count variables, one can assign Poisson distributions. With missing categorical and continuous covariates, it is recommended that the joint covariate distribution should be assigned by first specifying the distribution of the categorical covariates conditional on continuous covariates. In some special circumstances, when X_{ip} is strictly positive and continuous, a normal distribution can be specified on the transformed variable $log(X_{ip})$. If $log(X_{ip})$ is not approximately normal then other specifications such as an exponential or gamma distribution are recommended.

5.2 Simulation

A simulation study was conducted to determine the bias and mean squared error (MSE) of estimating $\boldsymbol{\beta}$ with various proportions of missing covariates under a MAR assumption. The data augmentation procedure for estimating $\boldsymbol{\beta}$ and MSEs was compared to a complete case analysis. In the simulation, at each iteration, three covariates (X_1, X_2, X_3), in which X_1 is binary and X_2, X_3 are continuous, were generated. X_1 was generated from the binomial distribution with a success probability of 0.3, X_2 was simulated from a normal distribution with mean $= 0$ variance $= 1$ and X_3 was generated from a normal distribution with mean $= 1$ and variance $= 2$. Furthermore, the response variable Y was generated from $\mathcal{N}\left(\beta_0 + \beta_1 x_1 + \beta_2 x_2 + \beta_3 x_3, \sigma^2 = 1\right)$ with $\beta_0 = 0$ and $\beta_1 = \beta_2 = \beta_3 = 1$. The sample size for each iteration was 100.

Various proportions of missing values were created in X_1 and X_2, which depended on non-missing X_3 in order to simulate the MAR missing mechanism. To model the missing probability for the variable X_1; $Pr\ (X_{i1} = \text{missing}|X_{i3})$, we consider the logistic model as follows;

$$Pr\ (X_{i1} = \text{missing}|X_{i3}) = \frac{exp\ (\gamma + X_{i3})}{1 + exp\ (\gamma + X_{i3})} = p_i.$$

Based on this probability, we generated the binary variable indicating whether X_{i1} is missing: $I_i \sim \text{Bernoulli}\ (1, p_i)$. If $I_i = 1$ then we deleted the corresponding value of X_{i1}. This process ensures that missingness in X_1 depends only on the X_3 which is fully observed. Similarly, we generated missing values in X_2, which depend on the fully observed covariate X_3. In addition to this, we have a fully observed response variable Y. Inferences about $\boldsymbol{\beta} = (\beta_0, \beta_1, \beta_2, \beta_3)$ from the complete data posterior $p\left(\boldsymbol{\beta}, \sigma^2, M_{i1}, M_{i2}|Y, \boldsymbol{X}\right) \propto p\left(\boldsymbol{\beta}, \sigma^2, M_{i1}, M_{i2}\right) L\left(Y_i|\boldsymbol{\beta}, \sigma^2, M_{i1}, M_{i2}, \boldsymbol{X}\right)$ can be obtained by using the random walk Metropolis algorithm (Metropolis and Ulam 1949; Metropolis et al. 1953; Hastings 1970). The joint missing covariate distribution, $p\ (M_{i1}, M_{i2}|\alpha_1, \alpha_2, \alpha_3, \alpha_4)$, is modeled as a sequence of one-dimensional distributions $p\ (M_{1i}|M_{2i}, \alpha_1, \alpha_2) \sim \text{Bernoulli}\ (1, C_i)$ where $\log\left(\frac{C_i}{1-C_i}\right) = \alpha_1 + \alpha_2 M_{i1}$ and $p\ (M_{2i}|\alpha_3, \alpha_4) \sim \mathcal{N}\ (\alpha_3, \alpha_4)$. In this simulation study, we placed the following prior distributions on the four regression parameters and hyper-parameters of the missing covariate distributions:

$$\begin{gathered} p\ (\beta_0)\ , p\ (\beta_1)\ , p\ (\beta_2)\ , p\ (\beta_3) \sim \mathcal{N}\ (0, var = 10) \\ p\ (\alpha_1)\ , p\ (\alpha_2)\ , p\ (\alpha_3) \sim \mathcal{N}\ (0, var = 10) \\ p\ (\alpha_4)\ , p\left(\sigma^2\right) \sim \text{gamma}\ (2, 2) \end{gathered}$$

Bayesian linear regression was fit by using the above described complete data likelihood, missing data model assumptions and prior distributions to estimate $\boldsymbol{\beta}$ and variances. Bayesian linear regression was also implemented for the complete cases to estimate $\boldsymbol{\beta}$ and variances. This process was repeated 500 times to evaluate the performance of the estimators in terms of bias and MSE for the regression parameters in order to compare complete case analysis to data augmentation. Tables 3 and 4 demonstrate the results from our simulation study comparing the bias and MSEs. Table 3 shows that complete case analysis has negligible biases for various combinations of the proportions of the missing data on X_1 and X_2, while with data augmentation the regression coefficients have slight biases away from the null when missingness in covariates is independent of the outcome given the covariates. However, the MSEs from Table 4, for both complete-case analysis and data augmentation in estimating the regression coefficients for missing covariates ($\beta_1 \& \beta_2$) are very similar and the MSE for the non-missing covariate, β_3, are significantly smaller for DA compared to complete-cases under various proportions of missing values, as shown in Table 4. Moreover, when the the missingness in covariates depends on the outcome, then complete-case estimates are biased towards the null, while data augmentation had negligible bias (Results are not shown here). For detailed discussion

Table 4 MSEs comparison of regression coefficients for complete-cases and data augmentation

Missing %		Complete cases			Data augmentation		
x_1	x_2	β_1	β_2	β_3	β_1	β_2	β_3
5	5	0.0484	0.0110	0.0044	0.0464	0.0109	0.0038
5	10	0.0610	0.0122	0.0053	0.0564	0.0123	0.0045
5	15	0.0627	0.0132	0.0078	0.0579	0.0146	0.0062
5	20	0.0730	0.0150	0.0092	0.0621	0.0170	0.0070
10	5	0.0527	0.0134	0.0060	0.0489	0.0136	0.0044
10	10	0.0610	0.0148	0.0078	0.0599	0.0150	0.0057
10	15	0.0642	0.0142	0.0089	0.0607	0.0137	0.0067
10	20	0.0689	0.0138	0.0119	0.0579	0.0155	0.0085
15	5	0.0557	0.0139	0.0071	0.0530	0.0136	0.0056
15	10	0.0659	0.0158	0.0094	0.0601	0.0160	0.0072
15	15	0.0663	0.0161	0.0110	0.0656	0.0157	0.0072
15	20	0.0719	0.0173	0.0151	0.0646	0.0187	0.0084
20	5	0.0644	0.0136	0.0083	0.0648	0.0124	0.0056
20	10	0.0772	0.0174	0.0133	0.0747	0.0155	0.0074
20	15	0.0764	0.0159	0.0162	0.0680	0.0156	0.0085
20	20	0.0915	0.0179	0.0198	0.0815	0.0188	0.0089

about the bias and efficiency of multiple imputation compared with complete-case analysis for missing covariate values, see (White and Carlin 2010; Chen et al. 2008).

5.3 BRFSS Data

We applied the FB approach to the Behavioral Risk Factor Surveillance System (BRFSS) data. BRFSS is the world's largest ongoing random-digit dialed (RDD) telephone survey, and is conducted by health departments in the United States. BRFSS collects state level data about US residents, inquiring about their health-related risk behaviors and events, chronic health conditions and use of preventive services brf (2015). The survey is conducted each year, having begun in 1984. For illustration purposes, we used the Georgia 2013 BRFSS data for the month of January. This dataset has 336 variables and 518 observations. To demonstrate the fully Bayesian approach for missing covariates, we selected the variable, Reported weight in pounds, as our response variable. Reported height in inches and participated in any physical activity or exercises in the past month (1 = Yes, 0 = No) were considered as covariates in our analysis. Because of non-response, some of the subjects are missing values for the physical activity (PA) variable. There were a total of 34 subjects who did not respond to the question of participating in physical activity in the past month. To fit the fully Bayes model, we used the random-walk Metropolis-Hastings algorithm for

Table 5 Regression coefficients and SEs for BRFSS data

Coefficients	Bayesian approach	Complete-cases
β_{PA}	–7.6668	–7.8990
	(2.5760)	(2.5062)
β_{ht}	3.4302	3.3865
	(0.1346)	(0.1379)

posterior sampling and use posterior means to estimate the regression coefficients for the covariates, reported height and participated in any physical activity in past months. The results are reported in Table 5.

6 Discussion

In this chapter, we have illustrated how a fully Bayesian regression modeling approach can be applied to incomplete data under an ignorability assumption. It is well known that the observed data are not sufficient to identify the underlying missing mechanism. Therefore, sensitivity analyses should be performed over various plausible models for the nonresponse mechanism (Little and Rubin 2014). In general, the stability of conclusions (inferences) over the plausible models gives an indication of their robustness to unverifiable assumptions about the mechanism underlying missingness. In linear regression, when missingness on Y depends on the fully observed Xs (MAR), DA has negligible bias and smaller MSEs compared to the complete-cases. When missingness in Xs depends on other Xs which are fully observed then the CC analysis has negligible bias and very similar MSEs compared to the DA for missing covariates. Furthermore, when missingness in covariates depends on the response, DA will perform better than CC. Because of these biases, the choice of the method should come from a substantive basis. In summary, a FB modeling approach enables coherent model estimation because missing data values are treated as parameters, which are easily sampled within MCMC simulations. The FB approach takes into account uncertainty about missing observations and offers a very flexible way to handle the missing data. In conclusion we remark that for missing covariates we have used a default class of priors to make inferences. However, for some studies, historical data may be available allowing for construction of informative priors that may further improve inference. For more details, Ibrahim et al. (2002) proposed a class of informative priors for generalized linear models with missing covariates.

References

Behavioral risk factor surveillance system. Retrieved July 5, 2015, from http://www.cdc.gov/brfss.

Chen, Q., Ibrahim, J. G., Chen, M. -H., & Senchaudhuri, P. (2008). Theory and inference for regression models with missing responses and covariates. *Journal of multivariate analysis, 99*(6), 1302–1331.

Daniels, M. J., & Hogan, J. W. (2008). *Missing data in longitudinal studies: Strategies for Bayesian modeling and sensitivity analysis*. CRC Press.

Etzioni, R., Pepe, M., Longton, G., Chengcheng, H., & Goodman, Gary. (1999). Incorporating the time dimension in receiver operating characteristic curves: A case study of prostate cancer. *Medical Decision Making, 19*(3), 242–251.

Geman, S., & Geman, D. (1984). Stochastic relaxation, gibbs distributions, and the bayesian restoration of images. *IEEE Transactions on Pattern Analysis and Machine Intelligence, 6*, 721–741.

Hastings, K. W. (1970). Monte-Carlo sampling methods using markov chains and their applications. *Biometrika, 57*(1), 97–109.

Ibrahim, J. G., Chen, M. -H., Lipsitz, & S. R., (1999a). Monte-Carlo em for missing covariates in parametric regression models. *Biometrics, 55*(2), 591–596.

Ibrahim, J. G., Lipsitz, S. R., & Chen, M. -H. 1999b. Missing covariates in generalized linear models when the missing data mechanism is non-ignorable. *Journal of the Royal Statistical Society: Series B (Statistical Methodology), 61*(1), 173–190.

Ibrahim, J. G., Chen, M. -H., & Lipsitz, S. R. (2002). Bayesian methods for generalized linear models with covariates missing at random. *Canadian Journal of Statistics, 30*(1), 55–78.

Lipsitz, S. R., & Ibrahim, J. G. (1996). A conditional model for incomplete covariates in parametric regression models. *Biometrika, 83*(4), 916–922.

Little, R. J. A., & Rubin, D. B. (2014). *Statistical analysis with missing data*. Wiley.

Pepe, M. S. (2003). *The statistical evaluation of medical tests for classification and prediction*. USA: Oxford University Press.

Metropolis, N., & Ulam, S. (1949). The Monte-Carlo method. *Journal of the American statistical association, 44*(247), 335–341.

Rosenbluth, M. N., Teller, A. H., & Teller, E. (1953). Equation of state calculations by fast computing machines. *The journal of Chemical Physics, 21*(6), 1087–1092.

Rubin, D. B. (1976). Inference and missing data. *Biometrika, 63*(3), 581–592.

Satten, G. A., & Carroll, R. J. (2000). Conditional and unconditional categorical regression models with missing covariates. *Biometrics, 56*(2), 384–388.

Schafer, J. L. (1997). *Analysis of incomplete multivariate data*. CRC press.

Tanner, M. A., & Wong, W. H. (1987). The calculation of posterior distributions by data augmentation. *Journal of the American statistical Association, 82*(398), 528–540.

White, I. R., & Carlin, J. B. (2010). Bias and efficiency of multiple imputation compared with complete-case analysis for missing covariate values. *Statistics in Medicine, 29*(28), 2920–2931.

Xie, F., & Paik, M. C. (1997). Multiple imputation methods for the missing covariates in generalized estimating equation. *Biometrics*, 1538–1546.

A Multiple Imputation Framework for Massive Multivariate Data of Different Variable Types: A Monte-Carlo Technique

Hakan Demirtas

Abstract The purpose of this chapter is to build theoretical, algorithmic, and implementation-based components of a unified, general-purpose multiple imputation framework for intensive multivariate data sets that are collected via increasingly popular real-time data capture methods. Such data typically include all major types of variables that are incomplete due to planned missingness designs, which have been developed to reduce respondent burden and lower the cost associated with data collection. The imputation approach presented herein complements the methods available for incomplete data analysis via richer and more flexible modeling procedures, and can easily generalize to a variety of research areas that involve internet studies and processes that are designed to collect continuous streams of real-time data. Planned missingness designs are highly useful and will likely increase in popularity in the future. For this reason, the proposed multiple imputation framework represents an important and refined addition to the existing methods, and has potential to advance scientific knowledge and research in a meaningful way. Capability of accommodating many incomplete variables of different distributional nature, types, and dependence structures could be a contributing factor for better comprehending the operational characteristics of today's massive data trends. It offers promising potential for building enhanced statistical computing infrastructure for education and research in the sense of providing principled, useful, general, and flexible set of computational tools for handling incomplete data.

1 Introduction

Missing data are a commonly occurring phenomenon in many contexts. Determining a suitable analytical approach in the presence of incomplete observations is a major focus of scientific inquiry due to the additional complexity that arises through missing data. Incompleteness generally complicates the statistical analysis in terms of biased

H. Demirtas (✉)
Division of Epidemiology and Biostatistics (MC923), University of Illinois at Chicago, 1603 West Taylor Street, Chicago, IL 60612, USA
e-mail: demirtas@uic.edu

D.-G. Chen and J.D. Chen (eds.), *Monte-Carlo Simulation-Based Statistical Modeling*, ICSA Book Series in Statistics, DOI 10.1007/978-981-10-3307-0_8

parameter estimates, reduced statistical power, and degraded confidence intervals, and thereby may lead to false inferences (Little and Rubin 2002). Advances in computational statistics have produced flexible missing-data procedures with a sound statistical basis. One of these procedures involves multiple imputation (MI), which is a stochastic simulation technique in which the missing values are replaced by $m > 1$ simulated versions (Rubin 2004). Subsequently, each of the simulated complete data sets is analyzed by standard methods, and the results are combined into a single inferential statement that formally incorporates missing-data uncertainty to the modeling process. MI has gained widespread acceptance and popularity in the last few decades. It has some well-documented advantages: First, MI allows researchers to use more conventional models and software; an imputed data set may be analyzed by literally any method that would be suitable if the data were complete. As computing environments and statistical models grow increasingly sophisticated, the value of using familiar methods and software becomes important. Second, there are still many classes of problems for which no direct maximum likelihood procedure is available. Even when such a procedure exists, MI can be more attractive due to fact that the separation of the imputation phase from the analysis phase lends greater flexibility to the entire process. Lastly, MI singles out missing data as a source of random variation distinct from ordinary sampling variability.

These days most incomplete data sets involve variables of many different types on a structural level; causal and correlational interdependencies are a function of a mixture of binary, ordinal, count, and continuous variables as well as nonresponse rates and mechanisms, all of which act simultaneously to characterize the data-analytics paradigms under consideration. The use of digital data is growing rapidly as researchers get more capable of collecting instantaneous self-reported data using mobile devices in naturalistic settings; and real-time data capture methods supply novel insights into determinants of how real-life processes are formed. Mobile phones are becoming ubiquitous and easy to use, and thus have the capacity to collect data quickly from large numbers of people and transfer this information to remote servers in an unobtrusive way. As these procedures yield relatively large number observations per subject, planned missing data designs have been developed to reduce respondent burden and lower the cost associated with data collection. Planned missing data designs include items or groups of items according to predetermined probabilistic sampling schemes. These designs are highly useful and will likely increase in popularity in the future.

The purpose of this work is to build theoretical, algorithmic, and implementation-based components of a unified, general-purpose multiple imputation framework, which can be instrumental in developing power analysis guidelines for intensive multivariate data sets that are collected via increasingly popular real-time data capture (RTDC hereafter) approaches. Such data typically include all major types of variables (binary, ordinal, count, and continuous) that are incomplete due to planned missingness designs. Existing MI methodologies are restrictive for RTDC data, because they hinge upon strict and unrealistic assumptions (multivariate normal model for continuous data, multinomial model for discrete data, general location model for a mix of normal and categorical data), commonly known as joint MI models (Schafer 1997). Some methods can handle all data types with relaxed assumptions, but lack theoretical justification (Van Buuren 2012; Raghunathan et al. 2001). For massive data as collected in RTDC studies, a novel imputation framework that can accommodate all four major types of variables is needed with a minimal set of assumptions. In addition, no statistical power (probability of correctly detecting an effect) and sample size (number of subjects, measurements, waves) procedures are available for RTDC data. The lack of these tools severely limits our ability to capitalize on the full potential of incomplete intensive data, and a unified framework for simultaneously imputing all types of variables is necessary to adequately capture a broad set of substantive messages that massive data are designed to convey. The proposed MI framework represents an important and refined addition to the existing methods, and has potential to advance scientific knowledge and research in a meaningful way; it offers promising potential for building enhanced statistical computing infrastructure for education and research in the sense of providing principled, useful, general, and flexible set of computational tools for handling incomplete data.

Combining our previous random number generation (RNG) work for multivariate ordinal data (Demirtas 2006), joint binary and normal data (Demirtas and Doganay 2012), ordinal and normal data (Demirtas and Yavuz 2015), count and normal data (Amatya and Demirtas 2015a), binary and nonnormal continuous data (Demirtas et al. 2012; Demirtas 2014, 2016) with the specification of marginal and associational parameters; our published work on MI (Demirtas 2004, 2005, 2007, 2008, 2009, 2010, 2017a; Demirtas et al. 2007, 2008; Demirtas and Hedeker 2007, 2008a, b, c; Demirtas and Schafer 2003; Yucel and Demirtas 2010), along with some related work (Demirtas et al. 2016a; Demirtas and Hedeker 2011, 2016; Emrich and Piedmonte 1991; Fleishman 1978; Ferrari and Barbiero 2012; Headrick 2010; Yahav and Shmueli 2012), a broad mixed data imputation framework that spans all possible combinations of binary, ordinal, count, and continuous variables, is proposed. This system is capable of handling the overwhelming majority of continuous shapes; it can be extended to control for higher order moments for continuous variables, and to allow over- and under-dispersion for count variables as well as the specification of Spearman's rank correlations as the measure of association. Procedural, conceptual, operational, and algorithmic details of the published, current, and future work will be given throughout the chapter.

The organization of the chapter is as follows: In Sect. 2, background information is provided on the generation of multivariate binary, ordinal, and count data with an emphasis on underlying multivariate normal data that form a basis for the subsequent discretization in the binary and ordinal cases, and correlation mapping using inverse cumulative distribution functions (cdfs) in the count data case. Then, different correlation types that are relevant to the work are described; a linear relationship between correlations before and after discretization is discussed; and multivariate power polynomials in the context of generating continuous data are elaborated on. In Sect. 3, operational characteristics of MI under normality assumption are articulated. In Sect. 4, an MI algorithm for multivariate data with all four major variable types is outlined under ignorable nonresponse by merging the available RNG and MI routines. Finally, in Sect. 5, concluding remarks, future research directions, and extensions are given.

2 Background on RNG

The proposed MI algorithm has strong impetus and computational roots derived from some ideas that appeared in the RNG literature. This section provides salient features of multivariate normal (MVN), multivariate count (MVC), multivariate binary (MVB), and multivariate ordinal (MVO) data generation. Relevant correlation structures involving these different types of variables are discussed; a connection between correlations before and after discretization is established; and the use of power polynomials, which will be employed at later stages to accommodate nonnormal continuous data, is explained.

MVN Data Generation: Sampling from the MVN distribution is straightforward. Suppose $Z \sim N_d(\mu, \Sigma)$, where μ is the mean vector, and Σ is the symmetric, positive definite, $d \times d$ variance-covariance matrix. A random draw from a MVN distribution can be obtained using the Cholesky decomposition of Σ and a vector of univariate normal draws. The Cholesky decomposition of Σ produces a lower-triangular matrix A for which $AA^T = \Sigma$. If $z = (z_1, ..., z_d)$ are d independent standard normal random variables, then $Z = \mu + Az$ is a random draw from the MVN distribution with mean vector μ and covariance matrix Σ.

MVC Data Generation: Count data have been traditionally modeled by the Poisson distribution. Although a few multivariate Poisson (MVP) data generation techniques have been published, the method in Yahav and Shmueli (2012) is the only one that reasonably works (allowing negative correlations) when the number of components is greater than two. Their method utilizes a slightly modified version of the NORTA (Normal to Anything) approach (Nelsen 2006), which involves generation of MVN variates with given univariate marginals and the correlation structure (R_N), and then transforming it into any desired distribution using the inverse cdf. In the Poisson case, NORTA can be implemented by the following steps:

1. Generate a k-dimensional normal vector $\boldsymbol{Z}_N$ from MVN distribution with mean vector $\boldsymbol{0}$ and a correlation matrix R_N.
2. Transform $\boldsymbol{Z}_N$ to a Poisson vector $\boldsymbol{X}_{POIS}$ as follows:
   (a) For each element z_i of $\boldsymbol{Z}_N$, calculate the Normal cdf, $\Phi(z_i)$.
   (b) For each value of $\Phi(z_i)$, calculate the Poisson inverse cdf with a desired corresponding marginal rate θ_i, $\Psi_{\theta_i}^{-1}(\Phi(z_i))$; where $\Psi_{\theta_i}(x) = \sum_{i=0}^{x} \frac{e^{-\theta}\theta^i}{i!}$.
3. $\boldsymbol{X}_{POIS} = \left[\Psi_{\theta_i}^{-1}(\Phi(z_i)), \ldots, \Psi_{\theta_k}^{-1}(\Phi(z_k))\right]^T$ is a draw from the desired MVP distribution with correlation matrix R_{POIS}.

An exact theoretical connection between R_{POIS} and R_N has not been established to date. However, it has been shown that a feasible range of correlation between a pair of Poisson variables after the inverse cdf transformation is within $[\underline{\rho} = Cor(\Psi_{\theta_i}^{-1}(U), \Psi_{\theta_j}^{-1}(1-U)), \overline{\rho} = Cor(\Psi_{\theta_i}^{-1}(U), \Psi_{\theta_j}^{-1}(U))]$, where θ_i and θ_j are the marginal rates, and $U \sim Uniform(0, 1)$. Yahav and Shmueli (2012) proposed a conceptually simple method to approximate the relationship between the two correlations. They have demonstrated that R_{POIS} can be approximated as an exponential function of R_N where the coefficients are the functions of $\underline{\rho}$ and $\overline{\rho}$. Once all the elements of R_N (that would correspond to R_{POIS} after correlation mapping) are approximated, the NORTA method can be used to generate draws from MVP distribution.

MVB Data Generation: Out of several correlated binary data simulation routines that have appeared in the literature (see Qaqish 2003 and references therein), the one that fits into our framework was proposed by Emrich and Piedmonte (1991) who introduced a method for generating correlated binary data. Let $X_1, ..., X_j$ represent binary variables such that $E[X_j] = p_j$ and $Cor(X_j, X_k) = \delta_{jk}$, where p_j $(j = 1, ..., J)$ and δ_{jk} $(j = 1, ..., J-1; k = 2, ..., J)$ are given, and where $J \geq 2$. Let $\Phi[t_1, t_2, \rho]$ be the cdf for a standard bivariate normal random variable with correlation coefficient ρ (tetrachoric correlation). Naturally, $\Phi[t_1, t_2, \rho] = \int_{-\infty}^{t_1}\int_{-\infty}^{t_2} f(z_1, z_2, \rho)dz_1dz_2$, where $f(z_1, z_2, \rho) = [2\pi(1-\rho^2)^{1/2}]^{-1} \times exp \Big[-(z_1^2 - 2\rho z_1 z_2 + z_2^2)/(2(1-\rho^2))\Big]$. We can generate multivariate normal outcomes (Z's) whose correlation parameters are obtained by solving the equation

$$\Phi[z(p_j), z(p_k), \rho_{jk}] = \delta_{jk}(p_j q_j p_k q_k)^{1/2} + p_j p_k, \tag{1}$$

for ρ_{jk} $(j = 1, ..., J-1; k = 2, ..., J)$ where $z(p)$ denotes the pth quantile of the standard normal distribution, and $q = 1 - p$. As long as δ_{jk} (phi coefficient) is within the feasible range, the solution is unique. Repeating this numerical integration process $J(J-1)/2$ times, one can obtain the overall correlation matrix (say Σ) for the d-variate standard normal distribution with mean 0. To create dichotomous outcomes (X_j) from the generated normal outcomes (Z_j), we set $X_j = 1$ if $Z_j \geq z(1 - p_j)$ and 0 otherwise for $j = 1, ..., J$. This produces a vector with the desired properties.

MVO Data Generation: A few multivariate ordinal data simulation routines have been published. The method proposed by Demirtas (2006) relies on simulating correlated binary variates as an intermediate step. After collapsing the specified ordinal levels to the binary ones, corresponding binary correlations are computed via simulation in a way to ensure that re-conversion to the ordinal scale delivers the specified distributional properties. In a similar operational logic to Demirtas (2006), Ferrari and Barbiero (2012) introduced a more direct routine in which an iterative computational procedure is implemented in an attempt to find the relationship between specified ordinal correlations and the correlations of the underlying normal variates that are assumed to be ordinalized through thresholds determined by the marginal ordinal proportions. Of note, unlike binary data, the correlations and odds ratios do not uniquely determine each other in the ordinal case, so marginal distributions and correlation structure constitute a partial specification (i.e., the first two moments do not fully specify MVO data), which is adequate in most applications. In what follows, the MVO data part of our algorithm will be based on Ferrari and Barbiero (2012), as it best conforms to our conceptual framework.

Different Correlation Types: A correlation between two continuous variables is usually computed as the common Pearson correlation. If one or both variables is/are dichotomized/ordinalized by a threshold concept of underlying continuous variables, different naming conventions are assigned to the correlations. A correlation between a continuous and a dichotomized/ordinalized variable is a biserial/polyserial and point-biserial/point-polyserial correlation before and after discretization, respectively. When both variables are dichotomized/ordinalized, the correlation between the two latent continuous variables is known as the tetrachoric/polychoric correlation. The phi coefficient is the correlation between two discretized variables; in fact the term phi coefficient is reserved for dichotomous variables, but for lack of a better term we also use it for ordinalized and count variables. In addition, we employ the same terminology for ordinal and count variables in what follows. All of these correlations are special cases of the Pearson correlation.

Relationships of Correlations in Discrete-Continuous and Continuous-Continuous Cases: Suppose that X and Y follow a bivariate normal distribution with a correlation of δ_{XY}. Without loss of generality, we may assume that both X and Y are standardized to have a mean of 0 and a variance of 1. Let X_D be the binary variable resulting from a split on X, $X_D = I(X \geq k)$, where k is the point of dichotomization. Thus, $E[X_D] = p$ and $V[X_D] = pq$ where $q = 1 - p$. The correlation between X_D and X, $\delta_{X_D X}$ can be obtained in a simple way, namely, $\delta_{X_D X} = \frac{Cov[X_D, X]}{\sqrt{V[X_D]V[X]}} = E[X_D X]/\sqrt{pq} = E[X|X \geq k]/\sqrt{pq}$. We can also express the relationship between X and Y via the following linear regression model:

$$Y = \delta_{XY} X + \epsilon \tag{2}$$

where ϵ is independent of X and Y, and follows $N \sim (0, 1 - \delta_{XY}^2)$. When we generalize this to nonnormal X and/or Y (both centered and scaled), the same relationship can be assumed to hold with the exception that the distribution of ϵ follows a nonnormal distribution. As long as Eq. 2 is valid,

$$\begin{aligned} Cov[X_D, Y] &= Cov[X_D, \delta_{XY} X + \epsilon] \\ &= Cov[X_D, \delta_{XY} X] + Cov[X_D, \epsilon] \\ &= \delta_{XY} Cov[X_D, X] + Cov[X_D, \epsilon] . \end{aligned} \tag{3}$$

Since ϵ is independent of X, it will also be independent of any deterministic function of X such as X_D, and thus $Cov[X_D, \epsilon]$ will be 0. As $E[X] = E[Y] = 0$, $V[X] = V[Y] = 1$, $Cov[X_D, Y] = \delta_{X_D Y}\sqrt{pq}$ and $Cov[X, Y] = \delta_{XY}$, Eq. 3 reduces to

$$\delta_{X_D Y} = \delta_{XY}\delta_{X_D X} \,. \tag{4}$$

In the bivariate normal case, $\delta_{X_D X} = h/\sqrt{pq}$ where h is the ordinate of the normal curve at the point of dichotomization. Equation 4 indicates that the linear association between X_D and Y is assumed to be fully explained by their mutual association with X (Demirtas and Hedeker 2016). The ratio, $\delta_{X_D Y}/\delta_{XY}$ is equal to $\delta_{X_D X} = E[X_D X]/\sqrt{pq} = E[X|X \geq k]/\sqrt{pq}$. It is a constant given p and the distribution of (X, Y). These correlations are invariant to location shifts and scaling, X and Y do not have to be centered and scaled, their means and variances can take any finite values. Once the ratio $(\delta_{X_D X})$ is found, one can compute the point-biserial or biserial correlation when the other one is specified.

When X is ordinalized to obtain X_O, the fundamental ideas remain unchanged. As long as the assumptions of Eqs. 2 and 4 are met, the method is equally applicable to the ordinal case in the context of the relationship between the polyserial (before ordinalization) and point-polyserial (after ordinalization) correlations. The easiest way of computing $\delta_{X_O X}$ is to generate X with a large number of data points, then ordinalize it to obtain X_O, and then compute the correlation between X_O and X. X could follow any continuous univariate distribution. However, for the purpose of the more general algorithm presented in Sect. 4, X is assumed to be a part of a MVN distribution before discretization. Similarly, in the Poisson case, $\delta_{X_{POIS} Y} = \delta_{XY}\delta_{X_{POIS} X}$ is valid. The only difference is that we use the inverse cdf method rather than discretization via thresholds as in the binary and ordinal cases.

Power Polynomials: Fleishman (1978) presented a moment-matching procedure that simulates nonnormal distributions often used in Monte-Carlo studies. It is based on the polynomial transformation, $Y = a + bZ + cZ^2 + dZ^3$, where Z follows a standard normal distribution, and Y is standardized (zero mean and unit variance). The distribution of Y depends on the constants a, b, c, and d, that can be computed for specified or estimated values of skewness ($\nu_1 = E[Y^3]$) and kurtosis ($\nu_2 = E[Y^4] - 3$). This procedure of expressing any given variable by the sum of linear combinations of powers of a standard normal variate is capable of covering a wide area in the skewness-elongation plane whose bounds are given by the general expression $\nu_2 \geq \nu_1^2 - 2$.

Assuming that $E[Y] = 0$, and $E[Y^2] = 1$, by using the moments of the standard normal distribution, the following set of equations can be derived:

$$a = -c \tag{5}$$

$$b^2 + 6bd + 2c^2 + 15d^2 - 1 = 0 \tag{6}$$

$$2c(b^2 + 24bd + 105d^2 + 2) - \nu_1 = 0 \tag{7}$$

$$24[bd + c^2(1 + b^2 + 28bd) + d^2(12 + 48bd + 141c^2 + 225d^2)] - \nu_2 = 0 \tag{8}$$

These equations can be solved by the Newton-Raphson method, or any other plausible root-finding or nonlinear optimization routine. More details for the Newton-Raphson algorithm in this particular setting is given by Demirtas et al. (2012), and a computer implementation can be found in Demirtas and Hedeker (2008a). Note that the parameters are estimated under the assumption that the mean is 0, and the standard deviation is 1; the resulting data set should be back-transformed to the original scale by reverse centering and scaling. It is well-known that linear transformations such as centering and scaling do not change the correlation value. The standardization does not affect the skewness and kurtosis values either, and hence it is merely a computational convenience for our purposes. Since $a = -c$, it reduces to solving the following equations:

$$g = \begin{bmatrix} g_1 = b^2 + 6bd + 2c^2 + 15d^2 - 1 \\ g_2 = 2c(b^2 + 24bd + 105d^2 + 2) - \nu_1 \\ g_3 = 24[bd + c^2(1 + b^2 + 28bd) + d^2(12 + 48bd + 141c^2 + 225d^2)] - \nu_2 \end{bmatrix} = \begin{bmatrix} 0 \\ 0 \\ 0 \end{bmatrix} .$$

The first derivative matrix is given by

$$H = \begin{bmatrix} g_1'(b) & g_1'(c) & g_1'(d) \\ g_2'(b) & g_2'(c) & g_2'(d) \\ g_3'(b) & g_3'(c) & g_3'(d) \end{bmatrix} ,$$

where $g_1'(b) = 2b + 6d$, $g_1'(c) = 4c$, $g_1'(d) = 6b + 30d$
$g_2'(b) = 2c(2b + 24d)$, $g_2'(c) = 2(b^2 + 24bd + 105d^2 + 2)$, $g_2'(d) = 2c(24b + 210d)$
$g_3'(b) = 24(d + 2bc^2 + 28c^2d + 48d^3)$, $g_3'(c) = 24(2c + 2b^2c + 56bcd + 282cd^2)$
$g_3'(d) = 24(b + 28bc^2 + 24d + 144bd^2 + 282c^2d + 900d^3)$.

Updating equations in Newton-Raphson are

$$\begin{bmatrix} b^{(t+1)} \\ c^{(t+1)} \\ d^{(t+1)} \end{bmatrix} = \begin{bmatrix} b^{(t)} \\ c^{(t)} \\ d^{(t)} \end{bmatrix} - H^{-1}g .$$

Fleishman's method has been extended in several ways in the literature. One such extension is a multivariate version proposed by Vale and Maurelli (1983), which plays a central role in this chapter. The procedure for generating multivariate continuous data begins with computation of the constants given in Eqs. 5–8, for each variable independently. The multivariate case can be formulated in matrix notation as shown below. First, let Z_1 and Z_2 be variables drawn from standard normal populations; let $\mathbf{w}'$ be the weight vector that contains the power function weights a, b, c, and d, $\mathbf{w}' = [a, b, c, d]$; and let $\mathbf{z}'$ be the vector of powers zero through three, $\mathbf{z}' = [1, Z, Z^2, Z^3]$. The nonnormal variable Y is then defined as the product of these two vectors, $Y = \mathbf{w}'\mathbf{z}$. Let $\delta_{Y_1Y_2}$ be the correlation between two nonnormal variables Y_1 and Y_2 corresponding to the normal variables Z_1 and Z_2, respectively. As the variables are standardized, meaning $E(Y_1) = E(Y_2) = 0$,

$\delta_{Y_1Y_2} = E(Y_1Y_2) = E(\mathbf{w}_1'\mathbf{z}_1\mathbf{z}_2'\mathbf{w}_2) = \mathbf{w}_1'\mathcal{R}\mathbf{w}_2$, where $\mathcal{R}$ is the expected matrix product of $\mathbf{z}_1$ and $\mathbf{z}_2'$:

$$\mathcal{R} = E(\mathbf{z}_1\mathbf{z}_2') = \begin{bmatrix} 1 & 0 & 1 & 0 \\ 0 & \delta_{Z_1Z_2} & 0 & 3\delta_{Z_1Z_2} \\ 1 & 0 & 2\delta^2_{Z_1Z_2}+1 & 0 \\ 0 & 3\delta_{Z_1Z_2} & 0 & 6\delta^3_{Z_1Z_2}+9\delta_{Z_1Z_2} \end{bmatrix},$$

where $\delta_{Z_1Z_2}$ is the correlation between Z_1 and Z_2. After algebraic operations, the following relationship between $\delta_{Y_1Y_2}$ and $\delta_{Z_1Z_2}$ in terms of polynomial coefficients ensues:

$$\delta_{Y_1Y_2} = \delta_{Z_1Z_2}(b_1b_2 + 3b_1d_2 + 3d_1b_2 + 9d_1d_2) + \delta^2_{Z_1Z_2}(2c_1c_2) + \delta^3_{Z_1Z_2}(6d_1d_2) \quad (9)$$

Solving this cubic equation for $\delta_{Z_1Z_2}$ gives the intermediate correlation between the two standard normal variables that is required for the desired post-transformation correlation $\delta_{Y_1Y_2}$. Clearly, correlations for each pair of variables should be assembled into a matrix of intercorrelations that is needed in multivariate normal data generation. Subsequently, the specified nonnormal variables are simulated through the respective normal components and polynomial coefficients for each variable. For a definitive source and in-depth coverage of the Fleishman polynomials, see Headrick (2010). We now briefly discuss our published RNG work that will serve as building blocks for the proposed approach.

Joint Generation of Different Types: In what follows, in addition to the abbreviations MVN, MVB, MVO, and MVC that were introduced earlier, NCT and MVNCT denote nonnormal continuous, and multivariate nonnormal continuous, respectively. In RNG for the MVB-MVN combination (Demirtas and Doganay 2012), the underlying mechanism comes from a combination of three well-known data generation routines. It assumes that all variables in the system jointly follow a multivariate normal density originally, but some of the components are dichotomized. The two sets of correlations are naturally altered with this operation: (1) the correlations among dichotomized variables, and (2) the correlations among normal and dichotomized variables. The magnitude of the first change needs to be computed through a series of double numerical integrations (Eq. 1), and that of the second change comes from Eq. 4. Once these transitions are performed, one can form an overall correlation matrix for a multivariate normal distribution that would lead to the specified correlations after dichotomizing some of the variables via thresholds that are determined by marginal proportions. In RNG for the MVO-MVN combination (Demirtas and Yavuz 2015) and the MVC-MVN combination (Amatya and Demirtas 2015a), the same logic applies with some operational differences. The connection of correlations before and after ordinalization/inverse cdf matching for the O-O/C-C pairs can be found by an iterative procedure (Demirtas 2006; Ferrari and Barbiero 2012) and by the method in Yahav and Shmueli (2012), respectively. The link between polyserial and point-polyserial correlations is established via the ordinal/count versions of Eq. 4

for the O-N/C-N pairs. Generation of MVN data with an overall correlation matrix that has three components (O-O, O-N, and N-N pairs) is followed by ordinalization via thresholds for the discrete part. In the count case, the inverse cdf matching replaces the threshold concept. In RNG for the MVB-MVNCT combination (Demirtas et al. 2012), the NCT-NCT pairs are handled through multivariate power polynomials. B-B correlations (phi coefficients) are transformed to tetrachoric correlations via Eq. 1. The procedure for carrying out the B-NCT part is a special case of the MVO-MVC-MVNCT combination we elaborate on in Sect. 4. These methods are capable of generating data that are consistent with the specified linear association structure for all variables, proportions of the binary and ordinal variables, Poisson rates for the count variables, and mean, variance, skewness, and peakedness behavior for the continuous variables.

3 Missing Data and MI

The properties of missing-data methods vary depending on the manner in which data became missing; every missing-data technique makes implicit or explicit assumptions about the missing-data mechanism. Many missing-data procedures in use today assume that missing values are missing at random (MAR) (Rubin 1976). Under MAR, missingness is related to the observed data, but conditionally independent of the missing data. A special case of MAR is missing completely at random (MCAR). Under MCAR, nonresponse is independent of observed and unobserved data; a weaker version of the MCAR assumption allows dependence on fully observed covariates. If the response probabilities depend on unobserved data; in this case, the missing values are said to be missing not at random (MNAR). A missing-data mechanism is said to be ignorable if the missing data are MAR or MCAR, together with a minor technical condition called distinctness. MI is a Monte-Carlo technique in which the missing values are replaced by a set of simulated versions of them. These simulated values are drawn from a Bayesian posterior predictive distribution for the missing values given the observed values and the missingness indicators. Carrying out MI requires two sets of assumptions. First, one must propose a model for the data distribution which should be plausible and should bear some relationship to the type of analysis to be performed. The second set of assumptions pertains to type of missingness mechanism. An assumption of MAR is commonly employed for MI. However, the theory of MI does not necessarily require MAR; MI may also be performed under nonignorable models. For the purposes of this chapter, ignorable nonresponse is assumed. The key idea of MI is that it treats missing data as an explicit source of random variability to be averaged over. The process of creating imputations, analyzing the imputed data sets, and combining the results is a Monte-Carlo version of averaging the statistical results over the predictive distribution of the missing data. In practice, a large number of multiple imputations is not required; sufficiently accurate results can often be obtained with several imputations. Once the imputations have been created, the completed data sets may be analyzed without regard for missing data;

all relevant information on nonresponse is now carried in the imputed values. Once the quantities have been estimated, the several versions of the estimates and their standard errors are combined by simple arithmetic. MI under normality assumption is central to this chapter as it forms a basis to MI for nonnormal continuous data as well as discrete data. Its operational details are given below. With the utility of MI, we can potentially reduce the number of observations per subject in RTDC studies with planned missingness (and subsequently reduce participant burden and study costs).

Let y_{ij} denote an individual element of Y, $i = 1, 2, ..., n$, $j = 1, 2, ..., p$. The ith row of Y is $y_i = (y_{i1}, y_{i2}, ..., y_{ip})^T$. Assume that $y_1, y_2, ..., y_n$ are independent realizations of a random vector, denoted as $(Y_1, Y_2, ..., Y_p)$, which has a multivariate normal distribution with mean vector μ and covariance matrix Σ; that is $y_1, y_2, ..., y_n|\theta \sim N(\mu, \Sigma)$, where $\theta = (\mu, \Sigma)$ is the unknown parameter and Σ is positive definite. The complete-data likelihood with this setting is proportional to $|\Sigma|^{-\frac{n}{2}} exp\big\{ -\frac{1}{2}\sum_{i=1}^{n}(y_i - \mu)^T\Sigma^{-1}(y_i - \mu)\big\}$. The maximum likelihood estimators for μ and Σ are well-known: $\hat{\mu} = \bar{y} = n^{-1}\sum\limits_{i=1}^{n} y_i$ and $\hat{\Sigma} = S = n^{-1}\sum\limits_{i=1}^{n}(y_i - \bar{y})$ $(y_i - \bar{y})^T$. When imputations are created under Bayesian arguments, MI has a natural interpretation as an approximate Bayesian inference for the quantities of interest based on the observed data. MI can be performed by first running an EM-type (expectation-maximization) algorithm (Dempster et al. 1977), and then by employing a data augmentation procedure (Tanner and Wong 1987). The EM algorithm is useful for two reasons: it provides good starting values for the data augmentation scheme, and it gives us an idea about the convergence behavior. Data augmentation using the Bayesian paradigm has been perceived as a natural tool to create multiply imputed data sets. A brief description of the MI process using data augmentation is as follows. When both μ and Σ are unknown, the conjugate class for the multivariate normal data model is the normal inverted-Wishart family. When a $p \times p$ matrix X has an inverted-Wishart density $(W^{-1}(k, \Gamma))$ with degrees of freedom parameter k and inverse-scale parameter Γ, the density is proportional to $|X|^{-(\frac{k+p+1}{2})} exp\big\{ -\frac{1}{2}tr(\Gamma^{-1}X^{-1})\big\}$ for $k \geq p$. Bayesian inference for $\theta = (\mu, \Sigma)$ proceeds with the formulation of prior distributions: Suppose that $\mu|\Sigma \sim N(\mu_0, \tau^{-1}\Sigma)$, where the hyperparameters μ_0 and $\tau > 0$ are fixed and known; and $\Sigma \sim W^{-1}(k, \Gamma)$, where $p \leq k$ and $\Gamma > 0$. The prior density for θ is then $f(\theta) \propto |\Sigma|^{-(\frac{k+p+2}{2})} exp\big\{ -\frac{1}{2}tr(\Gamma^{-1}\Sigma^{-1})\big\} exp$ $\big\{ -\frac{\tau}{2}(\mu - \mu_0)^T\Sigma^{-1}(\mu - \mu_0)\big\}$, and after some algebraic manipulations the complete-data likelihood can be re-expressed as $\propto |\Sigma|^{-\frac{n}{2}} exp\big\{ -\frac{n}{2}tr(\Sigma^{-1}S)\big\} exp\big\{ -\frac{n}{2}$ $(\bar{y} - \mu)^T\Sigma^{-1}(\bar{y} - \mu)\big\}$. Multiplying the prior and likelihood, the posterior distribution $P(\theta|Y)$ has also a normal inverted-Wishart form with new values for (τ, k, μ_0, Γ). In other words, the complete-data posterior is normal inverted-Wishart: $\mu|\Sigma, Y \sim N(\mu_0^*, (\tau^*)^{-1}\Sigma)$; and $\Sigma|Y \sim W^{-1}(k^*, \Gamma^*)$, where the updated hyperparameters are $\tau^* = \tau + n$, $k^* = k + n$, $\mu_0^* = \big(\frac{n}{\tau+n}\big)\bar{y} + \big(\frac{\tau}{\tau+n}\big)\mu_0$, and $\Gamma^* = \big[\Gamma^{-1} + nS + \big(\frac{\tau n}{\tau+n}\big)(\bar{y} - \mu_0)(\bar{y} - \mu_0)^T\big]^{-1}$. When no strong prior information is available about θ, one may apply Bayes' theorem with the improper prior $f(\theta) \propto |\Sigma|^{-\left(\frac{p+1}{2}\right)}$, which is the limiting form of the normal inverted-Wishart density as $\tau \to 0, k \to -1$ and $\Gamma^{-1} \to 0$, reflecting a state of relative ignorance. Initial estimates for θ are typically obtained by the EM algorithm. Let Y_{mis} and Y_{obs} denote the missing and observed parts of data, respectively. Then, data augmentation scheme is implemented as follows: First, a value of missing data from the conditional predictive distribution of Y_{mis}, $Y_{mis}^{(t+1)} \sim P(Y_{mis}|Y_{obs}, \theta^{(t)})$, is drawn. Then, conditioning on $Y_{mis}^{(t+1)}$, a new value of θ from its complete-data posterior, $\theta^{(t+1)} \sim P(\theta|Y_{obs}, Y_{mis}^{(t+1)})$ is drawn. Repeating these two steps from a starting value $\theta^{(0)}$ yields a stochastic sequence $(\theta^{(t)}, Y_{mis}^{(t)}) : t = 1, 2, ...$ whose stationary distribution is $P(\theta, Y_{mis}|Y_{obs})$, and the subsequences $\theta^{(t)}$ and $Y_{mis}^{(t)}$ have $P(\theta|Y_{obs})$ and $P(Y_{mis}|Y_{obs})$ as their respective distributions. For a reasonably large number of iterations, the convergence to these stationary distributions is achieved. Since the complete-data likelihood is assumed to follow a multivariate normal distribution, drawing from conditional distributions above is relatively straightforward and can be performed by applying sweep operators to subsets of the vector μ and the matrix Σ.

4 Connecting RNG and MI, Outline of a Unified MI Algorithm for Mixed Data

RNG and MI are intricately related under the assumption of ignorable nonresponse (MCAR and MAR), where an explicit model for missingness does not have to be posited for inferential purposes, and there are no residual relationships between missingness and outcomes once we take an appropriate account for observed data (i.e., missing values behave like a random sample of all values within subclasses defined by observed data). Both paradigms hinge upon a procedure that has substantially similar underlying rationale under ignorability. RNG/MI involves with producing/preserving marginal and associational trends using the specified/estimated quantities. Ignorability is known to hold for planned missingness designs (Schafer 1997). While there are typically other reasons for missingness (subject unable/unwilling to respond, device not with subject etc.), such missing values are assumed to be ignorable. The possibility of nonignorable missingness cannot be ruled out, however, MI literature favors rich imputation models, and the presence of many variables in the data sets, some of which are measured repeatedly, lead us to believe that MAR assumption is not implausible in the sense that causes or correlates of missingness and/or outcomes are likely to be included in the observed data. Let O, C, N, and NCT correspond to the ordinal, count, normal, and nonnormal continuous parts, respectively. As a binary variable is a special case of an ordinal variable, combinations that are involved with ordinal data inherently include binary data as well. First, the marginal features (proportions for binary and ordinal variables, the first four moments [mean, variance, skewness and kurtosis] of the continuous variables, and rate parameters for the count variables) and correlation matrix (based on all available information) among all variables should be found using the observed data (Y_{obs}). The skeleton of the MI algorithm is as follows.

1. Compute the polychoric correlations for the O-O pairs using the method in Ferrari and Barbiero (2012), separately for each ordinal pair.
2. Compute the polychoric correlations for the C-C pairs using the method in Yahav and Shmueli (2012), separately for each count pair.
3. Store the means and standard deviations of the continuous variables (needed in Step 21), and work with the centered and scaled versions of the continuous variables. Note that correlations remain unchanged with a linear transformation. Estimate the power coefficients (a, b, c, d) for each continuous variable by Eqs. 5–8 given corresponding ν_1 and ν_2 values.
4. For each NCT-NCT pair, using the constants in Step 3, find the intermediate correlation by solving Eq. 4.
5. For each O-NCT pair, assume that there are two identical standard normal (N) variables, one is the normal component of the continuous variable and the other underlies the ordinal variable before discretization. Compute $Cor(O, N)$ by the ordinal version of Eq. 3.
6. Solve for $Cor(NCT, N)$ assuming $Cor(O, NCT) = Cor(O, N) * Cor(NCT, N)$, so that the linear association between O and NCT is assumed to be fully explained by their mutual association with N. In this equation, $Cor(O, NCT)$ is specified, and $Cor(O, N)$ is found in Step 5.
7. Compute the intermediate correlation between NCT and N by Eq. 9. Notice that for standard normal variables, $b = 1$ and $a = c = d = 0$. So the intermediate correlation is the ratio, $Cor(NCT, N)/(b + 3d)$, where b and d are the coefficients of the nonnormal continuous variable.
8. For each C-NCT pair, assume that there are two identical standard normal (N) variables, one is the normal component of the continuous variable and the other underlies the count variable before the inverse cdf matching. Compute $Cor(C, N)$ by the count version of Eq. 4.
9. Solve for $Cor(NCT, N)$ assuming $Cor(C, NCT) = Cor(C, N) * Cor(NCT, N)$, so that the linear association between C and NCT is assumed to be fully explained by their mutual association with N. In this equation, $Cor(C, NCT)$ is specified, and $Cor(C, N)$ is found in Step 8.
10. Compute the intermediate correlation between NCT and N by Eq. 9, which is $Cor(NCT, N)/(b + 3d)$, as in Step 7.
11. For each O-C pair, suppose that there are two identical standard normal variables, one underlies the ordinal variable before discretization, the other underlies the count variable before the inverse cdf matching. Find $Cor(O, N)$ by Eq. 3. Then, assume $Cor(O, C) = Cor(O, N) * Cor(C, N)$. $Cor(O, C)$ is specified and $Cor(O, N)$ is calculated. Solve for $Cor(C, N)$. Then, find the underlying N-N correlation by Eq. 4.
12. Construct an overall correlation matrix, Σ^* using the results from Steps 1–11.
13. Check if Σ^* is positive definite. If it is not, find the nearest positive definite correlation matrix by the method of Higham (2002).

14. Randomly assign a normal score to each of the binary and ordinal measurement using normal quantiles in the appropriate range dictated by the marginal proportions.
15. Randomly assign a normal score to each count measurement based on the equivalence of cdfs of Poisson and normal distribution in the appropriate range dictated by the rate parameters.
16. Compute a normal score for each continuous measurement by finding the normal root in the Fleishman equation.
17. Perform MI under MVN assumption for the data that are formed in Steps 14–16 with the correlation matrix Σ^*.
18. Obtain ordinal variables using the thresholds determined by the marginal proportions using quantiles of the normal distribution.
19. Obtain count variables by the inverse cdf matching procedure.
20. Obtain continuous variables by the sum of linear combinations of standard normals using the corresponding (a, b, c, d) coefficients.
21. Transform back to the original scale for continuous variables by reverse centering and scaling.

In other words, the marginal and associational quantities are obtained by the examination of observed data; a new data set is formed whose marginals are normal via discretization, inverse cdf matching, and extracting the normal components for binary/ordinal, count, and nonnormal continuous parts, respectively; and finally an overall MVN correlation matrix is to be assembled. Subsequently, one can create imputed data sets by taking advantage of well-accepted Bayesian MI procedures with normalized scores and this intermediate correlation matrix, as explained in Sect. 3. Then, after MI, we can go back to the original types using the threshold concept for the binary and ordinal variables, and quantile transformation for the count variables, and as for the continuous variables, normal variates should be transformed to the original nonnormal continuous data by $Y = a + bZ + cZ^2 + dZ^3$, then reverse centering and scaling completes the process. Obviously, this whole procedure needs to be repeated $m > 1$ times to adequately account for missing data uncertainty. This novel imputation model encompasses all major variable types (binary, ordinal, count, continuous) to test the predictive effects of subject-level (time-static) and within-subject (time-varying) parameters. The proposed flexible and broad imputation approach allows us to address many research questions in RTDC studies, by adequately capturing the real incomplete-data trends.

The implementation of the above algorithm can be performed by adding Bayesian MI capabilities into our existing R packages **OrdNor**, **BinNor**, **MultiOrd**, **PoisNor** (Amatya and Demirtas 2015b, 2016a, b, c, respectively), **BinNonNor**, **PoisBinOrd**, **PoisBinNonNor** (Inan and Demirtas 2016a, b, c, respectively), **PoisBinOrdNor** (Demirtas et al. 2016b), **PoisBinOrdNonNor** (Demirtas et al. 2016c), **PoisNonNor** (Demirtas et al. 2016d), and **BinOrdNonNor** (Demirtas et al. 2016e).

5 Some Remarks and Discussion

Building upon our published and current work, the future steps are as follows: First, the third order polynomials can be extended to the fifth order system (Headrick 2002) for the continuous part in the spirit of controlling for higher order moments to cover a larger area in the skewness-elongation plane and to provide a better approximation to the probability density functions of the continuous variables; and the count data part will be augmented through the generalized Poisson distribution (Demirtas 2017b) that allows under- and over-dispersion, which is usually encountered in most applications, via an additional dispersion parameter. The computational pieces are to be put together for the general-purpose MI algorithm that can routinely be applied to intensive data with planned missingness. Second, the algorithm can be extended to accommodate the Spearman correlation. Although the Pearson correlation may not be the best association quantity in every situation, all correlations mentioned in this chapter are special cases of the Pearson correlation; it is the most widespread measure of association; and generality of the methods proposed herein with different kinds of variables requires the broadest possible framework. In an attempt to further broaden the scope and applicability of the ideas presented herein, the proposed MI approach can be extended to allow the specification of the Spearman's rho, which is more popular for discrete and heavily skewed continuous distributions. For the continuous-continuous pairs, the connection between the Pearson and Spearman correlations is given in Headrick (2010) through the power coefficients, and these two correlations are known to be equal for the binary-binary pairs. The relationship will be derived for all other variable type combinations. Inclusion of Spearman's rho as an option will allow us to specify nonlinear associations whose monotonic components are reflected in the rank correlation. Third, statistical power, sample size, and measurement frequency guidelines (currently unavailable to the best of our knowledge) can be developed. This may be very helpful in designing future RTDC studies with planned missingness as well as in many other contexts. Power is a function of the sample size, measurement frequency, standardized effect size, directionality of hypotheses, Type I error rate, mode of analysis, missing data rate, degree of associations among successive measurements and parameter(s) under consideration. For example, these power and sample size guidelines may be used to determine the minimal number of recruitable subjects for an RTDC study with a certain budget, proportion of missing prompts that can be accommodated and/or the number of prompts per day, the number of days are needed to attain a specified power level under a well-defined model with associated hypotheses.

The key advantages of the proposed tools are as follows: (1) Individual components are well-established; our published and current studies suggest that the goals are achievable. (2) Given their computational simplicity, flexibility, and generality, these methods are likely to be widely used by researchers, methodologists, and practitioners in a wide spectrum of scientific disciplines. (3) They could be very useful in graduate-level teaching of statistics courses that involve computation and simulation, and in training graduate students. (4) Having access to these methods is needed by potentially a large group of people. (5) Imputed variables can be treated as outcomes or predictors in subsequent statistical analyses as the variables are being imputed jointly. (6) Required quantities can either be specified or estimated from a real data set. (7) We allow the specification of the two prominent types of correlations (Pearson and Spearman correlations). This makes it feasible to generate linear and a broad range of nonlinear associations. (8) The continuous part can include virtually any shape (skewness, low or high peakedness, mode at the boundary, multimodality, etc.) that is spanned by power polynomials; the count data part can be under- or over-dispersed. (9) Ability to simultaneously imputing different types of data may facilitate comparisons among existing data analysis and computation methods in assessing the extent of conditions under which available methods work properly, and foster the development of new tools, especially in contexts where correlations play a significant role (e.g., longitudinal, clustered, and other multilevel settings). (10) The approaches presented here can be regarded as a variant of multivariate Gaussian copula-based methods as (a) the binary and ordinal variables are assumed to have a latent normal distribution before discretization; (b) the count variables go through a correlation mapping procedure via the anything-to-normal approach; and (c) the continuous variables consist of polynomial terms involving normals. (11) Availability of a general mixed data imputation algorithm can markedly facilitate simulated power-sample size calculations for a broad range of statistical models.

The following possible questions can be posed: (1) We have enough data in such studies, why do we want more through imputation? (2) In the presence of massive data, do we really need to be concerned about statistical power? Intensive data are typically analyzed with complicated statistical models with many parameters that are assumed to govern the process that differentiates population-averaged (similarities) and subject-specific (differences) effects, as well as cross-sectional (time-invarying) and longitudinal (time-varying) trends. These massive data are collected to potentially address a wide range of research questions; complex data are designed for complex set of hypotheses, a situation which is generally associated with a rich set of parameters for which more information is necessary for modeling purposes. In addition, researchers could recruit a larger group of participants with the same resources as subjects are asked to respond a subset of questions or prompts in planned missingness designs where MI can only help in drawing statistically sound conclusions in the sense that –more subjects, fewer observations per subject– type of settings, by which more questions can be answered, become more feasible. Same arguments equally apply to the statistical power and sample size context.

Examples from many application areas can be given. Modern data collection procedures, such as real-time data captures yield relatively large numbers of subjects and observations per subject, and data from such designs are sometimes referred to as intensive longitudinal data (Walls and Schafer 2006). For instance, in subjective well-being and quality of life studies, data are collected on people's material conditions such as income, health, education, environment, personal safety, and social connections, as well as subjects' momentary positive and negative affects as measured by individual mood items such as feeling happy, relaxed, cheerful, confident, accepted by others, sad, stressed, and irritable, etc. One such study that the author has been involved focused on adolescent smoking in the context of modeling the determinants of the variation in the adolescents' moods (Hedeker et al. 2008, 2009, 2012; Demirtas et al. 2012). Subjects carried the hand-held computers with them at all times during a seven consecutive day data collection period and were trained to both respond to multiple random prompts from the computers throughout the day and to event record smoking episodes. Questions included ones about place, activity, companionship, mood, and other subjective items. The proposed approach can be used to impute a large number of variables of all major types in a way that preserves the marginal and correlational features for such data.

All in all, the proposed MI framework could be a useful addition to the literature for the following major reasons: First, theoreticians, practitioners, data analysts, and methodologists across many different fields in social, managerial, behavioral, medical, and physical sciences will be able to multiply impute intensive multivariate data of mixed types with ease. Second, the proposed work can serve as a milestone for the development of more sophisticated data analysis, computation, and programming techniques in the digital information domains. Capability of accommodating many incomplete variables of different distributional nature, types, and dependence structures could be a contributing factor for better comprehending the operational characteristics of today's massive data trends. Third, the work can be helpful in promoting higher education in the form of training graduate students. Overall, it offers promising potential for building enhanced statistical computing infrastructure for education and research in the sense of providing principled, useful, general, and flexible set of computational tools for handling incomplete data.

References

Amatya, A., & Demirtas, H. (2015a). Simultaneous generation of multivariate mixed data with Poisson and normal marginals. *Journal of Statistical Computation and Simulation*, *85*, 3129–3139.

Amatya, A., & Demirtas, H. (2015b). *Concurrent generation of ordinal and normal data with given correlation matrix and marginal distributions*, R package **OrdNor**. http://CRAN.R-project.org/package=OrdNor.

Amatya, A., & Demirtas, H. (2016a). *Simultaneous generation of multivariate binary and normal variates*, R package **BinNor**. http://CRAN.R-project.org/package=BinNor.

Amatya, A., & Demirtas, H. (2016b). *Generation of multivariate ordinal variates*, R package **MultiOrd**. http://CRAN.R-project.org/package=MultiOrd.

Amatya, A., & Demirtas, H. (2016c). *Simultaneous generation of multivariate data with Poisson and normal marginals*, R package **PoisNor**. http://CRAN.R-project.org/package=PoisNor.

Demirtas, H. (2004). Simulation-driven inferences for multiply imputed longitudinal datasets. *Statistica Neerlandica*, *58*, 466–482.

Demirtas, H. (2005). Multiple imputation under Bayesianly smoothed pattern-mixture models for non-ignorable drop-out. *Statistics in Medicine*, *24*, 2345–2363.

Demirtas, H. (2006). A method for multivariate ordinal data generation given marginal distributions and correlations. *Journal of Statistical Computation and Simulation*, *76*, 1017–1025.

Demirtas, H. (2007). Practical advice on how to impute continuous data when the ultimate interest centers on dichotomized outcomes through pre-specified thresholds. *Communications in Statistics-Simulation and Computation*, *36*, 871–889.

Demirtas, H. (2008). On imputing continuous data when the eventual interest pertains to ordinalized outcomes via threshold concept. *Computational Statistics and Data Analysis*, *52*, 2261–2271.

Demirtas, H. (2009). Rounding strategies for multiply imputed binary data. *Biometrical Journal*, *51*, 677–688.

Demirtas, H. (2010). A distance-based rounding strategy for post-imputation ordinal data. *Journal of Applied Statistics*, *37*, 489–500.

Demirtas, H. (2014). Joint generation of binary and nonnormal continuous data. *Journal of Biometrics and Biostatistics*, *5*, 1–9.

Demirtas, H. (2016). A note on the relationship between the phi coefficient and the tetrachoric correlation under nonnormal underlying distributions. *American Statistician*, *70*, 143–148.

Demirtas, H. (2017a). Concurrent generation of binary and nonnormal continuous data through fifth order power polynomials, *Communications in Statistics—Simulation and Computation*, 46, 344–357.

Demirtas, H. (2017b). On accurate and precise generation of generalized Poisson variates, *Communications in Statistics—Simulation and Computation*, 46, 489–499.

Demirtas, H., Ahmadian, R., Atis, S., Can, F. E., & Ercan, I. (2016a). A nonnormal look at polychoric correlations: Modeling the change in correlations before and after discretization. *Computational Statistics*, *31*, 1385–1401.

Demirtas, H., Arguelles, L. M., Chung, H., & Hedeker, D. (2007). On the performance of bias-reduction techniques for variance estimation in approximate Bayesian bootstrap imputation. *Computational Statistics and Data Analysis*, *51*, 4064–4068.

Demirtas, H., & Doganay, B. (2012). Simultaneous generation of binary and normal data with specified marginal and association structures. *Journal of Biopharmaceutical Statistics*, *22*, 223–236.

Demirtas, H., Freels, S. A., & Yucel, R. M. (2008). Plausibility of multivariate normality assumption when multiply imputing non-Gaussian continuous outcomes: A simulation assessment. *Journal of Statistical Computation and Simulation*, *78*, 69–84.

Demirtas, H., & Hedeker, D. (2007). Gaussianization-based quasi-imputation and expansion strategies for incomplete correlated binary responses. *Statistics in Medicine*, *26*, 782–799.

Demirtas, H., & Hedeker, D. (2008a). Multiple imputation under power polynomials. *Communications in Statistics–Simulation and Computation*, *37*, 1682–1695.

Demirtas, H., & Hedeker, D. (2008b). An imputation strategy for incomplete longitudinal ordinal data. *Statistics in Medicine*, *27*, 4086–4093.

Demirtas, H., & Hedeker, D. (2008c). Imputing continuous data under some non-Gaussian distributions. *Statistica Neerlandica*, *62*, 193–205.

Demirtas, H., & Hedeker, D. (2011). A practical way for computing approximate lower and upper correlation bounds. *The American Statistician*, *65*, 104–109.

Demirtas, H., & Hedeker, D. (2016). Computing the point-biserial correlation under any underlying continuous distribution. *Communications in Statistics- Simulation and Computation*, *45*, 2744–2751.

Demirtas, H., Hedeker, D., & Mermelstein, J. M. (2012). Simulation of massive public health data by power polynomials. *Statistics in Medicine*, *31*, 3337–3346.

Demirtas, H., Hu, Y., & Allozi, R. (2016b). *Data generation with Poisson, binary, ordinal and normal components*, R package **PoisBinOrdNor**. https://cran.r-project.org/web/packages/PoisBinOrdNor.

Demirtas, H., Nordgren, R., & Allozi, R. (2016c). *Generation of up to four different types of variables*, R package **PoisBinOrdNonNor**. https://cran.r-project.org/web/packages/PoisBinOrdNonNor.

Demirtas, H., & Schafer, J. L. (2003). On the performance of random-coefficient pattern-mixture models for non-ignorable drop-out. *Statistics in Medicine*, *22*, 2553–2575.

Demirtas, H., Shi, Y., & Allozi, R. (2016d). *Simultaneous generation of count and continuous data*, R package **PoisNonNor**. https://cran.r-project.org/web/packages/PoisNonNor.
Demirtas, H., Wang, Y., & Allozi, R. (2016e). *Concurrent generation of binary, ordinal and continuous data*, R package **BinOrdNonNor**. https://cran.r-project.org/web/packages/BinOrdNonNor.
Demirtas, H., & Yavuz, Y. (2015). Concurrent generation of ordinal and normal data. *Journal of Biopharmaceutical Statistics*, *25*, 635–650.
Dempster, A. P., Laird, N. M., & Rubin, D. B. (1977). Maximum likelihood estimation from incomplete data via the EM algorithm. *Journal of Royal Statistical Society-Series B*, *39*, 1–38.
Emrich, J. L., & Piedmonte, M. R. (1991). A method for generating high-dimensional multivariate binary variates. *The American Statistician*, *45*, 302–304.
Ferrari, P. A., & Barbiero, A. (2012). Simulating ordinal data. *Multivariate Behavioral Research*, *47*, 566–589.
Fleishman, A. I. (1978). A method for simulating non-normal distributions. *Psychometrika*, *43*, 521–532.
Headrick, T. C. (2002). Fast fifth-order polynomial transforms for generating univariate and multivariate nonnormal distributions. *Computational Statistics and Data Analysis*, *40*, 685–711.
Headrick, T. C. (2010). *Statistical simulation: Power method polynomials and other transformations*. Boca Raton, FL: Chapman and Hall/CRC.
Hedeker, D., Demirtas, H., & Mermelstein, R. J. (2008). An application of a mixed-effects location scale model for analysis of ecological momentary mssessment (EMA) data. *Biometrics*, *6*, 627–634.
Hedeker, D., Demirtas, H., & Mermelstein, R. J. (2009). A mixed ordinal location scale model for analysis of ecological momentary assessment (EMA) data. *Statistics and Its Interface*, *2*, 391–402.
Hedeker, D., Mermelstein, R. J., & Demirtas, H. (2012). Modeling between- and within-subject variance in ecological momentary assessment (EMA) data using mixed-effects location scale models. *Statistics in Medicine*, *31*, 3328–3336.
Higham, N. J. (2002). Computing the nearest correlation matrix—a problem from finance. *IMA Journal of Numerical Analysis*, *22*, 329–343.
Inan, G., & Demirtas, H. (2016a). *Data generation with binary and continuous non-normal components*, R package **BinNonNor**. https://cran.r-project.org/web/packages/BinNonNor.
Inan, G., & Demirtas, H. (2016b). *Data generation with Poisson, binary and ordinal components*, R package **PoisBinOrd**. https://cran.r-project.org/web/packages/PoisBinOrd.
Inan, G., & Demirtas, H. (2016c). *Data generation with Poisson, binary and continuous components*, R package **PoisBinNonNor**. https://cran.r-project.org/web/packages/PoisBinNonNor.
Little, R. J. A., & Rubin, D. B. (2002). *Statistical analysis with missing data* (2nd ed.). New York, NY: Wiley.
Nelsen, R. B. (2006). *An introduction to copulas*. Berlin, Germany: Springer.
Qaqish, B. F. (2003). A family of multivariate binary distributions for simulating correlated binary variables with specified marginal means and correlations. *Biometrika*, *90*, 455–463.
Raghunathan, T. E., Lepkowski, J. M., van Hoewyk, J., & Solenberger, P. A. (2001). Multivariate technique for multiply imputing missing values using a sequence of regression models. *Survey Methodology*, *27*, 85–95.
Rubin, D. B. (1976). Inference and missing data. *Biometrika*, *63*, 581–592.
Rubin, D. B. (2004). *Multiple imputation for nonresponse in surveys* (2nd ed.). New York, NY: Wiley.
Schafer, J. L. (1997). *Analysis of incomplete multivariate data*. London, UK: Chapman and Hall.
Tanner, M. A., & Wong, W. H. (1987). The calculation of posterior distributions by data augmentation. *Journal of American Statistical Association*, *82*, 528–540.
Vale, C. D., & Maurelli, V. A. (1983). Simulating multivariate nonnormal distributions. *Psychometrika*, *48*, 465–471.
Van Buuren, S. (2012). *Flexible imputation of missing data*. Boca Raton, Florida: CRC Press.

Walls, T. A., & Schafer, J. L. (2006). *Models for intensive longitudinal data*. New York, NY: Oxford University Press.

Yahav, I., & Shmueli, G. (2012). On generating multivariate Poisson data in management science applications. *Applied Stochastic Models in Business and Industry*, *28*, 91–102.

Yucel, R. M., & Demirtas, H. (2010). Impact of non-normal random effects on inference by multiple imputation: A simulation assessment. *Computational Statistics and Data Analysis*, *54*, 790–801.

Hybrid Monte-Carlo in Multiple Missing Data Imputations with Application to a Bone Fracture Data

Hui Xie

Abstract In this chapter we introduce Hybrid Monte-Carlo (HMC) as an efficient method to sample from complex posterior distributions of many correlated parameters from a semi-parametric missing data model. The HMC enables a distribution-free likelihood-based approach to multiple imputation of missing values. We describe the modeling approach for modeling missing values that does not require assuming any specific distributional forms. We then describe the use of the HMC sampler to obtain inferences and generate multiple imputations under the model, and touch upon various implementation issues, such as choosing starting values, determining burn-in period, monitoring convergence, deciding stopping times. An R program is provided for analyzing missing data from a Bone Fracture study.

1 Introduction

Missing data occur often in health studies. Multiple imputation (MI) becomes a very popular approach to deal with missing data problems. Software programs implementing popular MI methods are widely available in SAS and R (Little and Rubin 2002; Harel and Zhou 2007; Horton and Lipsitz 2001; Ibrahim et al. 2005; Kenward and Carpenter 2007; Rubin 1987; Raghunathan et al. 2001; Yu et al. 2007). Chen et al. (2011) proposed a novel method for multiple imputation based on conditional semi-parametric odds ratio models. The approach overcomes some important limitations of the extant multiple imputation methods. By using the conditional semi-parametric odds ratio modeling framework, we can simultaneously address both the issue of the inflexibility of the MI methods employing the joint normal model and the issue of the possible incompatibility of the chained fully conditional models (Gelman and Raghunathan 2001; Gelman and Speed 1993, 1999; Buuren et al. 1999; Buuren 2007). As

H. Xie (✉)
Simon Fraser University, Burnaby, Canada
e-mail: huixie@uic.edu

H. Xie
The University of Illinois at Chicago, Chicago, USA

D.-G. Chen and J.D. Chen (eds.), *Monte-Carlo Simulation-Based Statistical Modeling*, ICSA Book Series in Statistics, DOI 10.1007/978-981-10-3307-0_9

a multiple imputation procedure, the approach of Chen et al. (2011) eliminates the need to evaluate intractable multiple integrations with respect to missing data. A major challenge in the approach is the difficulty to sample from the complex posterior distribution of many correlated parameters in the semi-parametric complete-data model. The task demands the utilization of the state-of-the-art Monte-Carlo methods for efficient posterior distribution sampling. Chen et al. (2011) address this issue by using the hybrid Monte-Carlo (HMC) method.

The purpose of this book chapter is not to propose a new methodology that is different from Chen et al. (2011), but to give a more detailed description of the issues when implementing the HMC to facilitate MI, and to illustrate the efficiency of HMC to sample from complex posterior distributions. We describe the use of the HMC sampler to obtain inference under the model, and touch upon various implementation issues, such as choosing starting values, determining burn-in period, monitoring convergence, and deciding stopping times. This chapter presents an application of the state-of-the-art Monte-Carlo sampling methods for statistical modeling and analysis. Data and computer programs are publicly available in order for readers to replicate model development and data analysis presented in the chapter by interested readers in their research.

2 The Bone Fracture Data

A hip fracture is a serious injury that can be life-shortening and costly. It frequently requires surgery, replacement or months of physical therapy. Barengolts et al. (2001) conducted a population-based case-control study to identify risk factors for hip fracture occurrence. The study collected data on the hip fracture outcome and a set of potential risk factors for hip fracture for a sample of male veterans. There were 218 cases (subjects with hip fracture) who are matched with controls based on age and race on a 1:1 ratio with a total sample size of 436 subjects.

Table 1 presents the two group comparisons between cases and controls for the nine predictors suggested as potentially important for predicting risk factors from preliminary exploratory data analysis. These predictors are a mixture of continuous and binary variables. Means and standard deviations are presented for continuous variables and proportions of "yes" are presented for the binary variables in Table 1. As is typical with such population studies, there exists various amounts of missingness for these variables as shown in Table 1.

Figure 1 plots the missing data pattern. The missing data pattern is arbitrary and non-monotone. Although the proportion of missingness ranges from 3.2% for the variable "Dementia" to 23.9% for the variable "Albumin", there were only about 50% complete cases (237 out of 436 subjects). Therefore the common logistic regression software used for analyzing this type of data will exclude about a half of the

Table 1 Summary statistics of predictors.

Variable	MissingPortion (%)	Control	Case	P-value
Etoh	10.1	31.5%	60.9%	$5.3e^{-9}$
Smoke	12.4	37.6%	68.6%	$1.7e^{-9}$
Dementia	3.2	4.6%	22.1%	$10.0e^{-8}$
Antiseiz	5.3	2.3%	16.8%	$4.0e^{-7}$
LevoT4	9.2	5.2%	5.9%	0.77
AntiChol	10.8	20.1%	5.4%	$1.5e^{-5}$
Albumin	23.9	3.83(0.54)	3.40(0.70)	$7e^{-10}$
BMI	10.3	27.07(5.62)	22.56(4.72)	$2e^{-16}$
log(HGB)	12.2	2.59(0.14)	2.47(0.18)	$7e^{-14}$

The table presents summary statistics for the nine predictors using observed data, stratified by the hip fracture outcomes (i.e. cases and controls). Proportions of "Yes" are presented for binary variables (the first six variables in the table) and means (standard deviations) are presented for the continuous variables (Albumin, BMI, log(HGB)). The column "P-value" presents P-values from the chi-square tests and two-sample t-test for the binary and continuous predictors, respectively. The column "MissingPortion" presents the proportion of missingness for each variable

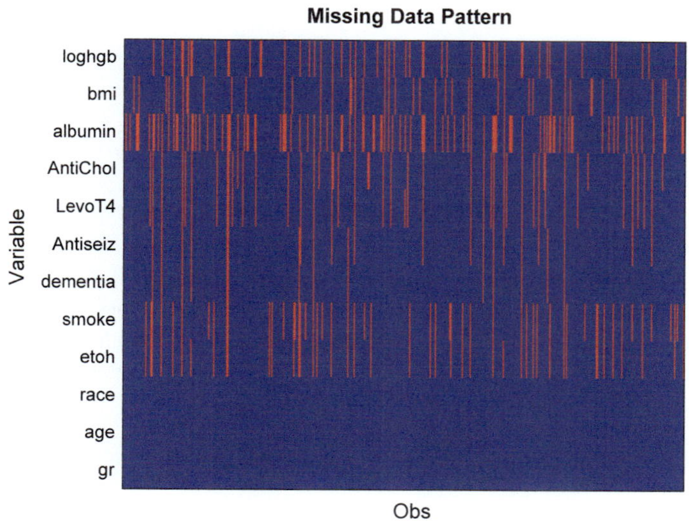

Fig. 1 Missing data pattern in the bone fracture data. The *x-axis* represents the observation index. The color "*red*" denotes the missing values

original sample. Such complete case analysis can substantially reduce effective sample size and introduce potentially significant selection bias for the identification of risk factors.

3 Imputation Modeling and Inference

3.1 Multiple Imputation to Missing Data Problems

We consider multiple imputation to the general missing data problems where the ideal complete data form a rectangular dataset with rows coming from independent and identically distributed draws from a multivariate probability distribution denoted as $f_\theta(Y)$ and the data for a row is denoted as $Y = (Y_1, \ldots, Y_t)$. Let $R = (R_1, \ldots, R_t)$ denote the missing data indicator for Y. $R_j = 1$ if Y_j is observed, and $R_j = 0$ otherwise. We consider arbitrary missing data patterns as illustrated in the bone fracture data. Let $Y = (Y^O, Y^M)$ where Y^O and Y^M denote the observed and missing components in Y, respectively. There are three types of missing data mechanisms as defined in Rubin (1976) and Little and Rubin (2002). They are missing completely at random (MCAR) when $f_\phi(R|Y)$ does not depend on Y, missing at random (MAR) when $f_\phi(R|Y) = f_\phi(R|Y^O)$, and missing not at random (MNAR) when $f_\phi(R|Y)$ depends on the missing values.

Multiple imputation replaces Y^M with $D(D > 1)$ plausible values from the posterior predictive distribution of missing data, i.e., $f(Y^M|Y^O, R)$. In practice, it is often easier to make draws from the joint distribution of all model unknowns $f(Y^M, \theta, \phi|Y^O, R)$, where

$$f(Y^M, \theta, \phi|Y^O, R) \propto \pi(\theta, \phi) f_\theta(Y^M, Y^O) f_\phi(R|Y^M, Y^O) \tag{1}$$

In practice, the standard analysis typically makes the assumptions of ignorability and prior independence of θ and ϕ, where the ignorability holds for likelihood-based inference when missingness is MAR and the parameters θ and ϕ are unrelated to each other (i.e., parameter distinctness). Under the above assumptions, $f_\phi(R|Y^M, Y^O)$ contains no information about Y^M and θ and can be ignored when the primary interest are on the posterior distribution of Y^M and θ. That is, the above posterior joint distribution can be factorized as the product of two parts:

$$f(Y^M, \theta, \phi|Y^O, R) \propto f(Y^M, \theta|Y^O) f(\phi|R, Y^O) \tag{2}$$

Although the MI approach of Chen et al. (2011) is general and can be applied to both ignorable and nonignorable missing data situations, a very challenging issue when dealing with potential nonignorable missing data involves positing models for missing data mechanisms and the associated model identifiability issues (Xie 2008, 2009). We therefore present the standard analysis of the bone fracture data under the ignorability assumption below.

3.2 Odds Ratio Models for Complete Data

As shown above, a key component in multiple imputation of missing data with arbitrary missingness pattern is to specify a joint model for the complete data. Let the density of Y under a product of Lebesgue measures and/or count measures be decomposed into consecutive conditional densities as

$$f(y_t, \ldots, y_1) = \prod_{j=1}^{t} f_j(y_j | y_{j-1}, \ldots, y_1).$$

Below we review the semi-parametric odds ratio models as employed in Chen et al. (2011) for imputation modeling of each consecutive conditional density function above. Let $(y_{t0}, \ldots, y_{10})$ be a fixed point in the sample. For a given conditional density $f_j(y_j | y_{j-1}, \ldots, y_1)$, define the odds ratio function relative to $(y_{j0}, \ldots, y_{10})$ as

$$\eta_j\{y_j; (y_{j-1}, \ldots, y_1) | y_{j0}, \ldots, y_{10}\} = \frac{f_j(y_j | y_{j-1}, \ldots, y_1) f_j(y_{j0} | y_{(j-1)0}, \ldots, y_{10})}{f_j(y_j | y_{(j-1)0}, \ldots, y_{10}) f_j(y_{j0} | y_{j-1}, \ldots, y_1)}.$$

For notational simplicity, we use $\eta_j\{y_j; (y_{j-1}, \ldots, y_1)\}$ to denote $\eta_j\{y_j; (y_{j-1}, \ldots, y_1) | y_{j0}, \ldots, y_{10}\}$. Chen (2004) shows that the conditional density can be rewritten as

$$f_j(y_j | y_{j-1}, \ldots, y_1) = \frac{\eta_j\{y_j; (y_{j-1}, \ldots, y_1)\} f_j(y_j | y_{(j-1)0}, \ldots, y_{10})}{\int \eta_j\{y_j; (y_{j-1}, \ldots, y_1)\} f_j(y_j | y_{(j-1)0}, \ldots, y_{10}) dy_j}.$$

In general, the choice of the fixed point can be arbitrary. In practice, certain choices, such as the center of the data points, may make computation numerically more stable.

In the following, we model the odds ratio function parametrically, which we denote by $\eta_j\{y_j; (y_{j-1}, \ldots, y_1), \gamma_j\}$, and we model $f_j(y_j | y_{(j-1)0}, \ldots, y_{10})$ nonparametrically, which we denote by $f_j(y_j)$. For notational convenience, we assume that $\eta_1(y_1) \equiv 1$. The joint model under this framework becomes

$$f(y_t, \ldots, y_1 | \gamma_2, \ldots, \gamma_t; f_1, \ldots, f_t) = \prod_{j=1}^{t} \frac{\eta_j\{y_j; (y_{j-1}, \ldots, y_1), \gamma_j\} f_j(y_j)}{\int \eta_j\{y_j; (y_{j-1}, \ldots, y_1), \gamma_j\} f_j(y_j) dy_j}. \quad (3)$$

Many different parametric models can be used for the odds ratio function. The most convenient model is perhaps the bilinear form for the logarithm of the odds ratio function. That is,

$$\log \eta_j\{y_j; (y_{j-1}, \ldots, y_1)\} = \sum_{k=1}^{j-1} \gamma_{jk}(y_j - y_{j0})(y_k - y_{k0}).$$

As illustrated in Chen et al. (2011), all generalized linear models with a canonical link function have this form of odds ratio function. In general, high-order terms may be introduced into the model as

$$\log \eta_j\{y_j;\,(y_{j-1},\ldots,y_1),\gamma_j\} = \sum_{k=1}^{j-1}\sum_{m_k=1}^{M_k}\sum_{l=1}^{L}\gamma_{jlkm_k}(y_j-y_{j0})^l(y_k-y_{k0})^{m_k}, \tag{4}$$

which reduces to the bilinear form when $M_k = L = 1$ for $k = 1,\ldots,j-1$. But we restrict ourselves to (4) with known L and M in this article. Note that if we allow L and M to be estimated, (4) can approximate any log-odds ratio function smoothly enough. We assume from now on that odds ratio functions are specified up to an unknown parameter γ, where $\gamma = (\gamma_2,\ldots,\gamma_t)$ and γ_j is the parameter for $\eta_j, j = 2,\ldots,t$.

To enhance modeling robustness, notice that $f_j(y_j) = f_j(y_j|y_{(j-1)0},\ldots,y_{10})$ conditions on the fixed reference point and thus behaves like a marginal distribution. By analogy to using the empirical distribution to estimate a marginal distribution, below $f_j(y_j)$ is modeled nonparametrically like a marginal distribution. Specifically, let $(y_{j1},\ldots,y_{jK_j})$ be the unique observed values in the dataset for Y_j. A nonparametric model for $f_j(y_j)$ assigns probability mass $p_j = (p_{j1},\ldots,p_{jK_j})$ on these unique data points as

$$Prob(Y_j = y_{jk}|y_{j-1,0},\ldots,y_{10}) = p_{jk},\;\; k = 1,\ldots,K_j, \tag{5}$$

$$\text{subject to} \sum_{k=1}^{K_j} p_{jk} = 1, \text{ and } 0 < p_{jk} < 1, \forall\, l. \tag{6}$$

Like the empirical marginal distribution estimates, a reasonably large sample size is needed so that the observed values cover the important range of the sample space. To relax the constraint in Eq. (6), we reparameterize p_j as $\lambda_j = (\lambda_{j1},\ldots,\lambda_{jK_j})$, such that $\lambda_{jk} = \ln(p_{jk}/p_{jK_j})$ for $k = 1,\ldots,K_j$. Thus, $p_{jk} = \exp(\lambda_{jk})/\sum_{u=1}^{K_j}\exp(\lambda_{ju})$.

Prior Specification

Since $\theta_j = (\gamma_j, f_j), j = 1,\ldots,t$, are, respectively, parameters from different conditional distributions, we assume that the priors for them are independent. Further, for any given j, the priors for (γ_j, f_j) can be reasonably assumed independent because γ_j resembles a location parameter and f_j resembles a scale parameter. More specifically, we assume the prior distribution for γ_j has the density $\psi_j(\gamma_j)$. For convenience, we set $\psi_1(\theta_1) \equiv 1$. The prior distribution for f_j is assumed to be a Dirichlet process as $\mathscr{D}_j(c_jF_j)$, where $c_j > 0$ and F_j is a probability distribution for Y_j for $j = 1,\ldots,t$. In practice, we choose c_j and F_j, and the hyperparameter in ψ_j such that they yield relatively noninformative priors. To ease the computation, we can use a Dirichlet process prior with the mean distribution having probability mass on the observed data points only, which becomes a Dirichlet prior $D(\alpha_{j1},\ldots,\alpha_{jK_j})$. As shown in Chen et al. (2011) the use of the Dirichlet prior is approximately correct for a relatively large sample size. Since we are dealing with a relatively large sample size, the difference in using different F_j disappears because the prior has little influence on the posterior in large samples (i.e. $n \gg c_j$).

3.3 Multiple Imputation Under the Framework

The joint posterior distribution of model unknowns (Y^M, θ) is sampled through the MCMC algorithm that iteratively samples the missing data Y^M and model parameters θ as described below.

1. Initialize the parameter vector $\theta = (\theta_1, \ldots, \theta_t)$, where $\theta_j = (\lambda_j, \gamma_j), j = 1, \ldots, t$. A set of readily available starting values can be obtained by assuming that the variables in the data matrix are independent of each other. That is, γ_j^0 are zeros, and the set of λ_j^0 values will then produce the empirical distribution of each variable.
2. Impute Y^{mis}. Given draws of the model parameters, we can impute the missing values in Y. One strategy is to impute missing values one component at a time. Suppose that the jth variable of Y_i, Y_{ij}, is missing for the ith observation. Then by the Bayes rule, Y_{ij}^{mis} is drawn from the following multinomial distribution on the unique observed values of this variable denoted as $(Y_{j1}, \ldots, Y_{jK_j})$:

$$Y_{ij}^{mis} | (Y_{i1}, \ldots, Y_{i(j-1)}, Y_{i(j+1)}, \ldots, Y_{ik}, Y_{iA}) \sim multinomial([P_{ij1}, \ldots, P_{ijK_j}]),$$

where the kth component in the multinomial probability vector $[P_{ij1}, \ldots, P_{ijK_j}]$, for $k = 1, \ldots, K_j$, is given as

$$P_{ijk} = \frac{f_\theta(y_{i1}, \ldots, y_{i(j-1)}, y_{jk}, y_{i(j+1)}, \ldots, y_{it})}{\sum_{k'=1}^{K_j} f_\theta(y_{i1}, \ldots, y_{i(j-1)}, y_{jk'}, y_{i(j+1)}, \ldots, y_{it})}$$

$$= \frac{f_{\theta_j}(y_{jk} | \mathcal{Y}_{ij}) \prod_{m=j+1}^{t} f_{\theta_m}(y_{im} | \mathcal{Y}_{im}(y_{jk}))}{\sum_{k'=1}^{K_j} f_{\theta_j}(y_{jk'} | \mathcal{Y}_{ij}) \prod_{m=j+1}^{t} f_{\theta_m}(y_{im} | \mathcal{Y}_{im}(y_{jk'}))},$$

where $\mathcal{Y}_{ij} = (y_{i1}, \ldots, y_{i(j-1)})$ denote the set of conditioning variables for modeling y_{ij}, $\mathcal{Y}_{im}(y_{jk}) = (y_{i1}, \ldots, y_{i(j-1)}, y_{ij} = y_{jk}, y_{i(j+1)}, \ldots, y_{i(m-1)})$ in which the missing value for y_{ij} is replaced with y_{jk}. When imputing data for Y_{ij}^{mis}, all the missing values in Y_i^M except the jth component take the imputed values in the previous iteration.
3. Draw $\theta = (\theta_1, \ldots, \theta_t)$. Once missing values in Y^M are imputed, we can make draws from the full conditional distributions of these model parameters. Note that when independent priors for $\theta_1, \ldots, \theta_t$ are assigned, their full conditional

distributions are also independent. Each set of parameters in $\theta_1, \ldots, \theta_t$ can then be sampled independently from each other using the hybrid Monte-Carlo (HMC) algorithm described below.

4. The iteration is then repeated until convergence and enough imputations are made.

4 Hybrid Monte-Carlo

The Random-Walk Metropolis-Hasting (RW-MH) is found to be inefficient to sample from the posterior distribution of the model parameters θ given the complete data since there are a relatively large number of parameters, which tend to be highly correlated. We adopt the Hybrid Monte-Carlo (HMC) method to sample for this posterior distribution. The HMC method is introduced by Duane et al. (1987), described in detail in Liu (2001), and adopted in Qian and Xie (2011) for handling missing covariates in marketing models. The HMC sampler uses the idea of Molecular Dynamic (MD) to propose new draws, which is followed by a Metropolis acceptance-rejection method to sample from a target distribution. Because the MD exploits the local dynamics of the target distribution, it suppresses the randomness of making proposal draws in the RW-MH algorithm. As a result, the HMC can substantially increase the acceptance rate of a Markov chain while maintaining a fast mixing of the chain. To sample from the posterior distribution, the HMC augments the parameter $\theta = (\lambda, \gamma)$ with a vector of invented momentum variables p and q, which has the same dimension as θ which defines the Hamiltonian function H as described below.

Note that the sampling distribution for (λ_j, γ_j) appears as in Chen et al. (2011)

$$\mathscr{P}(\gamma_j, \lambda_j) \propto \left\{ \prod_{i=1}^{n} \frac{\eta_j\{Y_j^i; (Y_{j-1}^i, \ldots, Y_1^i), \gamma_j\}}{\sum_{k=1}^{K_j} \eta_j\{y_{jk}; (Y_{j-1}^i, \ldots, Y_1^i), \gamma_j\} e^{\lambda_{jk}}} \right\} \psi_j(\gamma_j) \prod_{k=1}^{K_j} \exp\{(\delta_{jk} + a_{jk} - 1)\lambda_{jk}\}$$

where δ_{jk} denotes counts on y_{jk} and $\alpha_{jk} = c_j/n$. Let

$$U(\lambda_j, \gamma_j) = -\sum_{k=1}^{K_j} (\delta_{jk} + a_{jk} - 1)\lambda_{jk} - \log \psi_j(\gamma_j) - \sum_{i=1}^{n} \log \eta(Y_j^i; Y_{j-1}^i, \ldots, Y_1^i, \gamma_j)$$
$$+ \sum_{i=1}^{n} \log \left\{ \sum_{k=1}^{K_j} \eta_j\{y_{jk}; (Y_{j-1}^i, \ldots, Y_1^i), \gamma_j\} e^{\lambda_{jk}} \right\}.$$

and

$$H\{(\lambda_j, \gamma_j), (p_j, q_j)\} = U(\lambda_j, \gamma_j) + \frac{1}{2} \left\{ \sum_{k=1}^{K_j - 1} \frac{p_{jk}^2}{m_{jk}} + \sum_{k=1}^{D_j} \frac{q_{jk}^2}{n_{jk}} \right\}.$$

For given (p_j, q_j) and (m_j, n_j), an approximate molecular dynamic algorithm generating a candidate sample of (λ_j, γ_j) can be carried out by a given number of leap-frog approximations as follows (Duane et al. 1987). Let $(\lambda_j^{old}, \gamma_j^{old})$ be the current values of (λ_j, γ_j). Let $(\lambda_j^0, \gamma_j^0) = (\lambda_j^{old}, \gamma_j^{old})$. Draw p_j' from the normal distribution with mean 0 and variance $\text{diag}(m_{j1}, \ldots, m_{j(K_j-1)})$, and draw q_j' from the normal distribution with mean 0 and variance $\text{diag}(n_{j1}, \ldots, n_{jD_j})$. Then the initial momentum p_j^0 and q_j^0 have their elements given as follows:

$$p_{jk}^0 = p_{jk}' - \frac{\Delta}{2}\frac{\partial U}{\partial \lambda_{jk}}\left\{\lambda_j^0, \gamma_j^0\right\}$$
$$q_{jk}^0 = q_{jk}' - \frac{\Delta}{2}\frac{\partial U}{\partial \theta_{jk}}\left\{\lambda_j^0, \gamma_j^0\right\}$$

From the initial phase space $(\lambda_j^0, \gamma_j^0, p_j^0, q_j^0)$ of the system, we run the leap-frog algorithm in S steps to generate a new phase space $(\lambda_j^S, \gamma_j^S, p_j^S, q_j^S)$ where for the s step

$$\lambda_{jk}^s = \lambda_{jk}^{s-1} + \Delta\frac{p_{jk}^{s-1}}{m_{jk}}$$
$$\gamma_{jk}^s = \gamma_{jk}^{s-1} + \Delta\frac{p_{jk}^{s-1}}{m_{jk}}$$
$$p_{jk}^s = p_{jk}^{s-1} - \Delta_s\frac{\partial U}{\partial \lambda_{jk}}\left\{\lambda_{jk}^s, \gamma_{jk}^s\right\}$$
$$q_{jk}^s = q_{jk}^{s-1} - \Delta_s\frac{\partial U}{\partial \gamma_{jk}}\left\{\lambda_{jk}^s, \gamma_{jk}^s\right\}$$

where $s = 1, \ldots, S$, $\Delta_s = \Delta$ for $s < S$ and $\Delta_s = \frac{\Delta}{2}$ if $s = S$, Δ is the user-specified stepsize, and

$$\frac{\partial U}{\partial \lambda_{jk}} = -(\delta_{jk} + \alpha_{jk} - 1) + \sum_{i=1}^{n} \frac{\eta_j\{y_{jk}; (Y_{j-1}^i, \ldots, Y_1^i), \gamma_j\}e^{\lambda_{jk}}}{\sum_{k=1}^{K_j} \eta_j\{y_{jk}; (Y_{j-1}^i, \ldots, Y_1^i), \gamma_j\}e^{\lambda_{jk}}}$$

$$\frac{\partial U}{\partial \gamma_{jk}} = -\frac{\partial}{\partial \gamma_{jk}}\log \psi_j(\gamma_j) - \sum_{i=1}^{n}\frac{\partial}{\partial \gamma_{jk}}\log \eta_j\{Y_j^i; (Y_{j-1}^i, \ldots, Y_1^i), \gamma_j\}$$
$$+\sum_{i=1}^{n}\frac{\sum_{k=1}^{K_j}\frac{\partial}{\partial \gamma_{jk}}\eta_j\{y_{jk}; (Y_{j-1}^i, \ldots, Y_1^i), \gamma_j\}e^{\lambda_{jk}}}{\sum_{k=1}^{K_j} \eta_j\{y_{jk}; (Y_{j-1}^i, \ldots, Y_1^i), \gamma_j\}e^{\lambda_{jk}}}.$$

Let the current sample and the candidate sample obtained from S-iterations of the leap-frog approximation from the current sample be respectively denoted by $(\lambda_j^{old}, \gamma_j^{old})$ and $(\lambda_j^{new}, \gamma_j^{new})$. The candidate sample is then accepted with the probability

$$\min\left(1, \exp[-H\{(\lambda_j^{new}, \gamma_j^{new}), (p_j^{new}, q_j^{new})\} + H\{(\lambda_j^{old}, \gamma_j^{old}), (p_j^{old}, q_j^{old})\}]\right).$$

If the candidate sample is accepted, $(\lambda_j^{new}, \gamma_j^{new})$ is taken as the new current sample. Otherwise, $(\lambda_j^{old}, \gamma_j^{old})$ remains to be the current sample.

5 Implementing HMC for Model Fitting

5.1 Assigning Prior Distributions

We assign relatively noninformative priors for model parameters. We assign the prior for γ_j as $\psi_j(\gamma_j) = N(0, v_j^2 I_{n_j \times n_j})$ where n_j is the length of γ_j and v_j, the SD in the prior distribution for γ_j, is assigned a large value. As long as v_j is sufficiently large, the term $log\psi_j(\gamma_j)$ in the above equation for $U(\lambda_j, \gamma_j)$ will be negligible and the prior will become noninformative relative to data. In our analysis of bone fracture data, we assign $v_j = 10^3$. Further increasing values of v_j leads to negligible change in the results. In our analysis of bone fracture data, we choose $c_j = 1$ that corresponds to a uniform prior. As discussed in the "prior specification" paragraph in Sect. 3, even if c_j is not close to zero, as long as it is relatively small compared with the sample size n, the prior has little effects on the posterior inference.

5.2 Tuning Proposal Distribution

In the analysis of the bone fracture data, we set $m_{jk} = 1$ and $n_{jk} = 1$ for all j and k, the number of steps $S = 50$. Stepsize Δ was set to be 0.03 in the data analysis. Under these settings the resulting chain has about 60% acceptance rate. We tune the HMC sampler to have an acceptance rate at a level between 60 and 70%. The practice is supported by the recent theoretical work by Beskos et al. (2013) showing that the acceptance rate under the optimal tuning of a HMC sampler is 0.651. The relatively high acceptance rate under the HMC sampler while still maintaining a good mixing chain demonstrates the high efficiency of the HMC sampling method.

5.3 Starting Values

In order to speed the convergence, the starting parameter values were set as the maximum likelihood estimates (MLEs) of the semi-parametric odds ratio model parameters using one imputed dataset obtained from MICE, the R package for performing multiple imputations through chained equations. Using the set of starting

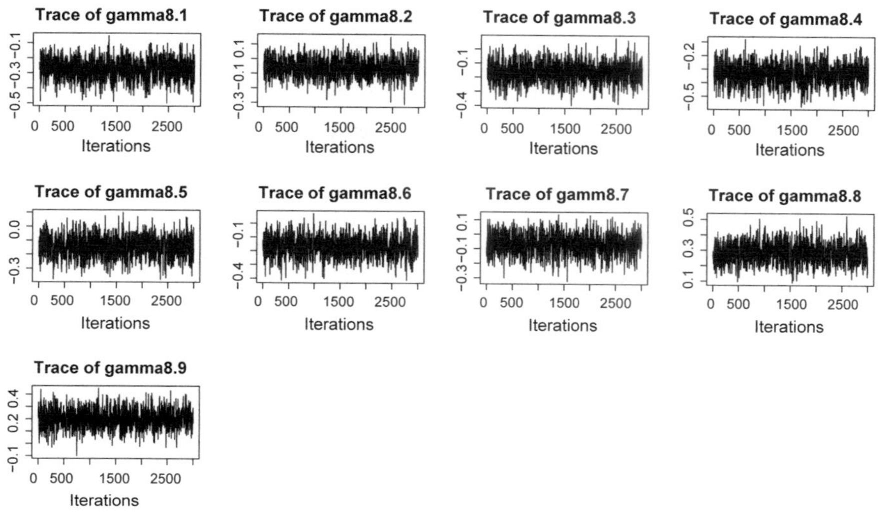

Fig. 2 Sample traceplots of the odds ratio function parameters for the BMI variable

values helps reduce the burn-in period relative to starting from parameter values estimated under the independence assumption.

5.4 Determining Burn-In

We use both the method by Geweke (1992) and the time series plots for convergence diagnosis on the Markov chains. The diagnosis suggests that the Markov chains converge to the stationary distributions within 1000 burn-in iterations for all test runs. As representative examples, Figs. 2 and 3 present sample trace plots and Geweke diagnostic plots for the odds ratio function parameters in the model for the variable "BMI". To be cautious, we use 2000 burn-in cycles in the analysis of bone fracture data.

5.5 Determining Iteration Intervals to Obtain Imputed Values

The autocorrelation plots show that the lag-50 draws are effectively uncorrelated. As representative samples, Fig. 4 provides autocorrelation plots for the odds ratio function parameters in the model for the variable "BMI". The plots shows that the lag-25 draws are effectively uncorrelated. To be cautious, we use 150 iterations between imputations in the data analysis.

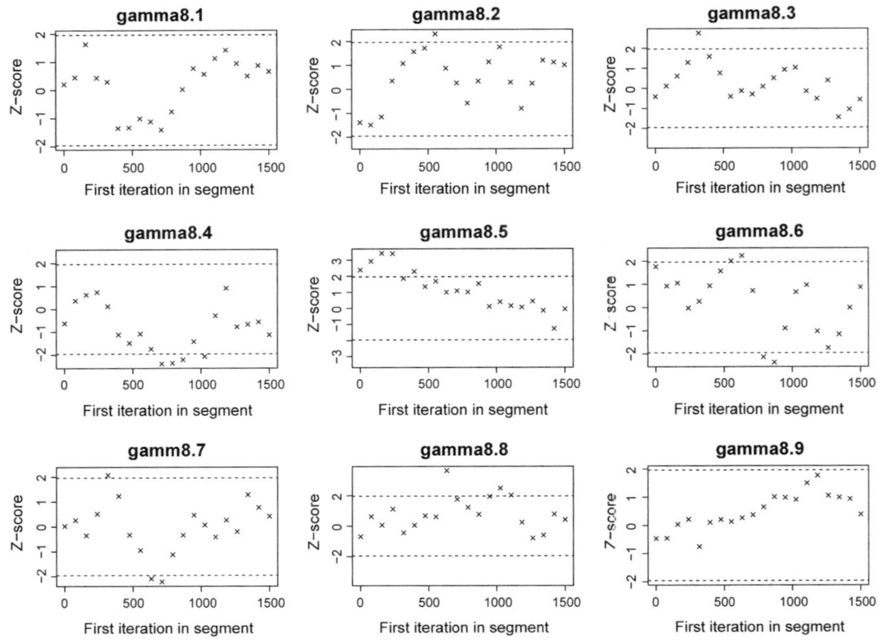

Fig. 3 Sample Geweke-Brooks diagnostic plots of the odds ratio function parameters for the BMI variable

5.6 Determining Stopping Time

After the chain reached the stationary distribution, when to stop the chain was determined primarily by the numbers of imputations and the iteration intervals between imputations. Therefore when we choose 150 iterations between imputations and 20 imputations, we stop the MCMC at 3000 iterations.

5.7 Output Analysis

At convergence, twenty imputed complete datasets were generated. When conducting MI for the bone fracture data, consecutive conditional odds ratio models are used in the imputation model. The order of the conditioning in the imputation model is *age, race, etoh, smoke, dementia, antiseiz, levoT4, antichol, albumin, bmi, log(hgb), fracture*. This means that we model *age*, *race* conditional on *age*, *etoh* conditional on *race* and *age*, and so on. In the analysis step, the imputed datasets are analyzed using the logistic regression model for *fracture*. Rubin's rule for result pooling is then applied to the estimates from the multiply imputed datasets. Note that by using the flexible odds ratio functions as described in Sect. 3 a semiparametric odds ratio

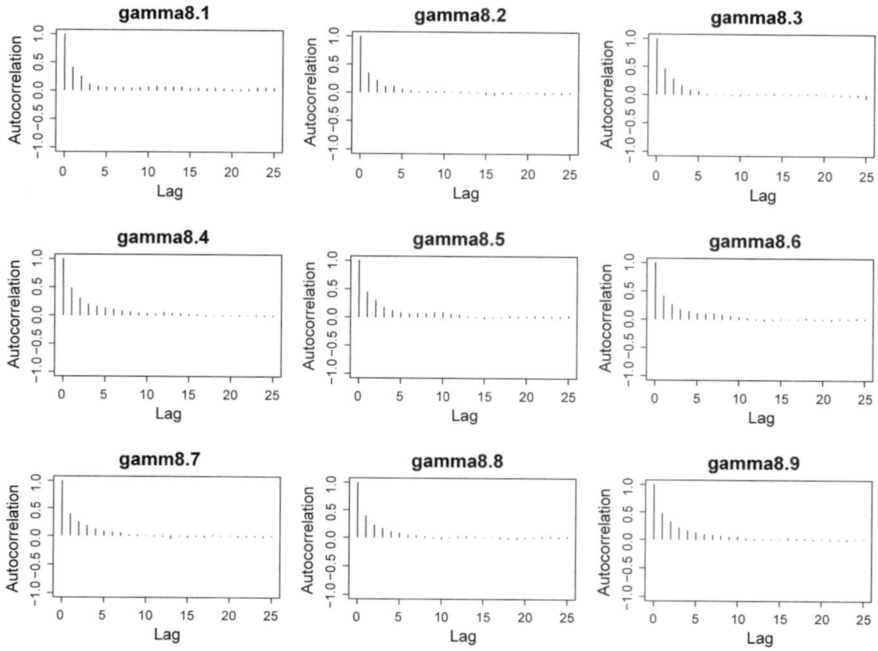

Fig. 4 Sample ACFs for the odds ratio parameters for the BMI variable

model compatible with the parametrically specified model for data analysis almost always exists.

For comparison purpose, we include the following alternative analysis. The first analysis is the complete case analysis. The second method is imputation based on the joint normal (JN) model, or its variant, termed the conditional Gaussian model (CGM). The CGM uses the location-scale model for a mixture of discrete and continuous outcomes. In the location-scale model, the log-linear model is used to model the categorical variables, and a conditional joint normal model is used to model the remaining continuous outcomes. We use the function *impCgm* in the missing data library of Splus 8.0 (Insightful Corp.), which implements the method. The third method is multiple imputation using fully conditional specifications. We use the R package MICE for this method. Given all the other variables at current values, the following default imputation methods are used to sequentially impute each variable in MICE: predictive mean matching for numeric data and logistic regression imputation for binary data.

The results on the analyses of the imputed datasets are listed in Tables 2 and 3 along with the results from complete-case analysis and the estimates from the other two MI methods. Table 2 reports the results without any interaction term among predictors in the model for hip fracture outcome. All multiple imputation methods show that the *LevoT4* is insignificant in contrast to the result from the complete-case

Table 2 Analysis of the imputed bone fracture data without Interaction

Variable	Method				
	CC	MICE	CGM	IMPA	IMPB
Etoh	1.39(0.39)	1.23(0.31)	1.18(0.30)	1.26(0.30)	1.22(0.29)
Smoke	0.93(0.40)	0.62(0.30)	0.51(0.29)	0.62(0.30)	0.70(0.29)
Dementia	2.51(0.72)	1.61(0.47)	1.54(0.45)	1.54(0.47)	1.54(0.45)
Antiseiz	3.31(1.06)	2.51(0.64)	2.44(0.60)	2.54(0.64)	2.38(0.63)
LevoT4	2.01(1.02)	0.92(0.64)	0.88(0.55)	0.95(0.60)	1.00(0.60)
AntiChol	−1.92(0.77)	−1.49(0.59)	−0.91(0.48)	−1.59(0.54)	−1.61(0.52)
Albumin	−0.91(0.35)	−1.03(0.28)	−1.01(0.26)	−1.07(0.27)	−1.03(0.29)
BMI	−0.10(0.04)	−0.10(0.03)	−0.11(0.03)	−0.11(0.03)	−0.11(0.03)
log(HGB)	−2.60(1.20)	−3.39(0.93)	−3.18(0.88)	−3.19(0.92)	−3.20(0.92)

CC Complete-case analysis; *MICE* multiple imputation using the Chained Equations; *CGM* multiple imputation using the conditional Gaussian model; *IMPA* imputation estimator based on 20 imputed datasets; *IMPB* imputation estimate based on 5 imputed datasets. Standard error estimates are in parentheses

Table 3 Analysis of the imputed bone fracture data with Interaction

Variable	Method				
	CC	MICE	CGM	IMPA	IMPB
Etoh	1.41(0.40)	1.13(0.29)	1.15(0.30)	1.20(0.32)	1.32(0.33)
Smoke	−9.21(5.69)	−5.32(4.34)	−3.05(4.52)	−3.41(4.48)	−3.84(4.47)
Dementia	2.80(0.79)	1.69(0.47)	1.54(0.47)	1.59(0.48)	1.60(0.47)
Antiseiz	4.12(1.29)	2.45(0.62)	2.51(0.63)	2.55(0.65)	2.54(0.66)
LevoT4	3.15(1.34)	0.41(0.65)	1.03(0.63)	0.97(0.69)	1.01(0.79)
AntiChol	5.08(4.15)	−0.72(1.99)	−1.26(2.34)	−2.42(2.39)	−2.50(2.27)
Albumin	5.90(4.04)	−3.07(3.40)	2.53(2.97)	2.85(3.23)	3.19(4.21)
BMI	−0.12(0.04)	−0.12(0.03)	−0.11(0.03)	−0.11(0.03)	−0.11 (0.03)
log(HGB)	4.60(5.99)	−7.56(4.80)	1.02(4.35)	1.60(4.69)	2.18(6.00)
Smoke * loghgb	4.05(2.28)	2.40(1.74)	1.40(1.79)	1.68(1.77)	1.84(1.76)
AntiChol*albumin	−2.36(1.40)	0.02(0.55)	0.07(0.62)	0.26(0.66)	0.30(0.63)
Albumin*loghgb	−2.67(1.67)	0.95(1.35)	−1.43(1.19)	−1.60(1.31)	−1.75 (1.71)

CC Complete-case analysis; *MICE* multiple imputation using the Chained Equations; *CGM* multiple imputation using the conditional Gaussian model; *IMPA* imputation estimator based on 20 imputed datasets; *IMPB* imputation estimate based on 5 imputed datasets. Standard error estimates are in parentheses

analysis. The results from all three MI methods are comparable to each other for this analysis.

Table 3 reports the analysis for logistic regression analysis with three interaction terms in the hip fracture outcome model. In order for the imputation model to be compatible with the analysis model, the three interaction terms are added into the odds ratio functions in our approach and to the logistic imputation model for the hip fracture in the MICE approach. In CGM, the interaction terms are combined with first-order terms to form a new data matrix, in which the missing values are multiply imputed in the imputation step. All multiple imputation methods show that the *LevoT4* is insignificant in contrast to the result from the complete-case analysis. The imputation estimates based on five imputed datasets are mostly close to those based on 20 imputed datasets. However, the estimates for some of the variables, such as *LevoT4*, have a relatively large change in magnitude, which may suggest that more than five imputed datasets are needed for this example. There are large numerical differences between some parameter estimates among the three multiple imputation methods. In particular, the signs of the estimates for *log(hgb)*, *albumin*, and *albumin*log(hgb)* are opposite for MICE when compared to those from CGM and our proposed MI method.

As noted in the literature and demonstrated through systematic simulation studies (Chen et al. 2011), imputations based on the JN model (or CGM) or the MICE can perform well in models without interactions. But they can perform poorly in accommodating interactions in the models. This explains that in the bone fracture data all three MI approach have similar results when no interactions are considered but substantial difference are found when interaction terms are considered in the analysis model. The bone fracture analysis results suggest that incorrect imputation models can induce substantial bias, and our approach provides a flexible and robust imputation method to correct for such bias.

6 Conclusion

With the advancement in computational power, the computationally-intensive Monte-Carlo methods has been increasingly finding their way to statistical methods and applications. Hybrid Monte-Carlo (HMC) has its unique advantage of increasing the efficiency of sampling from complex posterior distributions with many correlated parameters by utilizing the local information of likelihood functions. This chapter presents recent progress made in the application of state-of-the-art Monte-Carlo methods to powerful statistical models for missing data analysis.

One unique advantage of multiple imputation is that it separates the imputation stage from the analysis stage. Researchers who perform multiple imputations and who perform data analysis on the imputed datasets can be different and can have access to different levels of software and hardware capabilities. It is often the case that people who conduct the imputations are more statistically sophisticated and have access to more computational resources. Therefore more flexible models and computation-intensive methods are well suited for the MI approach to deal with missing data issues. We expect more research on the further development of flexible MI methods, enabled by the advancement in Monte-Carlo techniques.

References

Barengolts, E., Karanouh, D., Kolodny, L., & Kukreja, S. (2001). Risk factors for hip fractures in predominantly African-American veteran male population. *Journal of Bone and Mineral Research*, *16*, S170.

Beskos, A., Pillai, N., Roberts, G., Sanz-Serna, J., & Stuart, A. (2013). Optimal tuning of the hybrid Monte-Carlo algorithm. *Bernoulli*, *19*, 1501–1534.

Chen, H.Y., Xie, H., & Qian, Y. (2011). Multiple imputation for missing values through conditional semiparametric odds ratio models. *Biometrics*, 10–13.

Chen, H. Y. (2004). Nonparametric and semiparametric models for missing covariates in parametric regressions. *Journal of the American Statistical Association*, *99*, 1176–1189.

Duane, S., Kennedy, A. D., Pendleton, B. J., & Roweth, D. (1987). Hybrid Monte-Carlo. *Physics Letters B*, *195*(2), 216–222.

Gelman, A., & Raghunathan, T. E. (2001). Discussion on "conditionally specified distributions: An introduction". *Statistical Science*, *15*, 268–269.

Gelman, A., & Speed, T. P. (1993). Characterizing a joint probability distribution by conditionals. *Journal of the Royal Statistical Society. Series B*, *55*, 185–188.

Gelman, A., & Speed, T. P. (1999). Corrigendum: Characterizing a joint probability distribution by conditionals. *Journal of the Royal Statistical Society. Series B*, *61*, 483.

Geweke, J. (1992). Evaluating the accuracy of sampling–based approaches to calculating posterior moments. In J. M. Bernado, J. O. Berger, A. P. Dawid & A. F. M. Smith (Eds.), *Bayesian Statistics 4*. Oxford: Clarendon Press.

Harel, O., & Zhou, X. H. (2007). Multiple imputation: Review of theory, implementation and software. *Statistics in Medicine*, *26*, 3057–3077.

Horton, N. J., & Lipsitz, S. R. (2001). Multiple imputation in practice: Comparison of software packages for regression models with missing variables. *American Statistician*, *55*, 244–254.

Ibrahim, J. G., Chen, M. H., Lipsitz, S. R., & Herring, A. H. (2005). Missing-data methods for generalized linear models: A comparative review. *Journal of the American Statistical Association*, *100*, 332–346.

Kenward, M. G., & Carpenter, J. (2007). Multiple imputation: Current perspectives. *Statistical Methods in Medical Research*, *16*, 199–218.

Little, R. J. A., & Rubin, D. B. (2002). *Statistical analysis with missing values*. New York: Wiley.

Liu, J. S. (2001). *Monte-Carlo strategies in scientific computing*. New York: Springer.

Qian, Y., & Xie, H. (2011). No customer left behind: A distribution-free Bayesian approach to accounting for missing Xs in regression models. *Marketing Science*, *30*, 717–736.

Raghunathan, T. E., Lepkowski, J. M., van Hoewyk, J., & Solenberger, P. (2001). A multivariate technique for multiply imputing missing values using a sequence of regression models. *Survey Methodology*, *27*, 85–95.

Rubin, D. B. (1976). Inference and missing data. *Biometrika*, *63*, 92–581.

Rubin, D. B. (1987). *Multiple imputation for nonresponse in surveys*. New York: Wiley.

Schafer, J. L. (1997). *Analysis of incomplete multivariate data*. London: Chapman and Hall / CRC Press.

van Buuren, S. (2007). Multiple imputation of discrete and continuous data by fully conditional specification. *Statistical Methods in Medical Research*, *16*, 219–242.

van Buuren, S., Boshuizen, H. C., & Knook, D. L. (1999). Multiple imputation of missing blood pressure covariates in survival analysis. *Statistics in Medicine*, *18*, 681–694.

Xie, H. (2008). A local sensitivity analysis approach to longitudinal non-Gaussian data with non-ignorable dropout. *Statistics in Medicine*, *27*, 3155–3177.

Xie, H. (2009). Bayesian inference from incomplete longitudinal data: A simple method to quantify sensitivity to nonignorable dropout. *Statistics in Medicine*, *28*, 2725–2747.

Yu, L. M., Burton, A., & Rivero-Arias, O. (2007). Evaluation of Software for multiple imputation of semi-continuous data. *Statistical Methods for Medical Research*, *16*, 243–258.

Statistical Methodologies for Dealing with Incomplete Longitudinal Outcomes Due to Dropout Missing at Random

A. Satty, H. Mwambi and G. Molenberghs

Abstract Longitudinal studies are based on repeatedly measuring the outcome of interest and covariates over a sequences of time points. These studies play a vital role in many disciplines of science, such as medicine, epidemiology, ecology and public health. However, data arising from such studies often show inevitable incompleteness due to dropouts or even intermittent missingness that can potentially cause serious bias problems in the analysis of longitudinal data. In this chapter we confine our considerations to the dropout missingness pattern. Given the problems that can arise when there are dropouts in longitudinal studies, the following question is forced upon researchers: What methods can be utilized to handle these potential pitfalls? The goal is to use approaches that better avoid the generation of biased results. This chapter considers some of the key modelling techniques and basic issues in statistical data analysis to address dropout problems in longitudinal studies. The main objective is to provide an overview of issues and different methodologies in the case of subjects dropping out in longitudinal data for both the case of continuous and discrete outcomes. The chapter focusses on methods that are valid under the missing at random (MAR) mechanism and the missingness patterns of interest will be monotone; these are referred to as dropout in the context of longitudinal data. The fundamental concepts of the patterns and mechanisms of dropout are discussed. The techniques that are investigated for handling dropout are: (1) Multiple imputation (MI); (2) Likelihood-based methods, in particular Generalized linear mixed models (GLMMs); (3) Multiple imputation based generalized estimating equations (MI-GEE); and (4) Weighted estimating equations (WGEE). For each method, useful and important assumptions regarding its applications are presented. The existing literature in which we examine the effectiveness of these methods in the analysis of incomplete longitudinal data is discussed in detail. Two application examples are

A. Satty
Faculty of Mathematical Sciences and Statistics, Alneelain University, Khartoum, Sudan

H. Mwambi (✉)
School of Mathematics, Statistics and Computer Science, University of KwaZulu-Natal,
Private Bag X01 Scottsville 3209, Pietermaritzburg, South Africa
e-mail: MwambiH@ukzn.ac.za

G. Molenberghs
I-BioStat, Universiteit Hasselt & KU Leuven, Martelarenlaan 42, 3500 Hasselt, Belgium

D.-G. Chen and J.D. Chen (eds.), *Monte-Carlo Simulation-Based Statistical Modeling*, ICSA Book Series in Statistics, DOI 10.1007/978-981-10-3307-0_10

presented to study the potential strengths and weaknesses of the methods under an MAR dropout mechanism.

Keywords Multiple imputation GEE · Weighted GEE · Generalized linear mixed model (GLMM) · Likelihood analysis · Incomplete longitudinal outcome · Missing at random (MAR) · Dropout

1 Introduction

Longitudinal studies play a vital role in many disciplines of science including medicine, epidemiology, ecology and public health. However, data arising from such studies often show inevitable incompleteness due to dropouts or lack of follow-up. More generally, a subject's outcome can be missing at one follow-up time and be measured at the next follow-up time. This leads to a large class of dropout patterns. This chapter only pays attention to monotone dropout patterns that result from attrition, in the sense that if a subject drops out from the study prematurely, then on that subject no subsequent repeated measurements of the outcome are obtained. These commonly include studies done by the pharmaceutical industry as contained in protocols for many conditions where data are not collected after a study participant discontinues study treatment. This is highlighted in a recent report on the prevention and treatment of dropout by the National Research Council (Committee on National Statistics Division of Behavioral and Social Sciences and Education, http://www.nap.edu). A summary of the report was provided by Little et al. (2012). However, even in these studies, there typically is both unplanned and planned dropout. A predominately monotone pattern for missing outcome data is less common in clinical outcome studies and in publically-funded trials which are more of a pragmatic nature (e.g., trials in which the intention-to-treat estimand is the primary objective).

Given the problems that can arise when there are dropouts in longitudinal studies, the following question is forced upon researchers. What methods can be utilized to handle these potential pitfalls? The goal is to use approaches that better avoid the generation of biased results. The choice of statistical methods for handling dropouts has important implications on the estimation of the treatment effects, depending on whether one is considering a more of a pragmatic nature analysis or a more exploratory analysis. In case of a pragmatic analysis (intention-to-treat analysis), the goal of the clinical trial researchers is to produce a pragmatic analysis of the data. However, for incomplete longitudinal clinical trials, the dropouts complicate this process as most of the methods to be used when dealing with the dropout problem produce an exploratory analysis in nature rather than a pragmatic perspective. The literature presents various techniques that can be used to deal with dropout, and these range from simple classical ad hoc methods to model-based methods. These methods should be fully understood and appropriately characterized in relation to dropouts and should be theoretically proven before they are used practically. Further, each method is valid under some but usually not all dropout mechanisms, but one needs to realize that at the heart of the dropout problems it is impossible to identify the dropout mechanism (will be discussed later). Thus, it is important to address the mechanisms that govern dropouts. In this chapter, we present some of the various techniques to address the dropout problem in longitudinal clinical trials. The main objective is to investigate various techniques, and to discuss the most appropriate techniques for handling incomplete longitudinal data due to dropouts. The structure of the chapter is as follows. Section 2 presents the key notation and basic concepts used in the entire chapter but when new notation arises it will be explained at the point where it occurs. In Sects. 3 and 4, we give an overview of the various statistical methods in handling incomplete longitudinal studies due to dropout. Two application examples are provided for both cases, continuous and binary outcomes. The dropout generation schemes are also discussed. In addition, full analysis and results of the applications are also given. Finally, the chapter ends with a discussion and conclusion in Sect. 5.

2 Notation and Basic Concepts

Some notation is necessary to describe methods for analyzing incomplete longitudinal data with dropout. We will follow the terminology based on the standard framework of Rubin (1976), Little and Rubin (1987) in formulating definitions for data structure and missing data mechanisms. Let $Y_i = (Y_{i1}, ..., Y_{in_i})' = (Y_i^o, Y_i^m)'$ be the outcome vector of n_i measurements for subject i, $i = 1,...,n$, where Y_i^o represents the observed data part and Y_i^m denotes the missing data part. Let $R_i = (R_{i1}, ..., R_{in_i})'$ be the corresponding missing data indicator vector of the same dimension as Y_i, whose elements are defined as

$$R_{ij} = \begin{cases} 1 \text{ if } Y_{ij} \text{ is observed,} \\ 0 \qquad \text{otherwise.} \end{cases} \tag{1}$$

Complete data refers to the vector Y_i of planned measurements. This is the outcome vector that would have been recorded if no data had been missing. The vector R_i and the process generating it are referred to as the missingness process. In our case the R_i are here restricted to represent participant dropout, and so it has a monotone pattern (Verbeke and Molenberghs 2000). Thus the full data for the ith subject can be represented as (Y_i, R_i) and the joint probability for the data and missingness can be expressed as: $f(y_i, r_i \mid X_i, W_i, \theta, \xi) = f(y_i \mid X_i, \theta) f(r_i \mid y_i, W_i, \xi)$, where X_i and W_i are design matrices for the measurements and dropout mechanism, respectively, θ is the parameter vector associated with the measurement process and ξ is the parameter vector for the missingness process. According to the dependence of the missing data process on the response process, Little and Rubin (1987), Rubin (1976) classified missing data mechanisms as: missing completely at random (MCAR), missing at random (MAR) and not missing at random (MNAR). The missingness process is defined as MCAR if the probability of non-response is independent of the response; that is, $f(r_i \mid y_i, W_i, \xi) = f(r_i \mid W_i, \xi)$ and the missingness process is defined as MAR when the probability of non-response depends on the observed values of the response; that is, $f(r_i \mid y_i, W_i, \xi) = f(r_i \mid y_i^o, W_i, \xi)$. Finally, the missingness process is defined as MNAR if neither the MCAR nor the MAR assumptions hold, meaning that dependence on unobserved values of the response cannot be ruled out. That is, the probability of nonresponse depends on the missing outcomes and possibly on the observed outcomes. Our main focus is on the MAR mechanism for the dropout process.

When missingness is restricted to dropout or attrition, we can replace the vector R_i by a scalar variable D_i, the dropout indicator, commonly defined as

$$D_i = 1 + \sum_{j=1}^{n} R_{ij}. \tag{2}$$

For an incomplete dropout sequence, D_i denotes the occasion at which dropout occurs. In the formulation described above, it is assumed that all subjects are observed on the first occasion so that D_i takes values between 2 and $n + 1$. The maximum value $n + 1$ corresponds to a complete measurement sequence. If the length of the complete sequence is different for different subjects then we only need to replace n with n_i. However a common n holds where for example by design all subjects were supposed to be observed for an equal number of occasions or visits. Accordingly, an MCAR dropout mechanism implies $f(D_i = d_i \mid y_i, W_i, \xi) = f(D_i = d_i \mid W_i, \xi)$, MAR dropout mechanism, $f(D_i = d_i \mid y_i, W_i, \xi) = f(D_i = d_i \mid y_i^o, W_i, \xi)$ and MNAR dropout mechanism, $f(D_i = d_i \mid y_i, W_i, \xi) = f(D_i = d_i \mid Y_i^m, Y_i^o, W_i, \xi)$. There are parameters associated with the measurement process but suppressed for simplicity. Note that the MCAR mechanism can be seen as a special case of MAR. Hence the likelihood ratio test can be used to test the null hypothesis that the MCAR assumption holds. However it is not obvious to say a model based on the MAR mechanism is a simplification of a model based on the MNAR assumption. This assertion is supported by the fact that for any MNAR model there is a MAR counterpart that fits the data just as good as the MNAR model (Molenberghs et al. 2008).

3 Dropout Analysis Strategies in Longitudinal Continuous Data

Much of the literature involving missing data (or dropout) in longitudinal studies pertains to the various techniques developed to handle the problem. This section is devoted to providing an overview of the various strategies for handling missing data in longitudinal studies.

3.1 Likelihood Analysis

An appealing method for handling dropout in longitudinal studies is based on using available data, and these only, when constructing the likelihood function. This likelihood-based MAR analysis is also termed likelihood-based ignorable analysis, or direct likelihood analysis Molenberghs and Verbeke (2005). Direct likelihood analysis uses the observed data without the need of neither deletion nor imputation. In other words, no additional data manipulation is necessary when a direct likelihood analysis is envisaged, provided the software tool used for analysis is able to handle measurement sequences of unequal length (Molenberghs and Kenward 2007). To do so, under valid MAR assumption, suitable adjustments can be made to parameters at times when data are prone to incompleteness due to the within-subject correlation. Thus, even when interest lies in a comparison between two treatment groups at the last measurement time, such a likelihood analysis can be conducted without problems since the fitted model can be used as the basis for inference. When a MAR mechanism is valid, a direct likelihood analysis can be obtained with no need for modelling the missingness process. It is increasingly preferred over ad hoc methods, particularly when tools like the generalized linear mixed mixed effect models (Molenberghs and Verbeke 2005) are used. The major advantage of this method is its simplicity, it can also be fitted in standard statistical software without involving additional programming, using such tools as SAS software, procedures MIXED, GLIMMIX and NLMIXED. The use of these procedures has been illustrated by Verbeke and Molenberghs (2000), Molenberghs and Verbeke (2005). A useful summary for these procedures is presented by Molenberghs and Kenward (2007). Despite the flexibility and ease of implementation of direct likelihood method, there are fundamental issues when selecting a model and assessing its fit to the observed data, which do not occur with complete data. The method is sensible under linear mixed models in combination with the assumption of ignorability. Such an approach, tailored to the needs of clinical trials, has been proposed by Mallinckrodt et al. (2001a, b). For the incomplete longitudinal data context, a mixed model only needs missing data to be MAR. According to Verbeke and Molenberghs (2000), these mixed-effect models permit the inclusion of subjects with missing values at some time points for both missing data patterns, namely dropout and intermittent missing values. Since direct likelihood ideas can be used with a variety of likelihoods, in the first application example in this study we consider the general linear mixed-effects model for continuous outcomes that satisfy the Gaussian distributional assumption (Laird and Ware 1982) as a key modelling framework which can be combined with the ignorability assumption. For Y_i the vector of observations from individual i, the model can be written as follows

$$Y_i = X_i\beta + Z_i b_i + \varepsilon_i, \tag{3}$$

where $b_i \sim N(0, D)$, $\varepsilon_i \sim N(0, \Sigma_i)$ and $b_1, ..., b_N, \varepsilon_1, ..., \varepsilon_N$ are independent. The meaning of each term in (3) is as follows. Y_i is the n_i dimensional response vector for subject i, containing the outcomes at n_i measurement occasions, $1 \leq i \leq N$, N

is the number of subjects, X_i and Z_i are $(n_i \times p)$ and $(n_i \times q)$ dimensional matrices of known covariates, β is the p-dimensional vector containing the fixed effects, b_i is the q-dimensional vector containing the random effects and ε_i is a n_i dimensional vector of residual components, combining measurement error and serial correlation. Finally, D is a general $(q \times q)$ covariance matrix whose (i, j)th element is $d_{ij} = d_{ji}$ and Σ_i is a $(n_i \times n_i)$ covariance matrix which generally depends on i only through its dimension n_i, i.e., the set of unknown parameters in Σ_i will not depend upon i. This implies marginally $Y_i \sim N(X_i\beta, Z_iDZ_i' + \Sigma_i)$. Thus if we write $V_i = Z_iDZ_i' + \Sigma_i$ as the general covariance matrix of Y_i, then $f(y_i, \beta, V_i) = (2\Pi)^{\frac{-n}{2}} |V_i|^{\frac{-1}{2}} \exp\{-(y_i - X_i\beta)'V_i^{-1}(y_i - X_i\beta)/2\}$ from which a marginal likelihood involving all subjects can be constructed to estimate β. In the likelihood context, Little and Rubin (1987) and Rubin (1976) showed that when MAR assumption and mild regularity conditions hold, parameters θ and ξ are independent, and that likelihood based inference is valid when the missing data mechanism is ignored. In practice, the likelihood of interest is then based on the factor $f(y_i^o \mid \xi)$ (Verbeke and Molenberghs 2000). This is referred to as ignorability.

3.2 *Multiple Imputation (MI)*

Multiple imputation was introduced by Rubin (1978). It has been discussed in some detail in Rubin (1987), Rubin and Schenker (1986), Tanner and Wong (1987) and Little and Rubin (1987). The key idea behind multiple imputation is to replace each missing value with a set of M plausible values (Rubin 1996; Schafer 1997). The resulting complete data sets generated via multiple imputation are then analyzed by using standard procedures for complete data and combining the results from these analyses. The technique in its basic form requires the assumption that the missingness mechanism be MAR. Thus, multiple imputation process is accomplished through three distinct steps: (1) Imputation—create M data sets from M imputations of missing data drawn from a different distribution for each missing variable. (2) Analysis—analyze each of the M imputed data sets using standard statistical analysis. (3) Data pooling—combine the results of the M analyses to provide one final conclusion or inference. To discuss these steps in detail, we will follow the approach provided by Verbeke and Molenberghs (2000). Recall that we partitioned the planned complete data (Y_i) into Y_i^o and Y_i^m to indicate observed and unobserved data, respectively. Multiple imputation fills in the missing data Y_i^m using the observed data Y_i^o several times, and then the completed data are used to estimate ξ. If we know the distribution of $Y_i = (Y_i^o, Y_i^m)$ depends on the parameter vector ξ, then we could impute Y_i^m by drawing a value of Y_i^m from the conditional distribution $f(y_i^m \mid y_i^o, \xi)$. Because $\hat{\xi}$ is a random variable, we must also take its variability into account in drawing imputations. In Bayesian terms, $\hat{\xi}$ is a random variable of which the distribution depends on the data. So we first obtain the posterior distribution of ξ from the data, a distribution which is a function of $\hat{\xi}$. Given this posterior distribution, imputation algorithm can be used to draw a random ξ^* from the distribution of ξ, and to put this ξ^* in to draw a random Y_i^m from $f(y_i^m \mid y_i^o, \xi^*)$, using the following steps: (1) Draw ξ^* from the distribution of ξ, (2) Draw Y_i^{m*} from $f(y_i^m \mid y_i^o, \xi^*)$, and (3) Use the complete data (Y^o, Y^{m*}) and the model to estimate β, and its estimated variance, using the complete data, (Y^o, Y^{m*}):

$$\hat{\beta}_m = \hat{\beta}(Y) = \hat{\beta}(Y^o, Y^{m*}), \tag{4}$$

where the within-imputation variance is $U_m = V\hat{a}r(\hat{\beta})$. The steps described above are repeated independently M times, resulting in $\hat{\beta}_m$ and U_m, for $m = 1, ..., M$. Steps 1 and 2 are referred to as the imputation task, and step 3 is the estimation task. Finally, the results are combined using the following steps for pooling the estimates obtained after M imputations (Rubin 1987; Verbeke and Molenberghs 2000). With no missing data, suppose the inference about the parameter β is made using the distributional assumption $(\beta - \hat{\beta}) \sim N(0, U)$. The overall estimated parameter vector is the average of all individual estimates:

$$\hat{\beta}* = \frac{\sum_{m=1}^{M} \hat{\beta}_m}{M}, \tag{5}$$

with normal-based inferences for β based upon $(\hat{\beta} * - \beta) \sim N(0, V)$ (Verbeke and Molenberghs 2000). We obtain the variance (V) as a weighted sum of the within-imputation variance and the between-imputations variability:

$$V = W + \left(\frac{M+1}{M}\right) B, \tag{6}$$

where

$$W = \frac{\sum_{m=1}^{M} U_m}{M} \tag{7}$$

defined to be the average within-imputation variance, and

$$B = \frac{\sum_{m=1}^{M} (\hat{\beta}_m - \hat{\beta}*)(\hat{\beta}_m - \hat{\beta}*)'}{M - 1} \tag{8}$$

defined to be the between-imputation variance (Rubin 1987).

3.3 *Illustration*

To examine the performance of direct likelihood and multiple imputation methods, four steps were planned. The steps were as follow: First, a model was fitted to the full data (no data are missing), thus producing what we refer to as true estimates. Second, we generated a dropout rate of 10, 15 and 20% in the outcome (selected at random) variable using defined rules to achieve the required mechanism under MAR assumption. Third, the resulting incomplete data was analyzed using the two different methods using multiple imputation and direct likelihood. Fourth, results from the complete and incomplete data analysis were compared. The actual-data results were presented and used as references. The study aims to investigate how direct likelihood and multiple imputation compare to each other and to the true analysis.

Data Set—Heart Rates Trial

This data set was used in Milliken and Johnson (2009) to demonstrate analyses of repeated measures designs and to show how to determine estimates of interesting effects and provide methods to study contrasts of interest. The main objective was to investigate the effects of three treatments involving two active treatments and a control (AX23, BWW9 and CTRL) on heart rates, where each treatment was randomized to female individuals and each patient observed over four time periods. Specifically, each patient's heart rate was measured 5, 10, 15 and 20 min after administering the treatment. The only constraint is that the time intervals are not randomly distributed within an individual. In our case, we use the data to achieve a comparative analysis of two methods to deal with missing data. A model which is used to describe the data is similar to a split-plot in a completely randomized design. The model is

$$H_{ijk} = \mu + Time_j + \delta_i + Drug_k + (Time * Drug)_{jk} + \varepsilon_{ijk}, \tag{9}$$

where H_{ijk} is the heart rate of individual i at time j on drug k, $i = 1, ..., 24$, $j = 1, 2, 3, 4$ and $k = 1, 2, 3$. The model has two error terms: δ_i represents a subject random effect, and ε_{ijk} represents a time error component. The ideal conditions for a split-plot in time analysis is that: (1) the δ_i are independently and identically $N(0, \sigma_\delta^2)$, (2) the ε_{ijk} are independently and identically $N(0, \sigma_\varepsilon^2)$, and (3) the δ_i and ε_{ijk} are all independent of one another. The main purpose of this example is to investigate the effects of the three drugs. Thus, the type III tests of fixed effects and the differences between effects were the quantities of interest in the study. The primary null hypothesis (the difference between the drug main effects) will be tested. The null hypothesis is no difference among drugs. The significance of differences in least-square means is based on Type III tests. These examine the significance of each partial effect; that is, the significance of an effect with all the other effects in the model. In analysis results we present the significance of drug main effects, time main effects and the interaction of time and drug effects.

3.4 *Simulation of Missing Values*

Since there are no missing values in the example data set described above, it provides us with an opportunity to design a comparative study to compare the two methods to deal with missing data using the results from the complete data analysis as the reference. We carry out an application study to generate the data set with dropouts. In this application, we distinguish between two stages: (1) The dropout generation stage. (2) The analysis stage.

3.4.1 Generating Missing Data

In the first stage, we use the full data set to artificially generate missing values by mimicking the dropout at random mechanism. From the complete data, we draw 1000 random samples of size $N = 96$. The incomplete data was generated with 10, 15 and 20% dropout rate. We assume that the dropout depends only on the observed data. Furthermore, a monotone dropout pattern was imposed in the heart rate (outcome of interest); that is, if H_{ij} is missing, then H_{is} is missing for $s \geq j$. The explanatory variables drug, time and interaction between drug and time are assumed to be fully observed. In addition, in order to create the dropout model, we assume that dropout can occur only after the first two time points. Namely, dropout is based on values of H, assuming the H is fully observed in the first two time (time = 1, 2), while for the later times (time = 3, 4) some dropouts may occur. We assume an MAR mechanism for the dropout process and the dropout mechanism depends on individual previously observed values of one of the endpoints. For the MAR mechanism, H was made missing if its measurements exceeded 75 (the baseline mean for heart rate) the previous measurement occasion, beginning with the second post baseline observation. Thus in the generation, the missingness at time = 3, 4 was dependent on the most recently observed values. This was done to achieve the required mechanism under the MAR assumption.

3.4.2 Computations and Handling Missing Data

After generating the missing data mechanism and thus generating the data set with dropout, the next step was to deal with dropout. Handling dropout was carried out using direct likelihood analysis and multiple imputation methods with functions available in the SAS software package. Ultimately, likelihood, multiple imputation and analysis results from the fully observed data set can be compared in terms of their impact on various linear mixed model aspects (fixed effects and least squares means). The proposed methods dealt with the dropout according to the following:

- Imputing dropouts using multiple imputation techniques. This was achieved using procedures MI, MIXED and MIANALYZE with an LSMEANS option. The imputation model is based on model (9) which assumes normality of the variables. For the dropout under MAR, the imputation model should be specified (Rubin 1987). Thus, in the imputation model, we included all the available data (including the outcome, H) to predict the dropouts since they were potentially related to the imputed variable as well as to the missingness of the imputed variable. This means we used variables in the analysis model, variables associated with missingness of the imputed variable and variables correlated with the imputed variable. This was done to increase the plausibility of the MAR assumption, as well as to improve the accuracy and efficiency of the imputation. Once the multiple imputation model is chosen, the number of imputations must be decided. PROC MI was applied to generate M=5 complete data sets. We fixed the number of multiple imputations at M=5, since relatively small numbers are often deemed sufficient, especially for parameter estimation from normally distributed data (see, Schafer and Olsen 1998; Schafer 1999). PROC MIXED was used to set up effect parameterizations for the class variables and we used the ODS statement output to create output data sets that match PROC MIANALYZE for combining the effect mean estimates from the 5 imputed data sets. While PROC MIANALYZE cannot directly combine the least square means and their differences to obtain the effect means of drug and contrasts between drug groups from PROC MIXED, the LSMEANS table was sorted differently so that we enabled the use of the BY statement in PROC MIANALYZE to read it in.

- For comparison, the data was analyzed as they are, consistent with ignorability for direct likelihood analysis implemented with PROC MIXED with LSMEANS option. The REPEATED statement was used, in order to make sure the analysis procedure takes into account sequences of varying length and order of the repeated measurements. Parameters were estimated using Restricted Maximum Likelihood with the Newton-Raphson algorithm (Molenberghs and Verbeke 2005).

3.5 Results

A few points about the parameter estimates obtained by the proposed methods may be noted in the resulting tables. In Table 1, due to the similarities in the findings under the three dropout rates, the results for type III tests of fixed effects under 20 and 30% dropout rates are not presented but are available from the authors. Through the two evaluation criteria in Table 2, the largest bias, also the worst, are highlighted. For the efficiency criterion, the widest confidence interval, also the worst, 95% interval are highlighted.

The results that show the significance of the effects using direct likelihood and multiple imputation to handle dropout are presented in Table 1. Compared with the results based on the complete data set, we see that type III tests of fixed effects show that both direct likelihood and multiple imputation methods yielded statistically similar results. The analysis shows that the drug effect has significant p-values as its p-values, around 0.004, indicating a rejection of the null hypothesis of equal drug means. The p-value of the drug effect under multiple imputation (0.0043) was slightly higher in comparison to that of the direct likelihood analysis (0.0040), but both methods indicate strong evidence of significance compared to the p-value of 0.0088 for the original complete data set. Evidently, there are no extreme differences between the direct likelihood and multiple imputation methods. However, the p-value for the drug effect was significantly reduced by about 50% compared to the actual data

Table 1 Statistical test for drug, time and drug × time effects of complete data, direct likelihood and multiple imputation, under 15% dropout rate

	Effect	Type III tests of fixed effects			
		Num *df*	Den *df*	*F*-value	*Pr* > *F*
Actual-data	drug	2	21	5.99	0.0088
	time	3	63	12.96	<0.001
	drug × time	6	63	11.80	<0.001
Direct likelihood	drug	2	17.1	7.78	0.0040
	time	3	15.8	18.13	<0.001
	drug × time	6	15.8	25.74	<0.001
Multiple imputation	drug	2	21	7.14	0.0043
	time	3	447	84.15	<0.001
	drug × time	6	447	76.00	<0.001

Table 2 Bias and efficiency of MI and direct-likelihood, under different dropout rates: MIXED least squares means—(interaction terms are not shown)

Dropout rate	Effects	Bias		Efficiency	
		MI	Direct-likelihood	MI	Direct-likelihood
10%	AX23	0.08	**0.09**	0.97	**0.98**
	BWW9	−0.06	**−0.08**	0.95	**0.97**
	CTRL	**0.09**	0.05	0.88	**0.89**
	$time_1$	0.00	0.00	0.99	0.99
	$time_2$	0.00	0.00	0.99	0.99
	$time_3$	0.07	**0.09**	0.97	**0.98**
	$time_4$	**0.06**	0.04	0.94	**0.96**
20%	AX23	**0.11**	0.10	0.93	**0.94**
	BWW9	0.08	**0.10**	0.94	**0.97**
	CTRL	0.14	**0.16**	0.94	**0.97**
	$time_1$	0.00	0.00	0.98	0.98
	$time_2$	0.00	0.00	0.98	0.98
	$time_3$	0.09	**0.11**	1.27	**1.54**
	$time_4$	**0.08**	0.06	1.27	**1.34**
30%	AX23	0.24	**0.26**	1.08	**1.09**
	BWW9	0.14	**0.15**	1.08	**1.11**
	CTRL	0.19	**0.20**	1.09	**1.10**
	$time_1$	0.00	0.00	0.97	0.97
	$time_2$	0.00	0.00	**0.98**	0.97
	$time_3$	0.15	**0.17**	1.55	**1.68**
	$time_4$	**0.13**	0.12	1.58	**1.66**

Note The largest bias and efficiency for each given estimate presented in bold. MI = multiple imputation; Direct-likelihood

p-value. This indicates a real problem with dropout, both multiple imputation and direct likelihood may lead to rejection of the null hypothesis with a higher probability than would be the case if the data were complete. The test of significance for time effect in type III tests of fixed effects produced significant p-values of <0.0001 in both methods. The test for the interaction between drug and time effects gave a p-value of <0.0001 in both methods, indicating a strong evidence of time dependence on the drug effects. Generally, the proposed methods presented acceptable performance with respect to estimates of p-values in all cases when compared to that based on actual data. In two cases, namely p-values of time effect and interaction drug $\times$ time, the methods yielded the same results as those for complete data.

The results of MI and direct likelihood analysis in terms of bias and efficiency, under three dropout rates are presented in Table 2, which shows the results for the least square means. Note that, again due to similarities in the findings, we do not show full output, as the results of interactions terms are excluded. Examining these results we find the following. For 10% dropout rate, in terms of the biasedness of the estimates, the performance of both methods unsurprisingly yielded equally good performance. However, the benefits of MI over a direct likelihood are clearly evident. In some cases (estimates of time 1, time 2), the methods offered the same estimates as compared to the estimates from complete data. Such results are expected considering the fact that the first and second time points contained observed data for all patients considered in the analysis. An examination of the efficiency suggested that the estimates from MI were typically lower than those from direct likelihood. Nevertheless, the corresponding MI estimates estimates did not differ significantly from those of direct likelihood. Differences in efficiency estimates were never more than 0.03.

Considering the 20% dropout rate, the results revealed that direct likelihood slightly produces higher biases (the only exceptions to this rule occurred for estimate of AX23 and time 4). Regarding efficiency, both the MI and direct likelihood methods yielded estimates similar to each other, and in general, MI tends to have the smallest estimates. A comparison of 30% dropout rate again suggested that the estimates associated with MI were less biased than for direct likelihood, except for time 4. Efficiency results based on both methods were generally similar to their results with 10 and 20% dropout rates. Furthermore, between the two methods, the MI estimates were slightly different from those obtained by direct likelihood, although the degree of these differences was not very large. Overall, the performance of both methods appeared to be independent of the dropout rate.

One would ideally need to compare various means with each other. If there is no drug by time interaction, then we will often need to make comparisons between the drug main effect means and the time main effect means. Since the interaction effect mean is significant (as shown in Table 1), we need to compare the drugs with one another at each time point and/or times to one another for each drug. Comparisons of the time means within a drug are given in Fig. 1. Since the levels of time are quantitative and equally spaced, orthogonal polynomials can be used to check for linear and quadratic trends over time for each drug. The linear and quadratic trends in time for all drugs reveal that drug BWW9 shows a negative linear trend, and drug AX23

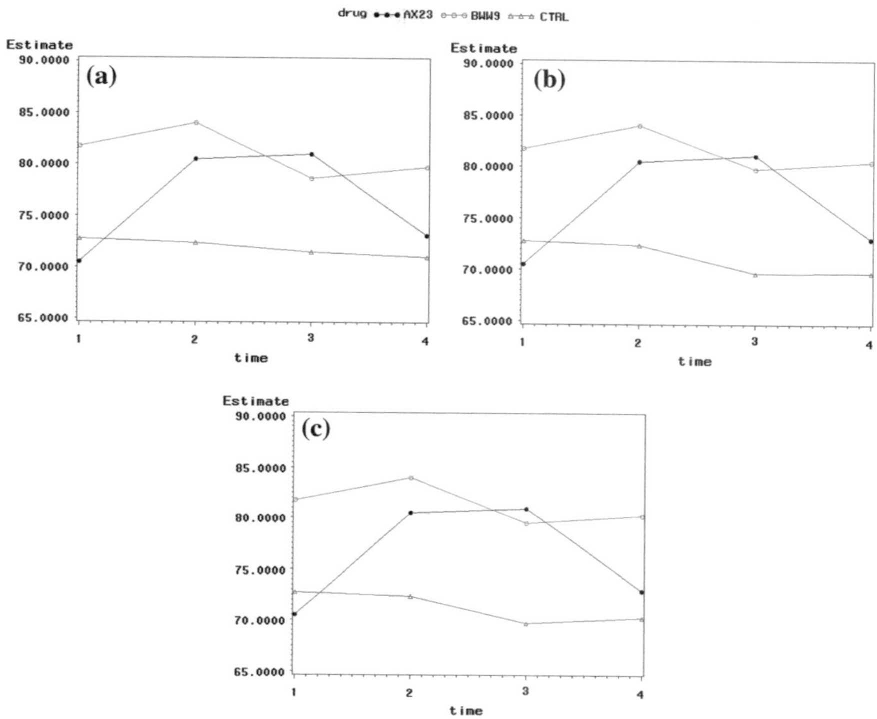

Fig. 1 **a** The heart rate data—Means over time for each drug for the heart rate data. **b** Direct likelihood—Means over time for each drug for the heart rate data. **c** Multiple imputation—Means over time for each drug for the heart rate data

shows a strong quadratic trend in all methods. Evidently, the differences occurred with drug CTRL in graphs (b) and (c) for direct likelihood and MI, respectively. Both methods yielded slightly different linear trends as compared to that from actual data. The graph in Fig. 1 displays these relationships.

4 Dropout Analysis Strategies in Longitudinal Binary Data

There is a wide range of statistical methods for handling incomplete longitudinal binary data. The methods of analysis to deal with dropout comprise three broad strategies: semi-parametric regression, multiple imputation (MI) and maximum likelihood (ML). In what follows, we utilize three common statistical methods in practice, namely WGEE, MI-GEE and GLMM. First, we compare the performance of the two versions or modifications of the GEE approach, and then show how they compare to the likelihood-based GLMM approach.

4.1 Weighted Generalized Estimating Equation (WGEE)

Next, we follow the description provided by Verbeke and Molenberghs (2005) in formulating the WGEE approach, thereby illustrating how WGEE can be incorporated into the conventional GEE implementations. Generally, if inferences are restricted to the population averages, exclusively the marginal expectations $E(Y_{ij}) = \mu_{ij}$ can be modelled with respect to covariates of interest. This can be done using the model $h(\mu_{ij}) = x_{ij}'\beta$, where $h(.)$ denotes a known link function, for example, the logit link for binary outcomes, the log link for counts, and so on. Further, the marginal variance depends on the marginal mean, with $Var(Y_{ij}) = v(\mu_{ij})\Omega$, where $v(.)$ and Ω denote a known variance function and a scale (overdispersion) parameter, respectively. The correlation between Y_{ij} and Y_{ik}, where $j \neq k$ for $i, j = 1, 2, ..., n_i$, can be given through a correlation matrix $C_i = C_i(\rho)$, where ρ denotes the vector of nuisance parameters. Then, the covariance matrix $V_i = V_i(\beta, \rho)$ of Y_i can be decomposed into the form $\Omega A_i^{1/2} C_i A_i^{1/2}$, where A_i is a matrix with the marginal variances on the main diagonal and zeros elsewhere. Without missing data, the GEE estimator for β is based on solving the equation

$$S(\beta) = \sum_{i=1}^{N} \frac{\partial \mu_i}{\partial \beta'} (A_i^{1/2} C_i A_i^{1/2})^{-1} (y_i - \mu_i) = 0, \tag{10}$$

in which the marginal covariance matrix V_i contains a vector ρ of unknown parameters. Now, assume that the marginal mean μ_i has been correctly modeled, then it can be shown that using Eq. (10), the estimator $\hat{\beta}$ is normally distributed with mean equal to β and covariance matrix equal to

$$Var(\hat{\beta}) = I_0^{-1} I_1 I_0^{-1}, \tag{11}$$

where

$$I_0 = \left(\sum_{i=1}^{N} \frac{\partial \mu_i'}{\partial \beta} V_i^{-1} \frac{\partial \mu_i}{\partial \beta'} \right), \tag{12}$$

and

$$I_1 = \left(\sum_{i=1}^{N} \frac{\partial \mu_i'}{\partial \beta} V_i^{-1} Var(y_i) V_i^{-1} \frac{\partial \mu_i}{\partial \beta'} \right). \tag{13}$$

For practical purposes, in (13), $Var(y_i)$ can be replaced by $(y_i - \mu_i)(y_i - \mu_i)'$ which is unbiased on the sole condition that the mean is correctly specified (Birhanu et al. 2011). Note that GEE arises from non-likelihood inferences, therefore "ignorability" discussed above, cannot be invoked to establish the validity of the method when dropout is under MAR hold (Liang and Zeger 1986). Only, when dropout is MCAR; that is, $f(r_i \mid y_i, X_i, \gamma) = f(r_i \mid X_i, \gamma)$ will estimating equation (10) yield consistent estimators (Liang and Zeger 1986). Under MAR, Robins et al. (1995) proposed the

WGEE approach to allow the use of GEE under MAR. The weights used in WGEE, also termed inverse probability weights, reflect the probability for an observation of subject to be observed (Robins et al. 1995). Therefore, the incorporation of these weights reduces possible bias in the regression parameter estimates. Based on Molenberghs and Verbeke (2005), we discuss the construction of these weights. According to them, such weights can be calculated as

$$\omega_{ij} \equiv P[D_i = j] = \prod_{k=2}^{j-1}(1 - P[R_{ik} = 0 \mid R_{i2} = \ldots = R_{i,k-1=1}]) \times P[R_{ij} = 0 \mid R_{i2} = \ldots = R_{i,j-1} = 1]^{I\{j \leq n_i\}}, \tag{14}$$

where $j = 2, 3, \ldots, n_i + 1$, $I\{\}$ is an indicator variable, and D_i is the dropout variable. The weights are obtained from the inverse probability provided the actual set of measurements are observed. In terms of the dropout variable D_i, the weights are written as

$$\omega_{ij} = \begin{cases} P(D_i = j \mid D_i \geq j) & \text{for} j = 2 \\ P(D_i = j \mid D_i \geq j)\prod_{k=2}^{j-1}[1 - P(D_i = k \mid D_i \geq k)] & \text{for} j = 3, \ldots, n_i \\ \prod_{k=2}^{n_i}[1 - P(D_i = k \mid D_i \geq k)] & \text{for} j = n_i + 1. \end{cases} \tag{15}$$

Now, from Sect. 2 recall that we partitioned Y_i into the unobserved components (Y_i^m) and the observed components (Y_i^o). Similarly, the mean μ_i can be partitioned into observed (μ_i^o) and missing components (μ_i^m). In the WGEE approach, the score equations to be solved are:

$$S(\beta) = \sum_{i=1}^{N}\sum_{d=2}^{n_i+1} \frac{I(D_i = d)}{\omega_{id}} \frac{\partial \mu_i}{\partial \beta'}(d)(A_i^{1/2}C_iA_i^{1/2})^{-1}(d)(y_i(d) - \mu_i(d)) = 0, \tag{16}$$

where $y_i(d)$ and $\mu_i(d)$ are the first $d - 1$ elements of y_i and μ_i respectively. In Eq. (16), $\frac{\partial \mu_i}{\partial \beta'}(d)$ and $(A_i^{1/2}C_iA_i^{1/2})^{-1}(d)$ are defined analogously, in line with the definitions of Robins et al. (1995). Provided that the ω_{id} are correctly specified, WGEE provides consistent estimates of the model parameters under a MAR mechanism (Robins et al. 1995).

4.2 Multiple Imputation Based GEE (MI-GEE)

An alternative approach that is valid under MAR is multiple imputation prior to generalized estimating equations, or, as we will term it in the remainder of this article, MI-GEE. The primary idea of the combination of MI and GEE comes from Schafer (2003). He proposed an alternative mode of analysis based on the following steps. (1) Impute the missing outcomes multiple times using a fully parametric model, such as a random effects type model. (2) After drawing the imputations, analyze the so-completed data sets using a conventional marginal model, such as the GEE method. (3) Finally, perform MI inference on the so-analyzed sets of data. As pointed out by Beunckens et al. (2008), MI-GEE comes down to first using the predictive distribution of the unobserved outcomes, conditional on the observed ones and covariates. Thereafter, when MAR is valid, missing data need no further attention during the analysis. In terms of the dropout mechanism, in the MI-GEE method, the imputation model needs to be specified. This specification can be done by an imputation model that imputes the missing values with a given set of plausible values (Beunckens et al. 2008). Details of this method can be found in Schafer (2003), Molenberghs and Kenward (2007) and Yoo (2009). In closely related studies, Beunckens et al. (2008) studied the comparison between the two GEE versions (WGEE and MI-GEE), and Birhanu et al. (2011) compared the efficiency and robustness of WGEE, MI-GEE and doubly robust GEE (DR-GEE). In this paper, however, we restrict attention to study how the two types of GEE (WGEE and MI-GEE) compared to the likelihood-based GLMM for analyzing longitudinal binary outcomes with dropout.

In the previous section, GEE, a special case of inverse probability weighting, was described as a useful device for the analysis of incomplete data, under an MAR mechanism. In this section, MI was described, and this suggests an alternative approach to handling MAR missingness when using GEE: use MAR-based MI together with a final GEE analysis for the substantive model. This emphasizes the valuable flexibility that this facility brings to MI, and can be considered as an example of using uncongenial imputation model. The term uncongenial was introduced by Meng (1994) for an imputation model that is not consistent with the substantive model, and it is for this reason that MI has much to offer in this setting. Further, Meng (1994) stated that it is one of the great strength of MI that these two models (substantive and imputation) do not have to be consistent in the sense that the two models need not be derived from an overall model for the complete data. GEE is one of the examples of situations in which such uncongenial imputation models might be of value (Molenberghs and Kenward 2007). As noted above GEE is valid under MCAR but not MAR. An alternative approach that is valid under MAR is MI prior to GEE, in which the imputation model is consistent with the MAR mechanism, but not necessarily congenial with the chosen substantive model. The population-averaged substantive model does not specify the entire joint distribution of the repeated outcomes, particularly the dependence structure is left unspecified, and so cannot be used as a basis for constructing the imputation model. Since we consider the MI-GEE method, the M imputed data sets combined with GEE on the imputed data is an alternative technique to likelihood inference and WGEE. It requires MAR for valid inferences.

4.3 Generalized Linear Mixed Model (GLMM)

An alternative approach to deal with dropout under MAR is to use likelihood-based inference (Verbeke and Molenberghs 2000). A commonly encountered random effects (or subject-specific) model for discrete longitudinal data is the generalized linear mixed model (GLMM) which is based on specifying a regression model for the responses conditional on an individual's random effects and assuming that within-subject measurements are independent, conditional on the random effects. The marginal likelihood in the GLMM is used as the basis for inferences for the fixed effects parameters, complemented with empirical Bayes estimation for the random effects (Molenberghs and Kenward 2007). As pointed out by Alosh (2010), the random effects can be included as a subset of the model for heterogeneity from one individual to another. Integrating out the random effects induces marginal correlation between the responses through the same individual (Laird and Ware 1982). Next, we briefly introduce a general framework for mixed effects models provided by Jansen et al. (2006) and Molenberghs and Kenward (2007). It is assumed that the conditional distribution of each Y_i, given a vector of random effects b_i can be written as follows

$$Y_i \mid b_i \sim F_i(\theta, b_i), \tag{17}$$

where Y_i follows a prespecified distribution F_i, possibly depending on covariates, and is parameterized via a vector θ of unknown parameters common to all individuals. The term b_i denotes the $(q \times 1)$ vector of subject-specific parameters, called random effects, which are assumed to follow a so-called mixing distribution Q. The distribution Q depends on a vector of unknown parameter, say ψ; that is, $b_i \sim Q(\psi)$. In terms of the distribution of Y_i, the b_i reflect the between unit-heterogeneity in the population. Further, given the random effects b_i, it is assumed that the components Y_{ij} in Y_i are independent of one another. The distribution function (F_i) provided in model (17) becomes a product over the n_i independent elements in Y_i. Inference based on the marginal model for Y_i can be obtained across their distribution $Q(\psi)$, provided one is not following a fully Bayesian approach. Now, assume that the $f_i(y_i \mid b_i)$ represents the density function and corresponds to the distribution F_i, while $q(b_i)$ represents the density function and corresponds to the distribution Q. Thus, the marginal density function of Y_i can be written as follows

$$f_i(y_i) = \int f_i(y_i \mid b_i) q(b_i) db_i. \tag{18}$$

The marginal density is dependent on the unknown parameters θ and ψ. By assuming the independence of the units, the estimates of $\hat{\theta}$ and $\hat{\psi}$ can be obtained using the maximum likelihood function that is built into model (18). The inferences can be obtained following the classical maximum likelihood theory. The distribution Q is assumed to be of a specific parametric form, for example a multivariate normal distribution. The integration in model (18), depending on both F_i and Q_i, may or may not be analytically possible. However, there are some proposed solutions based on Taylor series expansions of either $f_i(y_i \mid b_i)$ or on numerical approximations of the integral, for example, adaptive Gaussian quadrature. Verbeke and Molenberghs (2000) noted that for the classical linear mixed model, $E(Y_i)$ equals $X_i\beta$, meaning that the fixed effects have a subject-specific as well as a population-averaged interpretation. However, for nonlinear mixed models, the interpretation of random effects has important ramifications for the interpretation of the fixed effects regression parameters. The fixed effects only reflect the conditional effect of covariates, and the marginal effect is difficult to obtain, as $E(Y_i)$ is given by

$$E(Y_i) = \int y_i \int f_i(y_i \mid b_i) q(b_i) db_i dy_i. \tag{19}$$

In GLMM, a general formulation is as follows. It assumes that the elements Y_{ij} of Y_i are conditionally independent, given a $(q \times 1)$ vector of random effects b_i, with density function based on a classical exponential family formulation with conditional mean depending on both fixed and random effects. This leads to the conditional mean $E(Y_{ij} \mid b_i) = a'(\eta_{ij}) = \mu_{ij}(b_i)$, and the conditional variance is assumed to depend on the conditional mean according to $Y_{ij} \mid b_i = \Theta a''(\eta_{ij})$. One needs a link function, say h (for binary data, a canonical link is the logit link), and typically uses a linear regression with parameters β and b_i for the mean, i.e., $h(\mu_i(b_i)) = X_i\beta + Z_i b_i$. Here, we note that the linear mixed model is a special case with an identity link function. The random effects b_i are again assumed to be sampled from a multivariate normal distribution, with mean 0 and $(q \times q)$ covariance matrix. The canonical link function is usually used to relate the conditional mean of Y_{ij} to η_i; that is, $h = a'^{-1}$, such that $\eta_i = X_i\beta + Z_i b_i$. In principle, any suitable link function can be used (Fitzmaurice et al., 2004). In considering the link function of the logit form and assuming the random effects to be normally distributed, the familiar logistic-linear GLMM follows. For a more detailed overview, see, Jansen et al. (2006) and Molenberghs and Verbeke (2005).

4.4 Simulation Study

Note that the parameters in a marginal model, such as GEE, and a hierarchical model, such as GLMM, do not have the same interpretation. Indeed, the fixed effects in the latter are to be interpreted conditional upon the random effect. While there is no difference between the two in the linear mixed model, this is not the case for non-Gaussian outcomes, in particular for binary data. Fortunately, as stated in Molenberghs and Verbeke (2005) and references therein, the GLMM parameters can be approximately transformed to their marginal counterpart. In particular, when the random-effects structure is confined to a random intercept b_i, normally distributed with mean 0 and variance σ^2, then the ratio between the marginal and random effects parameter is approximately equal to $\sqrt{1 + c^2\sigma}$, where $c = 16\sqrt{3}/(15\pi)$. This ratio will be used in our simulation study to make the parameters comparable.

4.4.1 Design

The main objective of this study was to compare WGEE, MI-GEE and GLMM for handling dropout missing at random in longitudinal binary data. To do so, we used the following steps: (1) A complete longitudinal binary data set was generated, and the marginal logistic regression was fitted to the data to derive the parameter estimators. (2) Once the complete dataset was generated, 100 random samples of $N = 250$ and 500 subjects were drawn. (3) MAR dropout was created, for various dropout rates. (4) The above methods were applied to each simulated data set. The results from the simulated data were then compared with those obtained from the complete data. (5) The performances of WGEE, MI-GEE and GLMM were evaluated in terms of bias, efficiency and mean square error (MSE). The GLMM estimates were first adjusted for comparability before this evaluation of performance.

4.4.2 Data Generation

Simulated data were generated in order to emulate data typically found in longitudinal binary clinical trials data. The longitudinal binary data with dropout were simulated by first generating complete data sets. Then, 100 random samples of sizes $N = 250$ and 500 subjects were drawn. We assumed that subjects were assigned to two arms (Treatment = 1 and Placebo = 0). We also assumed that measurements were taken under four time points ($j = 1, 2, 3, 4$). The outcome (Y_{ij}) which is the measurement of subject i, measured at time j, was defined as 1 if the measurement is positive, and 0 if otherwise. The two levels of the outcome can represent a specific binary health outcome, but generally we labeled one outcome "success, i.e., 1" and the other "failure, i.e., 0". Then, we looked at logistic regression as modeling the success probability as a function of the explanatory variables. The main interest here is in the marginal model for each binary outcome Y_{ij}, which we assumed follows a logistic regression. Consequently, longitudinal binary data were generated according to the following logistic model with linear predictor

$$logitE(y_ij = 1|T_j, trt_i, b_i) = \beta_0 + b_i + \beta_1 T_j + \beta_2 trt_i + \beta_3(T_j * trt_i), \qquad (20)$$

where $\beta = (\beta_0, \beta_1, \beta_2, \beta_3)$, and the random effects b_i's are assumed to account for the variability between individuals and assumed to be $i.i.d.$ with a normal distribution, i.e., $b_i \sim N(0, \sigma^2)$. In this model, fixed categorical effects include treatment (trt), times (T) and treatment-by-time interaction ($T * trt$). For this model, throughout, we fixed $\beta_0 = -0.25$, $\beta_1 = 0.5$, $\beta_2 = 1.0$ and $\beta_4 = 0.2$. We also set a random intercept $b_i \sim N(0, 0.07)$. For each simulated data set, dropouts were created in the response variable, Y_{ij}, chosen stochastically. We assumed that the dropout can occur only after the second time point. Consequently, there are three possible dropout patterns. That is, dropout at the third time point, dropout at the fourth time point, or no dropout. The dropouts were generated at time j and the subsequent times were assumed to be dependent on the values of outcome measured at time $j-1$. Under model (20), we simulated a case where the MAR specification was different for the two outcomes (positive and negative). In particular, for time point, $j = 3$, we retained the criterion that if the dependent variable (Y_{ij}) was positive (i.e., $Y_{ij} = 1$), then the subject dropped out at the next time point, i.e., $j+1$. Dropouts were selected to yield approximate rates of 10, 20 and 30%. A monotone missing pattern (i.e., data for an individual up to a certain time) was considered, thus simulating a trial where the only source of dropout was an individual's withdrawal.

4.5 Analysis

In the analysis, different strategies were used to handle dropout: by weighting, by imputation and by analyzing the data with no need to impute or weight, consistent with MAR assumption, for WGEE, MI-GEE and GLMM, respectively.

4.5.1 WGEE

As discussed above, the WGEE method requires a model for the dropout mechanism. Consequently, we first fitted the following dropout model using a logistic regression,

$$logitP(D_i = j \mid D_i \geq j) = \gamma_0 + \gamma_1 y_{i,j-1} + \gamma_2 trt_i, j = 3, 4. \qquad (21)$$

where the predictor variables were the outcomes at previous occasions ($y_{i,j-1}$), supplemented with genuine covariate information. Model (21) is based on logistic regression for the probability of dropout at occasion j for individual i, conditional on the individual still being in the study (i,e., the probability of being observed is modeled). Note that mechanism (21) allows for the one used to generate the data and described in above only as a limiting case. This is because our dropout generating mechanism has a deterministic flavor. Strictly speaking, the probabilities of observation in WGEE are required to be bounded away from zero, to avoid issues with the weight. The effect of our choice is that WGEE is subjected to a severe stress test. It will be seen in the results section that, against this background, WGEE performs rather well. To estimate the probabilities for dropout as well as to pass the weights (predicted probabilities) to be used for WGEE, we used the "DROPOUT" and "DROPWGT" macros described in Molenberghs and Verbeke (2005). These macros could be used without modification. The "DROPOUT" macro is used to construct the variables *dropout* and *previous*. The outcome *dropout* is binary and indicates if individual had dropped out of the study before its completion, whereas, the *previous* variable refers to the outcome at previous occasions. After fitting a logistic regression, the "DROP-WGT" macro is used to pass the weights to the individual observations in WGEE. Such weights, calculated as the inverse of the cumulative product of conditional probabilities, can be estimated as $w_{ij} = 1/(\lambda_{i1} \times ... \times \lambda_{ij})$, where λ_{ij} represents the probability of observing a response at time j for the ith individual, conditional on the individual being observed at the time $j - 1$. Once the dropout model (21) was fitted and the weight distribution was checked, we merely included the weights by means of the WEIGHT statement in SAS procedure GENMOD. As mentioned earlier, the marginal measurement model for WGEE should be specified. Therefore, the model that we considered takes the form of

$$logitE(y_{ij}) = \beta_0 + \beta_1 T_j + \beta_2 trt_i + \beta_3 (T_j * trt_i). \qquad (22)$$

Here, we used the compound symmetry (CS) working correlation matrix. A random intercept b_i was excluded when considering WGEE.

4.5.2 MI-GEE

The analysis was conducted by imputing missing values using the SAS procedure MI, which employs a conditional logistic imputation model for binary outcomes. For the specification of the imputation model, an MAR mechanism is considered; that is, the imputation model comprises two-level covariate (i.e., treatment versus placebo classification) as well as longitudinal binary outcomes values at times $j =$ 1; 2; 3; 4. To be precise, for the imputation model, we used a logistic regression with measurements at the second time point as well as the two-level covariate to fill in the missing values that occur at the third time point. In a similar way, the imputation at the fourth time point is done using the measurements at the third time point including both imputed and observed, as predictors, as well as the measurements at the second time point which is always observed and the two-level covariate. Note that we describe here multiple imputation in a sequential fashion, making use of the time ordering of the measurements. Therefore, the next value is imputed based on the previous values, whether observed or already imputed. This is totally equivalent to an approach where all missing values are imputed at once based on the observed sub-vector. This implies that the dropout process was accommodated in the imputation model. It appears that there is potential for misspecification here. However, multiple imputation is valid under MAR. Whether missingness depends on one or more earlier outcomes, MAR holds, so the validity of the method is guaranteed (Molenberghs and Kenward 2007). In terms of the number of the imputed data sets, we used $M = 5$ imputations. GEE was then fitted to each completed data set using SAS procedure GENMOD. The GEE model that we considered is based on (22). The results of the analysis from these 5 completed (imputed) data sets were combined into a single inference using Eqs. (5),

(6), (7) and (8). This was done by using SAS procedure MIANALYZE. Details of implementation of this method are given in Molenberghs and Kenward (2007) and Beunckens et al. (2008).

4.5.3 GLMM

Conditionally on a random intercept b_i, the logistic regression model is used to describe the mean response, i.e., the distribution of the outcome at each time point separately. Specifically, we considered fitting model (20). This model assumed that there is natural heterogeneity across individuals and accounted for the within-subject dependence in the mean response over time. Model (20) was fitted using the likelihood method by applying the NLMIXED procedure in SAS software. This procedure relies on numerical integration and includes a number of optimization algorithms (Molenberghs and Verbeke 2005). Given that the evaluation and maximization of the marginal likelihood for GLMM needs integration, over the distribution of the random effects, the model was fitted using maximum likelihood (ML) together with adaptive Gaussian quadrature (Pinheiro and Bates 2000) based on numerical integration which works quite well in procedure NLMIXED. This procedure allows the use of Newton-Raphson instead of a Quasi-Newton algorithm to maximize the marginal likelihood, and adaptive Gaussian quadrature was used to integrate out the random effects. The adaptive Gaussian quadrature approach makes Bayesian approaches quite appealing because it is based on numerical integral approximations centered around the empirical Bayes estimates of the random effects, and permits maximization of the marginal likelihood with any desired degree of accuracy (Anderson and Aitkin 1985). An alternative strategy to fitting the GLMM is the penalized quasi-likelihood (PQL) algorithm (Stiratelli et al. 1984). However, in this study this algorithm is not used as it often provides highly biased estimates (Breslow and Lin 1995). Also, we ought to keep in mind that the GLMM parameters need to be re-scaled in order to have an approximate marginal interpretation and to become comparable to their GEE counterparts.

4.5.4 Evaluation Criteria

In the evaluation, inferences are drawn on the data before dropouts are created and the results used as the main standard against those obtained from applying WGEE, GLMM and MI-GEE approaches. We evaluated the performance of the methods using bias, efficiency, and mean square error (MSE). These criteria are recommended in Collins et al. (2001) and Burton et al. (2006). (1) Evaluation of bias: we defined the bias as the difference between the average estimate and the true value; that is, $\pi = (\bar{\hat{\beta}} - \beta)$ where β is the true value for the estimate of interest, $\bar{\hat{\beta}} = \Sigma_{i=1}^{S}\hat{\beta}_i/S$, S is the number of simulation performed, and $\hat{\beta}_i$ is the estimate of interest within each of the $i = 1, \ldots, S$ simulations. (2) Evaluation of efficiency: we defined the efficiency as the variability of the estimates around the true population coefficient. In this chapter, it was calculated by the average width of the 95% confidence interval. The 95% interval is approximately four times the magnitude of the standard error. Therefore, a narrower interval is always desirable because it leads to more efficient methods. (3) Evaluation of accuracy: the MSE provides a useful measure of the overall accuracy, as it incorporates both measures of bias and variability (Collins et al. 2001). It can be calculated as follows: MSE=$(\bar{\hat{\beta}} - \beta)^2$+$(SE(\hat{\beta}))^2$, where $SE(\hat{\beta})$ denotes the empirical standard error of the estimate of interest over all simulations (Burton et al. 2006). Generally, small values of MSE are desirable (Schafer and Graham 2002).

4.5.5 Simulations Results

The simulations results of WGEE, MI-GEE and GLMM in terms of bias, efficiency and MSEs, under N=250 and 500 sample sizes are presented in Table 3. A few points about the parameter estimates obtained by the proposed methods through the three evaluation criteria may be noted for each estimate in Table 3. First, the largest bias, also the worst, are highlighted. Second, for the efficiency criterion, the widest confidence interval, also the worst, 95% interval are highlighted. Third, for the evaluation of MSEs, the greatest values, also the worst, are highlighted. As we will see, the findings in general favoured MI-GEE over both WGEE and GLMM, regardless of the dropout rates.

By looking at this table, we observed that for 10% dropout rate, bias was least in the estimates of MI-GEE than in both WGEE and GLMM. In particular, the worst performance of WGEE and GLMM on bias permeated through the estimates of β_2 and (β_0, β_1, β_3), respectively, indicating a discrepancy between the average and the true parameter (Schafer and Graham 2002). Between the two MI-GEE and WGEE methods, the WGEE estimates were slightly different from those obtained by MI-GEE, although the degree of these differences was not very large. The efficiency performance was acceptable for both methods and comparable to each other, but low for most parameters under WGEE. The efficiency estimates associated with GLMM were larger than with WGEE and MI-GEE. In terms of MSEs, both WGEE and MI-GEE outperformed GLMM as they tend to have smallest MSEs. Overall, they yielded MSEs much closer to each other, however under 500 sample size, MI-GEE gave smallest MSEs.

Considering the 20% dropout rate, the results revealed that in most cases, GLMM consistently produced the most biased estimates. The only exception occurred for estimates of β_2 under 250 sample size as well as β_2 and β_3, under 500 sample size. For estimating all parameters, efficiency estimates by WGEE and MI-GEE were similar to each other and smaller than GLMM's estimates, except for β_3 under 500 sample size. In comparison with WGEE and MI-GEE, GLMM gave larger MSEs in magnitude than the two, except for estimate of β_0 and β_2 under 250 and 500 sample sizes, respectively. Comparing WGEE and MI-GEE, the MSEs associated with both methods were closer to each other and in one case—MSE of β_3 under 250 sample

Table 3 Bias, efficiency and mean square error of the WGEE, MI-GEE and GLMM Methods, under MAR mechanism over 100 samples: $N = 250$ and 500 subjects.

Dropout rate	Parameter	Bias			Efficiency			MSE		
		WGEE	MI-GEE	GLMM	WGEE	MI-GEE	GLMM	WGEE	MI-GEE	GLMM
						$N = 250$				
10%	β_0	0.094	0.061	**0.099**	0.005	0.012	**0.018**	0.028	0.018	**0.041**
	β_1	−0.099	−0.030	**−0.107**	0.003	0.013	**0.084**	0.018	0.086	**0.097**
	β_2	**0.053**	0.039	0.050	0.004	0.004	**0.011**	0.051	0.093	**0.107**
	β_3	0.018	0.012	**0.023**	0.002	0.004	**0.005**	0.007	0.008	**0.015**
20%	β_0	0.047	0.006	**0.052**	0.012	0.012	**0.031**	0.027	**0.060**	0.031
	β_1	0.033	0.139	**0.141**	0.012	0.014	**0.028**	0.048	0.020	**0.052**
	β_2	**0.131**	0.122	0.130	0.005	0.011	**0.017**	0.051	0.091	**0.102**
	β_3	−0.076	−0.038	**0.080**	0.006	0.007	**0.009**	0.008	0.008	**0.016**
30%	β_0	−0.065	−0.036	**−0.085**	0.026	0.003	**0.041**	0.071	0.072	**0.087**
	β_1	0.167	0.143	**0.169**	**0.023**	0.011	0.013	**0.089**	0.035	0.044
	β_2	0.178	0.171	**0.182**	0.015	0.005	**0.019**	0.069	0.032	**0.073**
	β_3	0.033	**0.104**	0.079	0.013	0.005	**0.016**	0.025	0.014	**0.047**
						$N = 500$				
10%	β_0	0.043	0.011	**0.051**	0.156	0.144	**0.162**	0.019	0.016	**0.059**
	β_1	−0.179	−0.242	**−0.249**	0.057	0.054	**0.068**	0.048	0.044	**0.053**
	β_2	**0.221**	0.211	0.220	0.093	0.086	**0.129**	0.097	0.082	**0.101**
	β_3	0.047	0.010	**0.056**	**0.036**	0.032	0.034	0.009	0.009	**0.017**
20%	β_0	0.080	0.078	**0.091**	0.154	0.138	**0.161**	0.130	0.111	**0.145**
	β_1	−0.195	−0.139	**−0.201**	0.068	0.053	**0.073**	0.052	0.037	**0.082**
	β_2	0.265	**0.293**	0.289	0.099	0.089	**0.153**	**0.120**	0.118	0.119
	β_3	**0.067**	0.020	0.064	**0.041**	0.032	0.034	0.009	0.007	**0.014**
30%	β_0	**0.136**	0.117	0.121	0.131	0.164	**0.173**	0.139	0.193	**0.198**
	β_1	−0.232	−0.218	**−0.243**	0.072	0.048	**0.074**	0.066	0.061	**0.091**
	β_2	0.342	0.184	**0.351**	0.084	0.093	**0.107**	0.186	0.136	**0.193**
	β_3	**0.067**	0.066	0.064	**0.097**	0.029	0.068	**0.012**	0.010	**0.012**

Note The largest bias, efficiency and mean square error for each given estimate presented in bold. MI-GEE = multiple imputation based generalized estimating equation; WGEE = weighted generalized estimating equation; LMM = linear mixed model; GLMM = generalized linear mixed model; MSE = mean square error

size—they gave the same values. As was the case for 10% under 500 sample size, MSEs by WGEE tended to be larger than those obtained by MI-GEE.

A comparison of 30% dropout rate again suggested that the results based on GLMM typically displayed greater estimation bias than did WGEE and MI-GEE, indicating a difference between the average estimate and the true values. Efficiency by MI-GEE appeared to be independent of the sample size in most cases, meaning the MI-GEE method yielded more efficient estimates across both sample sizes. Thus, MI-GEE was more efficient than WGEE, yet more efficient than GLMM. The latter yielded the largest values in most cases. With respect to MSEs, results that are computed by GLMM yielded largest values, showing no substantial improvement over GLMM under different sample sizes when compared with the results computed by WGEE and MI-GEE. Under 500 sample size, it can also be observed that in terms of the estimate of β_3, the MSE value for WGEE was equal to that based on GLMM, and they gave larger MSEs than did MI-GEE, whereas compared to WGEE, the MI-GEE still resulted in smaller MSEs. Generally, with increasing sample size, the performance of MI-GEE was better than that for WGEE and GLMM.

4.6 Application Example: Dermatophyte Onychomycosis Study

These data come from a randomized, double-blind, parallel group, multi-center study for the comparison of two treatments (we will term them in the remainder of this article, active and placebo) for toenail dermatophyte onychomycosis (TDO). Toenail dermatophyte onychomycosis is a common toenail infection, difficult to treat, affecting more than 2% of population. Further background details of this experiment are given in De Backer et al. (1996) and in its accompanying discussion. In this study, there were 2×189 patients randomized under 36 centers. Patients were followed 12 weeks (3 months) of treatment. Further, patients were followed 48 weeks (12 months) of total follow up. Measurements were planned at seven time points, i.e., at baseline, every month during treatment, and every 3 months afterward for each patient. The main interest of this experiment was to study the severity of infection relative to treatment of TDO for the two treatment groups. At the first occasion, the treating physician indicates one of the affected toenails as the target nail, the nail that will be followed over time. We restrict our analyses to only those patients for which the target nail was one of the two big toenails. This reduces our sample under consideration to 146 and 148 patients, in active group and placebo group, respectively. The percentage and number of patients that are in the study at each month is tabulated in Table 4 by treatment arm. Due to a variety of reasons, the outcome has been measured at all 7 scheduled time points for only 224 (76%) out of the 298 participants. Table 5 summarizes the number of available repeated measurements per patient, for both treatment groups separately. We see that the occurrence of missingness is similar in both treatment groups. We now apply the aforementioned methods to this data set. Let Y_{ij} be the severity of infection, coded as yes (severe) or no (not severe), at occasion j for patient i. We focus on assessing the difference between both treatment arms for onychomycosis. An MAR missing mechanism is assumed. For the WGEE and MI-GEE methods, we consider fitting Model (22). For the GLMM method, the above mentioned ratio is used. A random intercept b_i will be included in Model (22) when considering the random effects models. The results of the three methods are listed in Table 6. It can be seen from the analysis that the associated p-values for the main variable of interest, i.e., treatment are all nonsignificant, their p-values being all greater than 0.05. Such results should be expected considering the fact both marginal and random effect model may present similar results in terms of hypothesis testing (Jansen et al. 2006). However, when compared to WGEE and MI-GEE, the

Table 4 Number and percentage of patients with severe toenail infection at each time point, for each treatment arm separately

		Baseline	1 month	2 month	3 month	6 month	9 month	12 month
Active group	Number severe	54	49	44	29	14	10	14
	N%	146	141	138	132	130	117	133
	(%)	37.0	34.7	31.9	22.0	10.8	8.5	10.2
Placebo group	Number severe	55	48	40	29	8	8	6
	N%	148	147	145	140	133	127	131
	(%)	37.2	32.6	27.6	20.7	6.0	6.3	4.6

Table 5 Toenail data: Number of available repeated measurements for each patient, by treatment arm separately

Number of observed	Active group		Placebo group	
	N	%	*N*	%
1	4	2.74	1	0.68
2	2	1.37	1	0.68
3	4	2.74	3	2.03
4	2	1.37	4	2.70
5	2	1.37	8	5.41
6	25	17.12	14	9.46
7	107	73.29	117	79.05
Total	146	100	148	100

Table 6 Toenail data: (parameter estimates; standard errors) and *p*-values for WGEE, MI-GEE and GLMM

Effect	Parameter	WGEE	MI-GEE	GLMM
Intercept	β_0	(−0.301; 0.216) (0.4613)	(−0.051; 0.233) (0.4016)	(0.421; 3.981) (0.5400)
trt_i	β_1	(−0.201; 0.069) (0.1211)	(−0.309; 0.039) (0.0998)	(0.432; 0.251) (0.1312)
T_{ij}	β_2	(0.511; 0.442) (0.0073)	(0.025; 0.301) (0.0008)	(0.705; 0.487) (0.0410)
$trt_i * T_{ij}$	β_3	(−0.118; 0.164) (0.8004)	(−0.044; 0.063) (0.7552)	(0.401; 0.222) (0.6602)

GLMM method provided different results. Namely, its estimates were much bigger in magnitude. This in line with previous study conducted by Molenberghs and Verbeke (2005). In addition, the parameter estimates as well as the standard errors are more varied for GLMM than in the WGEE and MI-GEE methods.

5 Discussion and Conclusion

In the first part of this chapter, we have compared two methods applied to incomplete longitudinal data with continuous outcomes. The findings of our analysis in general suggest that both direct likelihood and multiple imputation performed best under all three dropout rates, and they are more broadly similar in results. This is to be expected as both approaches are likelihood based and Bayesian analysis, respectively, and therefore valid under the assumption of MAR (Molenberghs and Kenward 2007). The result of direct likelihood are in line with the findings that likelihood-based analyses are appropriate for the ignorability situation (Verbeke and Molenberghs 2000; Molenberghs and Verbeke 2005; Mallinckrodt et al. 2001a, b). Because of simplicity, and ease of implementation using many statistical tools such as the SAS software procedures MIXED, NLMIXED and GLIMMIX, direct likelihood might be adequate to deal with dropout data when the MAR mechanism holds, provided appropriate distributional assumptions for a likelihood formulation of the data also hold. Moreover, a method such as multiple imputation can be conducted without problems using statistical software such as the SAS procedures MI and MIANALYZE, and if done correctly, is a versatile, powerful and reliable technique to deal with dropouts that are MAR in longitudinal data with continuous outcomes. It would appear that the recommendation of Mallinckrodt et al. (2003a), Mallinckrodt et al. (2003b) to use direct likelihood and multiple imputation for dealing with incomplete longitudinal data with continuous outcomes is supported by the current analysis. At this point, we have to make it clear that the scope of this study is limited to direct likelihood and multiple imputation strategies. We note that there are several other strategies available to deal with incomplete longitudinal data with continuous outcome under ignorability assumption, however these methods are not covered in this study.

From the second part of the chapter that is based on dealing with binary outcomes, the results in general favoured MI-GEE over both WGEE and GLMM. This MI-GEE advantage is well documented in Birhanu et al. (2011). However, the current analysis differs from that based on Birhanu et al. (2011) as their analysis compared MI-GEE, WGEE and Doubly robust GEE in terms of the relative performance of the singly robust and Doubly robust versions of GEE in a variety of correctly and incorrectly specified models. Furthermore, the bias for MI-GEE based estimates in this study was fairly small, demonstrating that the imputed values did not produce markedly more biased results. This was to be expected as many authors, for example, Beunckens et al. (2008) noted that the MI-GEE method may provide less biased estimates than a WGEE analysis when the imputation model is correctly specified. From an extensive small and high sample sizes (i.e., N=250 and 500) simulation study, it emerged that MI-GEE is rather efficient and more accurate than other methods investigated in the current paper, regardless of dropout rate which also shows that the method does well as the dropout rate increases. Overall, the MI-GEE performance appeared to be independent of the sample sizes. However, in terms of efficiency, in some cases, it was less efficient than WGEE, yet more efficient and accurate than GLMM. This was specially true for WGEE when the rate of dropout was small and the sample size was small as well. In summary, the results further recommended MI-GEE over WGEE. However, both MI-GEE and WGEE methods may be selected as the primary analysis methods for handling dropout under MAR in longitudinal binary outcomes, but convergence of the analysis models may be affected by the discreteness or sparseness of the data.

Molenberghs and Verbeke (2005) stated that the parameter estimates from the GLMM are not directly comparable to the marginal parameter estimates, even when the random effects models are estimated through a marginal inference. They also transformed the GLMM parameters to their approximate GEE counterparts, using a ratio that makes the parameter estimates comparable. Therefore, an appropriate adjustments need to be applied to GLMM estimates in order to have an approximate marginal interpretation and to become com- parable to their GEE counterparts. Using this ratio in the simulation study, the findings showed that, although all WGEE, MI-GEE and GLMM are valid under MAR, there were slight differences between the parameter estimates and never differed by a large amount, in most cases. As a result, it appeared that for both sample sizes, the GLMM based results were characterized by the larger estimates for nearly all cases, although the degree of the difference in magnitude was not very large. In addition, it did not appear that the magnitude of this difference differed between the three dropout rates.

Although there was a discrepancy between the GLMM results on the one hand, and both the WGEE and MI-GEE results on the other, there are several important points to consider in the GLMM analysis of incomplete longitudinal binary data. The fact is that the GLMM may be applicable in many situations and offers an alternative to the models that make inferences about the overall study population when one is interested in making inferences about individual variability to be included in the model (Verbeke and Molenberghs 2000; Molenberghs and Verbeke 2005). Furthermore, it is important to realize that GLMM relies on the assumption that the data are MAR, provided a few mild regularity conditions hold, and it is as easy to implement and represent as it would be in contexts where the data are complete. Consequently, when this condition holds, valid inference can be obtained with no need for extra complication or effort, and the GLMM assuming an MAR process, is more suitable (Molenberghs and Kenward 2007). In addition, the GLMM is very general and can be applied for various types of discrete outcomes when the objective is to make inferences about individuals rather than population averages, and is more appropriate for explicative studies.

As a final remark, recall that MI-GEE has been the preferred method for analysis as it outperformed both the WGEE and GLMM estimations in the simulation study results. Despite this, the current study has focussed on handling dropout in the outcome variable, the MI-GEE can be well conducted in terms of the missingness in the covariates in the context of real-life, and can yield even more precise and convincing results since the choice for the WGEE method is not that straightforward. This can be justified by the fact that in the imputation model, the covariates that are conditioned on the analysis model are not included. The other available covariates can be included in the imputation model without being of interest in the analysis model, therefore yielding better imputations as well as wider applicability. Additionally, multiple imputation methods such as MI-GEE avoid some severe drawbacks encountered using direct modeling methods such as the excessive impact of the individual weights in the WGEE estimation or the poor fit of the random subject effect in the GLMM analysis. For further discussion, see Beunckens et al. (2008).

Lastly, we submit that the scope of the second part of thus chapter is limited to three approaches. This work is not intended to provide a comprehensive account of analysis methods for incomplete longitudinal binary data. We acknowledge that there are several methods available for incomplete longitudinal binary data under the dropouts that are MAR. However, these methods are beyond the scope of the study. This article exclusively deals with the WGEE, MI-GEE and GLMM paradigms that represent different strategies to deal with dropout under MAR.

References

Alosh, M. (2010). Modeling longitudinal count data with dropouts. *Pharmaceutical Statistics*, *9*, 35–45.

Anderson, J. A., & Aitkin, M. (1985). Variance component models with binary response: Interviewer variability. *Journal of the Royal Statistical Society, Series B*, *47*, 203–210.

Beunckens, C., Sotto, C., & Molenberghs, G. (2008). A simulation study comparing weighted estimating equations with multiple imputation based estimating equations for longitudinal binary data. *Computational Statistics and Data Analysis*, *52*, 1533–1548.

Birhanu, T., Molenberghs, G., Sotto, C., & Kenward, M. G. (2011). Doubly robust and multiple-imputation-based generalized estimating equations. *Journal of Biopharmaceutical Statistics*, *21*, 202–225.

Breslow, N. E., & Lin, X. (1995). Bias correction in generalised linear models with a single component of dispersion. *Biometrika*, *82*, 81–91.

Burton, A., Altman, D. G., Royston, P., & Holder, R. (2006). The design of simulation studies in medical statistics. *Statistics in Medicine*, *25*, 4279–4292.

Carpenter, J., & Kenward, M. (2013). *Multiple imputation and its application*. UK: Wiley.

Collins, L. M., Schafer, J. L., & Kam, C. M. (2001). A comparison of inclusive and restrictive strategies in modern missing data procedures. *Psychological Methods*, *6*, 330–351.

De Backer, M., De Keyser, P., De Vroey, C., & Lesaffre, E. (1996). A 12-week treatment for dermatophyte toe onychomycosis: terbinafine 250mg/day vs. itraconazole 200mg/day? a double-blind comparative trial. *British Journal of Dermatology*, *134*, 16–17.

Dempster, A. P., Laird, N. M., & Rubin, D. B. (1977). Maximum likelihood from incomplete data via the EM algorithm (with discussion). *Journal of Royal Statistical Society: Series B*, *39*, 1–38.

Jansen, I., Beunckens, C., Molenberghs, G., Verbeke, G., & Mallinckrodt, C. (2006). Analyzing incomplete discrete longitudinal clinical trial data. *Statistical Science*, *21*, 52–69.
Laird, N. M., & Ware, J. H. (1982). Random-effects models for longitudinal data. *Biometrics*, *38*, 963–974.
Liang, K. Y., & Zeger, S. L. (1986). Longitudinal data analysis using generalized linear models. *Biometrika*, *73*, 13–22.
Little, R. J. A., & Rubin, D. B. (1987). *Statistical analysis with missing data*. New York: Wiley.
Little, R. J. A. (1995). Modeling the drop-out mechanism in repeated-measures studies. *Journal of the American Statistical Association*, *90*, 1112–1121.
Little, R. J., & DAgostino, R., Cohen, M. L., Dickersin, K., Emerson, S. S., Farrar, J., Frangakis, C., Hogan, J. W., Molenberghs, G., Murphy, S. A., Neaton, J. D., Rotnitzky, A., Scharfstein, D., Shih, W, J., Siegel, J. P., & Stern, H., (2012). The prevention and treatment of missing data in clinical trials. *The New England Journal of Medicine*, *367*, 1355–1360.
Mallinckrodt, C. H., Clark, W. S., & Stacy, R. D. (2001a). Type I error rates from mixedeffects model repeated measures versus fixed effects analysis of variance with missing values imputed via last observation carried forward. *Drug Information Journal*, *35*, 1215–1225.
Mallinckrodt, C. H., Clark, W. S., & Stacy, R. D. (2001b). Accounting for dropout bias using mixed-effect models. *Journal of Biopharmaceutical Statistics*, *11*, 9–21.
Mallinckrodt, C. H., Clark, W. S., Carroll, R. J., & Molenberghs, G. (2003a). Assessing response profiles from incomplete longitudinal clinical trial data under regulatory considerations. *Journal of Biopharmaceutical Statistics*, *13*, 179–190.
Mallinckrodt, C. H., Sanger, T. M., Dube, S., Debrota, D. J., Molenberghs, G., Carroll, R. J., et al. (2003b). Assessing and interpreting treatment effects in longitudinal clinical trials with missing data. *Biological Psychiatry*, *53*, 754–760.
Milliken, G. A., & Johnson, D. E. (2009). *Analysis of messy data. Design experiments* (2nd ed., Vol. 1). Chapman and Hall/CRC.
Molenberghs, G., Kenward, M. G., & Lesaffre, E. (1997). The analysis of longitudinal ordinal data with non-random dropout. *Biometrika*, *84*, 33–44.
Molenberghs, G., & Verbeke, G. (2005). *Models for discrete longitudinal data*. New York: Springer.
Molenberghs, G., & Kenward, M. G. (2007). *Missing data in clinical studies*. England: Wiley.
Molenberghs, G., Beunckens, C., Sotto, C., & Kenward, M. (2008). Every missing not at random model has got a missing at random counterpart with equal fit. *Journal of Royal Statistical Soceity: Series B*, *70*, 371–388.
Pinheiro, J. C., & Bates, D. M. (2000). *Mixed effects models in S and S-Plus*. New York: Springer.
Rubin, D. B. (1976). Inference and missing data. *Biometrika*, *63*, 581–592.
Rubin, D. B. (1978). Multiple imputations in sample surveys. In *Proceedings of the Survey Research Methods Section* (pp. 20–34). American Statistical Association.
Rubin, D. B., & Schenker, N. (1986). Multiple imputation for interval estimation from simple random samples with ignorable nonresponse. *Journal of the American Statistical Association*, *81*, 366–374.
Rubin, D. B. (1987). *Multiple imputation for nonresponse in surveys*. New York: Wiley.
Rubin, D. B. (1996). Multiple imputation after 18+ years (with discussion). *Journal of the American Statistical Association*, *91*, 473–520.
Schafer, J. L. (1997). *Analysis of incomplete multivariate data*. New York: Champan and Hall.
Schafer, J. L., & Olsen, M. K. (1998). Multiple imputation for multivariate missing-data problems: A data analysts perspective. *Multivariate Behavioral Research*, *33*, 545–571.
Schafer, J. L. (1999). Multiple imputation: A primer. *Statistical Methods in Medical Research*, *8*, 3–15.
Schafer, J. L., & Graham, J. W. (2002). Missing data: Our view of the state of the art. *Psychological Methods*, *7*, 147–177.
Schafer, J. L. (2003). Multiple imputation in multivariate problems when the imputation and analysis models differ. *Statistica Neerlandica*, *57*, 19–35.

Stiratelli, R., Laird, N., & Ware, J. (1984). Random effects models for serial observations with dichotomous response. *Biometrics*, *40*, 961–972.

Tanner, M. A., & Wong, W. H. (1987). The calculation of posterior distributions by data augmentation (with discussion). *Journal of the American Statistical Association*, *82*, 528–550.

Verbeke, G., & Molenberghs, G. (2000). *Linear mixed models for longitudinal data*. New York: Springer.

Yoo, B. (2009). The impact of dichotomization in longitudinal data analysis: A simulation study. *Pharmaceutical Statistics*, *9*, 298–312.

Applications of Simulation for Missing Data Issues in Longitudinal Clinical Trials

G. Frank Liu and James Kost

Abstract Missing data handling in longitudinal clinical trials has gained considerable interest in recent years. Although a lot of research has been devoted to statistical methods for missing data, there is no universally best approach for analysis. It is often recommended to perform sensitivity analyses under different assumptions to assess the robustness of the analysis results from a clinical trial. To evaluate and implement statistical analysis models for missing data, Monte-Carlo simulations are often used. In this chapter, we present a few simulation-based approaches related to missing data issues in longitudinal clinical trials. First, a simulation-based approach is developed for generating monotone missing data under a variety of missing data mechanism, which allows users to specify the expected proportion of missing data at each longitudinal time point. Secondly, we consider a few simulation-based approaches to implement some recently proposed sensitivity analysis methods such as control-based imputation and tipping point analysis. Specifically, we apply a delta-adjustment approach to account for the potential difference in the estimated treatment effects between the mixed model (typically used as the primary model in clinical trials) and the multiple imputation model used to facilitate the tipping point analysis. We also present a Bayesian Markov chain Monte-Carlo method for control-based imputation which provides a more appropriate variance estimate than conventional multiple imputation. Computation programs for these methods are implemented and available in SAS.

1 Introduction

Handling missing data in longitudinal clinical trials has gained considerable interest in recent years among academic, industry, and regulatory statisticians alike. Although substantial research has been devoted to statistical methods for missing data, no single analytic approach has been accepted as universally optimal. A common recommendation is to simply conduct sensitivity analyses under different assumptions to assess

G.F. Liu (✉) · J. Kost
Merck & Co. Inc., 351 N. Sumneytown Pike, North Wales, PA, USA
e-mail: Guanghan_frank_liu@merck.com

D.-G. Chen and J.D. Chen (eds.), *Monte-Carlo Simulation-Based Statistical Modeling*, ICSA Book Series in Statistics, DOI 10.1007/978-981-10-3307-0_11

the robustness of the analysis results from a clinical trial. For decades, the gold standard for longitudinal clinical trials has been using mixed-models for repeated measures (MMRM, see e.g., Mallinckrodt et al. 2008). However, the MMRM analysis requires the assumption that all missing data are missing at random (MAR), an assumption which cannot be verified, and might even be considered as unlikely for some study designs and populations. The current expectation is that regulatory agencies will require sensitivity analyses to be conducted to evaluate the robustness of the analytic results to different missing data assumptions (European Medicines Agency 2010; National Academy of Sciences 2010). Clinical trial statisticians are thus well-advised to understand the ramifications that various missing data mechanisms (MDMs, see e.g., Little and Rubin 1987) have on their proposed analyses, most notably on bias, type I error control, and power. To that end, Monte-Carlo simulations are often used to conduct trial simulations under different MDMs for evaluating statistical analysis models.

In this chapter, we present three simulation-based approaches related to missing data issues in longitudinal clinical trials. First, a simulation-based approach is developed for generating monotone missing multivariate normal (MVN) data under a variety of MDMs (i.e., missing-completely-at-random [MCAR], MAR, and missing-not-at-random [MNAR]), which allows users to specify the expected proportion of missing data at each longitudinal time point. Second, a simulation-based approach is used to implement a recently proposed "tipping-point" sensitivity analysis method. Specifically, a delta-adjustment is applied to account for the potential difference in the estimated treatment effects between the mixed model (typically used as the primary model in clinical trials) and the multiple imputation model used to facilitate the tipping point analysis. Last, a Bayesian Markov chain Monte-Carlo (MCMC) method for control-based imputation is considered that provides a more appropriate variance estimate than conventional multiple imputation. Computation programs for some of these methods are made available in SAS.

In practice, there are two types of missing data. The first type is *intermittent missing* where the missing data of a subject is followed by at least one timepoint at which data are observed. The second type is *monotone missing,* which is typically caused by study attrition (i.e., early drop out of a subject). Common reasons for intermittent missing data include missed subject-visits, data collection errors, or data processing (e.g., laboratory) errors. Because these errors are unlikely to be related to the value of the data itself (had that value been observed), an assumption that the intermittent missing data are MAR, or even MCAR, is often appropriate. Therefore, it may be considered reasonable to first impute the intermittent missing data under MAR before performing the analysis (e.g., Chap. 4, O'Kelly and Ratitch 2014). With this consideration, the discussion that follows focuses primarily on monotone missing data. Although caution should be exercised because intermittent data might be MNAR for some studies in which the disease condition is expected to fluctuate over time.

2 Generation of Study Data with a Specified MDM and Cumulative Drop-Out Rates

Whether the clinical trial statistician wants to investigate novel approaches in the analysis of missing data or simply wants to compute power for an upcoming study, it is often useful to generate MVN data (given mean μ and covariance matrix Σ) under a specific MDM, with specified expected cumulative drop-out rates at each longitudinal timepoint. This section presents a method for generating monotone missing data, with a simple process outlining how to add intermittent (i.e., nonmonotone) missing data provided at the end of this section.

It is of interest to generate longitudinal MVN data, given μ and Σ, with expected cumulative drop-out rates (CDRs) over time under a given monotone MDM. For the MCAR MDM, this step is easily accomplished by first generating a subject-specific U(0,1) random-variate, and then comparing that variate to the target CDR at each timepoint. Starting with the first postdose timepoint, if the random variate is smaller than the target CDR, then that subject can be considered as having dropped out, with the data at that timepoint and all subsequent timepoints set to missing.

The approach for the MAR and MNAR MDMs is more complicated because the missingness for these MDMs depends on the data itself. Analytic or closed-form solutions are not yet available for a general MDM specification. As opposed to the unconditional approach used for the MCAR MDM, the subjects who have already dropped out need to be accounted for. Specifically, the conditional probability needs to be calculated for each individual who will drop out at Time t given that the subject is still in the study at Time t–1. Defining CDR_t, t = 1 to T, as the desired postdose expected cumulative droprate at Time t, this conditional proabability is expressed as $(CDR_t\text{-}CDR_{t-1})/(1\text{-}CDR_{t-1})$. For the purposes of this chapter, the baseline time point is assumed to be nonmissing (i.e., $CDR_0 = 0$).

Let $Y_{tjk} \sim N(\mu_k, \Sigma)$, with t = 0 to T, j = 1 to n, and k = 1 to K, for T total timepoints, n total observations, and K total groups (e.g., treatment arms). As noted, the baseline measurement, Y_{0jk}, is assumed to be nonmissing. Let p_{tj} represent the estimated conditional probability of dropping out at postdose time t for subject j, conditioned on subject j not having already dropped out. Finally, let Ψ represent one or more tuning parameters governing the effect of Y values on p_{tj}, with the designation of a positive (negative) value of Ψ indicating that higher (lower) values of Y are more likely to result in drop out.

Specifically, a logistic model, $\text{logit}(p_{tj}) = f(Y_j, \boldsymbol{\alpha}, \Psi)$ is considered, with the tuning parameter(s) Ψ pre-specified by users based on the desired MDM. The vector $\boldsymbol{\alpha} = (\alpha_1, \ldots, \alpha_t)$ is then estimated based on Monte-Carlo simulations such that the resulting missing data are sufficiently close to the specified CDRs (per the user-defined tolerance parameter ϵ). Without loss of generality, consider the following MDM, which follows a simple MAR process, where missingness at a given timepoint is solely a function of the observation at the previous timepoint (conditioned on the subject having not already dropped out). To simplify notation, the subject indicator j for p_{tj} and y_{tj} is supressed in the following formulas:

$$\text{logit}(p_t) = \sum_{i=1}^{t} \alpha_i + \psi\, y_{t-1}, \quad t = 1, 2, \ldots, T. \qquad (1)$$

The α_i are solved in a stepwise manner, first solving for α_1 using $\text{logit}(p_1) = \alpha_1 + \psi\, y_0$, such that p_1 is sufficiently close to CDR_1.

Next, solve for α_2 using $\text{logit}(p_2) = \hat{\alpha}_1 + \alpha_2 + \psi\, y_1$, such that p_2 is sufficiently close to $(CDR_2 - CDR_1)/(1 - CDR_1)$, where $\hat{\alpha}_1$was estimated from the previous step. Prior to iteratively solving for α_2, it is required to identify and exclude data from those subjects in the simulated dataset that have dropped out. This step is accomplished by comparing a subject-specific (and timepoint-specific) U(0, 1) variate with the subject-specific value of p_1 (which is, in part, a function of the recently solved $\hat{\alpha}_1$). This process is continued through to time T.

Each α_t is solved using a bisectional approach in conjunction with a large random sample drawn from the specified MVN distribution, with convergence for α_t declared when

$$|\hat{p}_t - (\text{CDR}_t - \text{CDR}_{t-1})/(1 - \text{CDR}_{t-1})| < \epsilon,$$

where ϵ is a user-defined convergence criterion and $\hat{p}_t$ is a function of Y, Ψ, and $\hat{\alpha}_t$. The detailed steps for calculating these $\hat{\alpha}_t$ are provided in the section that follows.

General Algorithm to Solve for $\boldsymbol{\alpha}$

A bisectional approach is used to solve for each $\boldsymbol{\alpha}_t$ sequentially as follows:

0. Generate a large dataset of observations (e.g., 100000), Y, comprised of $Y_k \sim N(\boldsymbol{\mu}_k, \boldsymbol{\Sigma})$, with the proportion of observations following the distribution of Y_k equal to π_k, as determined by the treatment ratio per the study design.

Do Steps 1–9 for t = 1, …, T:

1. Initialize $\alpha_L = -10000$, $\alpha_C = 0$, $\alpha_U = 10000$, DONE = 0, COUNTER = 0
2. If $t \geq 2$, then simulate the missingness of observations at earlier timepoints by using the previously computed α. Delete any subjects who are simulated as having dropped out.
3. For each remaining observation in Y, compute an estimate of $\widehat{f(\alpha_{tj})}$ (= f(Y, $\boldsymbol{\alpha}$, Ψ)), which is a function of the previously estimated α, α_C, and some function of the y and Ψ (depending on the MDM model).
4. Compute $\hat{p}_{tj} = (1 + \exp(-\widehat{f(\alpha_{tj})}))^{-1}$ for each remaining observation in Y.
5. Compute $\hat{p}_t$ as the mean of the $\hat{p}_{tj}$.
6. Compute DIFF = $\widehat{p_t} - (\text{CDR}_t - \text{CDR}_{t-1})/(1 - \text{CDR}_{t-1})$, a measure of how accurate this guess at $\alpha_t (=\alpha_C)$ is.
7. (a) If $|\text{DIFF}| < \epsilon$ then DONE = 1 (We are satisfied with α_C).
   (b) else if DIFF > 0 then $\alpha_U = \alpha_C$ and $\alpha_C = (\alpha_L + \alpha_C)$ / 2 (i.e., search lower).
   (c) else if DIFF < 0 then $\alpha_L = \alpha_C$ and $\alpha_C = (\alpha_U + \alpha_C)$ / 2 (i.e., search higher).
8. COUNTER = COUNTER+1; If COUNTER = 50 then DONE = 1. (the value of COUNTER may be adjusted by users to avoid an endless loop due to nonconvergence, though COUNTER = 50 is likely sufficient.)
9. If Not DONE then Go to Step 3; if DONE, set $\hat{\alpha}_t = \alpha_C$

After the $\boldsymbol{\alpha}$ vector has been estimated, the actual missing data of interest can be simulated by randomly generating the complete MVN dataset (Y), followed by the determination of missingness. Based on $\boldsymbol{\alpha}$, Ψ, the MDM model, and the randomly generated complete dataset Y, subject-specific cutpoints $\hat{p}_{tj}$ are computed. These cutpoints represent the probability that Subject j will drop out of the study at timepoint t (conditioned on not already having dropped out). For each timepoint, a uniform variate is generated and compared to the appropriate cutpoint to determine whether the subject drops out at that timepoint. The process starts at the first postdose timepoint, and proceeds sequentially up to last time point T. As noted above, this process is actually required in the stepwise generation of the α values themselves (Step 2 in the algorithm above).

We note that the proposed algorithm is set up to handle a mixed distribution, with the CDRs at each timepoint defined over all treatments arms. Of course, the different treatment arms will presumably still have different CDRs as a function of the μ_k. If different defined CDRs are desired for each treatment arm (as opposed to defining the CDRs over all treatment arms and letting the μ_k provide differentiation between the treatment droprates), then the algorithm would need to be run once for each such arm (or group of arms), yielding a different α vector for each.

In the absence of historical treatment-specific drop-out information, one could take a two-step approach in specifying the CDRs. The first step involves running the algorithm assuming a CDR (perhaps corresponding to placebo data from literature) over all treatment arms and letting the assumed efficacy (i.e., μ_k) drive the treatment-specific drop-out rates. One could then use available safety information on the drug (and placebo) to fine tune those treatment-specific CDRs, thus generating a separate α vector for each treatment group.

The process described above can also be used to generate study data with MNAR MDM. For example, the model $\text{logit}(p_t) = \sum_{i=1}^{t}\alpha_i + \psi\, y_{t-1}$ can be replaced with two options: (a) $\text{logit}(p_t) = \sum_{i=1}^{t}\alpha_i + \psi\, y_t$, in which the missing probability depends on the missing data; or (b) $\text{logit}(p_t) = \sum_{i=1}^{t}\alpha_i + \psi_1 y_{t-1} + \psi_2\, y_t$, in which p_t depends on both observed y_{t-1} and the missing data y_t. In all these models, the tuning parameters ψ, ψ_1, and ψ_2 are prespecified by users.

One might also wish to add intermittent (i.e., nonmonotone) missing data. This can be accomplished by generating a vector (one for each postdose timepoint) of independent uniform variates for each subject and then comparing that vector to cutpoints (timepoint-specific, as desired) corresponding to the missing probability at each timepoint (e.g., 0.01). Presumably, this process would be applied in conjunction with the monotone process, with the monotone process meant to emulate missingness due to subject-drop out and the intermittent process meant to emulate an MCAR process. In such a case, the process used to generate the intermittent missing data is applied independently of the process used to generate monotone missing data, with a given observation designated as missing if either process determines that observation is missing. The overall probability of missing would clearly be higher than the level specified for either process on its own, thus requiring some downward adjustment of the defined drop-out probabilities of one or both processes.

Example: Power and Bias Evaluation for a Longitudinal Study with Missing Data

For sample size and power calculations, analytic approaches are available when missing data are MAR (e.g., Lu et al. 2008, 2009). In general, power loss stemming from missing data in a longitudinal trial depends on the proportion and timing of the missing data; that is, the cumulative drop-out rates (CDRs), as a function of the different effective information yielded from the observations over time. For example, one would expect that a study with drop outs occurring gradually over time would have less power than a study in which all of the drop outs occurred between the second-to-last and the last (presumably primary) timepoint.

Despite the available analytic approaches, power calculations for longitudinal clinical trials are often conducted via simulations, given the extreme flexibility that simulations afford. The use of simulations is especially common in trials with complicating factors such as (a) interim analyses for futility or for overwhelming efficacy, (b) multiplicity approaches covering multiple timepoints/endpoints, or (c) adaptations built into the designs (e.g., dropping an arm or adjusting the randomization ratio as a function of the accruing data). Of course, power calculations can also be simulated for relatively straightforward clinical trial designs.

The following simulation study investigates the effect that different methods of generating random MVN data, primarily with respect to MDMs, have on both the bias of the parameter estimates and the corresponding power calculations. The assumed parameters are based on data obtained from an actual clinical trial. The results from 10,000 simulation runs are summarized in Table 1.

The simulations used the following assumptions (four postdose timepoints):

$\alpha = 0.050$;

N/per arm = 120.

$\mu_{Pbo} = (0.00, 0.46, 0.92, 1.37, 1.83)$; $\mu_{Act} = (0.00, 0.23, 0.46, 0.69, 0.92)$; (Higher means represent lower efficacy).

A functional form was used for the variance-covariance matrix, with $\sigma_i = c^i * \sigma_0$, with $\sigma_0 = 0.860$, and $c = 1.288$; $i = 1, \ldots, 4$; and $\rho_{ij} = r{*}b^{|j-i|-1}$, with $r = 0.748$ and $b = 0.832$; $i, j = 0, \ldots, 4$.

Notably $\sigma_0^2 = 0.740$, $\sigma_4^2 = 5.616$, and $\rho_{0,4} = 0.431$, with $\mathrm{Var}(Y_4 - Y_0) = 4.6$.

CDR = (0.086, 0.165, 0.236, 0.300), with a 30 % CDR at Timepoint 4 (T4).

The following missing data patterns (MDPs) were considered (with $\Psi = \Psi_1 = \Psi_2 = 0.5$).

- MDP0: No missing data.
- MDP1: MCAR
- MDP2: Data are MCAR but only the baseline and last timepoint values are included in the analysis (Completers Analysis).
- MDP3: MAR with $\mathrm{logit}(p_t) = \sum_{i=1}^{t} \alpha_i + \psi\, y_{t-1}$
- MDP4: MNAR with $\mathrm{logit}(p_t) = \sum_{i=1}^{t} \alpha_i + \psi\, y_t$
- MDP5: Mixture of MAR and MNAR with $\mathrm{logit}(p_t) = \sum_{i=1}^{t} \alpha_i + \psi_1 y_{t-1} + \psi_2\, y_t$

Table 1 Summary of missing data, power, and estimated treatment effects from MMRM at time 4 (10,000 simulations)

MDP	Percentage of missing at T4		Estimated bias for mean at T4						Power (%)
			Completers			MMRM			
	Drug	Pbo	Drug	Pbo	Drug-Pbo	Drug	Pbo	Drug-Pbo	
MDP0	0.0	0.0	0.00	0.00	0.00	0.00	0.00	0.00	91.0
MDP1	30.1	30.0	0.00	0.00	0.00	0.00	0.00	0.00	81.1
MDP2	30.0	30.0	0.00	0.00	0.00	0.00	0.00	0.00	78.4
MDP3	28.1	31.9	−0.27	−0.33	0.06	0.00	0.00	0.00	81.1
MDP4	27.0	33.0	−0.41	−0.53	0.08	−0.18	−0.23	0.05	79.4
MDP5	25.9	33.9	−0.53	−0.73	0.20	−0.17	−0.24	0.07	79.0

Note MDP = missing data pattern. T4 = time point 4. MMRM = mixed-methods for repeated measures

Although the simulations were conducted by defining a 30 % drop-out rate over the two treatment arms, readers should note that the two treatment arms still have different drop-out rates for the MAR and MNAR MDMs. This difference is a result of higher efficacy over time in the drug test group as compared with the placebo group. In the data simulations, we use $\Psi = \Psi_1 = \Psi_2 = 0.5$ such that a higher value (worse) of observed (in MDP3) or unobserved (in MDP4 and MDP5) response will result in a high probability of drop out. This simulates common drop out in clinical trials due to lack of efficacy.

In going from MDP2 to MDP1, a modest gain in power is observed as a result of using the partial data from subjects that dropped out prior to T4. This gain underscores the importance of using the full longitudinal dataset when calculating power, as opposed to considering only the final timepoint.

Focusing on the completers only, clear bias can be observed for the MAR and MNAR scenarios (MDP3 to MDP5); this fact is important to note when powering a study based on simple summary statistics from completers, as is often done when using results from the literature. As expected, the MMRM is unbiased for all of the MCARscenarios (MDP0 to MDP2), as well as for the MAR scenario (MDP3), since MMRMs assume that all missing data is MAR. Conversely, bias is present in the MMRM analysis for both of the MNAR scenarios (MDP4 and MDP5). Since this bias is to the detriment of the drug under study (i.e., leads to a diluted estimate of the treatment effect), these scenarios result in roughly 2 % lower power. Compared with the results from completers, the MMRM analysis had relatively smaller bias.

3 Tipping Point Analysis to Assess the Robustness of MMRM Analyses

As mentioned in Sect. 1, the MMRM, which is often used as the primary analysis model, assumes that all missing data are MAR—an assumption that cannot be verified. Both the clinical trial sponsor and government regulatory agencies are interested in assessing the robustness of any conclusions coming from an MMRM analyses against deviations from the MAR assumption. Given this interest, many methods for sensitivity analysis have been proposed and developed (see e.g., NRC, 2010 and references therein). Some of the more notable proposed methods include selection models, pattern-mixture models, and controlled-imputation models (see, e.g., Carpenter et al. 2013; Mallinckrodt et al. 2013; O'Kelly and Ratitch 2014). Another method, which has recently gained attention is the so-called *tipping point* approach. At a high level, a tipping point analysis varies the imputed values for the missing data (usually for the treatment arms only) by the exact amount needed to make a significant result turn nonsignificant.

Ratitch et al. (2013) have proposed three variations of tipping point analyses using pattern imputation with a delta adjustment. Our discussion considers the variation in which standard multiple imputation is performed first, and then a delta-adjustment

(δ) is applied simultaneously to all imputed values in the treatment group. The goal is to find the smallest δ that will turn the significant p-value (as calculated from the primary MMRM model) to a nonsignificant value. In addition to being relatively straightforward to interpret, this approach has the attractive quality of returning a quantitative result that is directly comparable on the scale of interest, which can then be put into clinical context. The following steps provide details for a bisectional procedure to solve for this tipping point δ.

General Algorithm to Solve for Tipping Point δ

Note the definitions of the following algorithm variables:

- m: the number of imputations to be used in the multiple imputation procedure (a value should be prespecified in the study protocol).
- d: the difference between the maximum and minimum values for the variable/endpoint under investigation (i.e., the maximum allowable shift)
- df: degrees of freedom
- p_{target}: Target probability (e.g., type-I error)
- t_{target}: t-value corresponding to p_{target}. If lower values of Y represent higher levels of efficacy, then this value must be negated in the search algorithm because the target t-value needs to be negative. (Note that the corresponding degrees of freedom (df) are actually a function of the data, as defined in Step 5 below).
- ϵ: A tolerance level, on the t-scale, under which convergence can be declared (e.g. 0.001).
- p_{prim}: p-value from the primary model

Given a dataset with intermittent missing data, the basic algorithm to conduct the tipping-point analysis is outlined below; instructions for procedures conducted in SAS refer to SAS version 9.3 or later.

0. Initialize $\delta_L = -d$, $\delta_C = 0$, $\delta_U = d$, DONE = 0, COUNTER = 0.
1. Using a Markov Chain Monte-Carlo method (see e.g., Schafer 1997), make the observed dataset monotone-missing. This step can be accomplished for each treatment group using `proc mi` within SAS by using the option `mcmc chain=multiple impute=monotone` in conjunction with all covariates (excluding treatment) included in the primary analysis model. This step will generate m monotone-missing datasets. Note that the study protocol should specify the random seed used in this step.
2. Applying parametric regression to the m monotone-missing datasets, impute the missing values in a stepwise fashion starting with the first postdose timepoint. This step can be accomplished for each treatment group using `proc mi` in SAS using the `monotone reg` option in conjunction with all covariates (excluding treatment) included in the primary analysis model. This step will generate m complete datasets (one imputed dataset for each of the m monotone-missing datasets).

Do Steps 3–9 while Not DONE

3. Subtract δ_C from each of the imputed values of the test drug treatment arms (to the detriment of test drug).
4. Analyze each of the m post-imputation complete datasets using the primary model, obtaining point estimates for the parameter of interest (e.g., mean change-from-baseline treatment difference at the last timepoint) and the associated variance.
5. Using the `proc mianalyze` procedure in SAS, combine the m means and variances from the m analyses to obtain the final test statistic and p-value, t_{δ_c} and p_{δ_c}, respectively (Rubin 1987). The final test statistic $\hat{Q}\,/\,(V^{(1/2)})$ is approximately distributed as t_ν, where $\hat{Q}$. is the sample mean of the m point estimates, $V = \hat{U} + (m + 1)\,(B/m)$, $\hat{U}$ is the sample mean of the m variance estimates, and B is the sample variance of the m point estimates. The degrees of freedom, ν, are computed as follows (Barnard and Rubin 1999): $\nu = [(\nu_1)^{-1} + (\nu_2)^{-1}]^{-1}$, where $\nu_1 = (m - 1)\,[1 + (\hat{U}/(1 + m^{-1})\ B)]^2$ and $\nu_2 = (1 - \gamma)\ \ \nu_0(\nu_0 + 1)\,/\,(\nu_0 + 3)$, with $\gamma = (1 + m^{-1})\ B/V$ and where ν_0 represents the complete-data degrees of freedom.
6. Compute DIFF$=t_{\delta_c} - t_{\text{target}}$.
7. (a) if |DIFF| $< \epsilon$ then DONE $= 1$ (We are satisfied with δ_C)

   (b) else if (DIFF > 0) then $\delta_L = \delta_C$; $\delta_C = (\delta_C + \delta_U)/2$; (subtract larger δ)

   (c) else if (DIFF < 0) then $\delta_U = \delta_C$; $\delta_C = (\delta_C + \delta_L)/2$; (subtract smaller δ)
8. If Not DONE, then COUNTER =COUNTER+1;
9. If COUNTER = 50, then DONE = 1; (guard against endless loop due to non-convergence)

The final δ can be interpreted as the detrimental offset needed to apply to each imputed observation to change a significant result to a nonsignificant result. Confidence in the primary results stem from a large value of δ, relative to (a) the assumed treatment difference, (b) the observed treatment difference per the primary model, and/or (c) a widely accepted clinically meaningful difference. For example, in a trial of an anti-depressant drug in which a clinically meaningful difference might be around 2–3 points, a trial result could be considered as robust if we were to subtract $\delta \geq 3$ points from every imputed value in the treatment arm and still maintain a statistically significant result.

Conventionally, MMRM analysis is based on restricted maximum likelihood while the tipping point methodology is implemented using multiple imputation (MI) analysis. Ideally, applying $\delta = 0$ to the MI analysis would yield the p-value from the MMRM analysis (p_{prim}). More important, setting $p_{\text{target}} = p_{\text{prim}}$ would ideally yield a solution of $\delta = 0$. If the δ obtained does not equal 0, then the value of δ obtained when setting $p_{\text{target}} = \alpha$ will be biased, per the intended interpretation. Unfortunately, simulation results indicate that the above method will not always yield a δ value equal to 0 when setting $p_{\text{target}} = p_{\text{prim}}$. This inconsistency might be due to additional variation in the analysis of multiple imputation as compared to the restricted maximum likelihood analysis. To overcome this discrepancy, we advise running the above algorithm twice and in the following order: first, with the setting $p_{target} = p_{prim}$, and then with the setting $p_{target} = \alpha$, yielding δ_{prim} and δ_{α}. The final δ is then computed as $\delta = \delta_\alpha - \delta_{prim}$. The value δ_{prim} can be thought of as a calibration factor in going from the MMRM to the MI model, accounting for the methodological differences between the two, as well as for the inherent randomness in the MI process. Simulation results indicate that t-values and p-values arising from the MMRM and MI models ($\delta = 0$) are highly similar, providing reassurance that the δ (i.e., $\delta_\alpha - \delta_{prim}$) obtained using the MI model translates well to the MMRM model.

As might be expected, higher values of m will yield results with greater stability. This stability applies not only to the estimates produced by the MI approach, but also to the adjusted degrees of freedom (*df*). The adjustment for the *df* was first proposed by Barnard and Rubin (1999), and has subsequently gained widespread use (e.g., adopted in SAS). The major impetus for the adjustment, as compared to the initial proposal for a *df* adjustment as cited in Rubin (1987), was to guard against the possibility that the *df* used for the MI approach would exceed the *df* present in the original MMRM for the complete data. However, this *df* adjustment might be very conservative in certain situations, particularly for smaller sample sizes when low numbers of imputations are used. This characteristic of the *df* adjustment might have the effect of producing abnormally large δ values since the respective t-values from the original MMRM and from the MI approach will be based on different t-distributions. A simple fix is to ensure that a sufficient number of imputations are used in the MI.

The following simulation study investigates (a) the variation of the *df* for the MI approach at $\delta = 0$ for different values of m, (b) the differences between the MMRM and the MI approach (at $\delta = 0$) for the t-values and p-values, (c) the variation of δ_{prim} for different values of m, and (d) the variation of $\delta = \delta_\alpha - \delta_{prim}$ for various treatment effect sizes and CDRs.

Unless otherwise noted, the assumptions used in the simulation study for Sect. 2 were also used for all simulations in Sect. 3. For ease of interpretability, a simple MAR mechanism (MDP3 from Sect. 2) was assumed for the missing data.

Assessing the variation of the *df* at $\delta = 0$

Moderate-to-large differences in *df* between the original MMRM and the MI model could cause convergence issues or unreliable results when attempting to solve for the tipping point δ. Table 2 shows the summary of *df*s from the MMRM and MI model using 1,000 simulations. For the case of MDP3, the trial had about 80 % power with about 30 % missing data at Time 4. The tipping point analysis was performed for about 800 simulated cases for which the MMRM results were significant. The *df*s from the MMRM analyses varied between 150 and 204. However, the *df*s for MI varied from 7 to 232 for $m = 5$, and from 47 to 199 for $m = 20$. Since the adjusted *df* are a direct function of the data itself, it is challenging to provide absolute general guidance as to how many imputations, m, are enough. For the scenario considered in this section, it appears as though $m = 100$ is sufficient to have relatively small variation for the *df* under consideration.

Table 2 Summary of Variation for the *df* under MI approach (at $\delta = 0$) by m (based on 1,000 simulations)

	# Sig	*df*			
		MMRM		MI	
m		Mean	p00, p25, p50, p75, p100	Mean	p00, p25, p50, p75, p100
5	800	179	150, 174, 179, 184, 201	63	7, 27, 46, 84, 232
20	800	179	154, 173, 179, 184, 204	113	47, 95, 111, 129, 199
50	811	179	153, 174, 179, 184, 197	144	91, 133, 144, 155, 195
100	825	179	151, 173, 179, 185, 203	158	122, 150, 158, 166, 194
250	777	179	154, 174, 179, 184, 201	168	137, 162, 168, 174, 192

Note MMRM = mixed-methods for repeated measures. MI = multiple imputation

Table 3 Summary of differences between the MMRM and the MI approach (at $\delta = 0$) for the t-values and p-values by m (based on 1,000 simulations)

m	# sig	t-value difference (MI-MMRM)		p-value difference (MI-MMRM)	
		Mean	p00, p25, p50, p75, p100	Mean	p00, p25, p50, p75, p100
5	800	−0.05	−1.13, −0.30, −0.06, 0.19, 1.23	0.005	−0.034, 0.000, 0.000, 0.004, 0.114
20	800	−0.03	−0.59, −0.14, −0.03, 0.08, 0.58	0.000	−0.018, 0.000, 0.000, 0.000, 0.037
50	811	−0.01	−0.53, −0.07, −0.01, 0.06, 0.37	0.000	−0.015, 0.000, 0.000, 0.000, 0.027
100	825	0.01	−0.26, −0.04, 0.01, 0.06, 0.30	0.000	−0.010, 0.000, 0.000, 0.000, 0.017
250	777	0.01	−0.19, −0.02, 0.01, 0.04, 0.17	0.000	−0.011, 0.000, 0.000, 0.000, 0.000

Note MI = multiple imputation. MMRM = mixed-methods for repeated measures

Assessing the variation of δ_{prim}

As discussed above, an adjustment (δ_{prim}) is needed to account for the differences between the primary MMRM and the MI analysis used in the tipping point procedure. The following simulation study examines the variation of δ_{prim} as well as the differences of the t-values and p-values between the primary MMRM and the MI approach with $\delta = 0$ for various values of m. The tolerance level for convergence of the t-values was set at $\epsilon = 0.005$.

The simulation results in Table 3 indicate that the differences of both the t-values and the p-values between the MMRM and the MI model at $\delta = 0$ are typically small, particularly for $m \geq 100$. This finding provides general confidence that the MI model adequately approximates the MMRM.

Due to the extensive computation required to estimate δ_{prim}, only 100 simulations were conducted to investigate the variation of δ_{prim} as a function of m. Table 4 indicates that the variation of δ_{prim} generally decreases as m increases, with the pencentiles generally shrinking to values closer to 0 as m increases. However, this trend cannot be expected to continue as $m \to \infty$, because some differences due to methodology will persist. For the examined scenario, no clear improvement was seen moving from $m = 100$ to $m = 250$.

Table 4 Summary of δ_{prim} by m (based on 100 simulations)

m	# sig	δ_{prim}	
		Mean	p00, p25, p50, p75, p100
5	81	0.48	−0.93, −0.26, 0.30, 0.95, 3.67
20	79	0.20	−0.34, −0.08, 0.19, 0.46, 1.02
50	79	0.17	−0.30, −0.09, 0.21, 0.40, 0.79
100	81	0.14	−0.16, −0.05, 0.16, 0.32, 0.56
250	85	0.18	−0.16, −0.04, 0.21, 0.37, 0.79

Assessing the Variation of the Final Tipping Point $\delta = \delta_\alpha - \delta_{\mathrm{prim}}$

The effect of various treatment differences and CDRs on the distribution of δ was examined using the same simulation assumptions as before, but fixing $m = 100$. Note that $\mu_{\mathrm{Pbo}} = (0.00, 0.46, 0.92, 1.37, 1.83)$ is held constant, while μ_{Act} is set equal to $(1-\theta)\mu_{\mathrm{Pbo}}$, $\theta = 0.35$ and 0.50, with larger values of θ resulting in larger efficacy (since higher values of μ represent lower efficacy). Cumulative drop-out rates of (0.054, 0.106, 0.154, 0.200) and (0.086, 0.165, 0.236, 0.300) were considered.

As shown in Table 4, the offset δ_{prim} needed to align the results between the MMRM and the MI model is non-ignorable across the scenarios. Focusing on the scenario with a 30 % CDR at T4 and $\theta = 0.50$, we note that the mean value of δ_{prim} needed to calibrate the two models was estimated as 0.14, with observed values ranging from −0.16 to 0.56. As a frame of reference, the true treatment difference at T4 is $(1-\theta)\mu_{\mathrm{Pbo}} - \mu_{\mathrm{Pbo}} = -0.5(1.83) = -0.92$.

Staying with the same scenario, the mean value of δ is equal to −1.34. That is, on average, all of the missing values in the treatment arm would need to be detrimentally adjusted 1.34 points in order to make the significant p-value obtained become non-significant (i.e., equal to 0.05). Assuming these results were obtained for a single study, and in the context of an observed (or assumed) treatment difference of −0.92, such a δ can be considered as evidence of a fairly robust treatment effect.

Conclusions across scenarios are best drawn by focusing on the estimated mean and quartiles (as opposed to the more variable quantities of the simulated minimum and maximum values). As expected, Table 5 demonstrates that larger detrimental values have to be applied to the imputed data from the treatment arms as (a) the drop-out rate goes down and (b) the true treatment effect goes up.

Without going into great detail, one technical point bears mentioning. When applying δ to the imputed values, it seems reasonable to not allow adjusted values past the minimum or maximum allowable value of the endpoint. However, this restriction might need to be relaxed when applying the convergence algorithm.

Table 5 Summary of Variation of δ_{prim} and Final Tipping Point δ (based on 100 simulations)

CDR at T4 (%)	θ	Number sig.	δ_{prim}		δ	
			Mean	p00, p25, p50, p75, p100	Mean	p00, p25, p50, p75, p100
20	0.35	57	0.11	−0.19, −0.05, −0.03, 0.28, 1.02	−1.41	−4.41, −1.85, −1.13, −0.70, −0.01
20	0.50	83	0.12	−0.18, −0.05, −0.01, 0.35, 0.56	−2.21	−6.68, −3.24, −1.97, −1.19, −0.11
30	0.35	52	0.16	−0.17, −0.04, 0.22, 0.30, 0.70	−1.00	−4.34, −1.53, −0.80, −0.36, −0.01
30	0.50	81	0.14	−0.16, −0.05, 0.16, 0.32, 0.56	−1.34	−3.69, −1.88, −1.32, −0.69, −0.02

Note CDR = cumulative drop-out rates

4 Monte-Carlo Approaches for Control-Based Imputation Analysis

Control-based imputation (CBI) has recently been proposed as an approach for sensitivity analysis (Carpenter et al. 2013), in which different imputation methods are used for the treatment and control groups. The missing data in the control group are imputed under the assumption of MAR, while the missing data in the treatment group are imputed using the imputation model built from the control group. One of the primary assumptions in this CBI approach is that the true post-discontinuation efficacy response in the test drug group is similar to the efficacy response of those subjects continuing in the trial in the control group. This control-based imputation model might be reasonable when no rescue or other active medications are taken by patients who drop out (Mallinckrodt et al. 2013). In general, this CBI can provide a conservative estimate of the treatment effect in superiority trials. Recently, these methods have become more attractive because the assumptions are transparent and understandable for clinical trial scientists. The methods address an attributable treatment effect (estimand) under the intent-to-treat principle but exclude the potential confounding effect of rescue medications (Mallinckrodt et al. 2013). Thus, the estimand captures the causal-effect outcomes for the test therapy.

The three most commonly used CBI methods (Carpenter et al. 2013) are defined by specifying the mean profile after drop out in the treatment group using the profile in the control group as follows:

I. **Copy Increments in Reference** (CIR): The increment mean change from the time of drop out for a patient in the treatment group will be the same as the increment mean change for a patient in the control group. Namely, the mean profile after drop out for the treatment group will be parallel to the mean profile of the control group.
II. **Jump to Reference** (J2R): The mean profile after drop out for the test drug group will equal the mean profile of the control group. That is, the mean profile for the test drug group has a 'jump' from the mean of test drug before drop out to the mean of control after drop out.
III. **Copy Reference** (CR): the mean profile for a drop-out patient in test drug group will equal the mean profile for the control group for all time points, including the time points before drop out.

These CBI approaches can be implemented using multiple imputation. Several SAS macros to implement this methodology have been developed by the Drug Information Association (DIA) Missing Data Working Group (macros are available at www.missingdata.org.uk).

Consider a response vector for patient i, $\boldsymbol{Y}_i = \{Y_{ij}, j = 1, \ldots, t\}$, and assume

$$\boldsymbol{Y}_i|\boldsymbol{X}_i \sim N\left(\boldsymbol{\mu}_i, \Sigma\right).$$

Let μ_{ij} represent the mean for patient i at time$j$, with the MMRM specified as

$$\mu_{\mathrm{ij}} = \alpha_j + \beta_j D_i + \boldsymbol{\gamma}_j' \boldsymbol{X}_{i\cdot}, \tag{2}$$

where β_j is the mean treatment difference from control at time j after adjusting for the covariates X_i, D_i is an indicator for treatment (1 for treatment and 0 for control), and $\boldsymbol{\gamma}_i$ is a vector of coefficients for the covariates. The following steps can be used to implement the CBI analysis,

1. Fit the MMRM thus yielding the estimates $\hat{\alpha}_j$, $\hat{\beta}_j$, $\hat{\boldsymbol{\gamma}}_j$ and $\hat{\Sigma}$;
2. Assume non-informative priors for the parameters, and draw a sample for these parameters from their posterior distribution, denoted by α_j, β_j, $\boldsymbol{\gamma}_j$ and Σ. Note that the DIA Missing Data Working Group macros used SAS `PROC MCMC` to fit the MMRM model and draw these parameters.
3. For a patient who dropped out at time j, draw a sample from the conditional distribution to impute the missing vector, i.e.,
4.

$$\mathbf{y}_{\mathrm{mis}} | \mathbf{y}_{\mathrm{obs}}, \mathrm{X}, \boldsymbol{\mu}, \Sigma \sim \mathrm{N}\left(\boldsymbol{\mu}_{\mathrm{m}} + \Sigma_{\mathrm{mo}} \Sigma_{\mathrm{oo}}^{-1} \left(\mathbf{y}_{\mathrm{obs}} - \boldsymbol{\mu}_{\mathrm{o}}\right), \Sigma_{\mathrm{mm}} - \Sigma_{\mathrm{mo}} \Sigma_{\mathrm{oo}}^{-1} \Sigma_{\mathrm{om}}\right) \tag{3}$$

where

$$\boldsymbol{\mu} = \begin{pmatrix} \boldsymbol{\mu}_o \\ \boldsymbol{\mu}_{\mathrm{m}} \end{pmatrix}, \quad \Sigma = \begin{pmatrix} \Sigma_{\mathrm{oo}} & \Sigma_{\mathrm{om}} \\ \Sigma_{\mathrm{mo}} & \Sigma_{\mathrm{mm}} \end{pmatrix}$$

are split into sub-vectors and block matrices with dimensions corresponding to the observed (indicated with 'o') and missing data (indicated with m) portions of the response vector. The patient and time indicators i and j are omitted in the formulas for simplicity. To implement the CBI, if a patient is in placebo group, the $\boldsymbol{\mu}_o$ and $\boldsymbol{\mu}_m$ for the placebo group will be used. Otherwise, the means will be modified as specified per the chosen CBI approach. Specifically, a patient in the treatment group who dropped out after time j will have

a.

$$\boldsymbol{\mu}_{\mathrm{m}}^{\mathrm{d}} = \begin{cases} \boldsymbol{\mu}_{\mathrm{m}}^{\mathrm{p}} + \mu_{\mathrm{j}}^{\mathrm{d}} - \mu_{\mathrm{j}}^{\mathrm{p}} & \text{for CIR} \\ \boldsymbol{\mu}_{\mathrm{m}}^{\mathrm{p}} & \text{for JR} \\ \boldsymbol{\mu}_{\mathrm{m}}^{\mathrm{p}} + \Sigma_{\mathrm{mo}} \Sigma_{\mathrm{oo}}^{-1} \left(\boldsymbol{\mu}_{\mathrm{o}}^{\mathrm{d}} - \boldsymbol{\mu}_{\mathrm{o}}^{\mathrm{p}}\right) & \text{for CR} \end{cases} \tag{4}$$

b. where the superscripts d and p are used to indicate the mean vector for drug or placebo.

5. Repeat Steps 2 and 3 for the number of imputed datasets;
6. Analyze each imputed dataset using the primary model (e.g., ANCOVA model at last visit) to get estimated treatment difference and its standard error;
7. Combine the results using Rubin's rule for final statistical inference (Rubin 1987).

Simulation studies show the estimated variances for the treatment differences using the regular MI techniques are always larger than the corresponding empirical variances. This phenomenon has been noticed for the copy-reference imputation method by Lu (2014) and Ayele et al. (2014). Lu (2014) proposed an analytical approach to get the correct variance estimate using the delta method. However, that approach is available only for copy-reference imputation and requires special programming for each specific analysis model.

Liu and Pang (2015) proposed methods to get more appropriate variances for the CBI estimates. One of their approaches is a Bayesian MCMC-based method that accounts for the pattern of missing data and obtains the estimates for the treatment difference and its variance from empirical MCMC samples. Based on the mean profile specified in equation (3), the overall treatment difference at the last time point under CBI can be written as a weighted average over the missing data patterns, that is

$$\theta^{CBI} = \sum\nolimits_{j=1}^{t} \pi_j \mu_{\mathrm{tj}}^{d} - \mu_t^{p}$$

where μ_{tj}^{d} is the mean at last time point t under missing data pattern j as given in (3) and the $\{\pi_j, j = 1, \ldots, t\}$ are the proportions of patients in the missing data patterns for the drug group. As $\sum_{j=1}^{t}\pi_j = 1$, we have

$$\theta^{\mathrm{CBI}} = \begin{cases} \sum_{j=1}^{t}\pi_j(\mu_j^{d} - \mu_j^{p})\ for\ CIR \\ \pi_t(\mu_t^{d} - \mu_t^{p}) \qquad for\ J2R \\ \sum_{j=1}^{t}\pi_j(\mu_{\mathrm{tj}}^{d} - \mu_t^{p})\ for\ CR \end{cases} \tag{5}$$

where μ_j^{d} and μ_j^{p} are the means at time j for drug and placebo, respectively. Therefore, the treatment effect under CBI can be expressed as a linear combination of the parameters of MMRM and the proportions of patients in each pattern of missing data. We note this approach is a special pattern-mixture model (PMM) where the missing data are handled differently by the pattern of missing data only for the treatment group. The missing data in the placebo group are all sampled assuming a MAR process.

To account for the uncertainty of the proportions of missing data $\{\pi_j, j = 1, \ldots, t\}$, random proportions are also drawn from a Dirichlet distribution in the Bayesian MCMC process, which corresponds to a posterior distribution for the proportions with a Jefferys prior. The empirical distribution and statistical inference for θ^{CBI} are obtained from the MCMC samples. Specifically, the following steps are implemented in the process:

1. Specify flat priors for α_j, β_j, $\boldsymbol{\gamma}_j$ and Σ, for example, use N(0, σ, 2 = 10000) for α_j, β_j, and each of the element of $\boldsymbol{\gamma}_j$, and an inverse Wishart distribution IW ($t + 1$, I), where I is an identity matrix of dimension t. We used SAS `PROC MCMC` to fit the MMRM model, specifically:

a. Use conjugate sampling to draw a sample for Σ;
b. Use random walk Metropolis to draw samples for α_j, β_j, $\boldsymbol{\gamma}_j$;
c. For a patient who dropped out at time j, `PROC MCMC` will draw a sample for the missing vector $\mathbf{y}_{\text{mis}}$ from conditional distribution as specified in (2) with the parameters from above;
d. Draw $\{\pi_j, j = 1, \ldots, t\}$ from the Dirichlet $(m_1 + 0.5, \ldots, m_t + 0.5)$, where m_j is the number of drop-out patients at time j+1 in treatment group, and m_t is the number of completers;
e. Evaluate θ^{CBI} with the formula (4).

2. The above process will be run with a burn-in, then repeat Steps a through e to obtain an empirical posterior distribution for θ^{CBI}.

Note that this Bayesian MCMC process is a simulation based approach. It is important to check the convergence of the MCMC samples. Usually, the trace-plot can be examined visually, or some statistical measures can be checked such as Geweke or Raftery-Lewis that are provided by SAS `PROC MCMC` procedure.

We apply the regular MI analysis and Bayesian MCMC approach to an antidepressant drug trial dataset created by the DIA Missing Data Working Group (Mallinckrodt et al. 2013). The dataset was constructed from an actual clinical trial and made available by the Working Group (see www.missingdata.org.uk). The dataset contains 172 patients (84 in the treatment arm, 88 in the placebo control arm). Repeated measures for Hamilton Depression 17-item total scores were taken at baseline and Weeks 1, 2, 4, and 6, post-randomization. The Week 6 measurements were completed by about 76 % of the treatment group patients and about 74 % of the the control group patients. The analysis dataset included one patient record with intermittent missing data; in all analysis methods, the missing data for this patient were imputed under the assumption of MAR. The monotone missing data were imputed under CBI methods of CIR, CR, or J2R.

In the analysis of this dataset, we noticed that the MCMC sampling had high autocorrelation. To increase the stability of the results, we used 200 imputations in the conventional MI analysis, and used 2,000 iterations for turning, 2,000 iterations for burn-in, and 200,000 in the main sampling, keeping one from every 10 samples (with option `THIN=10` in `PROC MCMC`) to get a total of 20,000 samples for the posterior mean and standard deviation. Table 6 shows the analysis results. Compared with the mixed model analysis, the Bayesian MCMC under MAR produced very similar results. As compared to the results from the mixed model, the CBI analyses based on regular MI are conservative. With CBI, the point estimates are shrunk toward 0 but the standard errors (*SE*s) are very similar to the SEs from the MAR analysis. As such, the CBI analyses with regular MI have large $p-$values compared to the primary analysis under MAR. In fact, the result of the J2R analysis becomes insignificant. With the Bayesian MCMC approach, the CBI analysis results have similar point estimates as the CBI analysis with regular MI but have smaller *SE*s. As a result, the p-values from CIR, CR, and J2R all maintained significance.

Table 6 Primary and sensitivity analysis results for an anti-depressant trial

Method	$\hat{\theta}$ (SE)	95 % CI	p-value (%)
Primary analysis under MAR			
Mixed Model	−2.80(1.12)	[−5.01, −0.60]	1.3
Bayesian MCMC	−2.80(1.13)	[−5.03, −0.57]	1.4
CBI using regular MI			
CIR	−2.46(1.10)	[−4.65, −0.28]	2.7
CR	−2.38(1.11)	[−4.57, −0.20]	3.3
J2R	−2.12(1.13)	[−4.36, 0.12]	6.3
CBI using MCMC approach			
CIR w MCMC	−2.43(1.01)	[−4.39, −0.44]	1.7
CR w MCMC	−2.34(0.99)	[−4.25, −0.53]	1.9
J2R w MCMC	−2.10(0.86)	[−3.77, −0.42]	1.5

Note CI = confidence intervals for mixed model and regular multiple imputation (MI), credible interval for Bayesian MCMC (Markov chain Monte-Carlo method) approach. MAR = missing at random. CBI = control-based imputation. CIR = copy increments in reference. CR = copy reference. J2R = jump to reference

To check the convergence of the MCMC sampling, Fig. 1 shows the diagnostics plots for both the Bayesian MCMC samples for the primary analysis under MAR and the CR analysis. The trace-plots for both parameters show good mixing and stabilization. With the option of THIN=10, the autocorrelation decreases quickly. The posterior density curves are estimated well from the 20,000 samples.

5 Discussions and Remarks

In many clinical trials, missing data might be unavoidable. We have illustrated some applications of Monte-Carlo simulation methods for handling of missing data issues for longitudinal clinical trials. Simulation-based approaches to dealing with missing data can be extremely useful in the conduct of clinical trials; most notably in the design stage to calculate needed sample size and power, as well as in the final analysis stage to conduct sensitivity analyses. We described a method to generate MVN longitudinal data under different assumed MDMs with specified CDRs. As a sensitivity analysis, we applied a δ -adjustment approach to account for the potential difference between the MMRM (typically used as the primary model in clinical trials) and the MI model used to facilitate the tipping point analysis, and to propose an adjustment to the final tipping point calculation. Depending on the number of imputations used, the inferential statistics produced by the MMRM and MI models can differ, due in part to differences in the approximated degrees of freedom. A sufficient number of imputations should be used to reduce this variation. The appropriate number of imputations to be used should be confirmed via simulation

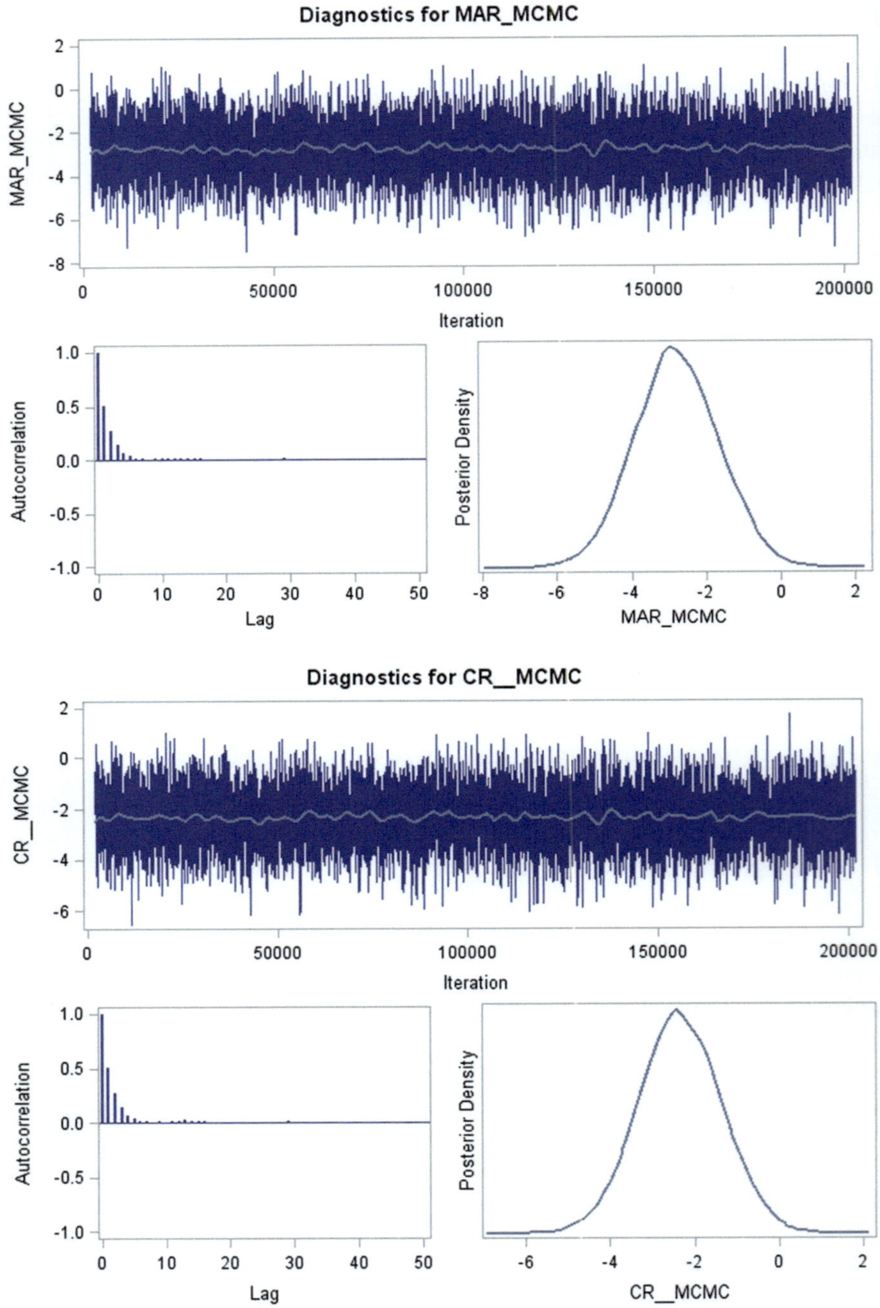

Fig. 1 Diagnostics plots for Bayesian MCMC under MAR and for Copy Reference Imputation Method

by the statistician during the analysis planning stage. We also presented a Bayesian MCMC method for CBI that provides a more appropriate variance estimate than regular multiple imputation.

The methods presented are only a few applications of simulation methods for missing data issues. Of course, many other simulation-based methods are available that can be used for missing data. For example, we considered only a logistic model for the MDM, noting that other models such as a probit model can also be used. In addition, the missing probabilities defined by the example MDMs depended only on the current time point and/or the next time point. Other MDMs may be defined allowing for the incorporation of additional time points. In the CBI approaches, we considered a Bayesian MCMC approach, although other avenues might also be used such as bootstrapping to obtain the appropriate variance for CBI methods. Although we considered only continuous endpoints, simulation-based methods can also be very useful in dealing with missing data for other types of endpoints such as binary, categorical, or time-to-event data.

One drawback for simulation-based methods is the random variation from the simulations. It is critical to assess the potential variation and/or monitor the convergence. When using simulation-based method for analysis of clinical trials, the analysis plan should pre-specify all the algorithms, software packages, and random seeds for the computation. Naturally, the analysis should use a sufficient number of imputations or replications in order to reduce the random variation. Of course, simulations examine only the statistical properties under the assumptions used in those simulations. Whenever possible, theoretical or analytic methods should be considered over simulations.

References

Ayele, B. T., Lipkovich, I., Molenberghs, G., & Mallinckrodt, C. H. (2014). A multiple imputation based approach to sensitivity analysis and effectiveness assessments in longitudinalm clinical trials. *Journal of Biopharmaceutical Statistics*, *24*, 211–228. doi:10.1080/10543406.2013.859148.

Barnard, J., & Rubin, D. B. (1999). Small-sample degrees of freedom with multiple imputation. *Biometrika*, *86*, 948–955. doi:10.1093/biomet/86.4.948.

Carpenter, J. R., Roger, J. H., & Kenward, M. G. (2013). Analysis of longitudinal trials with protocol deviation: A Framework for relevant, accessible assumptions, and inference via multiple imputation. *Journal of Biopharmaceutical Statistics*, *23*, 1352–1371. doi:10.1080/10543406.2013.834911.

European Medicines Agency. (2010). *Guideline on missing data in confirmatory clinical trials*. Retrieved from http://www.ema.europa.eu/docs/en_GB/document_library/Scientific_guideline/2010/09/WC500096793.pdf.

Little, R. J., & Rubin, D. B. (1987). *Statistical analysis with missing data*. New York: John Wiley.

Liu, G. F., & Pang, L. (2015). On analysis of longitudinal clinical trials with missing data using reference-based imputation. *Journal of Biopharmaceutical Statistics*. Advance online publication. http://dx.doi.org/10.1080/10543406.2015.1094810.

Lu, K., Luo, X., & Chen, P.-Y. (2008). Sample size estimation for repeated measures analysis in randomized clinical trials with missing data. *International Journal of Biostatistics*, *4*, 1–16. doi:10.2202/1557-4679.1098.

Lu, K., Mehrotra, D. V., & Liu, G. F. (2009). Sample size determination for constrained longitudinal data analysis. *Statistics in Medicine*, *28*, 679–699. doi:10.1002/sim.3507.

Lu, K. (2014). An analytic method for the placebo-based pattern mixture model. *Statistics in Medicine*, *33*, 1134–1145.

Mallinckrodt, C. H., Lane, P. W., Schnell, D., Peng, Y., & Mancuso, J. P. (2008). Recommendations for the primary analysis of continuous endpoints in longitudinal clinical trials. *Drug Information Journal*, *42*, 303–319. doi:10.1177/009286150804200402.

Mallinckrodt, C., Roger, J., Chuang-Stein, C., Molenberghs, G., Lane, P. W., O'Kelly, M., et al. (2013). Missing data: Turning guidance into action. *Statistics in Biopharmaceutical Research*, *5*, 369–382. doi:10.1080/19466315.2013.848822.

National Academy of Sciences. (2010). *The prevention and treatment of missing data in clinical trials. Panel on handling missing data in clinical trials* [Prepared by by the Committee on National Statistics, National Research Council]. Washington, DC: National Academics Press. Retrieved from http://www.nap.edu/catalog/12955/the-prevention-and-treatmentof-missing-data-in-clinical-trials.

O'Kelly, M., & Ratitch, B. (2014). *Clinical trials with missing data: A guide for practitioners*. West Sussex: Wiley. doi:10.1002/9781118762516.

Ratitch, B., O'Kelly, M., & Tosiello, R. (2013). Missing data in clinical trials: From clinical assumptions to statistical analysis using pattern mixture models. *Pharmaceutical Statistics*, *12*, 337–347. doi:10.1002/pst.1549.

Rubin, D. B. (1987). *Multiple imputationfor nonresponse in surveys*. New York: Wiley. doi:10.1002/9780470316696.

Schafer, J. L. (1997). *Analysis of incomplete multivariate data*. London: Chapman and Hall. doi:10.1201/9781439821862.

Application of Markov Chain Monte-Carlo Multiple Imputation Method to Deal with Missing Data from the Mechanism of MNAR in Sensitivity Analysis for a Longitudinal Clinical Trial

Wei Sun

Abstract Missing data in clinical trials could potentially arise from the mechanism of Missing Not At Random (MNAR). In order to understand the impact on a longitudinal clinical trial findings from missing data under the MNAR assumption, the sensitivity analyses could be carried out by multiple imputations. By progressively decreasing the treatment differences in those treated subjects who fell into an assumed MNAR pattern, the departure from Missing At Random (MAR) assumption could be investigated. This chapter aims to apply Markov Chain Monte-Carlo (MCMC) Multiple Imputation method to investigate that, under an MNAR assumption, that the missing data pattern of subjects who receive clinical trial treatment are similar to, worse than, or better than those of subjects who receive placebo with similar observed outcomes for two scenarios of early discontinuation: (1) discontinuation due to lack of efficacy and non-disease progression related Adverse Events (AEs); (2) discontinuation due to any reason. We also demonstrate how to apply MCMC multiple imputation method without assuming the data to have normal distribution.

Keywords Missing data · Multiple imputation · Markov Chain Monte-Carlo · Longitudinal clinical trial · Missing not at random · Early discontinuation · Sensitivity analysis · ANCOVA · Wilcoxon rank-sum test

1 Introduction

The reasons of missing data in randomized clinical trials include early discontinuations to the investigated medication (dropouts), skipped visits, and etc. A report from the National Research Council (2010) provides an excellent overview of the strengths and weaknesses of different methods to deal with missing data in a clinical trial (National Research Council 2010).

W. Sun (✉)
Manager Biostatistician at Otsuka America, Newyork, USA
e-mail: wei.sun@otsuka-us.com

D.-G. Chen and J.D. Chen (eds.), *Monte-Carlo Simulation-Based Statistical Modeling*, ICSA Book Series in Statistics, DOI 10.1007/978-981-10-3307-0_12

From a typical longitudinal clinical trial, a variable of interest is measured at baseline and fixed post-baseline time points from each randomized subject. To understand the impact on the findings in the trial from missing data under the MNAR assumption, we can analyze the incomplete longitudinal data as sensitivity analysis through multiple imputation (MI) (Rubin 1987). With this approach, multiple completed datasets are generated by imputing the missing values using an imputation model, and every completed dataset with imputed data is analyzed using the method that would have been used for a dataset without missing data. Then, parameter estimates and corresponding standard errors generated from each imputed dataset are combined in a final step to get the overall inference, which is accounted for uncertainty in the distribution of the imputed data.

There are several options for choosing the appropriate model to impute missing values. Markov Chain Monte-Carlo (MCMC) is a method to create pseudo-random samples from multidimensional and intractable probability distributions. Markov Chain Monte-Carlo of multiple imputation (MCMC MI) approach can be particularly useful for longitudinal data with missing values if the unknown missing data mechanism is missing not at random (MNAR), that is, when the missing depends on specific conditions related to the data observation or measurement. In the context of sensitivity analyses to evaluate how uncertainty in model inputs affects the model outputs, it has been acknowledged that models to deal with MNAR may be useful (Rubin 1987; Galbraith 2012; Verbeke and Molenberghs 2000).

This chapter aims to demonstrate the steps of applying MCMC MI to investigate that, under an MNAR assumption, that the missing data pattern of subjects who receive active treatment are similar to, worse than, or better than those of subjects who receive placebo with similar observed outcomes for two scenarios of early discontinuation: (1) discontinuation due to lack of efficacy and non-disease progression related Adverse Events (AEs); (2) discontinuation due to any reason, with the data from a simulated hypothetical longitudinal clinical trial.

Some continuous responses do not appear to follow a normal distribution. Near-normality can sometimes be obtained via a suitable data transformation (often logarithm), but that is not always possible, for example, the percent change from baseline of pain in a clinical trial for a medication to control pain. We probably observe negative percent changes. Importantly, regardless of whether the response of interest is intended to be analyzed on the original or a transformed scale, we often have outliers at the overall subject or single observation level in the analysis datasets from a longitudinal clinical trial. Therefore, later, after the section for the data of primary efficacy endpoint with normal distribution, we extend the demonstration, and also show how to apply MCMC multiple imputation method without assuming the data to have normal distribution.

The whole chapter is organized as follows. First, the approach suggested for dealing with the missing data with Multiple Imputation method is introduced. The details of imputation schemes employed to apply MCMC Multiple Imputation method to investigate the impact of missing data, under an MNAR assumption, are described in this section as well. Then, the design of a hypothetical longitudinal clinical trial is described. The results of two scenarios of early discontinuation are presented in this section as well. We also demonstrate how to apply MCMC multiple imputation method without assuming the data to have normal distribution in this section. Discussion for the research is provided in the last section.

2 Multiple Imputation to Deal with Missing Data

In a longitudinal clinical trial, let the underlying continuous measurement, collected at multiple time points, be denoted by Y_{it} for subject i at time point t for t = 1, …, T. Consider a randomized clinical trial with two treatment groups, A and P, where A denotes the active treatment and P the control. Let $A_i$denote the treatment (either A or P) for subject i.

Suppose that there are n subjects, then the matrix with n observations and T measurements can be written as:

$$D = \begin{pmatrix} X_1 \\ . \\ . \\ . \\ X_n \end{pmatrix} = \begin{pmatrix} x_1 & \cdots & x_{1T} \\ \vdots & \ddots & \vdots \\ x_n & \cdots & x_{nT} \end{pmatrix}$$

Missing values are an issue in a substantial number of statistical analyses in longitudinal clinical trials. It is common that some subjects may drop out of the trial before reaching the primary time point. While analyzing only complete cases has its simplicity, the information contained in the incomplete cases is lost. This approach also ignores possible systematic differences between the complete cases and the incomplete cases, and the resulting inference may be not applicable to the population of all cases, especially with a smaller number of complete cases. For longitudinal clinical trials with missing data, the sensitivity analyses are very important to assess how sensitive results are to reasonable changes in the assumptions that are made for the missing data.

There are several approaches to handling missing data in sensitivity analyses. One approach is single imputation, in which a value is substituted for each missing value. This approach treats missing values as if they were known in the complete-data analyses. Last Observation Carried Forward (LOCF) is one of the populous single imputation methods. The primary shortcoming associated with the single imputation schemes is the inability to accommodate variability/uncertainty. Single imputation does not reflect the uncertainty about the predictions of the unknown missing values, and the resulting estimated variances of the parameter estimates are biased toward zero.

Multiple imputation is another populous approach to handle missing data for continuous responses in a longitudinal clinical trial. MI was proposed by Rubin (1987), and it accounts for missing data not only by restoring the natural variability in the missing data, but also by incorporating the uncertainty caused by the estimation process.

There are general two methods: (1) the classic MI method, proposed by Rubin (1987) and based on a Bayesian paradigm, in which the model parameters are independently drawn from the posterior distribution for each imputed data set; (2) the frequentist MI method (Wang and Robins 1998) with the fixed model parameters at the maximum likelihood estimates for all imputed data sets. Once the missing data on the underlying continuous responses have been imputed, the completed data sets are then analyzed separately by standard methods, and the results are later combined to produce estimates and confidence intervals that incorporate missing data uncertainty.

Multiple imputation does not attempt to estimate each missing value through simulated values, but rather to represent a random sample of the missing values. This process results in valid statistical inferences that properly reflect the uncertainty that results from missing values, such as valid confidence intervals for parameters.

The general procedure of the MI scheme consists of:

- The missing data are imputed using an appropriate model to generate m complete data sets.
- The m complete data sets are analyzed.
- The results from the m complete data sets are combined for the inference using Rubin's Rules Rubin (1987).

2.1 Multiple Imputation via MCMC

There are some options for choosing the appropriate model to generate a complete data set. The MCMC method is one of the popular one, which is useful in both Bayesian and frequentist statistical inference. It consists of a class of algorithms for sampling from probability distributions based on constructing a Markov chain that has the desired distribution as its stationary distribution. It combines the Monte-Carlo method for sampling randomness and the Markov chain method for sampling independence with its stationary distribution.

MCMC can be used to generate pseudo-random samples from multidimensional and otherwise intractable probability distributions via Markov chains (Schafer 1997). A Markov chain, which converts the sampling schema into a time-series sequence, is a sequence of random values whose probabilities in a time interval depend upon the value of the number from a previous time point. The controlling factor in a Markov chain is the transition probability. It is a conditional probability for the system to go to a particular new state, given the current state of the system. Since the Markov chain is in a time-series format, the sample independence by examination of sample autocorrelation can be checked. As time interval increases toward infinite, the Markov chain converges to its stationary distribution. Assuming a stationary distribution exists, it is unique if the chain is irreducible. Irreducible means any set of states can be reached from any other state in a finite number of moves.

MCMC imputation is one of the features provided in the SAS Procedure MI in SAS/STAT® system. Procedure MI in SAS can be used for arbitrary missing data imputation or random sample data set simulation based on the complete input data set as prior information.

Assuming that the data are from a multivariate normal distribution, data augmentation is applied with missing data by repeating the two steps: the imputation step and the posterior step.

The imputation step simulates the missing values for each observation independently with the estimated mean vector and covariance matrix. That is, if the variables with missing values for observation i are denoted by $Y_{i(mis)}$ and the variables with observed values are denoted by $Y_{i(obs)}$, then the imputation step draws values for $Y_{i(mis)}$ from a conditional distribution $Y_{i(mis)}$ given $Y_{i(obs)}$. The posterior step simulates the posterior population mean vector and covariance matrix from the complete sample estimates. These new estimates are then used in the imputation step.

The two steps are iterated long enough for the results to be reliable for a multiply imputed data set (Schafer 1997). The goal is to have the iterates converge to their stationary distribution and then to simulate an approximately independent draw of the missing values. That is, with a current parameter estimate $\theta^{(t)}$ at t^{th} iteration, the Imputation step draws $Y_{mis}^{(t+1)}$ from $p(Y_{mis} \mid Y_{obs}, \theta^{(t)})$ and the Posterior step draws $\theta^{(t+1)}$ from $p(\theta \mid Y_{obs}, Y_{mis}^{(t+1)}.)$. This creates a Markov chain $(Y_{mis}^{(1)}., \theta^{(1)}), (Y_{mis}^{(2)}., \theta^{(2)})$, ..., which converges in distribution to $p(\theta \mid Y_{mis}, \theta \mid Y_{obs})$.

2.2 *Combining Inferences from Imputed Data Sets*

With m imputations, m different sets of the point and variance estimates for a parameter Q can be obtained. Let $\hat{Q}_i$ and $\hat{U}_i$ be the point and variance estimates from the ith imputed data set, i = 1, 2, …, m. Then the point estimate for Q from multiple imputations is the average of the m complete-data estimates:

$$\bar{Q} = \frac{1}{m}\sum\nolimits_{i=1}^{m} \hat{Q}_i$$

Let $\bar{U}$ be the within-imputation variance, which is the average of the m complete-data estimates

$$\bar{U} = \frac{1}{m}\sum\nolimits_{i=1}^{m} \hat{U}_i$$

and B be the between-imputation variance

$$B = \frac{1}{m-1}\sum\nolimits_{i=1}^{m} (\hat{Q}_i - \bar{Q})^2$$

Then the variance estimate associated with $\bar{Q}$ is the total variance

$$T = \bar{U} + (1 + \frac{1}{m})B$$

The statistic $(Q - \bar{Q})T^{-1/2}$ approximately distributed as a t-distribution with $\mathrm{v_m}$ degrees of freedom (Rubin 1987), where

$$\mathrm{v_m} = (\mathrm{m} - 1)[1 + \frac{\bar{U}}{\left(1 + m^{-1}\right) B}]^2$$

When the complete-data degrees of freedom $\mathrm{v_0}$ is small and there is only a modest proportion of missing data, the computed degrees of freedom, $\mathrm{v_m}$, can be much larger than $\mathrm{v_0}$, which is inappropriate. Barnard and Rubin (1999) recommended the use of an adjusted degrees of freedom, v_m^*.

$$v_m^* = [\tfrac{1}{v_m} + \tfrac{1}{\hat{v}_{obs}}]^{-1}$$

where

$$\hat{v}_{obs} = \frac{v_0 + 1}{v_0 + 3} + v_0(1 - {}'\Upsilon)$$

$${}'\Upsilon = \frac{(1 + m^{-1})}{T}\mathrm{B}$$

Similar to the univariate inferences, multivariate inferences based on Wald's tests can also be derived from the m imputed data sets.

3 Example of Clinical Trial and Sample Data

The data used in this paper are for illustration purposes only. In this chapter, to introduce the application of MCMC multiple imputation method to deal with missing data from the mechanism of MNAR, we assume the data are from a simulated hypothetical longitudinal clinical trial.

3.1 Introduction of a Simulated Longitudinal Clinical Trial

In our example, a double blind, randomized, placebo-controlled, parallel group clinical trial will be simulated for an active medication used in relieving persistent chronic pain. The clinical trial schema, presented in, Fig. 1 depicts the hypothetical design for this clinical trial. In the double blind randomization period, the eligible subjects will be randomly allocated to one of the investigational medicinal product (IMP; either active medication or placebo) using an IVRS and commence the treatment period.

Fig. 1 Study design and treatment schema

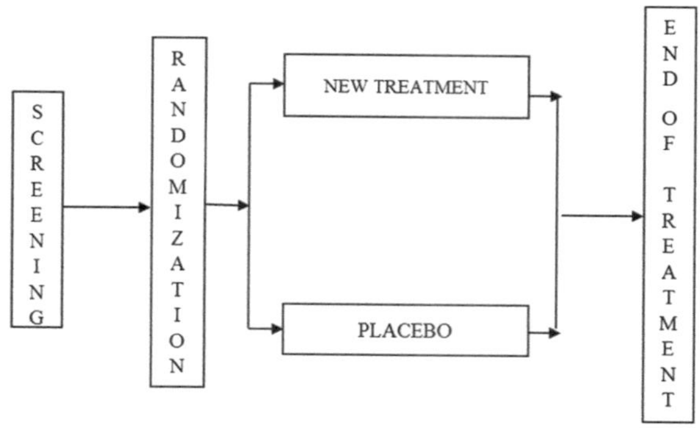

We assume two types of primary endpoints separately for the trial: one has the normal distribution, which is the change from the baseline to the end of treatment in a 0–10 numerical rating scale (NRS) average pain score in the simulated trial. And ANCOVA model will be applied with baseline measurement as covariate and treatment as main effect in the primary analysis. The other primary endpoint doesn't have normal distribution, which is the percent improvement from the baseline to the end of treatment in NRS average pain score in the simulated trial. And Wilcoxon rank-sum test will be applied as the primary analysis. No multiplicity issues will be considered in this research.

In the simulated analysis data set from this trial, we have 5 weekly measurements of NRS Average Pain Score (ranges from 0 to 10 with 0 = "no pain" and 10 = "pain as bad as you can imagine") in double blind treatment period after randomization from 122 subjects in the active medication arm and 126 subjects in the placebo arm (Tables 1 and 2).

The majority of missing data in this hypothetical clinical trial could be expected to be caused by subjects who dropped out of the clinical trial before completion. Based on the simulated data, the pattern of drop out were summarized below for each treatment group and primary reasons for discontinuation.

The simulated 0–10 NRS average pain scores at baseline and its change from baseline by week are presented in the table below.

3.2 *Assuming Data of Primary Efficacy Endpoint to Have Normal Distribution*

Let's first assume the primary endpoint is the change from the baseline to week 5 in NRS average pain score in the simulated clinical trial, and we don't need to consider multiplicity issues. The assumption that the data of the primary endpoint have the normal distribution is reasonable.

Table 1 Primary reasons for discontinuation

Number of:	ACT MED (N = 122)		Placebo (N = 126)		Total (N = 248)	
	n	(%)	n	(%)	n	(%)
Randomized	122	(100.0)	126	(100.0)	248	(100.0)
Discontinued	37	(30.3)	28	(22.2)	65	(26.2)
Adverse events	21	(17.2)	14	(11.1)	35	(14.1)
Sponsor discontinued study	0	(0.0)	0	(0.0)	0	(0.0)
Patient withdrew consent to participate	13	(10.7)	10	(7.9)	23	(9.3)
Lost to follow-up	0	(0.0)	0	(0.0)	0	(0.0)
Patient met (protocol specified) withdrawal criteria	1	(0.8)	1	(0.8)	2	(0.8)
Patient was withdrawn from participation by the investigator	2	(1.6)	3	(2.4)	5	(2.0)

Table 2 NRS average pain score by week

Change from baseline														
Base														
Week	Group	TRT						Line						
		N	Mean	SD	Min	MED	Max	N	Mean	Mean	SD	Min	MED	Max
Baseline	ACT MED	122	6.0	1.2	3.7	6.0	8.0							
	Placebo	126	6.1	1.2	3.8	6.0	8.5							
Week 1	ACT MED	122	5.6	1.5	2.0	5.7	8.9	122	6.0	−0.4	0.8	−3.7	−0.3	1.2
	Placebo	126	5.8	1.3	2.9	5.9	8.4	126	6.1	−0.3	0.6	−2.3	−0.1	1.6
Week 2	ACT MED	111	5.0	1.8	0.0	5.1	8.6	111	6.1	−1.1	1.4	−6.3	−0.6	1.0
	Placebo	116	5.4	1.6	1.6	5.5	8.7	116	6.0	−0.6	1.3	−5.3	−0.4	1.6
Week 3	ACT MED	102	4.8	1.8	0.0	4.9	8.0	102	6.0	−1.3	1.6	−7.6	−0.8	1.8
	Placebo	108	5.2	1.7	0.0	5.2	9.0	108	6.0	−0.8	1.4	−8.0	−0.5	1.3
Week 4	ACT MED	96	4.8	1.9	0.0	5.0	9.0	96	6.0	−1.2	1.6	−7.3	−0.9	2.0
	Placebo	102	5.2	1.6	1.7	5.1	8.3	102	6.0	−0.8	1.3	−5.0	−0.5	1.3
Week 5	ACT MED	91	4.6	1.8	0.0	4.4	10.0	91	5.9	−1.3	1.5	−6.7	−1.0	2.0
	Placebo	100	5.2	1.6	1.6	5.4	8.0	100	6.0	−0.8	1.5	−5.0	−0.3	2.1

ANCOVA was applied on the change of baseline at Week5 in NRS average pain scores (LOCF) with the baseline value as a covariate and treatment group as a factor. The mean change from baseline based on the Least Squares (LS) method, differences between least squares (LS) mean change, 95% CI for the difference and p-value were presented below in Table 3.

The primary analysis showed the significant treatment differences in the primary endpoint, the change from baseline to week 5 in NRS average pain scores ($p = 0.0051$).

Missing data in the double blind period in this trial could potentially arise from the mechanism of MNAR. In order to understand the impact on the trial findings from missing data under the MNAR assumption, multiple imputations will be carried out for the ITT Sample in this hypothetical trial on the endpoint, the mean difference at end of treatment visit in the randomized period. We assume the data with normal distribution first.

Rubin showed that in many applications as few as 5 times of imputation can provide efficient estimate based on multiple imputation. Given the advance in computation speed, 100 times of imputation have been used and considered appropriate.

Intermittent missing values for intermediate visit before the withdrawal visit will be imputed using the MCMC method in PROC MI with an IMPUTE = MONOTONE statement for 100 times for drug and placebo treatment groups, respectively. The resulting 100 partially imputed datasets will have a monotone missing pattern and will be further imputed under an MNAR assumption that the missing data pattern of subjects who receive drug are similar to, worse than, or better than those of subjects who receive placebo with similar observed outcomes for the following 2 scenarios:

(1) MNAR assumed for missing values resulting from discontinuation due to lack of efficacy (LOE) and non-disease progression related AEs (excluding those subjects who withdraw from treatment or die before the end of treatment visit of double blind treatment period for reasons related to underlying disease progression) from 10 subjects in the active medication arm and MAR for others, including the subjects in placebo arm;
(2) MNAR assumed for missing values resulting from discontinuation due to any reason (excluding those subjects who withdraw from treatment or die before the end of treatment visit of double blind treatment period for reasons related to underlying disease progression) from 24 subjects in the active medication arm and MAR for others, including the subjects in the placebo arm.

For each of these 2 scenarios, imputation will be carried out once on each of the 100 imputed datasets using PROC MI (with the 100 imputed datasets included in the BY statement of PROC MI) as follows:

- Step 1: Monotone missing data under the Missing At Random (MAR) assumption at time point t will be imputed by means and covariance from the observed endpoint at baseline and at all weekly post-baseline time points up to time point t (in chronological order) in their corresponding treatment groups (i.e., subjects in active medication arm whose missing data are assumed to be MAR and all subjects in placebo arm). The imputation will be realized using the PROC MI with the

Table 3 Change from baseline to week 5 (LOCF) in NRS average pain score

TRT group	N	Change from baseline									
		Mean	SD	Min	MED	Max	LSmean[a]	Estimated treatment effect[a]	95% CI[a]		
									Lower limit	Upper limit	P-val[a]
ACT MED	122	–1.2	1.6	–7.3	–0.8	2.0	–1.20	–0.54	–0.91	–0.16	0.0051
Placebo	126	–0.7	1.5	–5.3	–0.3	2.1	–0.66				

Note NRS score ranges from 0 to 10 with 0 = “no pain” and 10 = “pain as bad as you can imagine”
[a]derived from ancova model with treatment as factor, baseline value and study as covariates

MONOTONE REG option for each treatment group separately. The imputation model will include baseline value and its weekly measurement at all post-baseline time points up to time point t (in chronological order).

- Step 2: With the data imputed from Step 1, monotone missing data of subjects in the drug group under the MNAR assumption will be imputed. At each post-baseline time point t, the input dataset for PROC MI will include all placebo subjects and those subjects from the active medication arm that have values missing under MNAR at that time point. The imputation model will include baseline value and its weekly values at all post-baseline time points up to time point t (in chronological order). After the sequential imputation is completed for all time points, the imputed values at time point t plus a sensitivity parameter k $\times$ standard error of the observed change from baseline in weekly values in the placebo arm at the corresponding time point will then form the final imputed values. The sensitivity parameter k (where $k = 0, \pm 0.5, \pm 1.0, \pm 1.5$) will be used to explore the robustness of the estimated treatment difference to the degree of decrease or increase (positive values of k represent decrease and negative values represent increase) in efficacy from the placebo response pattern that is assumed for subjects with missing data under MNAR.

After missing values at all the time points are imputed, the change from baseline to Week 5 in the randomized period will be analyzed using the same ANCOVA model as specified for the primary efficacy analysis. For each value of k, results of the ANCOVA analysis on the 100 imputed datasets will be combined to derive an overall result using PROC MIANALYZE. The increment in the positive value of k will stop once the overall p-value is greater than a specified value. The decrease in the negative values of k will continue until the overall p-value becomes smaller than the p-value from the primary efficacy analysis.

Since the primary analysis in the simulated trial showed the significant treatment differences in the primary endpoint ($p = 0.0042$), the sensitivity parameter k was increased to make that the missing data pattern of subjects who receive drug were worse than those of subjects who receive placebo, which is assumed for subjects with missing data under MNAR. The treatment differences between active medication and placebo arms becomes smaller, and the increment of k stops once the overall p-value is greater than 0.05.

The final results from the 2 scenarios are presented in Tables 4 and 5.

The sensitivity analysis of the primary endpoint, the change from baseline to week 5, employing MNAR imputation for missing values due to AE related withdrawals (excluding withdrawals or deaths related to disease progression) in the active medication arm showed that without any alteration of the sensitivity parameter ($K = 0$) the mean change from baseline in NRS average pain score difference was negative; the active medication was significantly different from placebo ($p = 0.0052$). From the results in Table 4, we also know that the study conclusion from ANCOVA (LOCF) is reversed when the sensitivity parameter is 14 ($p = 0.0500$). Thus, if the sensitivity parameter 14, for MNAR assumption for missing values from discontinuation due

Table 4 Change from baseline—ITT (multiple imputation with MNAR assumed for missing values from discontinuation due to AEs in ACT MED group)

Sensitivity parameter K[a]	Treatment group	n	Mean	SD	Estimate	SE	95% CI		P-value
							Lower limit	Upper limit	
0	ACT MED	122	–1.293	1.653	–0.5837	0.208	–0.992	–0.175	0.0052
	Placebo	126	–0.737	1.547					
1	ACT MED	122	–1.281	1.649	–0.5716	0.208	–0.980	–0.163	0.0061
	Placebo	126	–0.737	1.547					
2	ACT MED	122	–1.269	1.645	–0.5595	0.208	–0.968	–0.151	0.0073
	Placebo	126	–0.737	1.547					
3	ACT MED	122	–1.257	1.643	–0.5473	0.208	–0.955	–0.139	0.0086
	Placebo	126	–0.737	1.547					
4	ACT MED	122	–1.245	1.642	–0.5352	0.208	–0.943	–0.127	0.0102
	Placebo	126	–0.737	1.547					
……									
13	ACT MED	122	–1.137	1.676	–0.4262	0.210	–0.839	–0.013	0.0431
	Placebo	126	–0.737	1.547					
14	ACT MED	122	–1.125	1.684	–0.4141	0.211	–0.828	0.000	0.0500
	Placebo	126	–0.737	1.547					

[a]Sensitivity parameter represents the degree of decrease or increase (positive values represent decrease and negative values represent increase, in the unit of S.E.) in nrs score from the imputed missing values based on the placebo response.
[b]number of randomized subjects.
Note Pain score ranges from 0 to 10 with higher score for more severe pain (larger reduction from baseline represents greater improvement).
sd: standard deviation; se: standard error

Table 5 Change from baseline to week 5 (multiple imputation with MNAR assumed for missing values from discontinuation due to any reasons in ACT MED group)

Sensitivity parameter K[a]	Treatment group	n	Mean	SD	Estimate	SE	95% CI		P-value
							Lower limit	Upper limit	
0	ACT MED	122	–1.217	1.677	–0.5089	0.216	–0.933	–0.085	0.0187
	Placebo	126	–0.737	1.547					
0	ACT MED	122	–1.188	1.681	–0.4799	0.216	–0.904	–0.055	0.0268
	Placebo	126	–0.737	1.547					
2	ACT MED	122	–1.160	1.688	–0.4509	0.216	–0.876	–0.025	0.0378
	Placebo	126	–0.737	1.547					
3	ACT MED	122	–1.131	1.696	–0.4219	0.217	–0.848	0.005	0.0526
	Placebo	126	–0.737	1.547					

[a]Sensitivity parameter represents the degree of decrease or increase (positive values represent decrease and negative values represent increase, in the unit of S.E.) in NRS score from the imputed missing values based on the placebo response.
[b]number of randomized subjects.
SD Standard Deviation; *SE* Standard Error

to AEs in the active medication arm, is plausible, the conclusion from ANCOVA (LOCF) is questionable.

Conducting the same analysis but imputing MNAR for withdrawals for any reason (excluding withdrawals or deaths related to disease progression) also showed a negative, and significant, mean difference without alteration of the sensitivity parameter (K = 0) (p = 0.0187). From the results in Table 5, the study conclusion from ANCOVA (LOCF) is reversed when the sensitivity parameter is 4 (p = 0.0526). Thus, if the sensitivity parameter 4, for MNAR assumption for missing values from discontinuation due to any reasons in the active medication arm, is plausible, the conclusion from ANCOVA (LOCF) is questionable.

3.3 Not Assuming Data of Primary Efficacy Endpoint to Have Normal Distribution

If we change the primary endpoint to the percent improvement from the baseline to the end of treatment in NRS average pain score in the hypothetical clinical trial, the data of the primary endpoint don't have the normal distribution. In the primary analysis, they will be compared between treatment groups using Wilcoxon rank-sum test and estimate of the median difference between new treatment and placebo, together with approximate 95% CI, will be calculated using the Hodges-Lehmann approach.

The results from the primary analysis are presented in Table 6 as median difference between groups, where a positive value would indicate a treatment difference in favor of the active medication, and 95% CIs; both determined using Hodges-Lehmann. Statistical significance was determined using Wilcoxon rank-sum test to yield p-value.

Subjects in the active medication arm had a median percent improvement from baseline of 13.3% compared with 5.4% in the placebo arm; the difference between groups was statistically significant (median difference = 8.00; CI: 2.93, 13.4; p = 0.0020).

The impact of missing data under the MNAR assumption will be assessed via sensitivity analyses with multiple imputations on weekly percent improvement in NRS average pain score from baseline.

We will apply the same steps introduced before for the data to have normal distribution to get imputed data for missing data. After missing values at all the time points are imputed, percent improvement from baseline to Week 5 from the fully imputed datasets will be calculated as 100% × (baseline—Week 5 mean)/baseline and it will then analyzed using the same model as specified for the primary efficacy analyses. For each value of k, the Hodges-Lehmann estimates of the median difference and asymptotic standard error from 100 imputed datasets will be analyzed using PROC MIANALYZE to derive overall median difference and 95% confidence interval.

Table 6 Percent improvement from baseline to end of treatment in NRS average pain score

Treatment group	N	Mean	Min	Q1	Median	Q3	Max	Hodges-Lehmann estimate		Wilcoxon rank sum test P-value
								Median diff.	95% CI	
ACT MED	122	19.5	–32.6	1.82	13.3	35.7	100.0	8.00	(2.93, 13.4)	0.0020
Placebo	126	10.4	–43.9	–2.86	5.4	20.2	72.2			

Note PAIN score ranges from 0 to 10 with higher score for more severe pain. % improvement = (baseline pain NRS mean—end of treatment pain NRS mean)/baseline pain NRS mean X 100

The SAS code to utilize PROC NPAR1WAY to get the median difference and asymptotic standard error from each imputation and use PROC MIANALYZE to derive overall median difference and 95% confidence interval will be as below:

```
proc npar1way data=indata wilcoxon hl(refclass=1 or 2);
        class TRT;
        by _imputation_;
        var PIMP;
        ods output WilcoxonTest=wilcox HodgesLehmann=hl;
run;
proc mianalyze data=hl;
  modeleffects shift;
  stderr stderr;
  ods output ParameterEstimates=Overall_hl;
run;
```

The results of the Wilcoxon rank-sum test on the 100 imputed datasets will be combined to derive an overall p value. The test statistic for making inference will be based on the method provided by Rubin and a modified macro from Mogg and Mehrotra (2007). The increment in the positive value of k will stop once the overall p-value is greater than 0.05. The decrease in the negative values of k will continue until the overall p-value becomes smaller than the p-value from the primary efficacy analysis.

As what we introduced in before, since the primary analysis in the simulated trial showed the significant treatment differences in the primary endpoint ($p = 0.0042$), the increment of sensitivity parameter k will stop once the overall p-value is greater than 0.05.

The sensitivity analysis of the primary endpoint, the percent improvement from the baseline to the end of treatment in NRS average pain score, using the same Wilcoxon ranksum test, and employing MNAR imputation for missing values due to AE related withdrawals (excluding withdrawals or deaths related to disease progression) in the active medication arm, showed that without any alteration of the sensitivity parameter ($K = 0$) the median difference was positive; active medication was significantly different from placebo (median difference = 9.51; 95% CI: 3.26, 15.76; $p = 0.0032$) (Table 7). From the results in this table, we also know that the study conclusion from Wilcoxon rank-sum test is reversed when the sensitivity parameter is 33 ($p = 0.0512$). Thus, if the sensitivity parameter 33, for MNAR assumption for missing values from discontinuation due to AEs in the active medication arm, is plausible, the conclusion from Wilcoxon rank-sum test is questionable.

Conducting the same analysis but imputing MNAR for withdrawals for any reason (excluding withdrawals or deaths related to disease progression) also showed a positive, significant, median difference without alteration of the sensitivity parameter ($K = 0$) (median difference = 7.93; 95% CI: 1.55, 14.32; $p = 0.0158$) (Table 8). Significant treatment differences were noted in weekly percent improvement from baseline in average pain 0–10 NRS score when analyzed using MMRM (Table 8). From the results in Table 8, the study conclusion from Wilcoxon rank-sum test is reversed when the sensitivity parameter is 4 ($p = 0.0538$). Thus, if the sensitivity

Table 7 Percent improvement from baseline to end of treatment—ITT (multiple imputation with mnar assumed for missing values from discontinuation due to AEs in active treatment group)

Sensitivity parameter K[a]	Treatment group	N[b]	Mean	Min	Q1	Median	Q3	Max	Hodges-Lehmann estimate		Wilcoxon rank sum test P-value
									Median diff.	95% CI	
0	ACT MED	122	21.2	–45.3	3.9	16.0	37.9	101.1	9.51	(3.26, 15.76)	0.0032
	Placebo	126	11.0	–52.1	–4.2	6.5	22.3	72.6			
1	ACT MED	122	21.0	–46.4	3.9	16.0	37.8	100.7	9.36	(3.12, 15.60)	0.0035
	Placebo	126	11.0	–52.1	–4.2	6.5	22.3	72.6			
2	ACT MED	122	20.8	–47.9	3.4	15.8	37.7	100.4	9.21	(2.95, 15.47)	0.0039
	Placebo	126	11.0	–52.1	–4.2	6.5	22.3	72.6			
3	ACT MED	122	20.6	–49.8	3.0	15.6	37.7	100.2	9.06	(2.76, 15.36)	0.0045
	Placebo	126	11.0	–52.1	–4.2	6.5	22.3	72.6			
4	ACT MED	122	20.4	–52.1	2.5	15.4	37.7	100.0	8.91	(2.63, 15.20)	0.0050
......											
32	ACT MED	122	14.8	–144.7	0.0	13.8	36.2	100.0	6.52	(–0.05, 13.09)	0.0492
	Placebo	126	11.0	–52.1	–4.2	6.5	22.3	72.6			
33	ACT MED	122	14.6	–148.3	0.0	13.8	35.8	100.0	6.48	(–0.10, 13.05)	0.0512
	Placebo	126	11.0	–52.1	–4.2	6.5	22.3	72.6			

[a]Sensitivity parameter represents the degree of decrease or increase (positive values of K represent decrease and negative values represent increase) in efficacy
[b]Number of ITT subjects.
Scenario 1 MNAR assumed for missing values resulting from discontinuation due to AES (excluding those subjects withdraw from treatment or died before the end of week 5 for reasons related to underlying disease progression) in the act medgroup and MAR for others

Table 8 Percent improvement from baseline to end of treatment—ITT (multiple imputation MNAR assumed for missing values from discontinuation due to any reasons(*) in active treatment group)

Sensitivity parameter K[a]	Treatment group	N[b]	Mean	Min	Q1	Median	Q3	Max	Hodges-Lehmann estimate		Wilcoxon rank sum test P-value
									Median diff.	95% CI	
0	ACT MED	122	19.7	–42.9	1.7	14.5	36.6	102.2	7.93	(1.55, 14.32)	0.0158
	Placebo	126	11.0	–52.1	–4.2	6.5	22.3	72.6			
1	ACT MED	122	19.2	–44.6	1.1	14.1	36.2	101.4	7.47	(1.11, 13.84)	0.0215
	Placebo	126	11.0	–52.1	–4.2	6.5	22.3	72.6			
2	ACT MED	122	18.7	–46.6	0.8	13.6	36.1	100.7	7.06	(0.66, 13.47)	0.0295
	Placebo	126	11.0	–52.1	–4.2	6.5	22.3	72.6			
3	ACT MED	122	18.2	–48.8	0.6	13.4	35.9	100.2	6.67	(0.25, 13.10)	0.0417
	Placebo	126	11.0	–52.1	–4.2	6.5	22.3	72.6			
4	ACT MED	122	17.7	–51.4	–0.1	13.2	35.9	100.0	6.28	(–0.20, 12.75)	0.0553
	Placebo	126	11.0	–52.1	–4.2	6.5	22.3	72.6			

[a]Sensitivity parameter represents the degree of decrease or increase (positive values of K represent decrease and negative values represent increase) in efficacy
[b]number of ITT subjects
*MNAR assumed for missing values resulting from discontinuation due to any reasons excluding subject died or discontinued due to disease progression before the end of week 5 in the act med group and MAR for others

parameter 4, for MNAR assumption for missing values from discontinuation due to any reasons in the active medication arm, is plausible, the conclusion from Wilcoxon rank-sum test is questionable.

4 Discussion

Missing data are common in clinical trials. To understand the impact on the findings from missing data under the MNARSignificant treatment differences were noted assumption, the sensitivity analyses could be carried out by applying MCMC Multiple Imputation method. To investigate the departure from MAR assumption, we can conduct sensitivity analysis by progressively decreasing the treatment differences in those treated subjects who fell into an assumed MNAR pattern, That is, making the missing data pattern of subjects who receive active medication similar to, worse than, or better than those of subjects who receive placebo for different scenarios of early discontinuation.

The sensitivity parameter is used to explore the robustness of the estimated treatment difference to the degree of decrease or increase in efficacy from the placebo response pattern that is assumed for subjects with missing data under MNAR. It starts from 0 and increases or decreases, until conclusion from the primary analysis is overturned (it is called tipping point analysis), or it becomes clinically meaningless to go even higher. Note that when 0% is used, the MI procedure would produce an analysis which is essentially MAR.

The example in this chapter demonstrated that the sensitivity analyses with MCMC MI method to deal with missing data from the mechanism of MNAR are appropriate to evaluate robustness of conclusions to a range of conditions in a longitudinal clinical trial.

References

Barnard, J., & Rubin, D. B. (1999). Small-sample degrees of freedom with multiple imputation. *Biometrika*, *86*, 948–955.

Galbraith, S. (2012). Applied missing data analysis by Craig K Enders. *Australian & New Zealand Journal of Statistics*, *54*, 251. doi:10.1111/j.1467-842X.2012.00656.x.

Mogg, R., & Mehrotra, D. V. (2007). Analysis of antiretroviral immunotherapy trials with potentially non-normal and incomplete longitudinal data. *Statistics in Medicine*, *26*(3), 484–497.

National Research Council. (2010). *The prevention and treatment of missing data in clinical trials*. Washington, DC: National Academies Press. http://www.nap.edu/catalog.php?record_id-12955.

Rubin, D. B. (1987). *Multiple imputation for nonresponse in surveys*. New York: Wiley.

Schafer, J. L. (1997). *Analysis of incomplete multivariate data*. New York: Chapman and Hall.

Verbeke, G., & Molenberghs, G. (2000). *Linear mixed models for longitudinal data*. New York: Springer.

Wang, N., & Robins, J. M. (1998). Large sample theory for parametric multiple imputation procedures. *Biometrika*, *85*, 935–948.

Part III
Monte-Carlo in Statistical Modellings and Applications

Monte-Carlo Simulation in Modeling for Hierarchical Generalized Linear Mixed Models

Kyle M. Irimata and Jeffrey R. Wilson

Abstract It is common to encounter data that have a hierarchical or nested structure. Examples include patients within hospitals within cities, students within classes within schools, factories within industries within states, or families within neighborhoods within census tracts. These structures have become increasingly common in recent times and include variability at each level which must be taken into account. Hierarchical models which account for the variability at each level of the hierarchy, allow for the cluster effects at different levels to be analyzed within the models (Shahian et al. in Ann Thorac Surg, 72(6):2155–2168, 2001). This chapter discusses how the information from different levels can be used to produce a subject-specific model. However, there are often cases when these models do not fit as additional random intercepts and random slopes are added to the model. This addition of additional parameters often leads to non-convergence. We present a simulation study as we explore the cases in these hierarchical models which often lead to non-convergence. We also used the 2011 Bangladesh Demographic and Health Survey data as an illustration.

1 Introduction

Hierarchical logistic regression models consist of inherent correlation due to different sources of variation. At each level of the hierarchy, we have random intercepts and sometimes random slopes as well as the appropriate fixed effects. We have done extensive work with the GLIMMIX and NLMIXED procedures in fitting hierarchical models and have noted the trials and tribulations in computing regression estimates and covariance estimates associated with hierarchical models in SAS, as attested by

K.M. Irimata
School of Mathematical and Statistical Sciences, Arizona State University, Tempe, AZ 85287, USA
e-mail: kirimata@asu.edu

J.R. Wilson (✉)
W.P. Carey School of Business, Arizona State University, Tempe, AZ 85287, USA
e-mail: Jeffrey.wilson@asu.edu

D.-G. Chen and J.D. Chen (eds.), *Monte-Carlo Simulation-Based Statistical Modeling*, ICSA Book Series in Statistics, DOI 10.1007/978-981-10-3307-0_13

others. We have had several occasions when our models do not converge. In some cases, we found that the convergence criterion was satisfied, but the standard error for the covariance parameters was given as "." This problem has gained the attention of many (Hartzel et al. 2001; Wilson and Lorenz 2015 to name a few). We do not know with certainty why certain convergence problems exist. As such we provide some understanding and make some suggestions based on our own work as well as work done by others. We also provide the steps and results of a simulation study which can be expanded upon for further exploration of the problem and its remedies.

In this chapter, we discuss the use of two-level and three-level hierarchical models for binary data, although it is possible to analyze higher level data. We discuss the use of models with effects at level 2 and level 3 representing random intercepts and random slopes. These random effects are added into the model to account for unobservable effects that are known to exist but were not measured or cannot be measured. We also discuss the use of simulations as a means of investigating issues or irregularities. This process is presented as an exercise in simulating hierarchical binary data, which for simplicity is restricted to the two-level case, although the techniques discussed can be readily expanded for higher levels. These simulated models have incorporated a random intercept and a random slope at level 2. We implement a hierarchical model using the GLIMMIX procedure in SAS, to identify factors that contribute to AIDS knowledge in Bangladesh and investigate models that do and do not converge based on the number of fixed effect predictors.

2 Generalized Linear Model

The birth of the generalized linear models unified many methods (Nelder and Wedderburn 1972). These models consist of a set of n independent random variables $Y_1 \ldots . . Y_n$, each with a distribution from the exponential family. We define a generalized linear model as having three components: the random component, the systematic component, and the link component. We define the log-likelihood function based on unknown mean parameters, a dispersion parameter, and a weight parameter, denoted by θ_i, φ, and ω_i respectfully, and of the form (Smyth 1989),

$$l(\phi_i^{-1}, \omega_i : y_i) = \sum_i \{\omega_i \phi_i^{-1} [y_i \theta_i - b(\theta_i)] - c(y_i, \omega_i \phi_i^{-1})$$

with ϕ_i unknown and assume that

$$c(y_i, \omega_i \phi_i^{-1}) = \omega_i \phi_i^{-1} a(y_i) - \frac{1}{2} s(-\omega_i \phi_i^{-1}) + t(y_i)$$

Thus we have a generalized linear model for the mean such that

$$\mu_i = E(Y_i) = b'(\theta_i) = \mathbf{x}_i'\boldsymbol{\beta}$$

where $\mathbf{x}_i' = (x_1, \ldots.., x_p)'$ is the vector of covariates and $\boldsymbol{\beta}$ is the vector of regression parameters. The functions a (y) and b (θ_i) are known functions. We also present the generalized linear model as

$$\mathbf{Y} = \mathbf{X}\beta + \varepsilon$$

where the random component belongs to the exponential family of distributions, while in the marginal form we present $g(E(Y)) = \mathbf{X}\beta$. However, when the set of outcomes from the outcomes Y_i are not independent, then the generalized linear model in its pure form is no longer appropriate and we must use generalized linear mixed models.

3 Hierarchical Models

It is common in fields such as public health, education, demography, and sociology to encounter data structures where the information is collected based on a hierarchy. For instance, in health studies, we often see patients nested within doctors and doctors nested within hospitals. In these types of cases, there is variability at each level of the hierarchy, resulting in intraclass correlation due to the clustering. As a result of the correlation at each level inherent from these hierarchical structures, the standard logistic regression is inappropriate (Rasbash et al. 2012). Ignoring these levels of design while researching the outcome is sure to lead to erroneous results unless the intraclass correlation is of an insignificant size (Irimata and Wilson 2017). Others have demonstrated that ignoring a level of nesting in the data can impact variance estimates and the available power to detect significant covariates (Wilson and Lorenz 2015). When seeking to appropriately analyze these types of correlated data, we must extend the generalized linear models by accounting for the association among the responses.

Hierarchical models, also referred to as nested models or mixed models are statistical models that extend the class of generalized linear models (GLMs) to address and account for the hierarchical (correlated) nesting of data (Hox 2002; Raudenbush and Bryk 2002; Snijders and Bosker 1998). We will refer to these as the hierarchical generalized linear models (HGLMs). This approach incorporates a random effect, usually according to the normal distribution, although non-normal random effects can also be used. The extension required in HGLMs is not as involved when the outcomes follow a conditional normal distribution and the random effects are normally distributed. However, when dealing with outcomes that are not normally distributed (i.e. binary, categorical, ordinal), the extension is not as straightforward. In these cases, we often use a link other than the identity and must specify an appropriate error distribution for the response at each level. We thus present the conditional mean explanation rather than the marginal mean.

While most work have concentrated on random intercepts, we have often been confronted with data requiring multiple random intercepts and even random slopes. When using the GLIMMIX procedure in SAS, we often find that models which include multiple random intercepts or even one random intercept with one random slope may not converge. Therefore, this chapter introduces the reader to hierarchical models with dichotomous outcomes (i.e., hierarchical generalized linear models), and provides concrete examples of non-convergence and possible remedies in these situations.

We present hierarchical models as

$$\mathbf{Y} = \mathbf{X}\beta + \mathbf{Z}\theta + \varepsilon$$

where the random effects θ have a multivariate normal distribution with mean vector zero and covariance matrix G, with the distribution of the errors ε as normal with mean vector 0 and covariance matrix R. The **X** matrix consists of the fixed effects with vector of regression parameters β while the Z matrix consists of columns, each representing the random effects with vector of parameters θ. Researchers refer to this as compensating for the correlation through the systematic component. Thus we often write in the conditional response form as

$$g(E[\mathbf{Y}|\,\theta]) = \mathbf{X}\beta + \mathbf{Z}\theta$$

where $\theta \sim \mathcal{N}(0, G)$. The unconditional covariance matrix for Y, is

$$\text{var}(\mathbf{Y}) = \mathbf{A}^{1/2}\mathbf{R}\mathbf{A}^{1/2} + \boldsymbol{G}$$

and the conditional covariance matrix, given the random effects is given by

$$\text{var}(\mathbf{Y}|\theta) = \mathbf{A}^{1/2}\mathbf{R}\mathbf{A}^{1/2} = \mathbf{V}.$$

Thus, it is common in literature to refer to the G-side and R-side effects, which refer to the covariance matrix of the random effects, and the covariance matrix of the residual effects, respectfully.

In SAS, the GLIMMIX procedure distinguishes between the G-side and R-side effects and can model the random effects as well as correlated errors. This procedure fits generalized linear mixed models based on linearization and relies on a restricted pseudo-likelihood method of estimation. We revisit the method here as it helps us to understand the problems regarding non-convergence. This estimation is essentially based on the following.

Consider the conditional mean as

$$E[Y|\,\theta] = g^{-1}(\mathbf{X}\beta + \mathbf{Z}\theta)$$

and using Taylor series expansion we linearize $g^{-1}\,(\mathbf{X}\beta + \mathbf{Z}\theta)$ about the points $\tilde{\beta}$ and $\tilde{\theta}$ which gives

$$g^{-1}\,(\mathbf{X}\beta + \mathbf{Z}\theta) \cong g^{-1}\left(\mathbf{X}\tilde{\beta} + \mathbf{Z}\tilde{\theta}\right) + \frac{\partial g^{-1}\,(\mathbf{X}\beta + \mathbf{Z}\theta)}{\partial\beta}\left(\beta - \tilde{\beta}\right) + \frac{\partial g^{-1}\,(\mathbf{X}\beta + \mathbf{Z}\theta)}{\partial\theta}(\theta - \tilde{\theta})$$

$$g^{-1}\,(\mathbf{X}\beta + \mathbf{Z}\theta) \cong g^{-1}\left(\mathbf{X}\tilde{\beta} + \mathbf{Z}\tilde{\theta}\right) + \Omega_{|\tilde{\beta}\tilde{\theta}}\mathbf{X}\left(\beta - \tilde{\beta}\right) + \Omega_{|\tilde{\beta}\tilde{\theta}}\mathbf{Z}(\theta - \tilde{\theta})$$

where $\Omega_{|\tilde{\beta}}$ and $\Omega_{|\tilde{\theta}}$ denote the matrix of derivatives evaluated at $\tilde{\beta}$ and $\tilde{\theta}$ respectively. Thus

$$\Omega_{|\tilde{\beta}\tilde{\theta}}\{g^{-1}\,(\mathbf{X}\beta + \mathbf{Z}\theta)\} \cong \Omega_{|\tilde{\beta}\tilde{\theta}}{}^{-1}\{g^{-1}\left(\mathbf{X}\tilde{\beta} + \mathbf{Z}\tilde{\theta}\right)\} + \mathbf{X}\left(\beta - \tilde{\beta}\right) + \mathbf{Z}(\theta - \tilde{\theta})$$

So

$$\Omega_{|\tilde{\beta}\tilde{\theta}}\{g^{-1}\,(\mathbf{X}\beta + \mathbf{Z}\theta)\} - \Omega_{|\tilde{\beta}\tilde{\theta}}{}^{-1}\{g^{-1}\left(\mathbf{X}\tilde{\beta} + \mathbf{Z}\tilde{\theta}\right)\} \cong \mathbf{X}\beta + \mathbf{Z}\theta - (\mathbf{X}\tilde{\beta} + \mathbf{Z}\tilde{\theta})$$

and

$$\mathbf{X}\beta + \mathbf{Z}\theta \cong \left(\mathbf{X}\tilde{\beta} + \mathbf{Z}\tilde{\theta}\right) + \Omega_{|\tilde{\beta}\tilde{\theta}}{}^{-1}\{g^{-1}\,(\mathbf{X}\beta + \mathbf{Z}\theta)\} - \Omega_{|\tilde{\beta}\tilde{\theta}}{}^{-1}\{g^{-1}\left(\mathbf{X}\tilde{\beta} + \mathbf{Z}\tilde{\theta}\right)\}$$

$$\mathbf{X}\beta + \mathbf{Z}\theta \cong \left(\mathbf{X}\tilde{\beta} + \mathbf{Z}\theta\right) + \Omega_{|\tilde{\beta}\tilde{\theta}}{}^{-1}\{(E[\mathbf{Y}|\,\theta]) - \{g^{-1}\left(\mathbf{X}\tilde{\beta} + \mathbf{Z}\tilde{\theta}\right)]\}$$

Hence we consider the approximation and use the similar structure denoted by $\mathbf{X}\tilde{\beta} + \mathbf{Z}\tilde{\theta}$ to represent the matrix of fixed effects multiplied by a beta-like term and Z matrix of random effects multiplied by a theta-like term and we denote $\Omega_{|\tilde{\beta}\tilde{\theta}}{}^{-1}\left\{(E[\mathbf{Y}|\,\theta]) - \{g^{-1}\left(\mathbf{X}\tilde{\beta} + \mathbf{Z}\tilde{\theta}\right)\right\} = \zeta$ as an error-like term. So we can think of the approximation as a linear term and defined as

$$Y_{\text{approx}} = \mathbf{X}\beta + \mathbf{Z}\theta + \zeta$$

with the variance

$$\mathrm{var}[\mathrm{Y}_{\mathrm{approx}} | \theta] = \mathrm{var}[\{(\mathrm{E}\,[\mathbf{Y} | \theta])\}] = \Omega_{|\tilde{\beta}\tilde{\theta}}{}^{-1}\mathbf{A}^{1/2}\mathbf{R}\mathbf{A}^{1/2}\Omega_{|\tilde{\beta}\tilde{\theta}}{}^{-1}$$

As such this can be seen as a linear approximation, given by $\mathrm{Y}_{\mathrm{approx}}$ with fixed effects β, and random effects θ and variance of ζ given by $\mathrm{var}[\mathrm{Y}_{\mathrm{approx}} | \theta]$.

3.1 Approaches with Binary Outcomes

Binary outcomes are very common in healthcare research, amongst many other fields. For example, one may investigate whether a patient has improved or recovered after discharge from the hospital or not. For healthcare and other types of research, the logistic regression model is one of the preferred methods of modeling data when the outcome variable is binary. In its standard form, it is a member of a class of generalized linear models specific to the binomial random component. As is customary in regression analysis, the model makes use of several predictor variables that may be either numerical or categorical. However, a standard logistic regression model assumes that the observations obtained from each unit are independent. If we were to fit a standard logistic regression to nested data, the assumption of independent observations is seriously violated. This violation could lead to an underestimation of the standard errors, which in turn can lead to conclusions of a significant effect, when in fact it is not.

Multilevel approaches for nested data can also be applied to analysis of dyadic data to take into account the nested sources of variability at each level (Raudenbush 1992). Many researchers have explored the use of these two-level approaches with binary outcomes (see for example McMahon et al. 2006).

4 Three-Level Hierarchical Models

In the analysis of multilevel data, each level provides a component of variance that measures intraclass correlation. For instance, consider a hierarchical model at three levels for the kth patient seeing the jth doctor in the ith hospital. The patients are at the lower level (level 1) and are nested within doctors (level 2) which are nested within hospitals at the next level (level 3). We consider the hospital as the primary unit, doctors as secondary unit, and patients as the observational unit. These clusters are treated as random effects. We make use of random effects as we believe there are some non-measurable influences on patient outcomes based on the doctor and also based on the hospital. Some effects may be positive and some effects may be negative, but overall we assume their average effects are zero.

4.1 With Random Intercepts

At level 1, we may take responses from different patients, while noting their age (Age) and length of stay (LOS). The outcomes are modeled through a logistic regression model

$$\log\left[\frac{p_{ijk}}{1-p_{ijk}}\right] = \gamma_{oij} + \gamma_{1ij}\mathrm{Age}_{ijk} + \gamma_{2ij}\mathrm{Los}_{ijk} \tag{4.1}$$

where γ_{oij} is the intercept, γ_{1ij} is the coefficient associated with the predictor Age_{ijk}, and γ_{2ij} is the coefficient associated with the predictor Los_{ijk} (length of stay) for $k = 1, 2, \ldots, n_{ij}$ patients; $j = 1, 2, \ldots, n_i$ doctors and $i = 1, \ldots, n$; hospitals. Each doctor has a separate logistic model. If we allow the effects of Age and LOS on the outcome to be the same for each doctor, but allow the intercept to be different on the logit scale, we have parallel planes for their predictive model. The γ_{oij} intercept represents those differential effects among doctors.

At level 2, we assume that the intercept γ_{oij} (which allows a different intercept for doctors within hospitals) depends on the unobserved factors specific to the ith hospital, the covariates given as associated with the doctors within the ith hospital, and a random effect u_{oij} associated with doctor j within hospital i. Thus,

$$\gamma_{oij} = \gamma_{oi} + \gamma_{1i}\mathrm{Experience}_{ij} + u_{oij} \tag{4.2}$$

where $\mathrm{Experience}_{ij}$ is the experience for the jth doctor within the ith hospital. Similarly, hospital administration policies may have different effects on doctors. At level 3, the model assumes that differential hospital policies depend on the overall fixed intercept β_0 and the random intercept u_{oi} associated with the unmeasurable effect for hospital i. Thus,

$$\gamma_{oi} = \beta_0 + u_{oi} \tag{4.3}$$

By successive substitution into the expression for γ_{oi} in (4.3) into (4.2), and then by substituting the resulting expression for γ_{oij} into (4.1), we obtained

$$\log\left[\frac{p_{ijk}}{1-p_{ijk}}\right] = \beta_0 + \gamma_{1i}\mathrm{Experience}_{ij} + \gamma_{1ij}\mathrm{Age}_{ijk} + \gamma_{2ij}\mathrm{Los}_{ijk} + u_{oi} + u_{oij} \tag{4.4}$$

The combination of random and fixed terms results in a generalized linear mixed model with two random effects; hospitals denoted by $u_{oi} \sim \mathcal{N}(0, \sigma^2_{u_i})$ and doctors denoted by $u_{oij} \sim \mathcal{N}(0, \sigma^2_{u_{ij}})$ with covariance $\sigma_{u_{oi}, u_{oij}}$. From Eq. (4.4), the model consists of the overall mean plus experience of doctors plus age of patient, length of stay plus effects due to hospitals and effects due to doctors for each individual. Hence, we have a subject-specific model.

4.2 Three-Level Logistic Regression Models with Random Intercepts and Random Slopes

Consider the three-level random intercept and random slope model consisting of a logistic regression model at level 1,

$$\log\left[\frac{p_{ijk}}{1-p_{ijk}}\right] = \gamma_{oij} + \gamma_{1ij}\mathrm{Age}_{ijk} + \gamma_{2ij}\mathrm{Los}_{ijk} \tag{4.5}$$

where both γ_{oij} and γ_{2ij} are random, for $k = 1, 2, \ldots, n_{ij}$; $j = 1, 2, \ldots, n_i$; and $i = 1, \ldots, n$. So each doctor has a different intercept and the rates of change with respect to length of stay are not the same for all the doctors. However, there are some unobserved effects related to LOS that impact remission. There are factors associated with LOS and the doctors' impacts on patients vary as LOS varies. The intercept represents a group of unidentifiable factors that impact the overall effect of the doctor on the patient's success, while the slope represents the differential impact that the particular variable (LOS) has that results in differences among patients.

So, at level 2, γ_{oij} and γ_{2ij} are treated as response variables within the model,

$$\gamma_{oij} = \gamma_{oi} + \gamma_{1i}\mathrm{Experience}_{ij} + u_{oij} \tag{4.6}$$

$$\gamma_{2ij} = \gamma_{2i} + u_{2ij} \tag{4.7}$$

where γ_{oi} and γ_{2i} are random effects. Equation (4.6) assumes the intercept γ_{oij} for doctors nested within hospital j, depends on the unobserved intercept specific to the ith hospital, the effects associated with the doctor's experience in the hospital, and a random term u_{oij} associated with doctor j within hospital i. The slope γ_{2ij} depends on the overall slope γ_{2i} for hospital i and a random term u_{2ij}.

At level 3, the model shows that the hospitals vary based on random effects

$$\gamma_{oi} = \beta_{00} + u_{oi} \tag{4.8}$$

$$\gamma_{2i} = \beta_{22} + u_{2i} \tag{4.9}$$

The intercept γ_{oi} depends on the overall fixed intercept β_{00} and the random term u_{oi} associated with the hospital i, while the hospital slope γ_{2i} depends on the overall fixed slope β_{22} and the random effect u_{2i} associated with the slope for hospital i. By substituting the expression for γ_{oi} and γ_{2i} into (4.7) and (4.8), and then substituting the resulting expression for γ_{oij} and γ_{2ij} into (4.9), we obtained

$$\log\left[\frac{p_{ijk}}{1-p_{ijk}}\right] = \beta_{00} + \gamma_{1ij}\mathrm{Age}_{ijk} + \gamma_{1i}\mathrm{Experience}_{ij} + u_{oi} + u_{oij} + \left(\beta_{22} + u_{2i} + u_{2ij}\right)\mathrm{Los}_{ijk} \tag{4.10}$$

Thus, we have a generalized linear mixed model with random effects u_{oi}, u_{oij}, γ_{1i} and γ_{1ij}. Therefore, Los_{ijk} is associated with both a fixed and random part. We take advantage of this regrouping of terms to incorporate the random effects and their variance-covariance matrix, so that u_{oi}, u_{oij}, γ_{1i} and γ_{1ij} are jointly distributed normally with a mean of zero and a covariance matrix reflecting the relationships between the random effects.

4.3 Nested Higher Level Logistic Regression Models

For higher than three level nested we can easily present the model, though executing the necessary computations may be tedious. Imagine if we had the data with another level, hospitals nested within cities (level 4 denoted by h). Cities may have their own way of monitoring healthcare within their jurisdiction. We also believed that the number of beds within the hospital is a necessary variable. For such data, we will have the kth patient nested within the jth doctor which is nested within ith hospital which is nested in the hth city. Then the model is:

$$\log\left[\frac{p_{hijk}}{1-p_{hijk}}\right] = \beta_{00} + \gamma_{1hij}\text{Age}_{hijk} + \gamma_{1hi}\text{Experience}_{hij} +$$
$$\gamma_{1h}\text{Bed}_{hi} + u_{oh} + u_{ohi} + u_{ohij} + \left(\beta_{22} + u_{2hi} + u_{2hij}\right)\text{Los}_{hijk} \quad (4.11)$$

5 Possible Problems with Hierarchical Model

5.1 Issues in Hierarchical Modeling

We found that convergence of parameter estimates can sometimes be difficult to achieve, especially when fitting models with random slopes or higher levels of nesting. Some researchers have found that convergence problems may occur if the outcome is skewed for certain clusters or if there is quasi or complete separation. Such phenomena destroy the variability within clusters which is essential to obtaining the solutions. In addition, including too many random effects may not be computationally possible (Schabenberger 2005).

We also found what other researchers did; for hierarchical logistic models for nested binary data, it is often not feasible to estimate random effects for both intercepts and slopes at the same time in a model. Newsom (2002) showed that we can have models with too many parameters to be estimated given the number of covariance elements included. Others found that such models can lead to severe convergence problems, which can limit the modeling. Before fitting these conditional models, McMahon et al. (2006) suggested that one should determine whether there is significant cluster interdependence to justify the use of multilevel modeling. Irimata and Wilson (2017) through simulation gave some further guidance.

Regardless of the number of clusters, Austin (2010) found that for all statistical software procedures, the estimation of variance components tended to be poor when there were only five subjects per cluster. The number of clusters on the mean number of quadrature points was negligible. However, when the random effects were large, Rodriquez and Goldman (1995) found substantial decreases in the estimation of fixed effects and/or variance components. They also found that there was bias in the estimation when the number of subjects per cluster was small.

These hierarchical models can be fitted through SAS with the GLIMMIX or NLMIXED procedure as well as in SPSS and R. Maas and Hox (2004) claimed that only one random statement is supported in the NLMIXED procedure so that nonlinear mixed models cannot be assessed at more than two levels. However, Hedeker et al. (2008), Hedeker et al. (2012) showed how more than one random statement can be used for continuous data in the NLMIXED procedure with more than two-levels.

5.2 Parameter Estimations

The conditional joint distribution of the responses and the distribution of the random effects provide a joint likelihood which cannot necessarily be readily written down in closed form. However, we still need to estimate the regression coefficients and the random components. In so doing, it is imperative for us to use some form of approximations. Sometimes researchers have used the quasi-likelihood approach through a Taylor series expansion to approximate the joint likelihood. The approximate likelihood is maximized to produce maximized quasi-likelihood estimates. The disadvantage which many researchers have pointed out with this approach is the bias involved with quasi-likelihoods (Wedderburn 1974). Other researchers have resorted to numerical integration, split up into quadratures, to obtain approximations of the true likelihood. More integration points will increase the number of computations and thus impede the speed to convergence, although it increases the accuracy. Each added random component increases the integral dimension. A random intercept is one dimension (one added parameter), a random slope makes that two dimensions. Our experience is that the three-level nested models with random intercepts and slopes often create problems regarding convergence.

5.3 Convergence Issues in SAS

We spent considerable time overcoming the challenges of the GLIMMIX procedure. We reviewed available literature and discussed with those with experience using SAS. Although there are by no means guarantees that there will not be challenges, we provide in this chapter our experiences, underscored by others, as well as suggestions for improving the performance of this procedure.

Non-convergence in the GLIMMIX procedure can be identified by looking at the output and the log. The most obvious indication of issues is in the convergence criterion, which is provided below the iteration history. When convergence is not obtained, SAS will provide the following warning: "DID NOT CONVERGE".

A successful convergence message does not itself necessarily guarantee that the model converged. In some cases, the convergence criterion will be satisfied, but the standard error for one or more of the (non-zero) covariance parameters will be missing. When this occurs, the standard error will be given by a "." instead of an actual estimate. In these cases, the output may look similar to the following:

Covariance Parameter Estimates			
Cov Parm	*Subject*	*Estimate*	*Standard Error*
Intercept	div	0.09097	.
urban	div	0.01127	.

When there is non-convergence, there are a number of possible remedies. Many authors, such as Kiernan et al. (2012) have offered a number of possible solutions. Researchers using the GLIMMIX procedure may choose to:

- Drop certain variables
- Relax the convergence criterion
- Increase the value of ABSCONV =
- Change the covariance structure using TYPE =
- Adjust the quadrature using QUAD =
- Utilize different approximation algorithms such as TECH = NRRIDG or TECH = NEWRAP, in the NLOPTIONS statement.
- Increase the number of iterations using MAXITER = in the NLOPTIONS statement
- Control the number of outer iterations using the INITGLM option
- Increase the number of optimizations using the MAXOPT = option
- Rescale data values to reduce issues relating to extreme values
- Utilize an alternative approach, such as the %HPGLIMMIX MACRO (Xie and Madden 2014)

For a more thorough discussion of the procedure itself, Ene et al. (2015) provided a thorough introduction to the use and interpretation of the GLIMMIX procedure in SAS.

6 Simulation of Data

The IML procedure in SAS was used to simulate two-level data following a generalized linear mixed model with random intercepts and random slopes. In this example, we explored the effects of including an increasing number of fixed effects when using the GLIMMIX procedure to fit a logistic regression model with one random intercept and one random slope. The approaches discussed in this section can readily be expanded to simulate data with more than two levels, although only two levels are discussed for ease of interpretation and understanding.

6.1 Simulation Setup

Here we set the parameters for the simulation. We will assume that our random intercept has variance σ^2_{INT}=7 and that the random slope has variance σ^2_{SLOPE}=15. We also assume that there are six continuous fixed effects. Each of the fixed effects has a mean of 1, with some random noise added such that the means are not all equal. The fixed effects are assumed to independent of one another and also pairwise independent of the random slope. The simulated data will include 15 clusters of observations, each with a randomly chosen number of observations between 2 and 40.

```
proc iml;
*Set the variance of the random slope;
sigInt = 7;
sigSlope = 15;
*Set the coefficients;
Bcont = 0.09;
*Set the observation level parameters;
*Set the means for 6 continuous fixed effects, and one
random slope;
means = {1,1,1,1,1,1,6};
*Slightly alter the means;
noise = normal(j(7,1));
noise[7]=0;
means = means+noise;
*Set the covariance of the fixed and random predictors;
R=I(7);
*Select the number of clusters;
b = 15;
*Randomly select the number of observations in each clus-
ter;
randobs = j(b,1);
call randgen(randobs, "Uniform");
*Transform to be between 2 and 40;
n = 2 + floor((41-2)*randobs);
*Calculate the overall total number of observations;
ntot = sum(n);
```

Once the parameters for the simulation are chosen, the cluster level data are created. Each of the random (cluster) intercepts are chosen according to independent random standard normal distributions with mean of 0 and standard deviation of 1.

The random (cluster) slope coefficients are also chosen according to independent random normal distributions with our specified variance and a mean of -1. In effect, each of the 15 clusters is assigned a unique cluster level intercept and slope term. Our design matrix is created using these random values.

```
*Create cluster level data;
*Cluster IDs;
cid = (1:b)`;
*Cluster random intercepts;
cint = randnormal(b,0,1);
*Cluster random slopes;
cslope = randnormal(b,-1,sigSlope);
*Loop through to create the design matrix;
cluster = j(ntot, 3);
startindex = 1;
do i=1 to b;
      endindex = startindex + n[i] - 1;
      cluster[startindex:endindex,1] = cid[i];
      cluster[startindex:endindex,2] = cint[i];
      cluster[startindex:endindex,3] = cslope[i];
      startindex = endindex + 1;
end;
```

Once the cluster level data are created, we can generate the observation level data. We create a matrix of independent normal realizations to serve as the observations for each of the six continuous fixed as well as the random slope variables. The realizations of each variable are created using a multivariate random normal. The fixed effect predictors are also transformed for better model fitting.

```
*Create observation level data;
X = randnormal(ntot, means, R);
*Apply some changes to the observation level data;
X[,1:6] = (X[,1:6]/1.6 + 5.1)*10;2] = bin(X[,2],cuts) - 1;
```

We combine our simulated data to create two matrices. The first matrix is used to combine all fixed and random effects information, while the second matrix provides a reduced set of information for use in simulation of the response. This second matrix removes information on the true random slope coefficient and the true cluster ID and thus contains information on the six fixed effects and the random intercept term.

```
*Create matrix of both cluster and observation level data;
alldat = X || cluster;
*Final data for simulation, excluding the random slope
predictor and cluster ID;
keepind = {1,2,3,4,5,6,9};
simdat = alldat[,keepind];
```

The coefficients for the fixed effect predictors are set according to those specified at the start of the simulation. The cluster level (random) intercept is assigned a coefficient equal to the square root of the random intercept variance term; since the random intercepts were originally simulated from a standard normal distribution, this coefficient introduces the specified variance into the simulation. These coefficients are also standardized based on the standard deviation of the respective observations.

```
*Set coefficients;
beta = j(7,1);
beta[1:6]=Bcont;
beta[7] = sqrt(sigInt);
*Standardize betas by the standard deviation;
datadev = STD(simdat);
beta = beta / datadev`;
```

We create our response as a function of these covariates. The simulated data are multiplied by the coefficients and the effect of the random slope is added in. The resulting value is then converted into a probability and used to create a binary response according to the Bernoulli distribution. This response is then combined with a "blinded" data matrix which has the value of the cluster intercept and the random slope coefficients removed. The final matrix is then output to a SAS data set with specified variable names.

```
*Create the response with the random slope effect added;
xb = simdat * beta + cluster[,3] # alldat[,7];
probs = 1 / (1 + exp(-xb));
y = rand("Bernoulli",probs);
*Create the final data with the cluster intercept removed;
outdat = y || alldat[,1:8];
*Output to a data set;
create SimData from
      outdat[colname={"Y" "X1" "X2" "X3" "X4" "X5" "X6"
"Xclust" "CID"}];
append from outdat;
close SimData;
*Quit IML;
quit;
```

The outputted data set is then analyzed using the GLIMMIX procedure in SAS. Each of the fixed effect predictors is added to the model one by one to determine the point at which this procedure will fail, if at all. A partial example of these analyses are shown below.

```
*Analyze the data using glimmix;
*One fixed effect;
proc glimmix data=SimData1;
       class CID;
       model Y(event="1") = X1 Xclust / dist=binary
link=logit;
       random intercept Xclust / type=vc subject=CID;
run;

*Two fixed effects;
proc glimmix data=SimData;
       class CID;
       model Y(event="1") = X1 X2 Xclust / dist=binary
link=logit;
       random intercept Xclust / type=vc subject=CID;
run;

[...]

*Six fixed effects;
proc glimmix data=SimData;
       class CID;
       model Y(event="1") = X1 X2 X3 X4 X5 X6 Xclust /
dist=binary link=logit;
       random intercept Xclust / type=vc subject=CID;
run;
```

6.2 *Simulation Results*

Although the GLIMMIX procedure is a powerful tool for fitting generalized linear models, it is not uncommon to find that the procedure fails to provide results. We utilized a simulation study similar to the one utilized in the previous section to investigate the effect of the number of predictors on the failure rates in the GLIMMIX procedure. A SAS macro was implemented to run the simulation across a variety of conditions and the GLIMMIX procedure was used to analyze the data under each condition for 1000 replications per condition. Each simulated data set contained information on a binary outcome, an identifier label for cluster number, one (random) cluster level predictor, and six fixed effect predictors. For each simulated data set, the GLIMMIX procedure was used to analyze the data set six times, where each call to the procedure included one additional fixed effect predictor.

Table 1 Failure rates for the GLIMMIX procedure (three clusters)

	Variances		Number of predictors					
Beta	Intercept	Slope	1	2	3	4	5	6
Weak	Low	Low	1	0.773	0.783	0.786	0.799	0.844
Moderate	Low	Low	1	0.760	0.784	0.804	0.798	0.815
Strong	Low	Low	1	0.780	0.785	0.807	0.817	0.821
Weak	Low	Medium	0	0.418	0.638	0.743	0.858	0.941
Moderate	Low	Medium	0	0.501	0.706	0.856	0.903	0.934
Strong	Low	Medium	0	0.325	0.506	0.675	0.788	0.893
Weak	Low	High	0	0.423	0.577	0.684	0.793	0.877
Moderate	Low	High	0	0.549	0.689	0.820	0.898	0.911
Strong	Low	High	0	0.450	0.629	0.703	0.826	0.865
Weak	Medium	Low	0	0.492	0.644	0.751	0.823	0.912
Moderate	Medium	Low	0	0.399	0.518	0.675	0.811	0.885
Strong	Medium	Low	0	0.322	0.488	0.641	0.745	0.817
Weak	Medium	Medium	0	0.459	0.604	0.698	0.820	0.899
Moderate	Medium	Medium	0	0.423	0.607	0.760	0.856	0.923
Strong	Medium	Medium	0	0.428	0.537	0.648	0.761	0.846
Weak	Medium	High	0	0.422	0.610	0.716	0.844	0.902
Moderate	Medium	High	0	0.367	0.557	0.725	0.846	0.910
Strong	Medium	High	0	0.393	0.543	0.636	0.788	0.831
Weak	High	Low	0	0.463	0.565	0.712	0.791	0.877
Moderate	High	Low	0	0.392	0.515	0.662	0.845	0.885
Strong	High	Low	0	0.364	0.509	0.664	0.743	0.851
Weak	High	Medium	0	0.529	0.701	0.777	0.880	0.956
Moderate	High	Medium	0	0.413	0.602	0.684	0.854	0.915
Strong	High	Medium	0	0.356	0.511	0.669	0.769	0.845
Weak	High	High	0	0.324	0.519	0.661	0.779	0.839
Moderate	High	High	0	0.327	0.484	0.656	0.831	0.869
Strong	High	High	1	0.376	0.581	0.738	0.808	0.842

In particular, the conditions examined were the number of data clusters, the strength of the variance of both the random intercept and slope and strength of fixed effect coefficients. The simulation took into account data sets with either 3, 15 or 45 clusters of data. The random effect variances we investigated included all combinations of low, medium and high variances for the random intercept and random slope—yielding a total of nine different variance combinations. The fixed effects also took three levels of strength—weak, moderate or strong.

The results of this simulation are given in Tables 1, 2 and 3 and are also displayed graphically in Fig. 1. These displays provide the failure rates for the 1000 simulations conducted for each of the specified conditions, thus higher values indicate poorer

Table 2 Failure Rates for the GLIMMIX Procedure (fifteen clusters)

	Variances		Number of predictors					
Beta	Intercept	Slope	1	2	3	4	5	6
Weak	Low	Low	0	0.413	0.400	0.404	0.414	0.396
Moderate	Low	Low	0	0.415	0.424	0.433	0.434	0.422
Strong	Low	Low	0	0.441	0.430	0.431	0.418	0.428
Weak	Low	Medium	0	0.138	0.180	0.337	0.484	0.610
Moderate	Low	Medium	0	0.182	0.272	0.387	0.493	0.579
Strong	Low	Medium	0	0.121	0.199	0.273	0.378	0.459
Weak	Low	High	0	0.152	0.281	0.401	0.509	0.635
Moderate	Low	High	0	0.200	0.269	0.355	0.459	0.545
Strong	Low	High	0	0.131	0.239	0.332	0.370	0.473
Weak	Medium	Low	0	0.171	0.251	0.321	0.449	0.567
Moderate	Medium	Low	0	0.148	0.258	0.445	0.570	0.613
Strong	Medium	Low	0	0.093	0.157	0.222	0.328	0.411
Weak	Medium	Medium	0	0.148	0.243	0.325	0.411	0.530
Moderate	Medium	Medium	0	0.167	0.294	0.406	0.514	0.573
Strong	Medium	Medium	0	0.118	0.189	0.284	0.348	0.455
Weak	Medium	High	0	0.214	0.304	0.396	0.459	0.627
Moderate	Medium	High	0	0.220	0.295	0.399	0.478	0.590
Strong	Medium	High	0	0.158	0.238	0.305	0.404	0.489
Weak	High	Low	0	0.092	0.191	0.324	0.404	0.531
Moderate	High	Low	0	0.157	0.252	0.366	0.440	0.552
Strong	High	Low	0	0.085	0.141	0.249	0.351	0.446
Weak	High	Medium	0	0.129	0.217	0.347	0.446	0.532
Moderate	High	Medium	0	0.122	0.227	0.335	0.503	0.605
Strong	High	Medium	0	0.133	0.194	0.272	0.359	0.437
Weak	High	High	0	0.140	0.232	0.329	0.470	0.541
Moderate	High	High	0	0.093	0.173	0.258	0.369	0.505
Strong	High	High	0	0.129	0.199	0.266	0.332	0.465

performance as a higher proportion of the calls to the GLIMMIX procedure failed to provide results. Tables 1, 2 and 3 divide the results of the simulations based on the number of clusters in each simulation, where Table 1 summarizes the simulations with 3 data clusters each, Table 2 summarizes the simulations with 15 data clusters each and Table 3 summarizes the simulations with 45 data clusters each. The first column in each of the tables provides the strength of the fixed effects predictor (weak, moderate or strong). The second and third columns denote the simulation settings for the variance of the random intercept and slope, respectively, where each variance term

Table 3 Failure Rates for the GLIMMIX Procedure (forty-five clusters)

	Variances		Number of predictors					
Beta	Intercept	Slope	1	2	3	4	5	6
Weak	Low	Low	0	0.243	0.247	0.245	0.236	0.235
Moderate	Low	Low	1	0.236	0.246	0.249	0.255	0.244
Strong	Low	Low	0	0.298	0.303	0.305	0.296	0.303
Weak	Low	Medium	0	0.055	0.090	0.123	0.159	0.184
Moderate	Low	Medium	0	0.051	0.098	0.127	0.163	0.243
Strong	Low	Medium	1	0.038	0.073	0.108	0.141	0.189
Weak	Low	High	1	0.035	0.073	0.099	0.140	0.197
Moderate	Low	High	1	0.038	0.061	0.101	0.157	0.214
Strong	Low	High	0	0.041	0.052	0.081	0.139	0.197
Weak	Medium	Low	0	0.060	0.089	0.134	0.169	0.223
Moderate	Medium	Low	0	0.061	0.093	0.127	0.189	0.215
Strong	Medium	Low	0	0.045	0.069	0.096	0.134	0.185
Weak	Medium	Medium	0	0.027	0.054	0.088	0.163	0.209
Moderate	Medium	Medium	0	0.062	0.104	0.139	0.168	0.208
Strong	Medium	Medium	0	0.071	0.099	0.118	0.146	0.187
Weak	Medium	High	0	0.041	0.088	0.130	0.177	0.209
Moderate	Medium	High	0	0.050	0.066	0.112	0.143	0.169
Strong	Medium	High	0	0.026	0.052	0.089	0.126	0.169
Weak	High	Low	0	0.031	0.057	0.080	0.141	0.209
Moderate	High	Low	0	0.031	0.061	0.104	0.139	0.187
Strong	High	Low	0	0.032	0.046	0.092	0.156	0.192
Weak	High	Medium	0	0.050	0.094	0.139	0.206	0.263
Moderate	High	Medium	0	0.068	0.111	0.161	0.182	0.259
Strong	High	Medium	0	0.030	0.068	0.101	0.129	0.173
Weak	High	High	0	0.038	0.086	0.139	0.170	0.235
Moderate	High	High	0	0.081	0.123	0.179	0.244	0.300
Strong	High	High	0	0.053	0.080	0.125	0.149	0.202

takes one of three levels (low, medium, high). The remaining six columns contain the failure rates as a proportion for the GLIMMIX procedure with a given number of fixed effects predictors. For instance, we can see from Table 1, in the first data row, in the last column that of the 1000 simulations with three clusters, weak fixed effects, low intercept variance and low slope variance, 84.4 % of the models with six fixed effect predictors failed to converge.

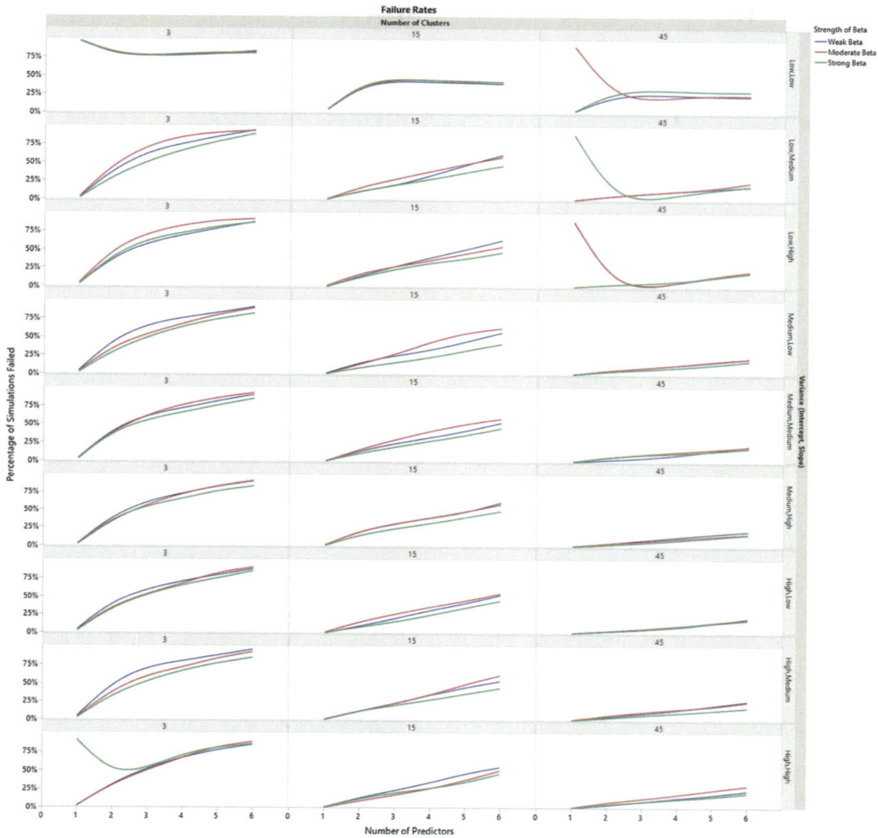

Fig. 1 Failure rates for the GLIMMIX procedure

Figure 1 provides a graphical representation of the same simulated data presented in Tables 1, 2 and 3. Each individual plot contains three lines representing the failure rates for each of the three strengths of the fixed effects. The blue line represents the simulations with weak predictors, the red line represents the simulations with moderate predictors and the green line represents the simulations with strong predictors. The vertical (Y) axis of each individual plot denotes the failure rates as a percentage, where higher values indicate higher rates of failure. The horizontal (X) axis within each of the individual plots represents the number of fixed effects included in the model for those simulations. The individual plots are also organized into three columns according to the number of data clusters in those simulations. The individual plots are further grouped into nine rows according to the strength of the random effects for those simulations. For example, in the individual plot in the last column of the first row contains information on the 1000 simulations in which there were 45 clusters, with weak fixed effects predictors, low random intercept variance and low random slope variance.

In general, as the number of predictors increased, the failure rates also increased. Notable exceptions include the case where there is very little variance in the random effects. For instance, in the case of low random intercept variance and low random slope variance, the failure rates may actually decrease, or increase only slightly. We can also see that the effect of increasing the number of predictors is also suppressed when there are more data clusters. In general, the GLIMMIX procedure is more successful in analyzing data with more clusters as illustrated by the lower failure rates. Similarly, data with overall stronger random effect variance is also less susceptible to failure as the number of predictors in the model increases. This holds true with respect to both the random intercept variance as well as the random slope variance.

7 Analysis of Data

7.1 Description

A subset of data from the 2011 Bangladesh Demographic and Health Survey is used in this study. This subset contains information on 1000 women between the ages of 10 and 49, living in Bangladesh. The data in this study are hierarchical in nature in that each of the women is nested within one of seven different districts, which correspond approximately to administrative regions in Bangladesh (NIPORT 2013). A simplified version of this structure is represented as Fig. 2.

The outcome of interest in this data set is a binary variable representing the woman's knowledge of AIDS. The variable takes one of two values representing knowledge of AIDS (1) or no knowledge of AIDS (0). In addition to this outcome, the data set also includes information on the woman's wealth index, age, number of living children as well as whether or not the woman lives in an urban or rural setting. Wealth index had five possible levels representing the quintile to which the woman belonged. Age represented the woman's age at the time of survey while number of living children represented how many living children the woman had at the time of survey. The urban/rural variable was a district level predictor as the value of this predictor were partially driven by the administrative region.

Please note to use the included DHS subset data, you must register as a DHS data user at: http://www.dhsprogram.com/data/new-user-registration.cfm. This subset data must not be passed on to others without the written consent of DHS (archive@dhsprogram.com). You are required to submit a copy of any reports/ publications resulting from using this subset data to: archive@dhsprogram.com.

Fig. 2 Hierarchical structure in 2011 DHS Study

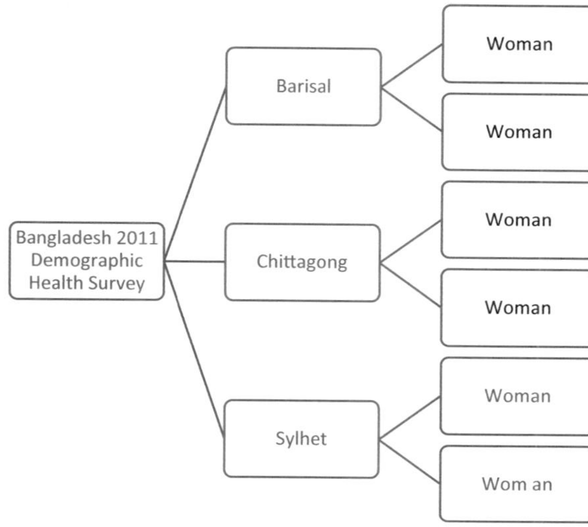

7.2 *Data Analysis*

We fit a logistic regression model with one random intercept and one random slope for the urban/rural variable. For these data, the random effects were used to address the clustering present due to districts. Each of these models was fitted using the GLIMMIX procedure in SAS. The first model included one fixed effect predictor for wealth index.

$$\log\left[\frac{p_{jk}}{1-p_{jk}}\right] = \beta_0 + \gamma_1 \mathrm{Urban}_j + \gamma_{1j} \mathrm{Wealth}_{jk} + u_{oj}$$

As in the data simulation section, these data can be analyzed in SAS using code similar to the example given below. Note that additional fixed effects predictors can be included in the model statement to fit additional models.

```
proc glimmix data=bang;
      class div urban wealth;
      model aids(event="1") = urban wealth / dist=binary
link=logit;
      random intercept urban /type=vc subject=div;
run;
```

The convergence criterion noted that the GLIMMIX procedure converged successfully and that we are also provided with standard errors for our random effects. Therefore, we see the procedure was successful in fitting the model.

Iteration History					
Iteration	*Restarts*	*Subiterations*	*Objective Function*	*Change*	*Max Gradient*
0	*0*	*4*	4562.1081534	2.00000000	0.00012
1	*0*	*3*	4679.3151727	0.37988319	0.000023
2	*0*	*2*	4718.2025244	0.05445086	0.000019
3	*0*	*1*	4720.2027844	0.00248043	0.000042
4	*0*	*1*	4720.2171672	0.00004268	1.244E-8
5	*0*	*0*	4720.2172745	0.00000000	5.905E-6

Convergence criterion (PCONV=1.11022E-8) satisfied.

Fit Statistics	
-2 Res Log Pseudo-Likelihood	4720.22
Generalized Chi-Square	973.41
Gener. Chi-Square / DF	0.98

Covariance Parameter Estimates			
Cov Parm	*Subject*	*Estimate*	*Standard Error*
Intercept	div	0.1149	0.1102
urban	div	0.05142	0.09565

Type III Tests of Fixed Effects				
Effect	*Num DF*	*Den DF*	*F Value*	*Pr > F*
urban	1	6	1.48	0.2692
wealth	4	982	21.92	<.0001

We also fit the model which included fixed effects for both wealth and age.

$$\log\left[\frac{\mathrm{p_{jk}}}{1-\mathrm{p_{jk}}}\right] = \beta_0 + \gamma_1 \mathrm{Urban_j} + \gamma_{1j} Wealth_{jk} + \gamma_{2j} \mathrm{Age_{jk}} + \mathrm{u_{oj}}$$

In this case, we can similarly see that the convergence criterion is satisfied and that estimates of the standard errors of the random effects are provided. Thus, we see that the GLIMMIX procedure was successful in fitting a model.

Iteration History					
Iteration	*Restarts*	*Subiterations*	*Objective Function*	*Change*	*Max Gradient*
0	*0*	*4*	4565.1551673	2.00000000	0.000164
1	*0*	*3*	4759.8377085	0.81042673	0.00017
2	*0*	*2*	4808.4720043	0.11542518	0.000114
3	*0*	*1*	4811.3231643	0.00580829	0.000136
4	*0*	*1*	4811.3435049	0.00011603	5.451E-8
5	*0*	*1*	4811.3434869	0.00000132	5.917E-9
6	*0*	*0*	4811.3434867	0.00000000	1.381E-7

Convergence criterion (PCONV=1.11022E-8) satisfied.

Fit Statistics	
-2 Res Log Pseudo-Likelihood	4811.34
Generalized Chi-Square	973.16
Gener. Chi-Square / DF	0.98

Covariance Parameter Estimates			
Cov Parm	*Subject*	*Estimate*	*Standard Error*
Intercept	div	0.1152	0.1059
urban	div	0.03594	0.08961

Type III Tests of Fixed Effects				
Effect	*Num DF*	*Den DF*	*F Value*	*Pr > F*
urban	1	6	1.70	0.2403
wealth	4	981	22.06	<.0001
age	1	981	43.47	<.0001

We added a third predictor for *number of living children* to our mixed model.

$$\log\left[\frac{p_{jk}}{1-p_{jk}}\right] = \beta_0 + \gamma_1 \mathrm{Urban}_j + \gamma_{1j}\mathrm{Wealth}_{jk} + \gamma_{2j}\mathrm{Age}_{jk} + \gamma_{3j}\mathrm{Children}_{jk} + u_{oj}$$

With the inclusion of this third predictor, we see that the GLIMMIX procedure fails to converge and consequently does not provide estimates of the standard errors of the random effects. Hence, we see that, although SAS is able to fit the model with two fixed effects, the inclusion of a third fixed effect leads to failure.

Iteration History					
Iteration	*Restarts*	*Subiterations*	*Objective Function*	*Change*	*Max Gradient*
0	*0*	*4*	4583.476774	2.00000000	3.121596
1	*0*	*3*	4788.3397625	2.00000000	6.129E-6
2	*0*	*2*	4842.5826312	0.36041184	0.000157
3	*0*	*1*	4846.0866446	0.19708062	0.000187
4	*0*	*1*	4847.0951336	0.16428366	1.49E-7
5	*0*	*1*	4848.0956362	0.14098626	2.589E-9
6	*0*	*1*	4849.0959182	0.12352732	7.98E-9
7	*0*	*1*	4850.096021	0.10993349	4.498E-8
8	*0*	*1*	4851.0960586	0.09904095	6.01E-11
9	*0*	*0*	4852.0960724	0.09011439	4.282E-6
10	*0*	*0*	4853.0960785	0.08266460	5.821E-6
11	*0*	*0*	4854.0960808	0.07635276	6.385E-6
12	*0*	*0*	4855.0960816	0.07093651	6.594E-6
13	*0*	*0*	4856.0960819	0.06623786	6.674E-6
14	*0*	*0*	4857.0960819	0.06212302	6.687E-6
15	*0*	*0*	4858.0960828	0.05848863	6.683E-6
16	*0*	*0*	4859.0960811	0.05525857	6.782E-6
17	*0*	*0*	4860.0960789	0.05236722	6.898E-6
18	*0*	*0*	4861.0960963	0.04974318	5.95E-6
19	*0*	*0*	4862.0961081	0.04737287	7.581E-6

Did not converge.

Covariance Parameter Estimates			
Cov Parm	*Subject*	*Estimate*	*Standard Error*
Intercept	div	0.09097	.
urban	div	0.01127	.

Although we do not explore its use in depth here, the %hpglimmix macro provides an alternative approach in SAS for fitting generalized linear mixed models (Xie and Madden 2014). This macro offers improvements in memory usage as well as processing time and supports the fitting of more complicated models as compared to the GLIMMIX procedure. Although this macro does not currently provide standard errors of the covariance parameter estimates or Type III test results, it can be useful when alternative approaches fail to resolve convergence issues in the GLIMMIX procedure. We fit the previously discussed model, which includes three fixed effects predictors as well as one random intercept and one random slope for the Bangladesh data. After loading the macro into the current SAS session, the model can be run using code similar to the following.

```
%hpglimmix(data=bang,
       stmts=%str(
              class div urban wealth children;
              model aids = urban wealth age children / solu-
tion ;
              random int urban / subject=div solution;
       ),
       error=binomial, maxit=50,
       link=logit
);
```

Though this model fails to converge in the GLIMMIX procedure, we see that %HPGLIMMIX provides results for the model which includes three fixed effect predictors.

Iteration History

Iteration	*Evaluations*	*Objective Function*	*Change*	*Max Gradient*
0	4	4892.189763	.	6.524357
1	5	4892.1622318	0.02753122	5.886723
2	3	4892.1566979	0.00553391	5.959108
3	3	4892.1564908	0.00020710	5.957165
4	5	4892.0850182	0.07147255	3.174453
5	4	4892.0847449	0.00027336	3.145538
6	4	4892.0726771	0.01206780	0.563614
7	4	4892.0726558	0.00002129	0.569235
8	4	4892.0726553	0.00000047	0.568549
9	5	4892.0724006	0.00025469	0.41386
10	*4*	4892.0721478	0.00025279	0.01987

Convergence criterion (GCONV=1E-8) satisfied.

Covariance Parameter Estimates

Cov Parm	*Subject*	*Estimate*
Intercept	div	0.09138
urban	div	0.01301
Residual		0.9905

Fit Statistics

-2 Res Log Likelihood	4892.07215
AIC (smaller is better)	4898.07215
AICC (smaller is better)	4898.09666
BIC (smaller is better)	4897.90988
CAIC (smaller is better)	4900.90988

Another possible remedy in this case is found in the NLMIXED procedure in SAS. This procedure utilizes likelihood-based approaches to fit mixed models for nonlinear outcomes (Wolfinger 1999). This procedure is readily available in SAS software and provides similar techniques to those available in the GLIMMIX procedure. Although the models that can be fit in both procedures are similar, it is worth noting that the two procedures use different techniques for estimation and thus the results may vary between the two approaches. However, because different estimation techniques are employed there are also cases in which the NLMIXED procedure will converge, while the GLIMMIX procedure will not.

The NLMIXED procedure is implemented differently as compared to many other procedures in SAS software. In particular, one must provide starting values for each of the parameters of interest, which can be estimated in a number of ways. In this example, we first used the logistic procedure to obtain estimates of the fixed effects parameters and specify a generic value of '1' for the variance of each of our random effects (intercept and slope). We also specified an equation with respect to our parameters and observed predictor values, and use this equation in the specification of our model statement through the calculation of our probability using the logit link. Finally, each of the random effects as well as the corresponding distribution is specified, and the subject assigned.

```
proc logistic data=bang;
      model aids_knowledge(event="1") = urban wealth age
children/ link=logit;
run;

proc nlmixed data=bang;
      parms b0=0.6916 b1=0.3335 b2=0.5921 b3=-0.0287 b4=-
0.2616 s2u = 1 s2r = 1;
      xb = b0 + u + (b1+rb1)*urban + b2*wealth + b3*age +
b4*children;
      p = exp(xb) / (1+exp(xb));
      model aids_knowledge ~binary(p);
      random u rb1 ~ normal([0,0],[s2u,0,s2r]) sub-
ject=div;
run;
```

We found that the NLMIXED procedure converges successfully and also provides solutions for both our fixed and random effects for the model which includes three fixed effects predictors. Wolfinger (1999) provides a good introduction to the NLMIXED procedure and its usage, as well as some of the underlying calculations.

NOTE: GCONV convergence criterion satisfied.

Fit Statistics

-2 Log Likelihood	972.1
AIC (smaller is better)	986.1
AICC (smaller is better)	986.2
BIC (smaller is better)	985.7

Parameter Estimates

Parameter	*Estimate*	*Standard Error*	*DF*	*t Value*	*Pr* > |*t*|	*Alpha*	*Lower*	*Upper*	*Gradient*
b0	0.6547	0.3339	5	1.96	0.1072	0.05	−0.204	1.5131	−0.0001
b1	0.3267	0.2801	5	1.17	0.2960	0.05	−0.393	1.0468	0.00032
b2	0.6156	0.06906	5	8.91	0.0003	0.05	0.4381	0.7931	−0.0004
b3	−0.0296	0.01113	5	−2.66	0.0448	0.05	−0.058	−0.001	0.00999
b4	−0.2567	0.06031	5	−4.26	0.0080	0.05	−0.412	−0.102	0.0018
s2u	0.03178	0.04894	5	0.65	0.5450	0.05	− 0.094	0.1576	0.00059
s2r	0.2798	0.2937	5	0.95	0.3846	0.05	−0.475	1.0349	−0.0001

In general, we found that the results of our data analysis are in agreement with our findings based on the simulation study. The GLIMMIX procedure was successful in analyzing the models with fewer fixed effects predictors. However, once we included additional fixed effects, we saw that the GLIMMIX procedure failed to converge. In these cases, we may choose to investigate only the smaller subset of predictors in order to get successful analyses. Alternatively, if the larger number of predictors is of interest, we can utilize the %HPGLIMMIX macro, which is able to achieve convergence, although the output is reduced. We may also utilize the NLMIXED procedure, which utilizes different methods for estimation.

8 Conclusions

Fitting hierarchical logistic regression models to survey binary data is common in a number of disciplines. These models are useful in analyzing survey data in the presence of clustering or correlation, which otherwise would make standard approaches inappropriate due to the lack of independence amongst the outcomes. Although there are a number of powerful approaches for fitting these models, such as the GLIMMIX and NLMIXED procedures in SAS, the computational complexity of the algorithms can often lead to failures in convergence.

Through the use of simulations, we obtained useful information for exploring the reasons for non-convergence, as well as steps to avoid these issues. In particular, when using the GLIMMIX procedure, researchers should be careful in selecting predictors to include in the model. The inclusion of too many predictors can lead to convergence issues, regardless of whether these predictors are fixed or random. When many predictors must be included due to research or knowledge constraints and if the GLIMMIX procedure failures to converge, other options can be explored to fit similar models. Because it utilizes different approaches, the NLMIXED procedure is a viable option for obtaining convergence in the mixed model setting when the GLIMMIX procedure fails. Recent advances, such as the %HPGLIMMIX macro can also be utilized as a remedy.

While we concentrated and presented results applicable only to the convergence issue in the GLIMMIX procedure for two-level hierarchical logistic regression models, we believe that these approaches can be readily adapted and expanded to explore different or more complex problems. In general, Monte-Carlo simulation offers a fast, and inexpensive avenue for investigating problems such as convergence, as well as appropriate solutions.

Acknowledgements This work is funded in part by the National Institutes of Health Alzheimer's Consortium Fellowship Grant, Grant No. NHS0007. The content in this paper is solely the responsibility of the authors and does not necessarily represent the official views of the National Institutes of Health.

References

Austin, P. C. (2010). Estimating multilevel logistic regression models when the number of clusters is low: A comparison of different statistical software procedures. *The International Journal of Biostatistics*, *6*(1), 1–20.

Austin, P. C., Manca, A., Zwarenstein, M., Juurlink, D. N., & Stanbrook, M. B. (2010). A substantial and confusing variation exists in handling of baseline covariates in randomized controlled trials: a review of trials published in leading medical journals. *Journal of Clinical Epidemiology*, *63*(2), 142–153.

Ene, M., Leighton, E. A., Blue, G. L., & Bell, B. A. (2015). Multilevel models for categorical data using SAS PROC GLIMMIX: The Basics. *SAS Global Forum 2015 Proceedings*.

Hartzel, J., Agresti, A., & Caffo, B. (2001). Multinomial logit random effects models. *Statistical Modelling*, *1*(2), 81–102.

Hedeker, D., Mermelstein, R. J., & Demirtas, H. (2008). An application of a mixed effects location scale model for analysis of Ecological Momentary Assessment (EMA) data. *Biometrics*, *64*(2), 627–634.

Hedeker, D., Mermelstein, R. J., & Demirtas, H. (2012). Modeling between- and within subject variance in Ecological Momentary Assessment (EMA) data using mixed-effects location scale models. *Statistics in Medicine*, *31*(27), 3328–3336.

Hox, J. J. (2002). *Multilevel analysis: Techniques and applications*. Mahwah: Lawrence Erlbaum Associates Inc.

Irimata, K. M., & Wilson, J. R. (2017). Identifying Intraclass correlations necessitating hierarchical modeling. *Journal of Applied Statistics*, accepted.

Kiernan, K., Tao, J., & Gibbs, P. (2012). Tips and strategies for mixed modeling with SAS/STAT procedures. *SAS Global Forum 2012 Proceedings*.

Kuss, O. (2002). Global goodness-of-fit tests in logistic regression with sparse data. *Statistics in Medicine*, *21*, 3789–3801.

Kuss, O. (2002). How to use SAS for logistic regression with correlated data. In *SUGI 27 Proceedings* (pp. 261–27).

Lesaffre, E., & Spiessens, B. (2001). On the effect of the number of quadrature points in a logistic random effects model: An example. *Journal of the Royal Statistical Society. Series C (Applied Statistics)*, *50*(3), 325–335.

Longford, N. T. (1993). *Random coefficient models*. Oxford: Clarendon Press.

Maas, C. J. M., & Hox, J. J. (2004). The influence of violations of assumptions on multilevel parameter estimates and their standard errors. *Computational Statistics & Data Analysis*, *46*(3), 427–440.

McMahon, J. M., Pouget, E. R., & Tortu, S. (2006). A guide for multilevel modeling of dyadic data with binary outcomes using SAS PROC NLMIXED. *Computational Statistics & Data Analysis*, *50*(12), 3663–3680.

National Institute of Population Research and Training (NIPORT). (2013). *Bangladesh demographic and health survey 2011*. NIPORT, MItra and Associates, ICF International: Dhaka Bangladesh, Calverton MD.

Nelder, J. A., & Wedderburn, R. W. M. (1972). Generalized linear models. *Journal of the Royal Statistical Society. Series A (General)135*(3), 370–384.

Newsom, J. T. (2002). A multilevel structural equation model for dyadic data. *Structural Equation Modeling: A Multidisciplinary Journal*, *9*(3), 431–447.

Rasbash, J., Steele, F., Browne, W. J., & Goldstein, H. (2012). *User's guide to WLwiN, Version 2.26*. Centre for Multilevel Modelling, University of Bristol. Retrieved from http://www.bristol.ac.uk/cmm/software/mlwin/download/2-26/manual-web.pdf.

Raudenbush, S. W. (1992). *Hierarchical linear models*. Newbury Park: Sage Publications.

Raudenbush, S. W., & Bryk, A. S. (2002). *Hierarchical linear models: Applications and data analysis methods*. Thousand Oaks: Sage Publications.

Rodriquez, G., & Goldman, N. (1995). An assessment of estimation procedures for multilevel models with binary responses. *Journal of the Royal Statistical Society, Series A (Statistics in Society)158*(1), 73–89.

SAS Institute Inc. (2013). *Base SAS® 9.4 Procedure guide: Statistical procedures* (2nd ed.). Cary, NC: SAS Institute Inc.

Schabenberger, O. (2005). Introducing the GLIMMIX procedure for generalized linear mixed models. *SUGI 30 Proceedings*, 196–30.

Shahian, D. M., Normand, S. L., Torchiana, D. F., Lewis, S. M., Pastore, J. O., Kuntz, R. E., et al. (2001). Cardiac surgery report cards: Comprehensive review and statistical critique. *The Annals of Thoracic Surgery*, *72*(6), 2155–2168.

Smyth, G. K. (1989). Generalized linear models with varying dispersion. *Journal of the Royal Statistical Society, Series B*, *51*, 47–60.

Snijders, T. A. B., & Bosker, R. J. (1998). *Multilevel analysis: An introduction to basic and advanced multilevel modeling*. Thousand Oaks: Sage Publications.

Three-level multilevel model in SPSS. (2016). UCLA: Statistical Consulting Group. http://www.ats.ucla.edu/stat/spss/code/three_level_model.htm.

Wedderburn, R. W. M. (1974). Quasi-likelihood functions, generalized linear models and the Gauss-Newton method. *Biometrika*, *61*(3), 439–447.

Xie, L., & Madden, L. V. (2014). %HPGLIMMIX: A high-performance SAS macro for GLMM Estimation. *Journal of Statistical Software*, *58*(8).

Wilson, J. R., & Lorenz, K. A. (2015). *Modeling Binary correlated responses using SAS, SPSS and R*. New York: Springer International Publishing.

Wolfinger, D. (1999). Fitting nonlinear mixed models with the new NLMIXED procedure. In *Sugi 24 Proceedings* (pp. 278–284).

Monte-Carlo Methods in Financial Modeling

Chuanshu Ji, Tao Wang and Leicheng Yin

Abstract The last decade has witnessed fast growing applications of Monte-Carlo methodology to a wide range of problems in financial economics. This chapter consists of two topics: market microstructure modeling and Monte-Carlo dimension reduction in option pricing. Market microstructure concerns how different trading mechanisms affect asset price formation. It generalizes the classical asset pricing theory under perfect market conditions by incorporating various friction factors, such as asymmetric information shared by different market participants (informed traders, market makers, liquidity traders, et al.), and transaction costs reflected in bid-ask spreads. The complexity of those more realistic dynamic models presents significant challenges to empirical studies for market microstructure. In this work, we consider some extensions of the seminal sequential trade model in Glosten and Milgrom (Journal of Financial Economics, 14(1), 71–100, 1985) and perform Bayesian Markov chain Monte-Carlo (MCMC) inference based on the trade and quote (TAQ) database in Wharton Research Data Services (WRDS). As more and more security derivatives are constructed and traded in financial markets, it becomes crucial to price those derivatives, such as futures and options. There are two popular approaches for derivative pricing: the analytical approach sets the price function as the solution to a PDE with boundary conditions and solves it numerically by finite difference etc.; the probabilistic approach expresses the price of a derivative as the conditional expectation under a risk neutral measure and computes it via numerical integration. Adopting the second approach, we notice the required integration is often performed over a high dimensional state space in which state variables are financial time series. A key observation is for a broad class of stochastic volatility (SV) models, the con-

C. Ji (✉)
Department of Statistics and Operations Research, University of North Carolina,
Chapel Hill, NC 27599-3260, USA
e-mail: cji@email.unc.edu

T. Wang
Bank of America Merrill Lynch,
Bank of America Tower, One Bryant Park, New York, NY 10036, USA

L. Yin
Exelon Business Services Company, Enterprise Risk Management,
Chase Tower 10 S. Dearborn St, Chicago, IL 60603, USA

D.-G. Chen and J.D. Chen (eds.), *Monte-Carlo Simulation-Based Statistical Modeling*, ICSA Book Series in Statistics, DOI 10.1007/978-981-10-3307-0_14

ditional expectations representing related option prices depend on high-dimensional volatility sample paths through only some 2D or 3D summary statistics whose samples, if generated, would enable us to avoid brute force Monte-Carlo simulation for the underlying volatility sample paths. Although the exact joint distributions of the summary statistics are usually not known, they could be approximated by distribution families such as multivariate Gaussian, gamma mixture of Gaussian, log-normal mixture of Gaussian, etc. Parameters in those families can be specified by calculating the moments and expressing them as functions of parameters in the original SV models. This method improves the computational efficiency dramatically. It is particularly useful when prices of those derivatives need to be calculated repeatedly as a part of Bayesian MCMC calibration for SV models.

1 Hierarchical Modeling in Market Microstructure Studies

The research in financial economics becomes more necessary after the financial crisis, with statistics playing an important role in such studies. Several milestones in modern finance, such as capital asset pricing model (CAPM), Black-Scholes-Merton derivatives pricing, hold under certain perfect market conditions, i.e. the market is fully efficient with no taxes, no transaction costs, no bid-ask spreads, unlimited short-selling, and all market participants sharing the same information, to name just a few. Those assumptions are clearly violated in real financial markets, evidenced by many empirical studies. Market microstructure concerns friction factors, aims to understand how asset price formation is affected by various trading mechanisms.

In this work, we will focus on two aspects of market microstructure that attract most attentions from financial economists: asymmetric information and bid/ask spreads. We will follow the model-based approach in the seminal work of Glosten and Milgrom (1985), referred to as G-M model in what follows. It is a sequential trade model assuming risk neutrality and a quote-driven protocol. The market maker posts bid and ask prices in every (discrete) time unit based on which traders place their orders. There are certain informed traders among other uninformed traders in the market, and the proportion of informed traders is represented by a parameter α. The information asymmetry induces adverse selection costs that force the market maker to quote different prices for buying and selling, leading to the bid-ask spread. The spread is a premium the market maker demands for trading with informed traders. A special feature in G-M model is to present explicitly how bid and ask prices change over time and are influenced by different trading orders.

Our research concerns empirical studies for G-M model and its extensions using real market data. Due to the complexity of many market microstructure models such as G-M, systematic model-based empirical studies are relatively lacking compared to the development of theoretical models and model-free descriptive data analysis. A noticeable contribution is Hasbrouck (2009) which considers an extension of Roll model (cf. Roll 1984) and uses the Gibbs sampler to estimate the effective trading cost and trading direction. To validate the method, a high correlation 0.965 is calculated between the Gibbs sampler estimates of the effective cost and the descriptive estimates based on high frequency TAQ data. The sophisticated hierarchical dynamic structure in G-M model presents a challenge to model-based inference using real market data. Little has been done in this direction. Das (2005) takes a useful step by presenting an algorithm for computing approximate solutions to the bid/ask prices and runs a simulation study under a modified G-M model. It helps us learn from the market maker's perspective, and paves a road for further studies.

In this work, we consider further extensions of G-M model and perform Bayesian MCMC inference based on the TAQ database in WRDS. Both the asymmetric information and bid-ask spread issues are addressed. To the best of our knowledge, our work is the first attempt at inference on market microstructure models of G-M type based on intra-day data. Since the main focus of this chapter is implementation of MCMC algorithms, some other useful results we developed along this line are not included here, such as incorporation of GARCH (1,1) model for the volatility of asset returns which furthermore improves the model fitting. More details are available in Tao Wang's Ph.D. thesis (cf Wang 2014) upon request.

1.1 The Model

The following setting for market microstructure is assumed:

- Let V_t denote the true underlying value (logarithmic share price) of a stock at time $t = 0, 1, \ldots$ The stock dynamics follows a random walk $V_t = V_{t-1} + \epsilon_t$, where the innovations $\epsilon_1, \epsilon_2, \ldots$ are iid $N(0, \sigma^2)$ random variables with parameter $\sigma > 0$.
- A single market maker sets the ask price A_t and the bid price B_t for one share of the stock at t.
- Traders enter the market sequentially. Each of them can buy the stock at the price A_t, or sell the stock at B_t. There are two types of traders: uninformed and informed. An uninformed trader (assumed not knowing V_t) will place a buy or sell order with equal probability η, or choose not to trade with probability $1 - 2\eta$. An informed trader, who is assumed to know V_t, will place a buy order if $V_t > A_t$ or a sell order if $V_t < B_t$, or no trade order if $B_t \leq V_t \leq A_t$.
- When setting A_t and B_t at each t, the market maker, knowing neither the type of the trader nor the true value V, will face an informed trader with probability α or an uninformed one with probability $1 - \alpha$, and receive an order placed by that trader, based on which he will update his belief in such a way as defining $A_t = E_t(V \mid \text{Buy})$ and $B_t = E_t(V \mid \text{Sell})$. Here $E_t(\cdot)$ denotes the conditional expectation given the market maker's information up to t, with "Buy/Sell" inserted in the condition to reflect the most recent order type.
- Denote the observed bid/ask prices by P_t^b and P_t^a respectively and assume $P_t^b \sim N(B_t, \delta^2)$ and $P_t^a \sim N(A_t, \delta^2)$, where $\delta > 0$ measures pricing errors (small perturbation).

Following Bayes' formula we get

$$
\begin{aligned}
E_t[V|\text{Sell}] &= \int v \; p_t(v|\text{Sell})dv \\
&= \int v \; \frac{P_t(\text{Sell}|v) \; f_t(v)}{P_t(\text{Sell})} dv, \qquad (1)
\end{aligned}
$$

where $f_t(v)$ is the normal density of V_t and $P_t(\text{Sell})$, the probability of receiving a sell order at time t, is given by

$$
P_t(\text{Sell}) = \alpha \int_{-\infty}^{B_t} f_t(v)dv + (1-\alpha)\eta. \qquad (2)
$$

Therefore,

$$
\begin{aligned}
B_t &= \frac{1}{P_t(\text{Sell})} \int_{-\infty}^{\infty} v P_t(\text{Sell}|v) f_t(v)dv \\
&= \frac{1}{P_t(\text{Sell})} \left(\alpha \int_{-\infty}^{B_t} v f_t(v)dv + (1-\alpha)\eta V_0 \right). \qquad (3)
\end{aligned}
$$

Similarly, the ask price is given by

$$
A_t = \frac{1}{P_t(\text{Buy})} \left(\alpha \int_{A_t}^{\infty} v f_t(v)dv + (1-\alpha)\eta V_0 \right), \qquad (4)
$$

where $P_t(\text{Buy})$ is given by

$$
P_t(\text{Buy}) = \alpha \int_{A_t}^{\infty} f_t(v)dv + (1-\alpha)\eta. \qquad (5)
$$

Note that determination of B_t amounts to solving (3) numerically since B_t appears on both sides of (3). Similarly, A_t is found by solving (4) numerically.

1.2 Bayesian Inference via MCMC Algorithms

The paradigm of Bayesian hierarchical modeling appears applicable for the market microstructure study in this work. The variables (called unknowns in Bayesian terms) can be classified in three layers: parameter $\theta = \{\alpha, \eta, \sigma^2, \delta^2\}$ (top layer); observed data $P^{a,b} = \{(P_t^a, P_t^b),\ t = 1, \ldots, T\}$ (bottom layer); unobserved latent variables $V = \{V_t,\ t = 0, 1\ldots, T\}$ (middle layer). The presence of latent variable V hinders the traditional maximum likelihood estimation (MLE) for θ, which could be tackled by E-M algorithms (cf. Dempster et al. 1977; Meng and van Dyk 1997). We adopt the Bayesian approach due to its flexibility of using conditional probabilities. The data $P^{a,b}$ is treated as given and the focus will be on the (joint) posterior distribution $\pi(\theta, V \mid P^{a,b})$ which is intractable analytically. Thanks to a rich class of MCMC computational algorithms, we can generate samples from a Markov chain with the state space for $\{\theta, V\}$ and the limiting distribution $\pi(\theta, V \mid P^{a,b})$. The key is to design the transition probability mechanism judiciously such that (i) the resulting chain converges to the target distribution $\pi(\theta, V \mid P^{a,b})$ rapidly (weak convergence issue); (ii) statistical parameters (as various functions of θ) can be estimated by corresponding sample statistics based on the observed MCMC sample paths accurately (variance reduction issue). There is a huge literature for MCMC. See Robert and Casella (2004), Brooks et al. (2011) for an in-depth coverage of basic MCMC theory and many related issues in applications.

In what follows, several elements in the proposed MCMC algorithm are described. See Appendix 1 for more details in implementation.

1.2.1 Priors

Assume the four components of θ are independent under the prior π. For α, a conjugate beta prior is adopted with its mode close to 0.1 because the proportion of informed traders in the market is relatively small. A uniform prior over the interval $(0, 1/2)$ for η is adopted, i.e. no information other than the restriction $0 < 2\eta < 1$ is used. For the volatility parameter σ^2, either a conjugate inverse gamma prior or uniform prior is used. For δ^2, we use a uniform prior over a small interval. Having specified the prior, the posterior distribution can be derived accordingly.

1.2.2 Metropolis-Hastings Within Gibbs

MCMC is a repertoire of algorithms among which Metropolis-Hastings algorithm (M-H) and the Gibbs sampler (GS) are the most popular ones. GS reflects a natural divide-and-conquer strategy when the state space is multi-dimensional. A one step transition of the MCMC chain amounts to cycling through a sequence of partitioning blocks of the state space, where each block can contain just a single variable or be a vector of several components. When updating one block, the states of other blocks remain fixed. M-H is an acceptance-rejection sampling scheme applied to Markov chains. It is very useful when direct sampling from a probability density becomes intractable, and it also has an advantage that we only need to know the density up to a normalizing constant factor. Although GS is shown to be a special case of M-H mathematically, most people in the MCMC community still consider them separately because they really represent very different ideas. We apply *Metropolis-Hastings within Gibbs* (MHwGS) to our setting. Using superscripts for MCMC iterations, the transition from step n to step $n + 1$ will follow

GS step: Partition the state space into k disjoint blocks and express the state variable as $x = \{x_1, ..., x_k\}$. Without loss of generality, let the updating follow a natural order $x_1 \to x_2 \to \cdots \to x_k$ in one iteration. Having done the updating $x_j^{(n)} \to x_j^{(n+1)}$ for blocks $j < i$, we are to update $x_i^{(n)}$ to $x_i^{(n+1)}$ by sampling from the conditional density $f(\cdot| \, x_j^{(n+1)}, j < i; x_j^{(n)}, j > i)$.

M-H step: When sampling directly from $f(\cdot| \, x_j^{(n+1)}, j < i; x_j^{(n)}, j > i)$ is difficult, we generate y from a proposal density $g(\cdot| \, x_j^{(n+1)}, j < i; x_j^{(n)}, j > i)$ first, then use the M-H ratio as a probability of accepting y and assigning $x_i^{(n+1)} = y$; otherwise stick to $x_i^{(n)}$ without a change and move forward to updating $x_{i+1}^{(n)}$, etc.

In this work, we simply let each of α, η, σ^2, δ^2; $V_1, ..., V_T$ be a block by itself (with $T + 4$ blocks in total). Choosing the proposal density g is an art. See Appendix 1 for more details.

1.2.3 Diagnostics for Convergence

Although the mathematical aspect of MCMC convergence is addressed by Marlov chain theory, an indispensable part of MCMC implementation in practice is to determine when should we stop running a chain and how should we use the samples to estimate various numerical characteristics of the target distribution. Here we only present two commonly used MCMC convergence diagnostic criteria. As many MCMC contributors commented in the literature, no single criterion can guarantee convergence and each one has its own pros and cons. A general suggestion is to use several criteria for each problem at hand.

Gelman-Rubin Method

The proposal by Gelman and Rubin (1992) consists of the following steps:

- Run $m > 1$ parallel chains of length $2n$ with over-dispersed starting values.
- Disregard the first n samples in each chain.
- Select a dynamic variable of interest, say x, and calculate its within-chain and between-chain sample variances.
- Compute the estimated variance as a weighted sum of within-chain and between-chain variances.
- Calculate the shrink factor.

The within-chain variance is given by $W = \frac{\sum_{j=1}^m s_j^2}{m}$, where $s_j^2 = \frac{\sum_{i=1}^n (x_{ij} - \bar{x}_j)^2}{n-1}$ is the sample variance for jth chain and $\bar{x}_j = \frac{\sum_{i=1}^n x_{ij}}{n}$. The between-chain variance is given by $B = \frac{n}{m-1} \sum_{j=1}^m (\bar{x}_j - \bar{\bar{x}})^2$, where $\bar{\bar{x}} = \frac{\sum_{j=1}^m \bar{x}_j}{m}$. B can be viewed as the variance of chain means multiplied by n. Then the estimated variance is $\widehat{Var}(x) = (1 - \frac{1}{n})W + \frac{1}{n}B$ and the shrink factor is $\hat{R} = \sqrt{\frac{\widehat{Var}(x)}{W}}$ whose value, if substantially above 1, would indicate lack of convergence. This criterion is easy to use but appears more necessary than sufficient in the sense that it may indicate convergence prematurely if the shrink factor happens to be close to 1 by chance. A remedy is to calculate the shrink factor at several points in time (gelman.plot in R package CODA) to see whether the shrink factor is really settled or still fluctuating.

Geweke method

The procedure proposed by Geweke (1992) is based on a test for equality between means of the first and last parts of a Markov chain (by default the first 10 % and the last 50 %). If the samples are drawn from the stationary distribution of the chain, the two means are equal and Geweke statistic follows the distribution $N(0, 1)$ asymptotically.

The test statistic is a standard Z-score: the difference between the two sample means divided by its estimated standard error. The standard error is estimated from the spectral density at zero so as to take into account any autocorrelation. Hence values of Z-score that fall in the extreme tails of $N(0, 1)$ suggest that the chain has not fully converged.

If Geweke's diagnostic indicates that the first and last parts sampled from a Markov chain are not drawn from the same distribution, it may be useful to discard the first few iterations to see if the rest of the chain has "converged". The geweke.plot in R package CODA shows what happens to Geweke's Z-score when successively larger numbers of iterations are discarded from the beginning of the chain. To preserve the asymptotic conditions required for Geweke's diagnostic, the plot never discards more than half the chain.

The first half of the Markov chain is divided into several segments, then Geweke's Z-score is repeatedly calculated. The first Z-score is calculated with all iterations in the chain, the second after discarding the first segment, the third after discarding the first two segments, etc. The last Z-score is calculated using only the samples in the second half of the chain.

1.3 Simulation Study

In order to test whether our MCMC algorithms work well, we do simulation study first. In the simulation study, we specify $\alpha = 0.1$, $\eta = 0.25$, $\sigma^2 = 0.25$, $\delta^2 = 0.09$. Based on the market model in Sect. 1.1, we calculated the bid and ask prices, and then use these synthetic data to estimate the four parameters in the model by our MCMC algorithms. We run two MCMC chains, each containing 50,000 samples, and discard 10,000 burn in samples. After the burn in stage, we retain one in every 20 samples as a new path. Table 1 examines the effectiveness of the estimation strategy, showing the true value, posterior summary statistics of those parameters. We could use the posterior mean or posterior median as an estimation of the parameter.

Table 1 Summary statistics of the posterior samples for all four parameters in simulation study

Parameter	True value	Min	Median	Mean	Max	Standard error
α	0.10	0.003	0.15	0.15	0.51	0.001
η	0.25	0.00	0.23	0.25	0.50	0.002
σ^2	0.25	0.05	0.21	0.25	1.67	0.002
δ^2	0.09	0.07	0.10	0.10	0.43	0.001

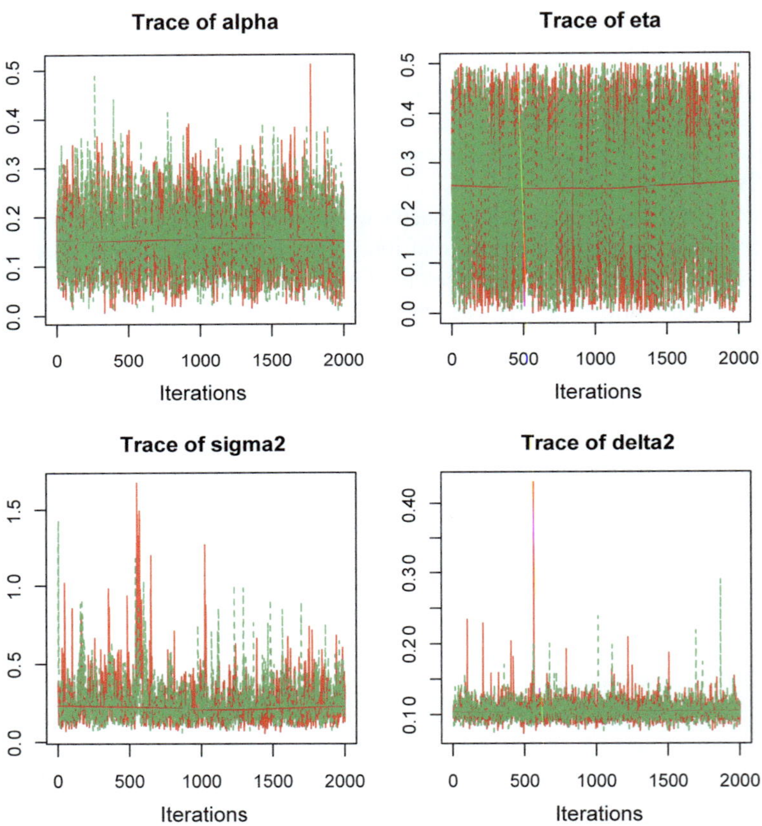

Fig. 1 Trace plot of the posterior samples in simulation study, *green* is chain 1 and *red* is chain 2

Figures 1, 2, 3 and 4 show the related convergence results of the MCMC algorithm.

Trace plots give us a direct insight of what values the posterior samples take at each iteration.

The autocorrelation plots show us how the autocorrelation changes with the increase of lag. From the autocorrelation plot, we can see except for σ^2, the autocor-

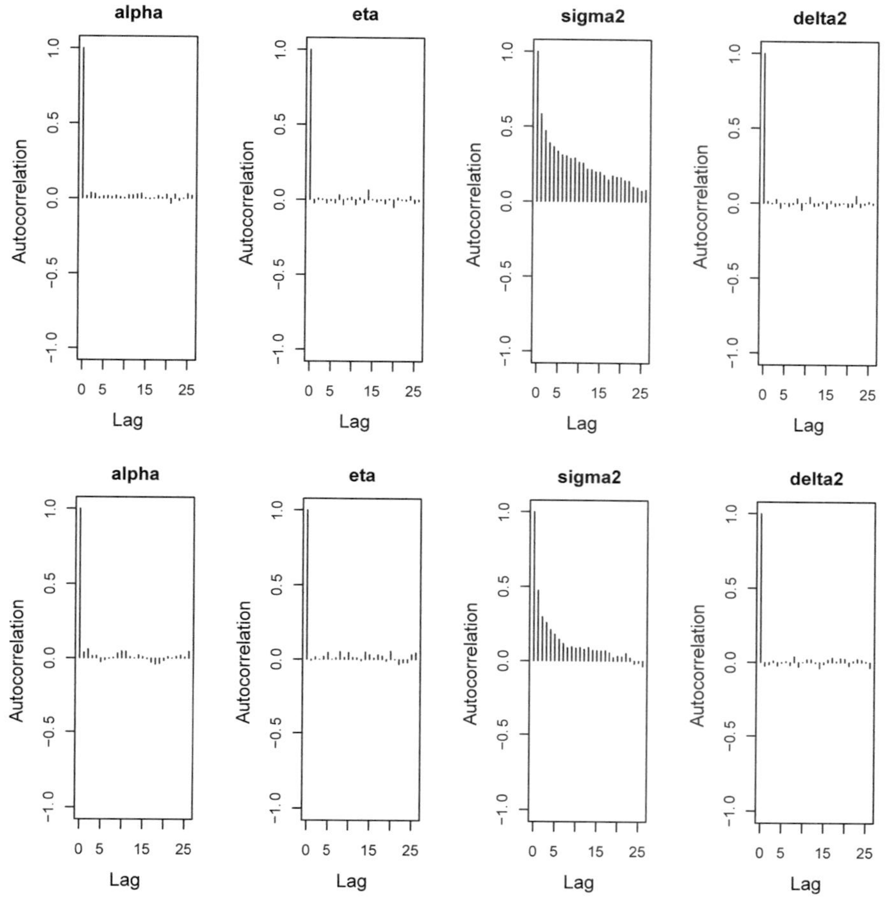

Fig. 2 Autocorrelation of the posterior samples in simulation study

relations for the other 3 parameters are near 0 at any lag. For σ^2, the autocorrelation decreases to 0 as the lag increases.

From Gelman-Rubin plot, we can see that the shrink factors for all four parameters converge to 1 after some iterations. Also, from Geweke plot, most of the Z-scores for all parameters are between -1.96 and 1.96. Both the Gelman-Rubin and Geweke plots show good MCMC convergence results.

The similarities between the posterior estimates and true values of the parameters and other convergence results indicate that our MCMC algorithm works well. Next step is to conduct empirical studies using real high frequency data.

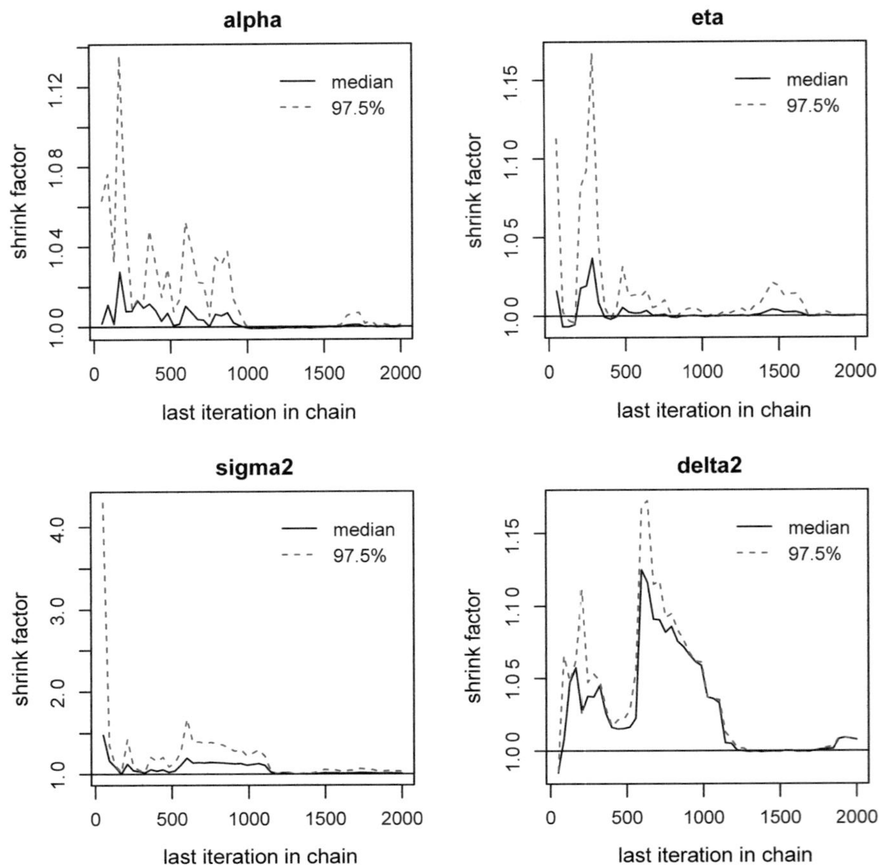

Fig. 3 Gelman-Rubin plots in simulation study. Convergence is suggested when the medians and the 97.5 percentiles approach 1

1.4 Empirical Study

1.4.1 Data

The data we used are the bid and ask prices of Microsoft stock in April, 2013 from TAQ database. TAQ contains intraday transactions data for all securities listed on the New York Stock Exchange (NYSE) and American Stock Exchange (AMEX), as well as Nasdaq National Market System (NMS) and SmallCap issues.

The data set has around 28,000,000 observations in total: over 900,000 observations on each trading day, and about 13 tradings at each time spot. There are a couple of major problems if we use raw bid and ask prices. One is computational budget constraint. The heavy computational burden would limit the sample to a relatively

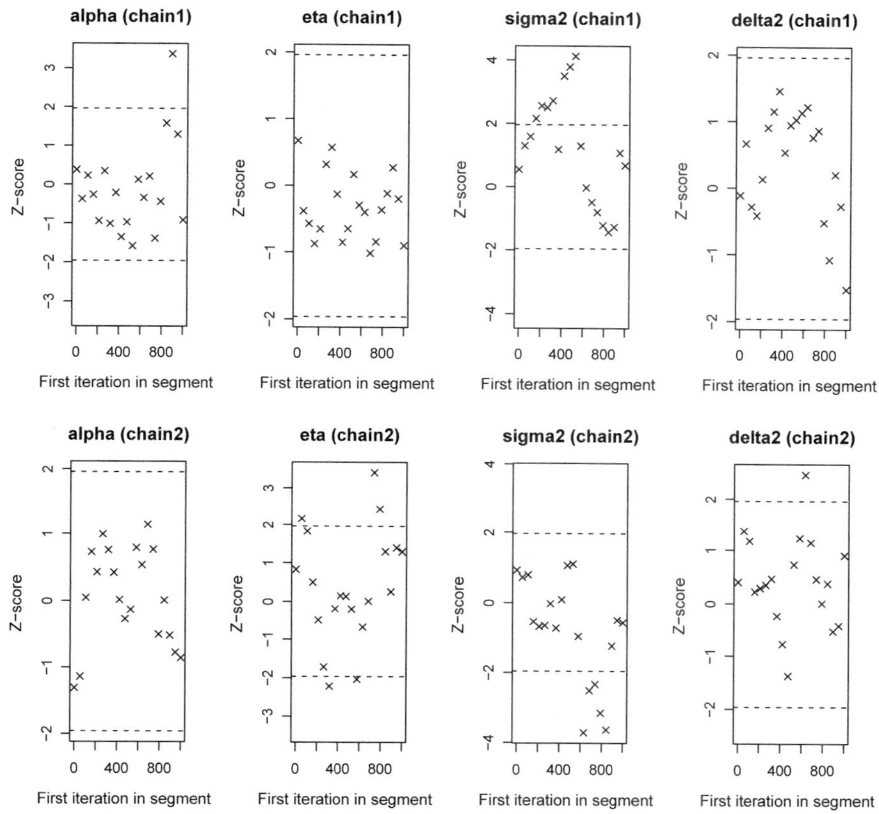

Fig. 4 Geweke plot in simulation study. Convergence is suggested when most of the Z-scores are between −1.96 and 1.96

short time horizon. Another issue is: too much noise in the original high frequency data would cause microstructure bias in inference. Therefore, our empirical study begins with some data processing:

- Missing data are deleted.
- The mean of all observations at the same trading time spot is used.
- since tradings are heavier at the beginning and the end of a trading day, while lighter around lunch time, we partition each day into 5 periods: 9:30 to 10:00, 10:00 to 11:30, 11:30 to 2:30, 2:30 to 3:30, 3:30 to 4:00 and use averages of bid/ask prices during each period.

Figure 5 shows the bid and ask prices for the cleaned data. From Fig. 1, it is hard to see the difference between the bid and ask prices since they only differ by 1 or 2 cents. Figure 6 shows a zoom-in version of Fig. 5, that plots only the first 10 bid and ask prices to help the visual effect.

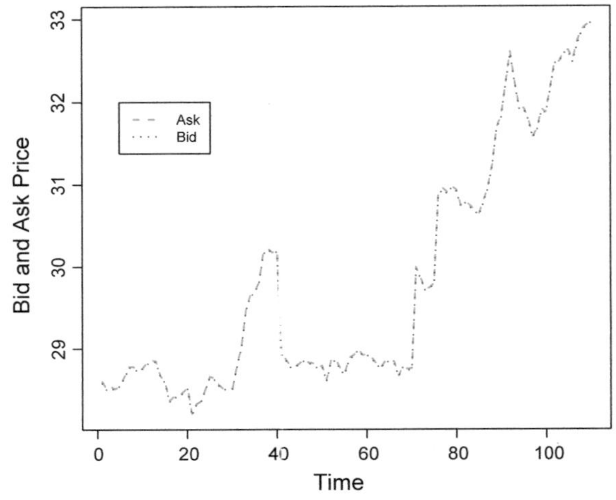

Fig. 5 Bid and ask prices for 22 consecutive trading days, x-axis represents the trading time after aggregation (5 trades per day after the aggregation of data)

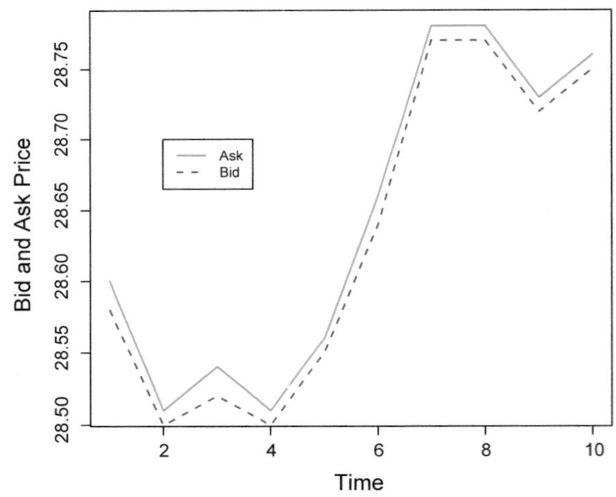

Fig. 6 Bid and ask prices for the first 2 consecutive trading days, x-axis represents the trading time after aggregation (5 trades per day after the aggregation of data)

1.4.2 Summary Statistics and MCMC Convergence

Table 2 shows the summary statistics of the posterior samples for the four parameters. Again, we could use posterior mean or posterior median as a point estimate for each parameter.

Figures 7, 8, 9 and 10 show the convergence results of the MCMC algorithm, similar to the analysis in the Simulation Study.

Table 2 Summary statistics of the posterior samples in empirical study

Parameter	Min	Median	Mean	Max	Standard error
α	0.0015	0.0856	0.0997	0.4415	0.0008
η	0.00004	0.2449	0.2477	0.5007000	0.0018
σ^2	0.0588	0.2141	0.2510	2.275	0.0019
δ^2	0.00006	0.2502	0.2509	0.4999	0.0018

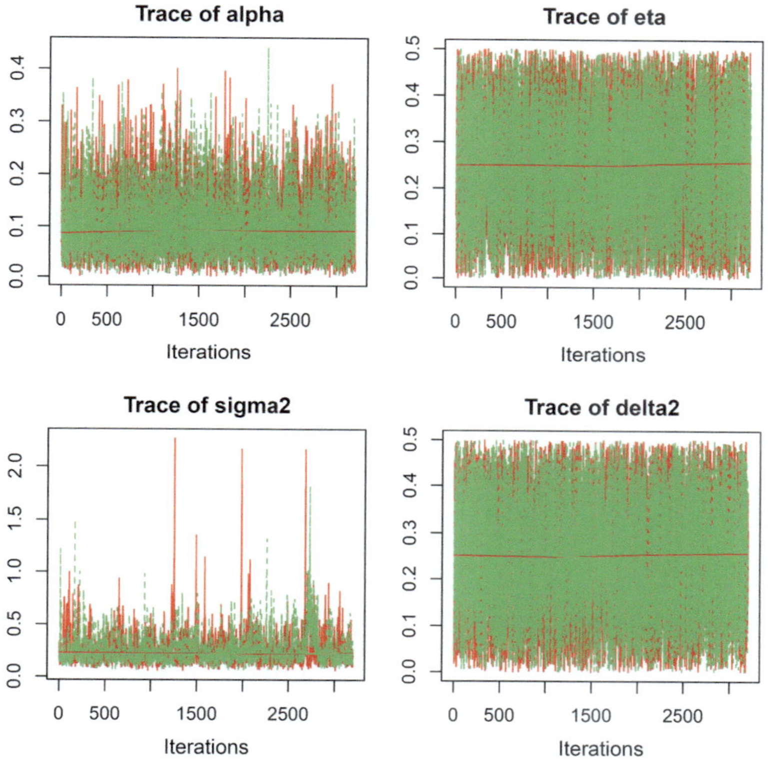

Fig. 7 Trace plot of the posterior samples in empirical study

1.5 *Economic Interpretation*

There are at least three sources for bid-ask spread: the adverse selection costs arising with asymmetric information, the inventory costs, and the order processing cost, which is associated with handling transactions. G-M model focuses on the first one. The extended G-M model we studied in this work also helps addressing the related issues. Besides the numerical evidence shown in Fig. 11, we also proved theoretically that the average bid-ask spread is an increasing function of α (see Wang 2014), i.e.

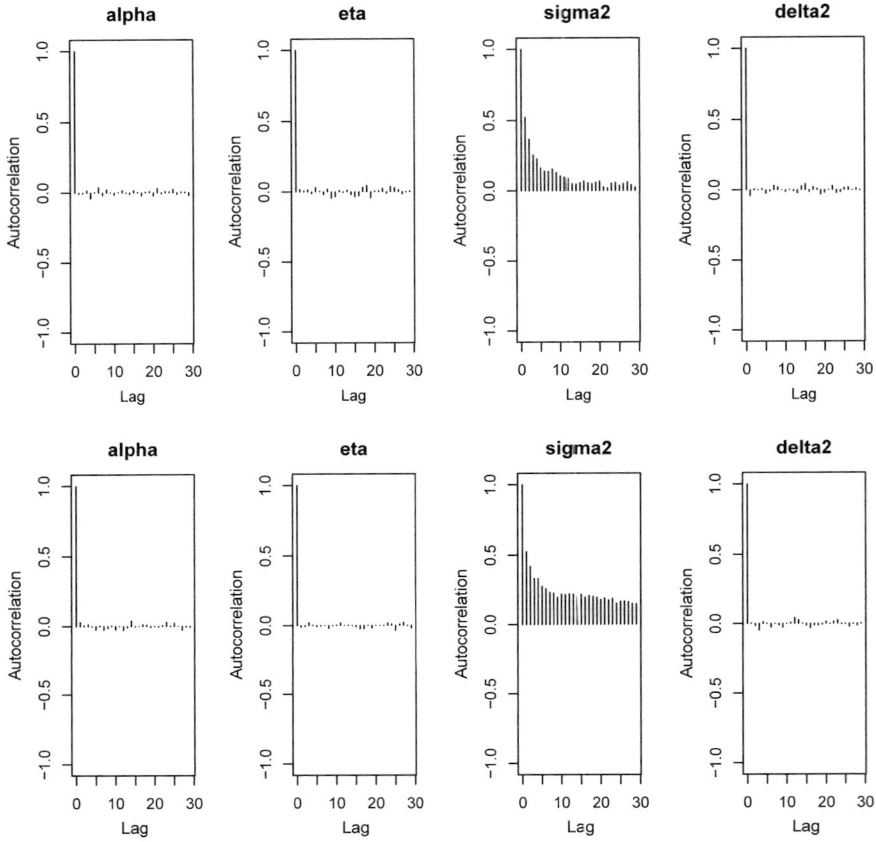

Fig. 8 Autocorrelation of the posterior samples in empirical study

the spread can be considered a premium that market makers demand for trading with agents with superior information. Another result developed in Wang (2014) is that under the extended G-M model, the bid-ask spread tends to zero at a certain rate as the number of trades go to infinity. However, this has not been shown in our empirical study, due to the limited time horizon used in the data. In addition, the bid-ask spread reflects the market maker's belief about asymmetric information. The degree of informed trading among market participants may not change in the short time period, at least from the market maker's viewpoint. This implies that the market maker makes no inference when he sees the total order imbalance at tick level. He may shift the whole bid-ask band rather than change the spread itself. Figure 12 shows the bid-ask spread for the cleaned data based on which we do not see much change of the spread over a short time period. More careful and thorough post-modeling analysis is still our ongoing project in which both in-sample (e.g. residual analysis) and out-of-sample (cross-validation) diagnostics are conducted.

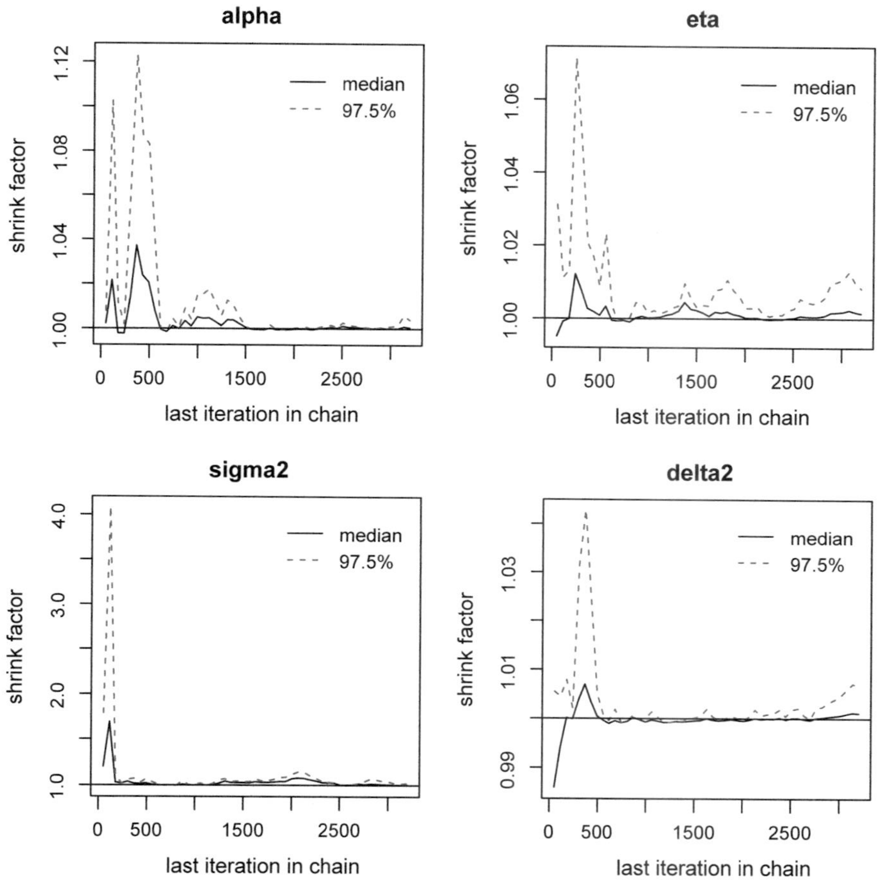

Fig. 9 Gelman-Rubin plots in empirical study. Convergence is suggested when the medians and the 97.5 percentiles approach 1

The extended G-M has conceivably made a number of unrealistic assumptions, such as the constant α for the proportion (or the impact) of informed traders in the market, and the constant trading volume associated with each trade. Modifications of those assumptions require more hard work in both theoretical and empirical studies.

1.6 Appendix 1

We provide more details of the MCMC algorithm here.

The hyperparameters in the prior distributions are set as follows:
in the beta prior for α: $\alpha_\alpha = 2,\ \beta_\alpha = 10$;

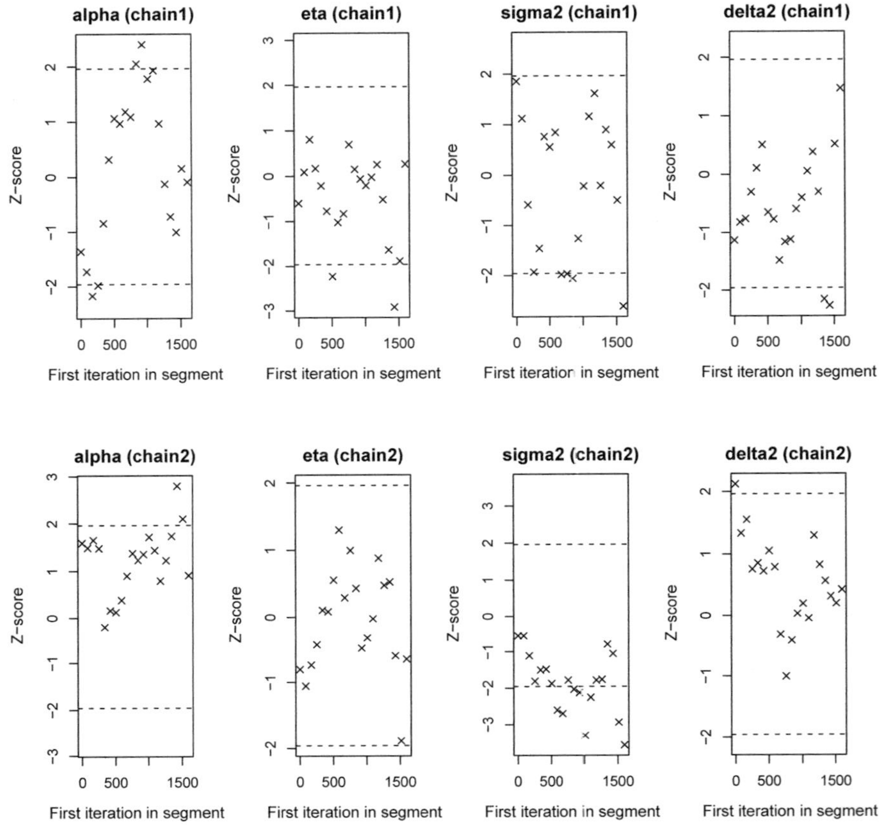

Fig. 10 Geweke plot in empirical study. Convergence is suggested when most of the Z-scores are between −1.96 and 1.96

in the inverse gamma prior for σ^2: $\alpha_\sigma = 3,\ \beta_\sigma = 1$;
in the uniform prior for σ^2: $u_{1\sigma} = 0,\ u_{2\sigma} = 0.5$;
in the uniform prior for δ^2: $u_{1\delta} = 0,\ u_{2\delta} = 0.5$.

Note that subscripts $t = 1, ..., T$ stand for real time in the model and superscripts $n = 1, ..., N$ represent the MCMC computational time.

Initialize $\alpha^{(0)}, \eta^{(0)}, \sigma^{2(0)}, \delta^{2(0)}, V_1^{(0)}, \ldots, V_T^{(0)}$, which can be assigned or sampled from the prior. A complete transition from the $(n-1)$th generation to the nth generation consists of the following steps:

Step 1: Update the latent variable $V_t^{(n)}$

For $t = 2, 3, \ldots, T-1$,

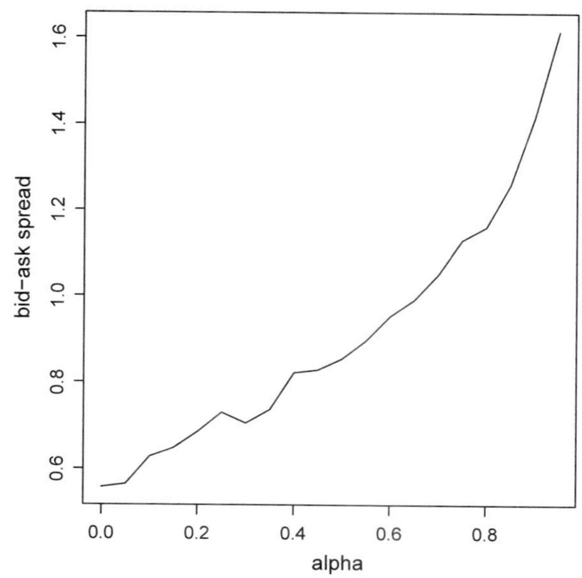

Fig. 11 The mean bid-ask spread versus α in the simulation study, the y-axis is the mean of bid-ask spreads across time using different α values

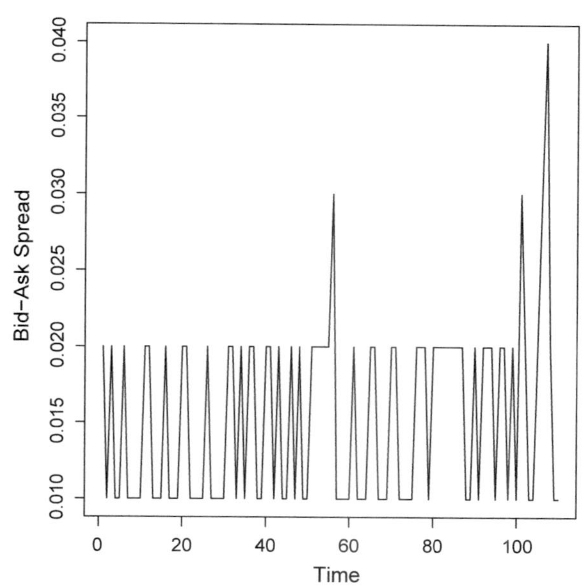

Fig. 12 The bid-ask spread versus time for the cleaned real data, the bid-ask spread does not change very much

$$
\begin{aligned}
& f(V_t^{(n)} | V^{(n-1)}, V_1^{(n)}, \ldots, V_{t-1}^{(n)}, \theta^{(n-1)}, P^a, P^b) \\
\propto \; & f(V_t^{(n)} | V_{t-1}^{(n)}, \theta^{(n-1)}) \, f(V_{t+1}^{(n-1)} | V_t^{(n)}, \theta^{(n-1)}) \\
\propto \; & \exp\left[-\frac{(V_t^{(n)} - V_{t-1}^{(n)})^2}{2\sigma^{2(n-1)}} - \frac{(V_{t+1}^{(n-1)} - V_t^{(n)})^2}{2\sigma^{2(n-1)}} \right] \\
\propto \; & \exp\left\{ -\frac{[V_t^{(n)} - \frac{V_{t-1}^{(n)} + V_{t+1}^{(n-1)}}{2}]^2}{2(\sigma^{(n-1)}/\sqrt{2})^2} \right\} \\
\sim \; & N\left(\frac{V_{t-1}^{(n)} + V_{t+1}^{(n-1)}}{2}, \frac{\sigma^{2(n-1)}}{2} \right)
\end{aligned}
$$

where $V^{(n)} = (V_1^{(n)}, \ldots, V_T^{(n)})$, and $\theta^{(n)} = (\alpha^{(n)}, \eta^{(n)}, \sigma^{2(n)}, \delta^{2(n)})$. Given all the other variables, we just sample from a normal distribution with mean $\frac{V_{t-1}^{(n)} + V_{t+1}^{(n-1)}}{2}$ and variance $\frac{\sigma^{2(n-1)}}{2}$ to get $V_t^{(n)}$.

For $t = 1$, generate $V_1^{(n)} \sim N(V_2^{(n-1)}, \sigma^{2(n-1)})$.

For $t = T$, generate $V_T^{(n)} \sim N(V_{T-1}^{(n)}, \sigma^{2(n-1)})$.

Step 2: Update $\sigma^{2(n)}$

If the prior is IG$(\alpha_\sigma, \beta_\sigma)$ (inverse gamma), then

$$
\begin{aligned}
& f(\sigma^{2(n)} | V^{(n)}, \alpha^{(n-1)}, \eta^{(n-1)}, \delta^{2(n-1)}, P^a, P^b) \\
& \propto \prod_{t=2}^{T} f(V_t^{(n)} | V_{t-1}^{(n)}, \alpha^{(n-1)}, \eta^{(n-1)}, \delta^{2(n-1)}, P^a, P^b) \, f(\sigma^{2(n)}) \\
& \propto \prod_{t=2}^{T} (\sigma^{2(n)})^{-\frac{1}{2}} \exp\left[-\frac{(V_t^{(n)} - V_{t-1}^{(n)})^2}{2\sigma^{2(n)}} \right] (\sigma^{2(n)})^{-\alpha_\sigma - 1} \, e^{-\frac{\beta_\sigma}{\sigma^{2(n)}}} \\
& \propto (\sigma^{2(n)})^{-(\frac{T-1}{2} + \alpha_\sigma) - 1} \exp\left[-\frac{\frac{\sum_{t=2}^{T} (V_t^{(n)} - V_{t-1}^{(n)})^2}{2} + \beta_\sigma}{\sigma^{2(n)}} \right] \\
& \sim \mathrm{IG}\left(\frac{T-1}{2} + \alpha_\sigma, \frac{\sum_{t=2}^{T} (V_t^{(n)} - V_{t-1}^{(n)})^2}{2} + \beta_\sigma \right).
\end{aligned}
$$

If the prior is Unif$(u_{1\sigma}, u_{2\sigma})$, then

$$
\begin{aligned}
& f(\sigma^{2(n)} | V^{(n)}, \alpha^{(n-1)}, \eta^{(n-1)}, \delta^{2(n-1)}, P^a, P^b) \\
& \propto \prod_{t=2}^{T} f(V_t^{(n)} | V_{t-1}^{(n)}, \alpha^{(n-1)}, \eta^{(n-1)}, \delta^{2(n-1)}, P^a, P^b) f(\sigma^{2(n)}) \\
& \propto (\sigma^{2(n)})^{-\frac{T-1}{2}} \exp\left[-\frac{\sum_{t=2}^{T} (V_t^{(n)} - V_{t-1}^{(n)})^2}{2\sigma^{2(n)}} \right] I_{\{\sigma^{2(n)} \in (u_{1\sigma}, u_{2\sigma})\}}.
\end{aligned}
$$

The posterior distribution is not in a closed form, hence we need MHwGS. The procedure is as follows:

- Simulate a sample y_σ from a proposal density q_σ, which is chosen to be the prior Unif$(u_{1\sigma}, u_{2\sigma})$ here.
- Denote the posterior distribution of σ^2 by f_σ, compute the acceptance probability (M-H ratio) $\rho_\sigma = min(1, \frac{f_\sigma(y_\sigma)q_\sigma(\sigma^{(n-1)})}{f_\sigma(\sigma^{(n-1)})q_\sigma(y_\sigma)})$.
- Let $\sigma^{2(n)} = \{\begin{matrix} y_\sigma & \text{with probability } \rho_\sigma \\ \sigma^{2(n-1)} & \text{with probability } 1-\rho_\sigma \end{matrix}$.

Step 3: Update $\alpha^{(n)}$

$$f(\alpha^{(n)}|V^{(n)}, \sigma^{2(n)}, \eta^{(n-1)}, \delta^{2(n-1)}, P^a, P^b)$$
$$\propto \prod_{t=1}^{T} \exp\left[-\frac{(P_t^a - A_t)^2}{2\delta^{2(n-1)}} - \frac{(P_t^b - B_t)^2}{2\delta^{2(n-1)}}\right] (\alpha^{(n)})^{\alpha_\alpha - 1}(1-\alpha^{(n)})^{\beta_\alpha - 1}$$

where B_t, A_t is given by (3) and (4) with η, α replaced by $\eta^{(n-1)}$ and $\alpha^{(n)}$ respectively. Again, we need MHwGS:

- Simulate a sample y_α from a proposal density q_α, which is chosen to be the prior Beta$(\alpha_\alpha, \beta_\alpha)$.
- Denote the posterior of α by f_α, compute the acceptance probability $\rho_\alpha = min(1, \frac{f_\alpha(y_\alpha)q_\alpha(\alpha^{(n-1)})}{f_\alpha(\alpha^{(n-1)})q_\alpha(y_\alpha)})$.
- Let $\alpha^{(n)} = \{\begin{matrix} y_\alpha & \text{with probability } \rho_\alpha \\ \alpha^{(n-1)} & \text{with probability } 1-\rho_\alpha \end{matrix}$.

Step 4: Update $\eta^{(n)}$

$$f(\eta^{(n)}|V^{(n)}, \sigma^{2(n)}, \alpha^{(n)}, \delta^{2(n-1)}, P^a, P^b)$$
$$\propto \prod_{t=1}^{T} \exp\left[-\frac{(P_t^a - A_t)^2}{2\delta^{2(n-1)}} - \frac{(P_t^b - B_t)^2}{2\delta^{2(n-1)}}\right] I_{\{\eta^{(n)}\in(0,1/2)\}}$$

where B_t, A_t is given by (3) and (4) with η, α replaced by $\eta^{(n)}$ and $\alpha^{(n)}$ respectively. Similarly, MHwGS is applied here:

- Simulate a sample y_η from a proposal density q_η, which is chosen to be the prior distribution Unif(0, 1/2).
- Denote the posterior distribution of η by f_η, compute the acceptance probability $\rho_\eta = min(1, \frac{f_\eta(y_\eta)q_\eta(\eta^{(n-1)})}{f_\eta(\eta^{(n-1)})q_\eta(y_\eta)})$.
- $\eta^{(n)} = \{\begin{matrix} y_\eta & \text{with probability } \rho_\eta \\ \eta^{(n-1)} & \text{with probability } 1-\rho_\eta \end{matrix}$.

Step 5: Update $\delta^{2(n)}$

$$f(\delta^{2(n)}|V^{(n)},\sigma^{2(n)},\alpha^{(n)},\eta^{(n)},P^a,P^b)$$
$$\propto (\delta^{2(n)})^{-T} \exp\left[-\sum_{t=1}^{T}\frac{(P_t^a-A_t)^2+(P_t^b-B_t)^2}{2\delta^{2(n)}}\right] I_{\{\delta^{2(n)}\in(u_{1\delta},u_{2\delta})\}}.$$

Here the needed MHwGS is given by

- Simulate a sample y_δ from a proposal density q_δ, which is chosen to be the prior Unif$(u_{1\delta},u_{2\delta})$.
- Denote the posterior density of δ as f_δ, compute the acceptance probability $\rho_\delta = min(1, \frac{f_\delta(y_\delta)q_\delta(\delta^{2(n-1)})}{f_\delta(\delta^{2(n-1)})q_\delta(y_\delta)})$.
- Let $\delta^{2(n)} = \begin{cases} y_\delta & \text{with probability } \rho_\delta \\ \delta^{2(n-1)} & \text{with probability } 1-\rho_\delta \end{cases}$.

Step 6: Now go to Step 1 for the updating in the next iteration of transition.

2 Monte-Carlo Strategies in Option Pricing for SABR Model

In finance, an option is a contract which gives the buyer the right, but not the obligation, to buy or sell an underlying asset or instrument at a specified strike price on or before a specified date, depending on the form of the option. Because valuation of option contracts depends on a number of other variables besides the underlying asset, it is a complex task and becomes a central topic in mathematical finance. For valuation of options, cases with closed-form pricing formulas are rare with exceptions of Black-Scholes-Merton model, Hestons model, just name a few. In general, numerical computation and approximation techniques are almost always required. There are basically two approaches. The analytical approach sets the price function as the solution to a PDE with boundary conditions, which are often solved by numerical methods such as finite difference etc. The probabilistic approach expresses an option price as the conditional expectation under a risk neutral measure which needs to be computed using numerical integration. Such integration is often performed over a high dimensional state space in which state variables are time series of underlying assets or volatilities. In this situation, Monte-Carlo simulation appears inevitable.

SABR (abbreviation for **stochastic** $\alpha\beta\rho$) model enjoys the popularity in the study of stochastic volatility with applications in asset pricing and risk management. Major references include Antonov and Spector (2012), Hagan et al. (2002) and (2005), Paulot et al. (2009), Rebonado et al. (2011), among others. The main feature of the SABR model, compared to some previous models, is its capability in reproducing the dynamic behavior of volatility smiles and skews, and thus in yielding more stable results for pricing and hedging. SABR assumes that the volatility of the asset (e.g. stock or forward) follows a geometric Brownian motion, and is correlated to the underlying forward price (leverage effect). So far almost all cited works for SABR have adopted the analytical approach using singular perturbation of the pricing function. In contrast, we take the probabilistic approach for pricing options under SABR.

The basic idea of our approach is Monte-Carlo dimension reduction using certain probability approximation schemes. Note that brute force Monte-Carlo for option pricing begins with Euler approximation that discretizes sample paths of the asset and volatility, followed by calculating the option payoff along each path then taking averages. In doing so, small steps in the discretization are needed to reduce the bias, but that would require a far greater number of simulated sample paths to reduce the variance. To alleviate the computational intensity, we observe that many pricing functionals of interest only depend on the volatility sample paths through two or three summary statistics. Therefore, it would suffice to know the joint distribution of those 2D or 3D summary statistics when calculating the option price. Our strategy will naturally be finding good approximations for the joint distribution when the exact form is not available.

After introducing SABR model and the option pricing formula, the probability approximation scheme boils down to expressing moments of some key summary statistics as functions of original model parameters. We have done this by exact analytical calculation, and obtained good results for all different ranges of parameter β, because three cases $\beta = 1$ (generalized Black-Scholes model), $\beta = 0$ (Gaussian model) and $0 < \beta < 1$ give rise to different pricing formulas. Since the basic idea is the same, we will only present the case $\beta = 1$ in this chapter and refer to the Ph.D. thesis Yin (2016) for the other two cases and the technical details.

2.1 SABR Model and Option Pricing for the Case $\beta = 1$

The risk-neutral dynamics of general SABR model is given by SDEs

$$dF(t) = r\ F(t)^{\beta}dt + \sigma(t)F(t)^{\beta}dW_1(t), \tag{6}$$

$$d\sigma(t) = \alpha\ \sigma(t)dW_2(t), \tag{7}$$

with the underlying asset value $F(t)$ (e.g. LIBOR forward rate, or forward swap rate, or stock price) and the volatility $\sigma(t)$, where r is the risk-free interest rate, $W_1(t)$ and $W_2(t)$ are two correlated standard Brownian motions with correlation coefficient ρ, i.e.

$$dW_1(t)\ dW_2(t) = \rho\ dt. \tag{8}$$

As we mentioned, the three model parameters $\alpha > 0$, $0 \le \beta \le 1$ and $-1 \le \rho \le 1$ give the reason for using the abbreviation SABR — stochastic $\alpha\beta\rho$ model, and by changing their values, a variety of interesting market behaviors can be mimicked.

The value of a European call option is defined by the expected value of discounted option payoff at maturity t_{ex}, i.e.

$$C(F_0, K) = e^{-rt_{ex}} E\left\{E_\sigma\left\{\max(F(t_{ex}) - K, 0)\right\}\right\}, \tag{9}$$

where $F_0 = F(0)$ is the present asset value, K is the strike price and $E_\sigma(\cdot)$ denotes the conditional expectation given the path $\sigma(t),\ 0 \le t \le t_{ex}$.

Consider the case $\beta = 1$ [called the *generalized BLack-Scholes* (*B-S*) *model* in th literature] and rewrite (6), (7), (8) as an equivalent form

$$dF(t) = rF(t)dt + \sigma(t)F(t)[\sqrt{1-\rho^2}dB_1(t) + \rho dB_2(t)] \tag{10}$$

$$d\sigma(t) = \alpha\sigma(t)dB_2(t) \tag{11}$$

where $B_1(t)$ and $B_2(t)$ are two independent standard Brownian motions and all other notations remain the same as defined before. We will use this form of SABR model in what follows.

Define

$$\Sigma^2 = \int_0^{t_{ex}} \sigma(u)^2 du, \tag{12}$$

$$X_1 = \int_0^{t_{ex}} \sigma(u)dB_1(u), \tag{13}$$

$$X_2 = \int_0^{t_{ex}} \sigma(u)dB_2(u). \tag{14}$$

Conditioning on the volatility path $\sigma(t),\ 0 \le t \le t_{ex}$, we have

$$E_\sigma\left\{\max\left(F(t_{ex}) - K, 0\right)\right\} = \\ F_0\ \exp\{rt_{ex} + \rho X_2 + \tfrac{1}{2}(1 - \rho - \sqrt{1-\rho^2} - \rho^2)\Sigma^2\}\Phi(d_1) \\ -K\ \Phi(d_2), \tag{15}$$

where Φ is the cdf of $N(0, 1)$, and d_1 and d_2 are given by

$$d_1 = d_2 + \sqrt{1-\rho^2}\ \Sigma, \tag{16}$$

$$d_2 = \frac{ln(\frac{F_0}{K}) - \frac{1}{2}(\rho + \sqrt{1-\rho^2})\ \Sigma^2 + \rho X_2 + rt_{ex}}{\sqrt{1-\rho^2}\Sigma}. \tag{17}$$

Therefore, it suffices to compute the option price $C(F_0, K)$ in (9) as the expected value under the joint distribution of (Σ^2, X_2) without simulating the entire volatility path on $[0, t_{ex}]$.

2.2 *Approximating the Distribution of* (Σ^2, X_2)

There are a couple of factors taken into account when choosing an approximate distribution for (Σ^2, X_2). Firstly, Σ^2 is positive and often has a skewed density. Secondly, X_2 is a martingale with respect to the time t_{ex}, and conditioning on $(\sigma(t), 0 \leq t \leq t_{ex})$, $X_2 \sim N(0, \Sigma^2)$. Hence we propose a gamma mixture of normals for the distribution of (Σ^2, X_2). Having decided the distribution family, we relate the two parameters of gamma density for Σ^2 to the first and second moments of (Σ^2, X_2), which in turn can be calculated analytically as closed forms of the original model parameters. That leads to specification of the parameters in gamma family as functions of parameters in SABR model. Similar results in using an inverse gamma mixture of normals and log-normal mixture of normals are given in Yin (2016).

Moments as functions of SABR model parameters

Denote $\sigma(0) = \sigma_0$, we have

$$
\begin{aligned}
E\{\Sigma^2\} &= \frac{\sigma_0^2}{\alpha^2}(e^{\alpha^2 t_{ex}} - 1),\\
E\{(\Sigma^2)^2\} &= \frac{2\sigma_0^4}{5\alpha^4}(\frac{1}{6}e^{6\alpha^2 t_{ex}} - e^{\alpha^2 t_{ex}} + \frac{5}{6}),\\
E\{(\Sigma^2)^3\} &= \frac{\sigma_0^6}{315\alpha^6}(e^{15\alpha^2 t_{ex}} - 7e^{6\alpha^2 t_{ex}} + 27e^{\alpha^2 t_{ex}} - 21).
\end{aligned}
$$

for every $n = 1, 2, ...,$

$$
E\{X_2^n\} = \frac{\sigma_0^n}{\alpha^n}\sum_{k=0}^{n}(-1)^{n-k}\binom{n}{k}\, e^{\frac{1}{2}k(k-1)\alpha^2 T}.
$$

in particular,

$$
\begin{aligned}
E\{X_2\} &= 0,\\
E\{X_2^2\} &= \frac{\sigma_0^2}{\alpha^2}(e^{\alpha^2 t_{ex}} - 1).
\end{aligned}
$$

moreover,

$$
Cov(\Sigma^2, X_2) = \frac{\sigma_0^3}{3\alpha^3}(e^{3\alpha^2 t_{ex}} - 3e^{\alpha^2 t_{ex}} + 2).
$$

Connections between parameters in the gamma mixture of normals and moments of (Σ^2, X_2)

For convenience, denote the needed moments by

$$\begin{aligned} E(X_2^2) &= E(\Sigma^2) \triangleq S, \\ E[(\Sigma^2)^2] &\triangleq \Delta, \\ Cov(\Sigma^2, X_2) &\triangleq \Gamma, \end{aligned}$$

where the first equality is due to Ito's isometry, and $S\Delta \geq \Gamma^2$ follows from Cauchy-Schwarz inequality.

Now we have

Proposition 14.1 *Let* $\Sigma^2 \sim Gamma(\kappa, \theta)$ *and* $X_2 \sim N(a_0 + a_1\Sigma^2, b\Sigma^2)$ *conditioning on the path* $\{\sigma(t), 0 \leq t \leq t_{ex}\}$ *where the constants* $a_0 \in I\!R$, $a_1 \in I\!R$, $b > 0$ *and the parameters* κ, θ *are given by*

$$\begin{aligned} \kappa &= \frac{S^2}{\Delta - S^2}, \\ \theta &= \frac{\Delta - S^2}{S}, \\ a_0 &= \frac{-S\Gamma}{\Delta - S^2}, \\ a_1 &= \frac{\Gamma}{\Delta - S^2}, \\ b &= 1 - \frac{\Gamma^2}{S(\Delta - S^2)}. \end{aligned}$$

Therefore, the gamma mixture of normals is fully specified by α, σ_0 and t_{ex} in the SABR.

2.3 Numerical Experiments and Empirical Calibration of SABR

All pricing methods require the SABR model parameters as inputs, which are set for empirical studies in what follows. Now let us introduce the data we use.

For different underlying asset types, there are different channels to obtain related option contracts information. Trading information for equity can be found on YAHOO! Finance channel. As for other underlying asset types such as energy, foreign exchange, interest rates and weather, option contracts in trading are often listed on CME Group website. In our study, we will use Microsoft stock, MSFT, as an example for equity, and iShare 20+ years treasury bond ETF, TLT, as an example of fixed income product. We will price the European options on these assets as of March 28th 2016 that will expire on May 20th 2016.

Figures 13 and 14 show option chains of MSFT and TLT on March 28th 2016 that expire on May 20th 2016. An option chain is simply a sequence of call and put

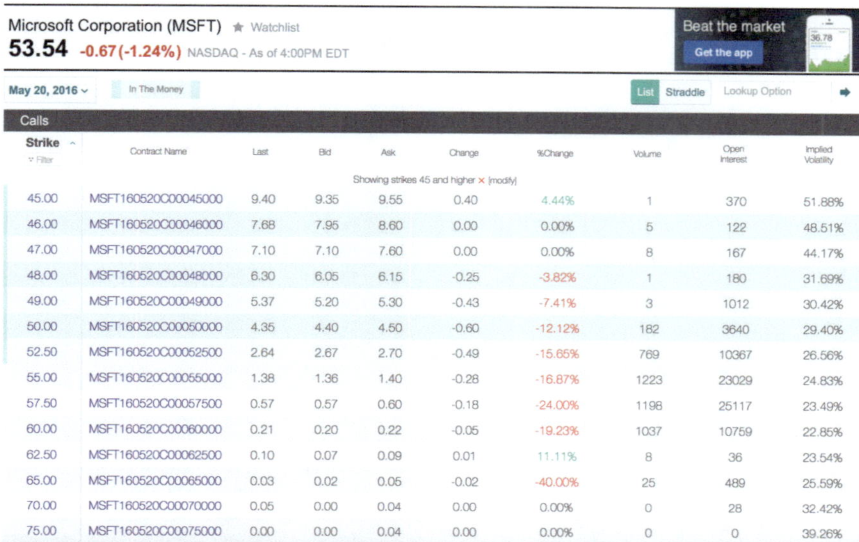

Microsoft Corporation (MSFT) ★ Watchlist
53.54 -0.67 (-1.24%) NASDAQ - As of 4:00PM EDT

May 20, 2016 · In The Money · List · Straddle · Lookup Option

Calls

Showing strikes 45 and higher

Strike	Contract Name	Last	Bid	Ask	Change	%Change	Volume	Open Interest	Implied Volatility
45.00	MSFT160520C00045000	9.40	9.35	9.55	0.40	4.44%	1	370	51.88%
46.00	MSFT160520C00046000	7.68	7.95	8.60	0.00	0.00%	5	122	48.51%
47.00	MSFT160520C00047000	7.10	7.10	7.60	0.00	0.00%	8	167	44.17%
48.00	MSFT160520C00048000	6.30	6.05	6.15	-0.25	-3.82%	1	180	31.69%
49.00	MSFT160520C00049000	5.37	5.20	5.30	-0.43	-7.41%	3	1012	30.42%
50.00	MSFT160520C00050000	4.35	4.40	4.50	-0.60	-12.12%	182	3640	29.40%
52.50	MSFT160520C00052500	2.64	2.67	2.70	-0.49	-15.65%	769	10367	26.56%
55.00	MSFT160520C00055000	1.38	1.36	1.40	-0.28	-16.87%	1223	23029	24.83%
57.50	MSFT160520C00057500	0.57	0.57	0.60	-0.18	-24.00%	1198	25117	23.49%
60.00	MSFT160520C00060000	0.21	0.20	0.22	-0.05	-19.23%	1037	10759	22.85%
62.50	MSFT160520C00062500	0.10	0.07	0.09	0.01	11.11%	8	36	23.54%
65.00	MSFT160520C00065000	0.03	0.02	0.05	-0.02	-40.00%	25	489	25.59%
70.00	MSFT160520C00070000	0.05	0.00	0.04	0.00	0.00%	0	28	32.42%
75.00	MSFT160520C00075000	0.00	0.00	0.04	0.00	0.00%	0	0	39.26%

Fig. 13 Microsoft stock option chain

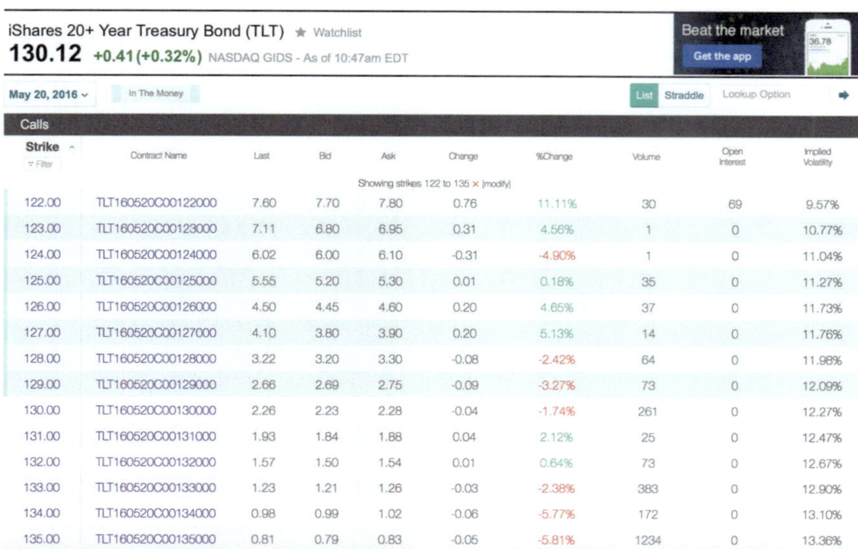

iShares 20+ Year Treasury Bond (TLT) ★ Watchlist
130.12 +0.41 (+0.32%) NASDAQ GIDS - As of 10:47am EDT

May 20, 2016 · In The Money · List · Straddle · Lookup Option

Calls

Showing strikes 122 to 135

Strike	Contract Name	Last	Bid	Ask	Change	%Change	Volume	Open Interest	Implied Volatility
122.00	TLT160520C00122000	7.60	7.70	7.80	0.76	11.11%	30	69	9.57%
123.00	TLT160520C00123000	7.11	6.80	6.95	0.31	4.56%	1	0	10.77%
124.00	TLT160520C00124000	6.02	6.00	6.10	-0.31	-4.90%	1	0	11.04%
125.00	TLT160520C00125000	5.55	5.20	5.30	0.01	0.18%	35	0	11.27%
126.00	TLT160520C00126000	4.50	4.45	4.60	0.20	4.65%	37	0	11.73%
127.00	TLT160520C00127000	4.10	3.80	3.90	0.20	5.13%	14	0	11.76%
128.00	TLT160520C00128000	3.22	3.20	3.30	-0.08	-2.42%	64	0	11.98%
129.00	TLT160520C00129000	2.66	2.69	2.75	-0.09	-3.27%	73	0	12.09%
130.00	TLT160520C00130000	2.26	2.23	2.28	-0.04	-1.74%	261	0	12.27%
131.00	TLT160520C00131000	1.93	1.84	1.88	0.04	2.12%	25	0	12.47%
132.00	TLT160520C00132000	1.57	1.50	1.54	0.01	0.64%	73	0	12.67%
133.00	TLT160520C00133000	1.23	1.21	1.26	-0.03	-2.38%	383	0	12.90%
134.00	TLT160520C00134000	0.98	0.99	1.02	-0.06	-5.77%	172	0	13.10%
135.00	TLT160520C00135000	0.81	0.79	0.83	-0.05	-5.81%	1234	0	13.36%

Fig. 14 iShare 20+ year treasury bond option chain

option strike prices along with their premiums for a given maturity period (cf. Harris 2003).

To fully describe the SABR model, we need an initial volatility and a risk-free interest rate. We use the historical volatility up to the as-of-date as a proxy of volatility

NOTE: futures symbols begin with the character @
index symbols begin with the character $

Symbol (option symbols)	hv20	hv50	hv100	DATE	curiv	Days/Percentile	Close

****
* Data generated by McMillan Analysis Corp. Copyright 2016 *
* www.optionstrategist.com 800-724-1817 *
****

Symbol (option symbols)	hv20	hv50	hv100	DATE	curiv	Days/Percentile	Close
$AUM	15	13	12	160324	12.86	597/ 97%ile	75.28
$AUX	15	13	12	160324	17.03	600/ 47%ile	132.86
$BKX	26	33	28	160208	34.55	426/ 99%ile	64.80
$BPX	11	12	10	160324	15.86	599/ 99%ile	70.66
$BRB	26	22	20	160119	29.09	346/ 71%ile	368.08
$BVZ	89	107	131	160324	45.21	251/ 55%ile	14.74
$CDD	13	13	11	160324	12.79	600/ 83%ile	132.49
$DJX	10	18	17	160324	12.89	600/ 66%ile	175.16
$EPX	89	87	72	150821	47.10	441/ 97%ile	192.89
$EUI	10	9	10	160324	10.80	597/ 50%ile	89.49
$EUU	9	9	10	160324	12.23	600/ 65%ile	111.74
$GBP	11	11	10	160324	12.94	598/ 98%ile	141.54
$GVZ	67	109	96	150402	136.07	573/ 95%ile	14.60
$HGX	16	26	25	160324	22.68	598/ 50%ile	224.78
$MNX	15	24	21	160324	19.26	585/ 80%ile	440.55
$NDO	17	15	13	160324	12.80	597/ 87%ile	67.01
$NDX	15	24	21	160324	13.06	595/ 49%ile	4405.53
$NZD	17	15	13	160324	16.05	600/ 24%ile	149.25
$OEX	12	19	18	160324	12.55	591/ 63%ile	904.06
$OSX	44	49	43	160324	36.06	480/ 90%ile	157.82
$OVX	58	70	72	150512	101.87	581/ 90%ile	33.70
$PZO	14	16	13	160324	19.67	600/ 72%ile	176.11
$RUI	13	19	18	151020	9.62	468/ 0%ile	1124.50
$RUT	20	24	21	160324	16.06	598/ 66%ile	1079.54
$RVX	84	87	96	160324	164.29	570/ 94%ile	19.40
$SFC	8	9	10	160324	10.10	597/ 37%ile	97.57
$SKA	9	8	9	160324	12.55	597/ 31%ile	82.95
$SOX	17	29	25	160324	27.23	569/ 74%ile	666.14
$SPX	12	19	18	160324	12.74	600/ 67%ile	2035.94
$VIX	89	107	131	160324	80.08	600/ 40%ile	14.74
$VXEEM	63	84	93	150331	77.40	486/ 23%ile	18.16
$VXEWZ	60	65	64	150406	70.48	494/ 38%ile	33.85
$VXST	132	157	193	150609	139.86	287/ 24%ile	15.22
$XAU	55	55	54	160324	49.05	598/ 86%ile	67.48

Fig. 15 Historical volatilities

initialization. This piece of information is provided by Option Strategist website to each traded underlying asset.

In Fig. 15, each underlying asset has three historical at-the-money volatilities based on the window length used to calculate that volatility. For example, there are three historical volatilities associated with MSFT, 24, 31 and 26%, which are calculated from 20, 50 and 100 days historical volatility respectively as of March 24th 2016. And the volatility of MSFT on that day is 20.82%.

Risk-free interest rate is the minimum rate of return an investor should expect for any investment. In practice, three-month U.S. Treasury bill is often used as a proxy of

Date	1 Mo	3 Mo	6 Mo	1 Yr	2 Yr	3 Yr	5 Yr	7 Yr	10 Yr	20 Yr	30 Yr
03/01/16	0.29	0.33	0.50	0.68	0.85	0.98	1.31	1.62	1.83	2.28	2.70
03/02/16	0.28	0.36	0.48	0.67	0.85	1.00	1.34	1.65	1.84	2.27	2.69
03/03/16	0.25	0.28	0.46	0.65	0.85	0.99	1.33	1.63	1.83	2.23	2.65
03/04/16	0.25	0.29	0.47	0.67	0.88	1.04	1.38	1.69	1.88	2.29	2.70
03/07/16	0.27	0.32	0.49	0.67	0.91	1.08	1.42	1.72	1.91	2.30	2.71
03/08/16	0.27	0.29	0.48	0.68	0.88	1.04	1.34	1.64	1.83	2.22	2.63
03/09/16	0.27	0.30	0.47	0.68	0.90	1.07	1.39	1.69	1.90	2.27	2.68
03/10/16	0.27	0.32	0.50	0.69	0.93	1.11	1.45	1.75	1.93	2.29	2.70
03/11/16	0.27	0.33	0.51	0.70	0.97	1.16	1.49	1.79	1.98	2.34	2.75
03/14/16	0.28	0.34	0.52	0.70	0.97	1.15	1.49	1.78	1.97	2.33	2.74
03/15/16	0.29	0.34	0.52	0.71	0.98	1.16	1.50	1.78	1.97	2.33	2.73
03/16/16	0.28	0.31	0.47	0.66	0.87	1.05	1.41	1.72	1.94	2.32	2.73
03/17/16	0.29	0.29	0.47	0.64	0.87	1.04	1.39	1.70	1.91	2.28	2.69
03/18/16	0.27	0.30	0.44	0.62	0.84	1.00	1.34	1.66	1.88	2.26	2.68
03/21/16	0.26	0.31	0.46	0.63	0.87	1.05	1.38	1.70	1.92	2.31	2.72
03/22/16	0.28	0.30	0.46	0.64	0.91	1.08	1.42	1.74	1.94	2.32	2.72
03/23/16	0.27	0.30	0.46	0.64	0.87	1.03	1.37	1.67	1.88	2.25	2.65
03/24/16	0.24	0.30	0.46	0.63	0.89	1.05	1.39	1.70	1.91	2.28	2.67
03/28/16	0.19	0.29	0.49	0.65	0.89	1.04	1.37	1.68	1.89	2.26	2.66
03/29/16	0.18	0.23	0.45	0.63	0.78	0.94	1.29	1.59	1.81	2.20	2.60
03/30/16	0.14	0.20	0.39	0.61	0.76	0.91	1.26	1.60	1.83	2.24	2.65

Fig. 16 Daily treasury yield curve rates

risk-free interest rate. U.S. Department of the Treasury releases daily treasury yield curves on its website, where we quote our risk-free interest rate. See Fig. 16.

Fitting SABR model parameters is not always straight forward because some of them are not observable from market. Therefore, we determine reasonable ranges for each of α and ρ, and simulate European call option prices in scenarios with different price-volatility correlation ρ and vol of vol α combinations.

Tables 3 and 4 specify SABR model parameters for MSFT and TLT respectively.

Table 5 shows MSFT option prices computed by the brute-force Monte-Carlo (Monte-Carlo column), and by dimension reduction methods we proposed using different distributions for (Σ^2, X_2), i.e. gamma mixture of normals (Gamma column), inverse-gamma mixture of normals (Inverse Gamma column) and log-normal mixture of normals (Log Normal column) respectively. As an option price is not known a priori at the time of pricing, we assume the brute-force MC gives a benchmark price. As for computational efficiency, the proposed approximation methods with different distributions all turn out to be much faster than the brute-force MC, and the results they produced also fall into a satisfactory range of accuracy. Figures 17 and 18 demonstrate pricing errors (see the caption of Fig. 17 for the definition) by using gamma mixture of normals. Note that the quality of the proposed approximation scheme depends on the parameter values. It is observed that the pricing is less accurate when ρ is near -1 than when ρ is near zero due to the leverage effect. And

Table 3 Model parameters microsoft stock

Item	Symbol	Value
Strike price	K	\$52.50
Closing price	F_0	\$53.54
Initial volatility	σ_0	26.56%
Time to maturity	t_{ex}	39
Risk-free interest rate	r	0.29%
Correlation	ρ	$[-1, 0]$
Vol of Vol	α	[0, 1]

Table 4 Model Parameters iShare 20+ years Treasury Bond ETF

Item	Symbol	Value
Strike price	K	\$125.00
Closing price	F_0	\$130.12
Initial volatility	σ_0	11.27%
Time to maturity	t_{ex}	39
Risk-free interest Rate	r	0.29%
Correlation	ρ	$[-1, 0]$
Vol of Vol	α	[0, 1]

Table 5 $\beta = 1$ SABR model prices comparison

α	ρ	Monte-Carlo	Gamma	Inverse gamma	Log normal
0.10	−0.05	2.86	2.86	2.85	2.86
0.10	−0.15	2.85	2.89	2.85	2.90
0.10	−0.25	2.83	2.90	2.92	2.86
0.10	−0.35	2.86	2.84	2.90	2.84
0.10	−0.45	2.84	2.92	2.89	2.89
0.10	−0.55	2.83	2.80	2.84	2.91
0.10	−0.65	2.84	2.79	2.87	2.99
0.10	−0.75	2.83	2.86	2.69	2.74
0.10	−0.85	2.85	2.77	2.83	2.69
0.10	−0.95	2.88	2.90	3.00	2.72
0.20	−0.05	2.85	2.86	2.86	2.86
0.20	−0.15	2.88	2.87	2.86	2.86
0.20	−0.25	2.85	2.89	2.92	2.97

(continued)

Table 5 (continued)

α	ρ	Monte-Carlo	Gamma	Inverse gamma	Log normal
0.20	−0.35	2.88	2.88	2.87	2.92
0.20	−0.45	2.87	2.94	2.90	2.83
0.20	−0.55	2.86	2.92	2.87	2.88
0.20	−0.65	2.86	2.82	2.92	2.83
0.20	−0.75	2.85	2.78	2.83	2.93
0.20	−0.85	2.86	2.87	2.92	2.79
0.20	−0.95	2.87	3.37	3.22	2.99
0.30	−0.05	2.84	2.87	2.87	2.87
0.30	−0.15	2.85	2.89	2.89	2.86
0.30	−0.25	2.87	2.89	2.92	2.83
0.30	−0.35	2.86	2.98	2.90	2.96
0.30	−0.45	2.86	2.90	2.93	2.87
0.30	−0.55	2.88	2.93	2.81	2.80
0.30	−0.65	2.87	2.93	2.83	2.93
0.30	−0.75	2.86	2.87	2.89	2.69
0.30	−0.85	2.86	2.63	2.87	2.95
0.30	−0.95	2.88	2.71	2.81	3.01
0.40	−0.05	2.86	2.87	2.87	2.87
0.40	−0.15	2.85	2.87	2.88	2.89
0.40	−0.25	2.86	2.87	2.91	2.89
0.40	−0.35	2.86	2.90	2.93	2.88
0.40	−0.45	2.84	2.88	2.88	3.00
0.40	−0.55	2.86	3.00	2.90	2.93
0.40	−0.65	2.86	2.99	3.03	2.89
0.40	−0.75	2.86	2.96	3.03	2.92
0.40	−0.85	2.89	2.77	3.00	3.07
0.40	−0.95	2.90	2.75	2.80	2.71
0.50	−0.05	2.87	2.88	2.86	2.88
0.50	−0.15	2.88	2.89	2.88	2.87
0.50	−0.25	2.86	2.88	2.86	2.89
0.50	−0.35	2.85	2.88	2.86	2.89
0.50	−0.45	2.87	2.89	2.92	2.86
0.50	−0.55	2.87	2.95	2.97	2.84
0.50	−0.65	2.84	2.77	2.84	2.89
0.50	−0.75	2.86	2.92	2.98	2.77
0.50	−0.85	2.91	2.82	2.80	2.96
0.50	−0.95	2.89	2.86	2.69	3.08
0.60	−0.05	2.87	2.86	2.88	2.87

(continued)

Table 5 (continued)

α	ρ	Monte-Carlo	Gamma	Inverse gamma	Log normal
0.60	−0.15	2.85	2.89	2.89	2.90
0.60	−0.25	2.87	2.90	2.90	2.92
0.60	−0.35	2.88	2.95	2.92	2.90
0.60	−0.45	2.88	2.81	2.95	3.00
0.60	−0.55	2.89	2.89	2.99	2.86
0.60	−0.65	2.87	2.92	2.88	2.88
0.60	−0.75	2.87	2.92	2.76	3.13
0.60	−0.85	2.89	2.80	2.83	2.80
0.60	−0.95	2.88	3.02	3.09	2.90
0.70	−0.05	2.86	2.86	2.88	2.89
0.70	−0.15	2.86	2.90	2.90	2.89
0.70	−0.25	2.89	2.92	2.93	2.90
0.70	−0.35	2.87	2.92	2.93	2.93
0.70	−0.45	2.88	3.03	2.89	2.95
0.70	−0.55	2.89	2.94	2.88	2.91
0.70	−0.65	2.88	2.82	2.91	2.91
0.70	−0.75	2.90	2.99	2.73	2.90
0.70	−0.85	2.87	2.93	2.91	2.97
0.70	−0.95	2.88	2.89	2.82	2.97
0.80	−0.05	2.89	2.91	2.87	2.91
0.80	−0.15	2.89	2.90	2.90	2.90
0.80	−0.25	2.90	2.94	2.88	2.91
0.80	−0.35	2.86	2.91	2.91	2.93
0.80	−0.45	2.88	2.94	2.86	2.87
0.80	−0.55	2.89	2.91	2.93	2.83
0.80	−0.65	2.88	2.97	2.95	2.89
0.80	−0.75	2.90	2.92	2.80	2.87
0.80	−0.85	2.89	3.03	2.98	2.87
0.80	−0.95	2.89	2.78	2.71	2.60
0.90	−0.05	2.88	2.87	2.90	2.87
0.90	−0.15	2.90	2.91	2.91	2.91
0.90	−0.25	2.89	2.93	2.93	2.92
0.90	−0.35	2.90	2.94	2.90	2.90
0.90	−0.45	2.91	2.98	2.98	2.94
0.90	−0.55	2.90	2.96	2.91	2.94
0.90	−0.65	2.89	3.11	2.97	3.00
0.90	−0.75	2.89	2.87	2.84	2.84
0.90	−0.85	2.90	2.83	2.75	2.89

(continued)

Table 5 (continued)

α	ρ	Monte-Carlo	Gamma	Inverse gamma	Log normal
0.90	−0.95	2.88	2.81	2.82	2.78
1.00	−0.05	2.89	2.90	2.87	2.88
1.00	−0.15	2.87	2.91	2.92	2.91
1.00	−0.25	2.86	2.94	2.92	2.92
1.00	−0.35	2.89	2.98	2.91	2.93
1.00	−0.45	2.90	3.02	2.90	2.93
1.00	−0.55	2.91	2.95	2.86	2.87
1.00	−0.65	2.90	2.88	2.85	2.92
1.00	−0.75	2.90	2.86	2.80	2.95
1.00	−0.85	2.91	2.91	2.83	2.87
1.00	−0.95	2.89	2.96	2.81	3.04

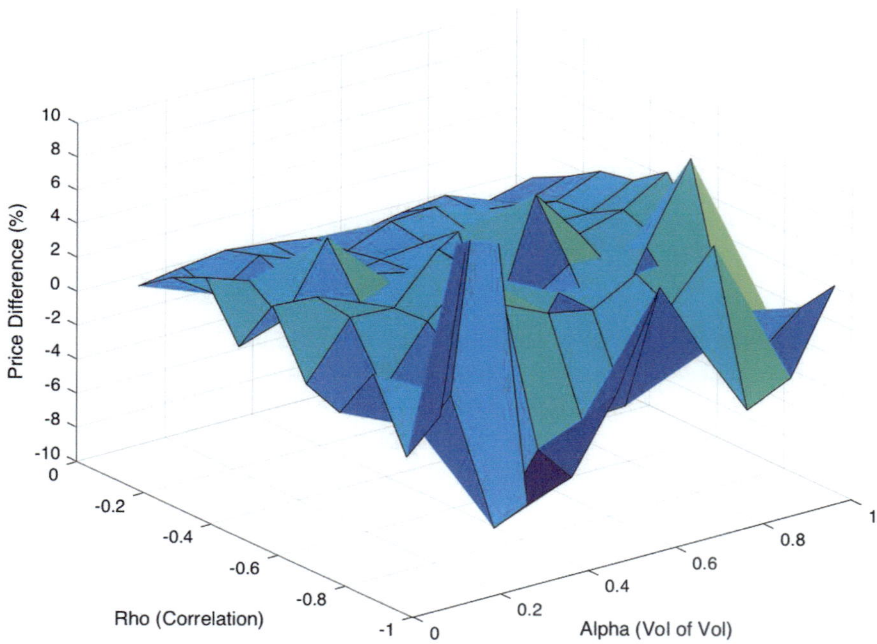

Fig. 17 Pricing error for the case $\beta = 1$, dimension reduction MC using gamma mixture of normal versus brute-force MC: let P_{DR} = option price computed using the dimension reduction MC, and P_{BF} = option price computed using the brute-force MC, then (y-coordinate)/100 = $\frac{P_{DR}-P_{BF}}{P_{BF}}$

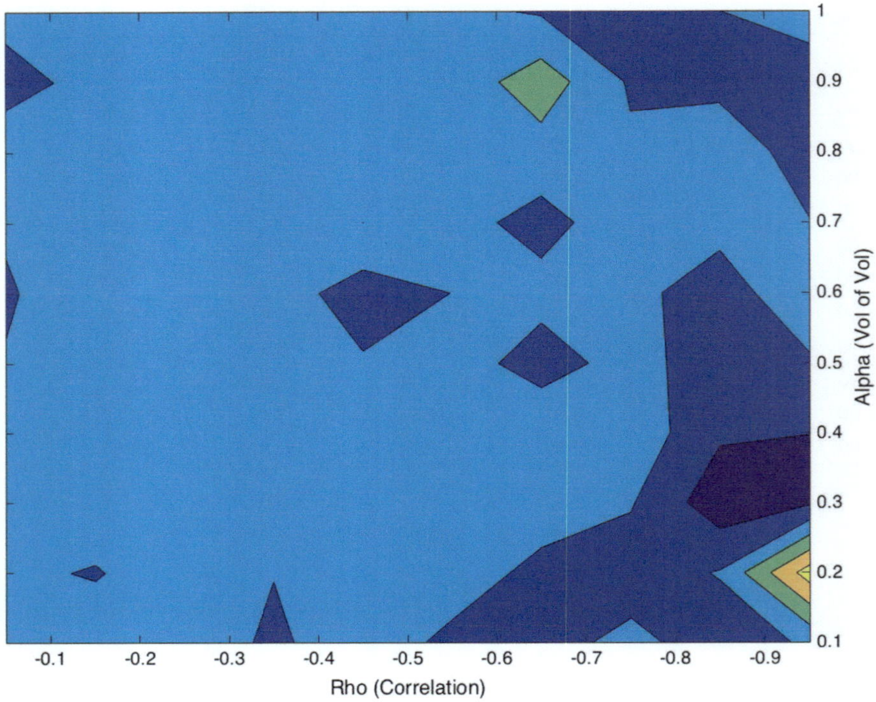

Fig. 18 *Top* view of Fig. 17: the large *light blue* area contains pairs (α, ρ) that yield small pricing errors compared to areas of other colors

greater values of α (volatility of volatility) tends to yield a better result, perhaps due to a greater degree of "mixing" in Monte-Carlo simulation. Table 5 covers a wide spectrum of combinations between α and ρ. Moreover, results using the other two approximate distributions, although not presented in this chapter, are in fact more stable. See Yin (2016) for detailed accounts.

References

Antonov, A., & Spector, M. (2012). Advanced analytics for the SABR model. *SSRN 2026350*.

Brooks, S., Gelman, A., Jones, G., & Meng, X.L. (2011). *Handbook of markov chain Monte-Carlo*. CRC Press.

Das, S. (2005). A learning market maker in the Glosten-Milgrom model. *Quantitative Finance*, *5*(2), 169–180.

Dempster, A. P., Laird, N. M., & Rubin, D. B. (1977). Maximum likelihood from incomplete data via EM algorithm (with discussion). *Journal of Royal Statistical Society, Series B*, *39*, 1–38.

Gelman, A., & Rubin, D. (1992). Inference from iterative simulation using multiple sequences. *Statistical Science*, *7*(4), 457–472.

Geweke, J. (1992). Evaluating the accuracy of sampling-based approaches to the calculation of posterior moments. *Bayesian Statistics*, 169–193.

Glosten, L., & Milgrom, P. (1985). Bid, ask and transaction prices in a specialist market with heterogeneously informed traders. *Journal of Financial Economics*, *14*(1), 71–100.

Hagan, P., Kumar, D., Lesniewski, A., & Woodward, D. (2002). Managing smile risk. *The Best of Wilmott*, 249–296.

Hagan, P., Lesniewski, A., & Woodward, D. (2005). Probability distribution in the SABR model of stochastic volatility. *Large deviations and asymptotic methods in finance* (pp. 1–35). Springer.

Harris, L. (2003). *Trading and exchanges*. Oxford University Press.

Hasbrouck, J. (2009). Trading costs and returns for US equities: estimating effective costs from daily data. *Journal of Finance*, *64*(3), 1–52.

Meng, X. L., & van Dyk, D. (1997). The EM algorithm: an old folk-song sung to a fast new tune (with discussion). *Journal of Royal Statistical Society, Series B*, *59*, 511–568.

Paulot, L. (2009). Asymptotic implied volatility at the second order with application to the SABR model. *SSRN 1413649*.

Rebonado, R., McKay, K., & White, R. (2011). *The SABR/LIBOR market model: Pricing, calibration and hedging for complex interest-rate derivatives*. Wiley.

Robert, C., & Casella, G. (2004). *Monte-Carlo statistical methods* (2nd ed.). Springer.

Roll, R. (1984). A simple implicit measure of the effective bid-ask spread in an efficient market. *Journal of Finance*, *39*(4), 1127–1139.

Wang, T. (2014). *Empirical analysis of sequential trade models for market microstructure*. Ph.D. thesis, University of North Carolina at Chapel Hill.

Yin, L. (2016). *Monte-Carlo strategies in option pricing for SABR model*. Ph.D. thesis, University of North Carolina at Chapel Hill.

Simulation Studies on the Effects of the Censoring Distribution Assumption in the Analysis of Interval-Censored Failure Time Data

Tyler Cook, Zhigang Zhang and Jianguo Sun

Abstract One problem researchers face when analyzing survival data is how to handle the censoring distribution. For practical convenience, it is often assumed that the observation process generating the censoring is independent of the event time of interest. This assumption allows one to effectively ignore the censoring distribution during the analysis, but it is clearly not always realistic. Unfortunately, one cannot generally test for independent censoring without additional assumptions or information. Therefore, the researcher is faced with a choice between using methods designed for informative or non-informative censoring without knowing the true nature of the censoring. This uncertainty creates a situation where the reliability of estimation and testing procedures is unknown as the assumptions are potentially violated. Fortunately, Monte-Carlo simulation methods can be very useful for exploring these types of questions under many different conditions. This chapter uses extensive simulation studies in order to investigate the effectiveness and flexibility of two methods developed for regression analysis of informative case I and case II interval-censored data under both types of censoring. The results of these simulation studies can provide guidelines for deciding between models when facing a practical problem where one is unsure about the informativeness of the censoring distribution.

T. Cook
University of Central Oklahoma, 100 North University Drive, Edmond, OK 73034, USA
e-mail: tcook14@uco.edu

Z. Zhang
Memorial Sloan Kettering Cancer Center, 1275 York Avenue, New York, NY 10065, USA
e-mail: zhangz@mskcc.org

J. Sun (✉)
University of Missouri, 146 Middlebush Hall, Columbia, MO 65211, USA
e-mail: sunj@missouri.edu

D.-G. Chen and J.D. Chen (eds.), *Monte-Carlo Simulation-Based Statistical Modeling*, ICSA Book Series in Statistics, DOI 10.1007/978-981-10-3307-0_15

1 Introduction

Many different disciplines deal with time-to-event data. Perhaps the most common example of survival data can be found in clinical trials, where doctors might be interested in studying something such as tumor onset time in a group of patients. Unfortunately, patients often drop out of the study for a variety of reasons. This causes a type of incomplete data because the exact time of tumor onset will be unknown for patients who leave the study. This problem of censoring is the hallmark characteristic of survival analysis, and it requires the development of special techniques.

Several unique types of censoring are seen in survival analysis. This chapter deals with a fairly general class of censoring called interval censoring. As the name suggests, one only knows that the event time of interest belongs to some interval of observation times with this type of data. Here, we will examine both case I and case II interval-censored data. With case I interval censoring, each subject is only observed once, and for this reason, the data are also often referred to as current status data. As a consequence of the data structure, the researcher only knows that the event time occurs before or after the observation time. Case II interval censoring usually involves multiple observation times. Here, at least one interval will be above zero and finite. This type of interval censoring is very frequently seen in medical studies where patients are scheduled for reoccurring follow-up visits.

Another aspect of censoring needing attention is the possible relationship between the observation time and the event time. It is often assumed that these two times are unrelated, and when this is the case, the censoring is called non-informative. It would be reasonable to assume that the censoring is non-informative if, for example, a patient drops out of a clinical trial because they need to move to a different city in order to start a new job. This assumption is attractive because it makes the analysis much less difficult. In fact, with current status data, non-informative censoring implies that the survival time and observation time are independent. However, this is clearly not always realistic. Say a patient drops out of a study due to illness. It is plausible that this censoring time has information related to tumor onset time, so it should not be ignored. This type of censoring is called informative censoring, and it is significantly more troublesome to handle.

When faced with a practical problem, the researcher will typically not know whether the censoring is informative or non-informative. Also, it is generally not possible to test this condition without additional knowledge or assumptions. This creates a slight dilemma when planning what method to use for data analysis. On the one hand, the researcher can choose to assume that the censoring is non-informative. This will make the analysis relatively straightforward, but it could tarnish the results when this seemingly naive assumption does not hold. On the other hand, one could assume that the censoring is informative. This might be more realistic, but increases the complexity of the analysis. Moreover, there is currently a lack of information for how these methods developed for informative censoring perform when the censoring is actually non-informative. This conundrum is the motivation for the research in this chapter. We are interested in exploring the flexibility of two methods developed for regression analysis of case I and case II interval-censored data under both informative and non-informative censoring. The goal of this research is to have a clearer picture of the performance of these methods when their assumptions are violated. This can provide insight into deciding between methods when working with a real dataset. To accomplish this, extensive Monte-Carlo simulation studies will be performed. The Monte-Carlo framework is ideal for investigated model performance and reliability because it allows one to investigate a wide range of specified conditions.

Survival analysis has a long history and a wide array of techniques for it are available in the literature. Many different approaches exist for handling problems involving case I interval-censored data. Several authors, including Peto (1973), Turnbull (1976) and Groeneboom and Wellner (1992), have proposed methods for nonparametric maximum likelihood estimation of the survival time distribution function. Nonparametric options also exist for treatment comparison, such as those described in Anderen and Ronn (1995) and Sun (1999). A wide range of models have also been suggested for regression analysis of current status data. Among them, the proportional hazards model was explored in Huang (1996), Lin et al. (1998) studied the use of the additive hazards model, and the proportional odds model was investigated in Rossini and Tsiatis (1996). Less work, however, has been done on current status data with informative censoring. Some examples exist in the context of tumorigenicity experiments. A three-state Markov model was discussed in Dewanji and Kalbfleisch (1986), and Lagakos and Louis (1988) proposed a method that utilizes known tumor lethality information. A more general process using latent random effects was described in Zhang et al. (2005). A similar number of options exist for case II interval-censored data. Many have considered the well-known proportional hazards model. Leading the way was the influential work of Finkelstein (1986), which used a maximum likelihood approach. Other models have also been explored. For example, the proportional odds model was examined by Huang and Wellner (1997) and Betensky et al. (2001) utilized the accelerate failure time model. Again, the informative censoring setup has attracted significantly less work. Notable examples include: Finkelstein et al. (2002) and Betensky and Finkelstein (2002) where general interval-censored data with informative censoring was considered. One limitation of the method of Betensky and Finkelstein (2002) is the requirement of follow-up after the event time. The method proposed by Zhang et al. (2007) for regression analysis of data with informative interval censoring avoids this issue and uses frailty terms to model dependence. An early comprehensive overview of theory and methodology for interval-censored data can be found in Sun (2006), and an up-to-date text on methodology and applications can be found in Chen et al. (2012).

The remainder of this chapter is outlined as follows. Section 2 describes the proposed models and parameter estimation techniques for the methods under investigation. Next, Sect. 3 presents the results from the extensive Monte-Carlo simulation studies. Finally, Sect. 4 summarizes and discusses the conclusions from the work in this chapter.

2 Methodology

In this section, we will describe the details for the methods under investigation for both case I and case II interval-censored data. First, the necessary notation and models will be introduced. Then, a summary of the parameter estimation procedure is given.

2.1 *Case I*

Here we outline the procedure proposed by Zhang et al. (2005) for regression analysis of informative current status data. This method was chosen for a number of reasons. It uses the additive hazards and proportional hazards models, which are two of the most well known and most often used models in survival analysis. A similar procedure for non-informative case I interval-censored data was proposed by Lin et al. (1998). That procedure had a relatively straightforward and simple estimation procedure, and the work of Zhang et al. (2005) shares many of those attractive characteristics. Also, Zhang et al. (2005) proposed to account for the informative censoring by using random effects, and this tool has been used in several other contexts including right-censored data (Huang and Wolfe 2002) and longitudinal data (Wang and Taylor 2001).

2.1.1 Notation and Models

Suppose that we have a survival study with n independent subjects, and define the following variables: the survival time of interest T_i, the observation time C_i, and a p-dimensional vector of possibly time-dependent covariates Z_i, for $i = 1, \ldots, n$. It will be assumed that the relationship between T_i and C_i can be modeled using an arbitrary mean zero random effect $b_i(t)$, which could also depend on time. The dependence between the survival and observation times can then be characterized by the specification of their respective hazard functions.

More specifically, we will assume that the T_i's follow the additive hazards frailty model, meaning the hazard function at time t is defined as:

$$\lambda_i(t|Z_i(s), b_i(s), s \leq t) = \lambda_1(t) + \beta' Z_i(t) + b_i(t) \tag{1}$$

given $\{Z_i(s), b_i(s), s \leq t\}$. Here $\lambda_1(t)$ is an unknown baseline hazard function, and the covariate effect on the survival time is represented by β, a p-dimensional vector of regression parameters. The C_i's will be assumed to follow a proportional hazards frailty model given $\{Z_i(s), b_i(s), s \leq t\}$. That is, the hazard function at time t is given by:

$$\lambda_i^c(t|Z_i(s), b_i(s), s \leq t) = \lambda_2(t) \exp(\gamma' Z_i(t) + b_i(t)) , \tag{2}$$

where $\lambda_2(t)$ is another unknown baseline hazard function, and γ represents the covariate effect on the observation times.

2.1.2 Parameter Estimation

Now we will consider the estimation of regression parameters. For simplicity, we will first examine the case where there are no informative censoring times. Assume that the survival study gives rise to the following observed data $\{(C_i, \delta_i = I(C_i \le T_i), Z_i(t), t \le C_i); i = 1, \ldots, n\}$. Define the counting process $N_i(t) = \delta_i I(C_i \le t) = I(C_i \le \min(T_i, t))$ and let $\Lambda_j(t) = \int_0^t \lambda_j(s)\mathrm{d}s,\ \ j = 1, 2$. Note that this counting process only jumps once if $C_i = t$ and $T_i \ge t$, and it can be shown (Zhang et al. 2005) that the probability for $\mathrm{d}N_i(t) = 1$ is given by

$$\begin{aligned} &\mathrm{d}\ \Pr\{T_i \ge t, C_i = t | Z_i(s), s \le t\} \\ &\quad = E\{\mathrm{e}^{-\Lambda_1(t) - B_i(t) - \beta' Z_i^*(t)} \lambda_2(t) \mathrm{e}^{\gamma' Z_i(t) + b_i(t)} \mathrm{d}t\} = \mathrm{e}^{\gamma' Z_i(t) - \beta' Z_i^*(t)} \mathrm{d}\Lambda_0^*(t), \quad (3) \end{aligned}$$

where $\mathrm{d}\Lambda_0^*(t) = \mathrm{e}^{-\Lambda_1(t)} E\{\mathrm{e}^{b_i(t) - B_i(t)}\}\mathrm{d}\Lambda_2(t)$, $B_i(t) = \int_0^t b_i(s)\mathrm{d}s$, and $Z_i^*(t) = \int_0^t Z_i(s)\mathrm{d}s$.

This is an interesting result because Eq. (3) is essentially what one would get from a standard proportional hazards model. Therefore, one can define the following martingales:

$$M_i^*(t) = N_i(t) - \int_0^t Y_i(s) \mathrm{e}^{\gamma' Z_i(s) - \beta' Z_i^*(s)} \mathrm{d}\Lambda_0^*(s),$$

where $Y_i(t) = I(C_i \ge t)$, and then use the well-known partial likelihood approach for estimation and inference concerning β and γ.

To be more specific, define

$$S^{(0)}(\beta, \gamma, t) = \frac{1}{n} \sum_{i=1}^{n} Y_i(t) \mathrm{e}^{\gamma' Z_i(t) - \beta' Z_i^*(t)},$$

$$S^{(1)}(\beta, \gamma, t) = \frac{1}{n} \sum_{i=1}^{n} Y_i(t) Z_i^*(t) \mathrm{e}^{\gamma' Z_i(t) - \beta' Z_i^*(t)},$$

and

$$S^{(2)}(\beta, \gamma, t) = \frac{1}{n} \sum_{i=1}^{n} Y_i(t) Z_i(t) \mathrm{e}^{\gamma' Z_i(t) - \beta' Z_i^*(t)}.$$

Now one can estimate β and γ by solving the estimating equations $U_\beta(\beta, \gamma) = 0$ and $U_\gamma(\beta, \gamma) = 0$, where

$$U_\beta(\beta,\gamma)=\sum_{i=1}^{n}\int_0^\infty\left\{Z_i^*(t)-\frac{S^{(1)}(\beta,\gamma,t)}{S^{(0)}(\beta,\gamma,t)}\right\}\mathrm{d}N_i(t)\,,$$

and

$$U_\gamma(\beta,\gamma)=\sum_{i=1}^{n}\int_0^\infty\left\{Z_i(t)-\frac{S^{(2)}(\beta,\gamma,t)}{S^{(0)}(\beta,\gamma,t)}\right\}\mathrm{d}N_i(t)\,.$$

Zhang et al. (2005) pointed out that $U_\beta(\beta,\gamma,t)$ and $U_\gamma(\beta,\gamma,t)$ are the partial score functions. Therefore, the estimates obtained from this approach are the maximum partial likelihood estimates. A main advantage of this method is that one is not required to estimate either of the two baseline hazard functions. Let $\hat{\theta}=(\hat{\beta}',\hat{\gamma}')'$ represent the obtained estimates of $\theta=(\beta',\gamma')'$. Then $\hat{\theta}$ is a consistent estimator and has asymptotically an approximately normal distribution.

This method can be easily extended to the case with censoring times on the C_i's. Suppose that there exists a censoring time C_i^c, which is independent of T_i and C_i, and $C_i^*=\min(C_i,C_i^c)$ is what is observed. Next, let $\xi_i=I(C_i^*=C_i)$ and define a new counting process $N_i^*(t)=\xi_i N_i(t)=I\{C_i\le\min(T_i,C_i^c,t)\}$. Estimation now proceeds as described above by solving the partial score functions, with the exception that we now define $Y_i(t)=I(C_i^*\ge t)$. The desirable results of consistency and asymptotic normality also hold for this situation.

2.2 *Case II*

Now we will discuss regression analysis of informative case II interval-censored data. For this data type, we will investigate the performance of the method proposed in Zhang et al. (2007). This method uses the popular proportional hazards model, and defines frailty terms to handle the dependence between the observation times and the failure time. As discussed in the previous section, similar approaches have proved successful in other contexts. Also, unlike Betensky and Finkelstein (2002), the method of Zhang et al. (2007) does not require follow-up times after the event time has occurred, which is a nice benefit.

2.2.1 Notation and Models

As before, consider a survival study with n independent subjects and let T_i denote the failure time for subject $i=1,\ldots,n$. Suppose that there exist two additional random variables U_i and V_i such that $U_i\le V_i$ and that one knows only whether T_i is less than or equal to U_i, between U_i and V_i, or greater than V_i. Also suppose that there exists a $p\times 1$ vector of covariates Z_i for each subject and define $W_i=U_i-V_i$, the gap time between the two observation times.

It will be assumed that the failure time and observation times are related through an unobserved random vector $b_i = (b_{1i}, b_{2i}, b_{3i})'$. The relationship among the variables is then modeled using the following hazard functions for T_i, U_i and W_i as

$$\lambda_i^{(T)}(t|Z_i, b_i) = \lambda_{t0}(t)\exp(\beta_t^{'} Z_i + b_{1i}), \quad (4)$$

$$\lambda_i^{(U)}(t|Z_i, b_i) = \lambda_{u0}(t)\exp(\beta_u^{'} Z_i + b_{2i}), \quad (5)$$

$$\lambda_i^{(W)}(t|Z_i, b_i) = \lambda_{w0}(t)\exp(\beta_w^{'} Z_i + b_{3i}), \quad (6)$$

respectively, where β_t, β_u and β_w are $p \times 1$ vectors of regression parameters, and $\lambda_{t0}(t)$, $\lambda_{u0}(t)$ and $\lambda_{w0}(t)$ are unknown baseline hazard functions.

In addition, it will be assumed that the latent random vector b_i follows a multivariate normal distribution such that $b_i \sim \mathrm{N}(0, \Sigma)$, where

$$\Sigma = \begin{pmatrix} \sigma_{11} & \sigma_{12} & \sigma_{13} \\ \sigma_{12} & \sigma_{22} & \sigma_{23} \\ \sigma_{13} & \sigma_{23} & \sigma_{33} \end{pmatrix}.$$

The values of the various σ's in the covariance matrix describe the relationship between the failure time and observation times. For example, if $\sigma_{12} = 0$, then the failure time T_i is independent of the first observation time U_i, given Z_i.

Next we will describe the construction of the likelihood function. Note that the exact value of T_i is not known since we have interval-censored data and the observed data can be expressed by $\{U_i, V_i, \delta_{1i}, \delta_{2i}, Z_i\}$ for $i = 1, \ldots, n$, where $\delta_{1i} = I(T_i \leq U_i)$ and $\delta_{2i} = I(U_i < T_i \leq V_i)$ are indicators identifying the interval containing T_i. Define

$$(L_i, R_i] = \begin{cases} (0, U_i] & \text{if } \delta_{1i} = 1 \\ (U_i, V_i] & \text{if } \delta_{2i} = 1 \\ (V_i, \infty] & \text{otherwise}, \end{cases}$$

and let $0 = s_0 < s_1 < \cdots < s_m = \infty$ be the set of times containing each L_i and R_i. Also, consider $\Lambda_{t0}(t) = \int_0^t \lambda_{t0}(u)\mathrm{d}u$ and $\gamma_j = \log \Lambda_{t0}(s_j)$. Finally, now also let $\gamma = (\gamma_1, \ldots, \gamma_{m-1})'$, $\theta = (\beta_t', \beta_u', \beta_w', \gamma', \sigma_{kl}, 1 \leq k \leq l \leq 3)'$, and $\Delta_i = (\delta_{1i}, \delta_{2i})$. One can now build the likelihood function of the observed data. Note that conditional on (U_i, W_i, Z_i, b_i), the likelihood for subject i is given by

$$L_{\Delta_i|U_i,W_i,b_i}(\theta) = \sum_{j=1}^{m} \alpha_{ij}[\exp\{-\exp(\beta_t' Z_i + b_{1i} + \gamma_{j-1})\} - \exp\{-\exp(\beta_t' Z_i + b_{1i} + \gamma_j)\}],$$

where $\alpha_{ij} = 1$ if $(s_{j-1}, s_j]$ is a subset of $(L_i, R_i]$ and 0 otherwise. The likelihood functions for U_i and W_i, conditional on (Z_i, b_i), have the forms

$$L_{U_i|b_i}(\theta) = \lambda_{u0}(U_i)\exp\{\beta_u' Z_i + b_{2i}\}\exp\{-\exp(\beta_u' Z_i + b_{2i})\Lambda_{u0}(U_i)\},$$

and

$$L_{W_i|b_i}(\theta) = \{\lambda_{w0}(W_i)\exp\{\beta_w' Z_i + b_{3i}\}\exp\{-\exp(\beta_w' Z_i + b_{3i})\Lambda_{w0}(W_i)\}\}^{\Psi_i},$$

where $\Lambda_{u0}(t) = \int_0^t \lambda_{u0}(u)\mathrm{d}u$, $\Lambda_{w0}(t) = \int_0^t \lambda_{w0}(u)\mathrm{d}u$, and $\Psi_i = I(W_i < \infty)$. Also define $O_i = \{\Delta_i, \Psi_i, U_i, W_i, Z_i\}$ to be the full observed data from subject i, and let $O = \{O_1, \ldots, O_n\}$ denote the combined data from all subjects. We can now write the full likelihood as

$$L_O(\theta) = \prod_{i=1}^{n} L_i(\theta; O_i) = \prod_{i=1}^{n} \int L_{\Delta_i|U_i,W_i,b_i}(\theta) L_{U_i|b_i}(\theta) L_{W_i|b_i}(\theta) f(b_i; \Sigma)\mathrm{d}b_i,$$

where $f(b_i; \Sigma)$ is the density function of b_i.

2.2.2 Parameter Estimation

The maximization of the likelihood function above is not straightforward since the b_i's are unknown. Therefore, the authors proposed using the EM algorithm in order to estimate the unknown parameters. The complete data is defined to be $\{(O_i, b_i), i = 1, \ldots, n\}$, and in typical fashion, one alternates between calculating the expectation of the log-likelihood, and then updating the estimate by maximizing the complete data likelihood. Variance estimation is achieved using Louis' formula. The procedure is quite complex, and a detailed discussion including recommendations for implementing the algorithm can be found in Zhang et al. (2007).

3 Simulation Studies

This section describes in detail the Monte-Carlo simulation studies that were performed in order to evaluate the performance of the models under a wide range of different conditions. As in the previous section, current status data will be examined first. Then follows the analysis of case II interval-censored data.

3.1 Case I

The majority of the simulations performed closely mirror those in the original work of Zhang et al. (2005). Several key model components change from simulation to simulation, but many remain constant. Specifically, the baseline hazard functions, $\lambda_1(t)$ and $\lambda_2(t)$, were set equal to one in all of the simulations. Also, each setup considered the situation with no censoring, 20% censoring, and 40% censoring.

This was achieved by setting $C_i^c = \tau$, where τ is a constant used to determine the percentage of censored observations. Each study used a sample size of $n = 200$ with 1000 replications.

Results are presented in tables that display the means of $\hat{\beta}$ and $\hat{\gamma}$ for several different combinations of true values for β and γ. Also, each table shows the means of the estimated standard deviations of $\hat{\beta}$ and $\hat{\gamma}$ (SEE) as well as the sample standard deviations of the point estimates (SE). Finally, the 95% empirical coverage probabilities are calculated.

3.1.1 Informative Censoring

First, we examine the case with informative censoring. This serves as a confirmatory analysis, and gives results that can be used for comparison with the other situations. The setup is the same as in the original paper except here we are only considering the discrete covariate case. Specifically, Z was generated from a Bernoulli distribution with success probability equal to 0.5. Exponential distributions were used for both the survival and observations times with hazards defined in (1) and (2), respectively. Time-independent random effects were generated from a standard normal distribution.

Tables 1, 2, and 3 show the results for these simulations. It is clear that as expected, the method seems to be performing well. The means of the parameter estimates are close to their true values, and the variance estimates are close, which suggests that the variance estimation procedure is valid. Also, the empirical 95% coverage probabilities are all fairly close to the desired level. The variance estimates grow as the censoring percentage increases. This can be expected since less information is observed with increased censoring.

Table 1 Case I informative censoring with discrete covariate and no censoring

γ	β	$\hat{\gamma}$				$\hat{\beta}$			
		Mean	SE	SEE	CP	Mean	SE	SEE	CP
0	0	0.003	0.291	0.282	0.950	0.028	0.407	0.389	0.958
	0.5	0.007	0.311	0.306	0.949	0.562	0.514	0.500	0.967
	1	0.035	0.341	0.331	0.946	1.145	0.661	0.647	0.966
0.2	0	0.181	0.282	0.276	0.948	–0.020	0.418	0.403	0.954
	0.5	0.219	0.291	0.297	0.948	0.533	0.513	0.511	0.961
	1	0.229	0.341	0.322	0.941	1.113	0.695	0.651	0.953
0.5	0	0.490	0.279	0.276	0.953	–0.026	0.471	0.442	0.951
	0.5	0.507	0.292	0.292	0.957	0.524	0.565	0.552	0.954
	1	0.522	0.326	0.313	0.945	1.084	0.735	0.685	0.946

Table 2 Case I informative censoring with discrete covariate and 20% censoring

γ	β	$\hat{\gamma}$				$\hat{\beta}$			
		Mean	SE	SEE	CP	Mean	SE	SEE	CP
0	0	0.015	0.313	0.316	0.957	0.025	0.569	0.560	0.953
	0.5	0.003	0.329	0.331	0.953	0.548	0.651	0.636	0.956
	1	0.007	0.355	0.348	0.960	1.033	0.766	0.740	0.957
0.2	0	0.194	0.318	0.313	0.953	–0.015	0.594	0.594	0.959
	0.5	0.209	0.333	0.325	0.950	0.532	0.691	0.668	0.953
	1	0.211	0.351	0.341	0.950	1.098	0.825	0.776	0.945
0.5	0	0.495	0.311	0.309	0.949	0.006	0.666	0.655	0.954
	0.5	0.509	0.313	0.320	0.962	0.546	0.728	0.729	0.960
	1	0.517	0.319	0.332	0.967	1.076	0.829	0.824	0.962

Table 3 Case I informative censoring with discrete covariate and 40% censoring

γ	β	$\hat{\gamma}$				$\hat{\beta}$			
		Mean	SE	SEE	CP	Mean	SE	SEE	CP
0	0	0.003	0.373	0.370	0.959	–0.027	0.961	0.928	0.949
	0.5	0.027	0.382	0.381	0.956	0.592	1.013	0.996	0.956
	1	–0.003	0.393	0.391	0.953	1.061	1.106	1.081	0.953
0.2	0	0.197	0.381	0.367	0.951	0.025	1.055	1.009	0.950
	0.5	0.191	0.375	0.375	0.955	0.509	1.089	1.065	0.946
	1	0.222	0.394	0.385	0.958	1.085	1.164	1.141	0.953
0.5	0	0.480	0.363	0.367	0.961	–0.041	1.145	1.142	0.961
	0.5	0.495	0.394	0.375	0.946	0.498	1.210	1.199	0.955
	1	0.508	0.390	0.381	0.953	1.069	1.346	1.272	0.951

3.1.2 Non-informative Censoring

The second situation considers the case with non-informative censoring. Using the proposed model, independent censoring is achieved by setting the latent random effects b_i's equal to zero. Then the survival and observation times were both generated from exponential distributions using the hazards defined in (1) and (2). Here we investigated the performance with both a discrete and continuous covariate. For the discrete case, it was assumed that Z followed a Bernoulli distribution with success probability 0.5, and a uniform distribution over [0, 1] was used for Z in the continuous case.

Tables 4 and 5 show the simulation results for a discrete and continuous covariate, respectively, with no censoring. These results display a number of important characteristics. Overall the point estimates for β and γ seem to be unbiased for both types of covariates. In general, the SE and SEE are reasonably close, which indicates that the variance estimates are sensible. Moreover, the coverage probabilities are largely

Table 4 Case I non-informative censoring with discrete covariate and no censoring

γ	β	$\hat{\gamma}$				$\hat{\beta}$			
		Mean	SE	SEE	CP	Mean	SE	SEE	CP
0	0	−0.006	0.290	0.290	0.952	−0.001	0.457	0.435	0.952
	0.5	0.019	0.332	0.313	0.946	0.550	0.585	0.537	0.947
	1	0.026	0.329	0.336	0.949	1.110	0.679	0.669	0.971
0.2	0	0.226	0.294	0.286	0.94	0.007	0.471	0.448	0.96
	0.5	0.217	0.304	0.306	0.953	0.547	0.556	0.553	0.968
	1	0.218	0.329	0.325	0.954	1.100	0.709	0.675	0.959
0.5	0	0.505	0.290	0.282	0.945	0.06	0.508	0.491	0.956
	0.5	0.512	0.305	0.299	0.954	0.553	0.624	0.593	0.955
	1	0.534	0.320	0.319	0.955	1.120	0.733	0.722	0.959

Table 5 Case I non-informative censoring with continuous covariate and no censoring

γ	β	$\hat{\gamma}$				$\hat{\beta}$			
		Mean $\hat{\gamma}$	SE	SEE	CP	Mean $\hat{\beta}$	SE	SEE	CP
0	0	−0.017	0.510	0.507	0.955	−0.012	0.784	0.754	0.947
	0.5	−0.002	0.504	0.536	0.967	0.521	0.863	0.888	0.960
	1	0.071	0.599	0.566	0.944	1.147	1.092	1.048	0.943
0.2	0	0.189	0.482	0.492	0.968	−0.013	0.756	0.763	0.968
	0.5	0.200	0.523	0.522	0.945	0.529	0.936	0.909	0.955
	1	0.216	0.554	0.550	0.952	1.050	1.080	1.057	0.953
0.5	0	0.502	0.489	0.480	0.953	0.004	0.835	0.823	0.956
	0.5	0.480	0.499	0.503	0.957	0.491	0.952	0.947	0.960
	1	0.480	0.540	0.526	0.940	0.970	1.134	1.082	0.946

accurate. Results with 20 and 40% censoring for both the discrete and continuous case can be found in Tables 6, 7, 8 and 9. These additional results mainly tell the same story. Estimates for both the discrete and continuous case tend to be unbiased, and the coverage probabilities are all quite close to the desired size. The important difference that can be seen is in the variance estimates. Specifically, the variance tends to increase for both types of covariates as the censoring percentage increases.

3.1.3 Multivariate Random Effects

The next simulation study investigated the performance when the structure of the random effects is misspecified. The proposed method assumes that the hazard functions for the survival and observation times share a random effect b_i. We examined a more general case where the relationship between the survival and observation times are characterized by an arbitrary random vector, $\mathbf{b}_i(t) = (b_i^1(t), b_i^2(t))$, with mean

Table 6 Case I non-informative censoring with discrete covariate and 20% censoring

γ	β	$\hat{\gamma}$				$\hat{\beta}$			
		Mean	SE	SEE	CP	Mean	SE	SEE	CP
0	0	–0.003	0.331	0.321	0.944	0.021	0.605	0.576	0.958
	0.5	–0.006	0.346	0.335	0.955	0.523	0.672	0.652	0.966
	1	0.036	0.362	0.352	0.95	1.110	0.772	0.759	0,962
0.2	0	0.211	0.308	0.316	0.961	0.027	0.608	0.609	0.961
	0.5	0.207	0.342	0.329	0.938	0.548	0.702	0.684	0.956
	1	0.238	0.346	0.344	0.949	1.140	0.835	0.790	0.953
0.5	0	0.513	0.321	0.213	0.945	0.037	0.694	0.668	0.949
	0.5	0.507	0.323	0.322	0.954	0.529	0.757	0.741	0.962
	1	0.525	0.344	0.336	0.952	1.090	0.871	0.841	0.956

Table 7 Case I non-informative censoring with continuous covariate and 20% censoring

γ	β	$\hat{\gamma}$				$\hat{\beta}$			
		Mean	SE	SEE	CP	Mean	SE	SEE	CP
0	0	–0.021	0.558	0.554	0.954	–0.026	1.024	0.986	0.944
	0.5	0.024	0.584	0.578	0.954	0.578	1.153	1.102	0.951
	1	–0.001	0.618	0.598	0.944	1.024	1.262	1.230	0.955
0.2	0	0.170	0.571	0.547	0.940	–0.039	1.110	1.051	0.950
	0.5	0.208	0.583	0.567	0.944	0.519	1.202	1.163	0.951
	1	0.239	0.602	0.586	0.954	1.118	1.361	1.288	0.951
0.5	0	0.496	0.541	0.537	0.959	–0.063	1.187	1.161	0.954
	0.5	0.541	0.558	0.552	0.956	0.600	1.289	1.260	0.956
	1	0.515	0.572	0.571	0.958	0.987	1.369	1.379	0.956

Table 8 Case I non-informative censoring with discrete covariate and 40% censoring

γ	β	$\hat{\gamma}$				$\hat{\beta}$			
		Mean	SE	SEE	CP	Mean	SE	SEE	CP
0	0	0.012	0.372	0.371	0.949	0.032	0.915	0.919	0.962
	0.5	0.001	0.396	0.382	0.949	0.479	1.060	0.994	0.945
	1	0.011	0.398	0.394	0.953	1.120	1.100	1.080	0.956
0.2	0	0.185	0.366	0.368	0.954	–0.037	1.010	0.996	0.959
	0.5	0.191	0.383	0.377	0.947	0.491	1.110	1.050	0.949
	1	0.210	0.391	0.386	0.963	1.050	1.180	1.140	0.955
0.5	0	0.509	0.388	0.369	0.947	0.005	1.190	1.140	0.942
	0.5	0.522	0.384	0.377	0.956	0.558	1.280	1.200	0.948
	1	0.517	0.379	0.384	0.962	1.060	1.280	1.26	0.950

Table 9 Case I non-informative censoring with continuous covariate and 40% censoring

γ	β	$\hat{\gamma}$				$\hat{\beta}$			
		Mean	SE	SEE	CP	Mean	SE	SEE	CP
0	0	–0.019	0.607	0.639	0.960	–0.077	1.609	1.590	0.956
	0.5	0.008	0.681	0.659	0.946	0.518	1.783	1.699	0.939
	1	–0.020	0.701	0.678	0.943	0.945	1.847	1.815	0.955
0.2	0	0.163	0.625	0.636	0.960	–0.108	1.769	1.719	0.943
	0.5	0.202	0.639	0.650	0.963	0.496	1.761	1.815	0.966
	1	0.214	0.659	0.667	0.957	1.043	1.953	1.940	0.940
0.5	0	0.522	0.643	0.633	0.948	0.003	1.976	1.950	0.946
	0.5	0.534	0.653	0.644	0.954	0.573	2.063	2.053	0.956
	1	0.540	0.714	0.658	0.936	1.078	2.296	2.152	0.944

zero. The hazard functions for the survival and observation times, respectively, are now defined as:

$$\lambda_i(t|Z_i(s), b_i^1(s), s \leq t) = \lambda_1(t) + \beta' Z_i(t) + b_i^1(t)\,, \tag{7}$$

$$\lambda_i^c(t|Z_i(s), b_i^2(s), s \leq t) = \lambda_2(t)\exp(\gamma' Z_i(t) + b_i^2(t))\,, \tag{8}$$

where $\lambda_1(t)$, $\lambda_2(t)$, β, and γ are the same as in (1) and (2).

For this case, we created current status data by first generating time-independent random effects assuming $\mathbf{b}_i \sim \text{MVN}(0, \Sigma)$ with

$$\Sigma = \begin{pmatrix} 1 & \rho \\ \rho & 1 \end{pmatrix},$$

where ρ, representing different levels of correlation between the random effects, was set to 0.3 and 0.5. For this situation only the discrete covariate was considered with Z coming from a Bernoulli distribution with success probability equal to 0.5. Finally, survival and observation times were generated from exponential distributions using the hazards defined in Eqs. (7) and (8).

Results for these simulation studies with no censoring are presented in Tables 10 and 11. Table 10 shows outcomes with $\rho = 0.3$ and Table 11 has the results when $\rho = 0.5$. Once again, additional simulations with 20 and 40% censoring can be found in Tables 12, 13, 14 and 15. It is clear from the results that the existence of a multivariate random effect causes serious problems for parameter estimation at both levels of correlation. The estimates for β and γ are biased in all cases. The coverage probabilities vary widely and are not close to the desired size. Similar conclusions can be seen when censoring is present, and once again the variance estimates increase as the censoring percentage increases.

Table 10 Case I multivariate random effect with $\rho = 0.3$ and no censoring

γ	β	$\hat{\gamma}$				$\hat{\beta}$			
		Mean	SE	SEE	CP	Mean	SE	SEE	CP
0	0	–0.001	0.249	0.247	0.964	0.005	0.348	0.315	0.979
	0.5	–0.074	0.260	0.257	0.942	0.169	0.411	0.359	0.740
	1	–0.129	0.282	0.272	0.915	0.450	0.512	0.448	0.602
0.2	0	0.170	0.261	0.244	0.932	0.047	0.392	0.340	0.968
	0.5	0.095	0.259	0.252	0.927	0.191	0.438	0.371	0.754
	1	0.036	0.257	0.267	0.913	0.483	0.534	0.462	0.630
0.5	0	0.401	0.245	0.239	0.914	0.076	0.413	0.364	0.973
	0.5	0.328	0.248	0.248	0.896	0.254	0.475	0.408	0.806
	1	0.302	0.260	0.261	0.867	0.571	0.561	0.501	0.731

Table 11 Case I multivariate random effect with $\rho = 0.5$ and no censoring

γ	β	$\hat{\gamma}$				$\hat{\beta}$			
		Mean	SE	SEE	CP	Mean	SE	SEE	CP
0	0	–0.006	0.263	0.238	0.940	0.052	0.400	0.298	0.942
	0.5	–0.176	0.275	0.255	0.900	0.061	0.383	0.343	0.643
	1	–0.224	0.267	0.265	0.840	0.366	0.417	0.397	0.541
0.2	0	0.142	0.274	0.236	0.920	–0.038	0.381	0.315	0.940
	0.5	–0.031	0.247	0.247	0.920	0.124	0.492	0.375	0.620
	1	0.035	0.244	0.260	0.900	0.562	0.419	0.434	0.664
0.5	0	0.492	0.230	0.232	0.940	0.150	0.334	0.352	0.981
	0.5	0.259	0.258	0.242	0.840	0.167	0.258	0.385	0.847
	1	0.301	0.271	0.256	0.860	0.654	0.673	0.490	0.643

Table 12 Case I multivariate random effect with $\rho = 0.3$ and 20% censoring

γ	β	$\hat{\gamma}$				$\hat{\beta}$			
		Mean	SE	SEE	CP	Mean	SE	SEE	CP
0	0	0.002	0.295	0.284	0.941	0.034	0.544	0.566	0.973
	0.5	–0.074	0.289	0.292	0.963	0.213	0.667	0.601	0.907
	1	–0.059	0.311	0.302	0.925	0.733	0.704	0676	0.932
0.2	0	0.196	0.297	0.281	0.952	0.119	0.638	0.618	0.991
	0.5	0.103	0.294	0.289	0.927	0.244	0.682	0.635	0.928
	1	0.073	0.294	0.297	0.933	0.610	0.750	0.681	0.874
0.5	0	0.427	0.251	0.278	0.965	0.106	0.670	0.679	0.956
	0.5	0.317	0.270	0.280	0.883	0.350	0.688	0.691	0.940
	1	0.339	0.310	0.289	0.938	0.681	0.722	0.743	0.925

Table 13 Case I multivariate random effect with $\rho = 0.3$ and 40% censoring

γ	β	$\hat{\gamma}$				$\hat{\beta}$			
		Mean	SE	SEE	CP	Mean	SE	SEE	CP
0	0	0.033	0.373	0.338	0.930	0.049	1.233	1.101	0.940
	0.5	–0.065	0.344	0.344	0.970	0.180	1.090	1.120	0.942
	1	–0.180	0.3877	0.349	0.900	0.243	1.218	1.147	0.883
0.2	0	0.169	0.396	0.341	0.910	0.162	1.279	1.232	0.971
	0.5	0.150	0.307	0.342	0.990	0.370	1.170	1.241	0.986
	1	0.119	0.363	0.347	0.940	0.742	1.325	1.250	0.940
0.5	0	0.409	0.311	0.333	0.980	–0.015	1.348	1.351	0.960
	0.5	0.389	0.305	0.339	0.950	0.500	1.271	1.358	0.991
	1	0.326	0.386	0.342	0.850	0.763	1.547	1.383	0.954

Table 14 Case I multivariate random effect with $\rho = 0.5$ and 20% censoring

γ	β	$\hat{\gamma}$				$\hat{\beta}$			
		Mean	SE	SEE	CP	Mean	SE	SEE	CP
0	0	–0.046	0.284	0.285	0.980	–0.029	0.666	0.587	0.926
	0.5	–0.029	0.327	0.295	0.940	0.247	0.614	0.624	0.921
	1	–0.013	0.326	0.302	0.920	0.588	0.711	0.657	0.867
0.2	0	0.142	0.289	0.282	0.960	–0.07	0.502	0.605	0.983
	0.5	0.090	0.260	0.283	0.940	0.352	0.534	0.630	0.982
	1	0.019	0.279	0.294	0.940	0.548	0.733	0.686	0.922
0.5	0	0.374	0.238	0.278	0.940	0.103	0.607	0.674	0.960
	0.5	0.356	0.338	0.282	0.880	0.218	0.747	0.699	0.901
	1	0.321	0.342	0.291	0.860	0.666	0.861	0.740	0.922

Table 15 Case I multivariate random effect with $\rho = 0.5$ and 40% censoring

γ	β	$\hat{\gamma}$				$\hat{\beta}$			
		Mean	SE	SEE	CP	Mean	SE	SEE	CP
0	0	–0.070	0.330	0.343	0.980	–0.185	1.057	1.110	0.961
	0.5	–0.058	0.349	0.348	0.960	0.200	1.131	1.152	0.942
	1	–0.043	0.372	0.353	0.940	0.781	1.239	1.154	0.943
0.2	0	0.139	0.364	0.336	0.940	–0.071	1.267	1.164	0.967
	0.5	0.137	0.435	0.341	0.860	0.238	1.503	1.219	0.925
	1	0.073	0.385	0.350	0.880	0.790	1.406	1.262	0.908
0.5	0	0.510	0.341	0.337	0.980	0.581	1.879	1.375	0.927
	0.5	0.415	0.263	0.337	0.960	0.494	1.165	1.331	0.981
	1	0.306	0.335	0.341	0.940	0.639	1.335	1.395	0.920

Table 16 Case I additive hazards model for observation times with no censoring

γ	β	$\hat{\gamma}$				$\hat{\beta}$			
		Mean	SE	SEE	CP	Mean	SE	SEE	CP
0	0	0.070	0.241	0.232	0.940	0.001	0.202	0.161	0.988
	0.5	0.031	0.280	0.281	0.952	0.624	0.438	0.415	0.953
	1	0.050	0.318	0.314	0.957	1.197	0.684	0.637	0.953
0.2	0	0.130	0.236	0.239	0.947	–0.206	0.234	0.223	0.899
	0.5	0.113	0.278	0.267	0.933	0.446	0.406	0.359	0.877
	1	0.130	0.293	0.302	0.946	1.032	0.603	0.580	0.919
0.5	0	0.314	0.260	0.248	0.876	–0.363	0.342	0.305	0.763
	0.5	0.177	0.238	0.239	0.694	0.072	0.280	0.230	0.406
	1	0.231	0.293	0.284	0.808	0.738	0.504	0.498	0.822

3.1.4 Model Misspecification

The final simulation study examined the case where the hazard function for the observation times is misspecified. Unlike Zhang et al. (2005), where a proportional hazards frailty model is assumed, we investigated the situation where the observation times follow an additive hazards frailty model, i.e. the hazard function for the observation times is given by:

$$\lambda_i^c(t|Z_i(s), b_i(s), s \leq t) = \lambda_3(t) + \gamma' Z_i(t) + b_i(t), \tag{9}$$

where $\lambda_3(t)$ is an unspecified baseline hazard function, and γ once again denotes the covariate effect on the observation times.

As with the last case, we focused only on the situation where Z was generated from a Bernoulli distribution with success probability 0.5. Here, current status data were created by first generating the b_i's from a standard normal distribution. Then, survival and observation times were produced from exponential distributions with hazards given by (1) and (9), with $\lambda_3(t) = 1$.

Table 16 shows the simulation results for this setup with no censoring. Upon inspection, it seems that the results in this case are mixed. The method performs adequately for certain parameter combinations and poorly for others. When γ is equal to zero, the bias for $\hat{\beta}$ and $\hat{\gamma}$ is small, and the coverage probabilities are fairly accurate. However, the results deteriorate as γ increases. Bias increases for both parameters and coverage probability drops. This could possibly be explained by the fact that the additive hazards model and proportional hazards model are similar for certain parameter values. Analysis of the results with censoring, which can be found in Tables 17 and 18, shows a similar outcome. The results are reasonable when γ is equal to zero and got worse as γ increases. Also it can be seen that the variance estimates increase as the censoring percentage increases.

Table 17 Case I additive hazards model for observation times with 20% censoring

γ	β	$\hat{\gamma}$				$\hat{\beta}$			
		Mean	SE	SEE	CP	Mean	SE	SEE	CP
0	0	0.012	0.301	0.303	0.956	0.010	0.552	0.536	0.966
	0.5	0.009	0.321	0.317	0.959	0.543	0.641	0.629	0.968
	1	0.020	0.356	0.335	0.944	1.131	0.839	0.765	0.953
0.2	0	0.162	0.318	0.303	0.950	–0.177	0.600	0.585	0.952
	0.5	0.099	0.336	0.313	0.922	0.420	0.684	0.638	0.940
	1	0.106	0.325	0.327	0.949	0.977	0.785	0.751	0.939
0.5	0	0.320	0.303	0.301	0.917	–0.352	0.659	0.641	0.918
	0.5	0.222	0.312	0.307	0.852	0.197	0.674	0.643	0.925
	1	0.259	0.332	0.321	0.881	0.822	0.784	0.749	0.933

Table 18 Case I additive hazards model for observation times with 40% censoring

γ	β	$\hat{\gamma}$				$\hat{\beta}$			
		Mean	SE	SEE	CP	Mean	SE	SEE	CP
0	0	–0.010	0.372	0.363	0.945	–0.023	1.065	1.042	0.952
	0.5	0.019	0.378	0.375	0.952	0.556	1.173	1.127	0.951
	1	0.003	0.406	0.385	0.945	1.015	1.306	1.213	0.943
0.2	0	0.149	0.396	0.364	0.929	–0.119	1.195	1.120	0.945
	0.5	0.089	0.393	0.370	0.928	0.360	1.228	1.150	0.940
	1	0.127	0.393	0.381	0.944	1.022	1.314	1.238	0.938
0.5	0	0.315	0.356	0.364	0.929	–0.344	1.269	1.230	0.939
	0.5	0.233	0.377	0.368	0.881	0.200	1.261	1.207	0.947
	1	0.268	0.378	0.376	0.906	0.851	1.350	1.290	0.951

3.1.5 Efficiency Comparison

It is also interesting to compare the efficiency of the method of Zhang et al. (2005) with one designed for independent censoring when the censoring is in fact non-informative. This could help a researcher determine which approach to use when it is either known, or assumed, that the censoring is non-informative. In order to accomplish this, another simulation study was conducted under the setup described in Lin et al. (1998). This method was chosen for comparison because the model specification is the same as that in Zhang et al. (2005) except for the frailty term. Specifically, the failure times and observation times were generated according to exponential distributions with the following hazards

$$\lambda_i(t|Z_i(s), b_i(s), s \leq t) = \lambda_1(t) + \beta' Z_i(t) , \tag{10}$$

Table 19 Case I efficiency comparison with method of Lin (1998)

	n = 100			n = 200		
	$\lambda_{c,0} = 0.5$	1.0	1.5	$\lambda_{c,0} = 0.5$	1.0	1.5
Bias	0.04	0.03	0.02	0.02	0.02	0.01
SE	0.38	0.42	0.50	0.25	0.29	0.33
SEE	0.38	0.41	0.49	0.25	0.28	0.33
95% CP	0.95	0.96	0.95	0.95	0.95	0.95
Bias	0.07	0.05	0.01	0.04	0.02	0.01
SE	0.65	0.74	0.92	0.42	0.49	0.60
SEE	0.61	0.72	0.87	0.40	0.48	0.58
95% CP	0.96	0.96	0.95	0.95	0.95	0.95

and

$$\lambda_i^c(t|Z_i(s), b_i(s), s \leq t) = \lambda_2(t) \exp(\gamma' Z_i(t)). \tag{11}$$

This represents the case where the censoring is independent. The baseline hazard function for the failure times was set to be $\lambda_1(t) = 1$, and the baseline hazard for the censoring times took the values $\lambda_2(t) = 0.5, 1.0,$ and 1.5. The true value of β was taken to be 0.5 and the covariate was generated from a uniform distribution over $(0, \sqrt{12})$. Samples sizes of $n = 100$ and $n = 200$ were investigated. 10,000 iterations were used for each combination of parameters.

The results for this simulation experiment can be found in Table 19. The top half of the table shows the original results from Lin et al. (1998). The bottom portion shows the outcomes using the method of Zhang et al. (2005). Bias and coverage probabilities are very close for both methods. However, the method of Lin et al. (1998) does have smaller SE and SEE. This indicates that it might be preferred to use their method when one believes the censoring is truly non-informative. This result makes sense since the approach of Lin et al. (1998) was developed for this type of censoring.

3.2 Case II

As was the case with current status data, many of the simulations for case II interval-censored data are similar. For simplicity, the baseline hazard functions $\lambda_{t0}(t)$, $\lambda_{u0}(t)$ and $\lambda_{w0}(t)$ were set equal to one in each setup. Also the diagonal elements of the covariance matrix were all set equal to 0.04 with the off-diagonal elements being 0.03. This produces correlation coefficients of 0.75 among b_{1i}, b_{2i} and b_{3i}. Covariates were generated for both the continuous and discrete case. The Z_i's were generated either from a uniform distribution over $[-1, 1]$ or a Bernoulli distribution with success

Table 20 Case II Informative censoring

Parameter	Continuous covariate				Discrete covariate			
	Bias	SE	SEE	CP	Bias	SE	SEE	CP
β_t	–0.0224	0.1846	0.1856	0.955	0.0210	0.1485	0.1473	0.948
β_u	–0.0266	0.1454	0.1401	0.939	–0.0185	0.1620	0.1570	0.941
β_w	–0.0251	0.1463	0.1400	0.937	–0.0133	0.1586	0.1570	0.948

probability of 0.5. Each simulation used a sample size of $n = 200$ with 1000 iterations unless otherwise noted.

Again, results are summarized in tables using several different statistics. The bias in estimating β_t, β_u and β_w is calculated by taking the mean of the parameter estimates minus the true value. Each table also shows the sample standard deviation of the point estimates (SE) as well as the mean of the estimated standard errors (SEE). Finally, 95% empirical coverage probabilities are calculated.

3.2.1 Informative Censoring

The first setup was a confirmatory simulation with dependent censoring. These simulations verify the original results and serve as a point of comparison for all the new cases. Here the data were generated according to the assumptions of the paper. Survival times were generated from an exponential distribution with hazard function given by (4). The first observation time and gap time were also generated from exponential distributions with hazards (5) and (6), respectively. In accordance with the original paper, we set $\beta_t = \beta_u = \beta_w = 1$.

The dependent censoring results for both a continuous covariate and a discrete covariate are in Table 20. Both results look very good. The bias is small for all three parameters. The variance estimates are all similar, and each coverage probability is around the specified 95%. Also, all of these outcomes closely match the corresponding results found in Zhang et al. (2007).

3.2.2 Non-informative Censoring

The next simulations investigated the performance when the censoring is independent. This setup can be obtained from the proposed model by setting the latent random effects equal to zero for each subject, i.e. $b_{1i} = b_{2i} = b_{3i} = 0$ for all i. Next, the survival time, first observation time, and gap time were generated from exponential distributions with hazards defined according to Eqs. (4), (5) and (6), respectively. Also, the true parameter values used here are $\beta_t = \beta_u = \beta_w = 1$.

Table 21 shows the outcomes for these simulations. The results seem to indicate that the proposed method does well for the case of independent censoring. Bias for all

Table 21 Case II non-informative censoring

Parameter	Continuous covariate				Discrete covariate			
	Bias	SE	SEE	CP	Bias	SE	SEE	CP
β_t	−0.0044	0.1836	0.1858	0.947	0.0238	0.1330	0.1463	0.965
β_u	0.0059	0.1430	0.1410	0.950	0.0105	0.1648	0.1583	0.940
β_w	0.0075	0.1442	0.1411	0.947	0.0216	0.1522	0.1585	0.957

three parameters is reasonably small for both the continuous and discrete covariate. The SE and SEE are always quite close, implying that the variance estimation is doing a good job, and the 95% coverage probabilities are also fairly close to the desired values. Moreover, these results are comparable to the dependent censoring setup in Table 20.

3.2.3 Model Misspecification

The next simulation studies examined cases where the hazard functions were misspecified. The proposed method assumes that all three hazard functions follow the proportional hazards model, and we were interested in evaluating the effectiveness of this approach when these assumptions are violated. To achieve this, the proportional hazards model was replaced with the additive hazards model in various combinations.

The first setup was a worst-case scenario where all three hazard functions were misspecified. That is, the hazard functions for T_i, U_i and W_i were defined as:

$$\lambda_i^{(T)}(t|Z_i, b_i) = \lambda_{t0}(t) + \beta_t^{'} Z_i + b_{1i} , \tag{12}$$

$$\lambda_i^{(U)}(t|Z_i, b_i) = \lambda_{u0}(t) + \beta_u^{'} Z_i + b_{2i} , \tag{13}$$

$$\lambda_i^{(W)}(t|Z_i, b_i) = \lambda_{w0}(t) + \beta_w^{'} Z_i + b_{3i} , \tag{14}$$

where once again β_t, β_u and β_w are $p \times 1$ vectors of regression parameters, and $\lambda_{t0}(t)$, $\lambda_{u0}(t)$ and $\lambda_{w0}(t)$ are unknown baseline hazard functions.

The values for T_i, U_i and W_i were again generated from exponential distributions, but the hazard functions were specified using Eqs. (12), (13) and (14). The true values for β_u and β_w were always equal to 1, and β_t took the values 0, 0.5, and 1.

The results for these simulations with $\beta_t = 1$ can be found in Table 22. It is clear from these simulations that the method performed poorly under these conditions. The bias is very large for all three parameters with both the continuous and discrete covariates, and the coverage probabilities are terrible. The results are similar when β_t is 0 and 0.5. The only exception is that the results for β_t improve as β_t decreases for the uniform covariate. This could possibly be explained by the fact that the proportional

Table 22 Case II additive hazard model for T, U, and W with $\beta_t = 1$

Parameter	Continuous covariate				Discrete covariate			
	Bias	SE	SEE	CP	Bias	SE	SEE	CP
β_t	–0.2424	0.1766	0.1866	0.738	–0.2796	0.1441	0.1463	0.496
β_u	–0.2916	0.2563	0.2514	0.795	–0.2817	0.1492	0.1510	0.510
β_w	–0.2652	0.2636	0.2518	0.795	–0.2651	0.1512	0.1515	0.549

hazards model and additive hazards model behave alike under these conditions. The full table with results for all parameter values are in Table 23.

Next, we considered the situation where only the hazards for U and W were misspecified. The survival times were generated from an exponential distribution using the correct proportional hazards model given by Eq. (4). The values for U and W are also from an exponential distribution, but the hazards are defined using the additive hazards model and Eqs. (13) and (14), respectively. Once again the true values for β_u and β_w are set equal to 1, and β_t takes the values 0, 0.5, and 1.

Table 24 summarizes the outcomes of these simulations with $\beta_t = 1$. There are still serious problems with the estimates for β_u and β_w in terms of bias and coverage probability. This can be expected since U and W were generated with incorrect hazard functions. However, the results are interesting because the estimation for β_t is quite good for both cases. The bias is small, the variance estimates are close, and the coverage probabilities are right around 95%. Results are nearly identical when β_t is equal to 0 and 0.5, and once again these additional results can be found in Table 25. This is promising since the primary goal is to estimate the covariate effect on the survival times. However, more research should go into explaining this curious outcome. It is possible that the variances and covariances of the random effects are small enough that they behave as if they are independent.

The last setup for these simulations tested the case where only the survival time hazard function is misspecified. Therefore, T_i was generated from an exponential distribution with the hazard function given by Eq. (12), while U_i and W_i were generated from exponential distributions using the hazards defined in (5) and (6). We investigated an extensive range of possible values for β_t, β_u, and β_w for both covariate types.

The results for the continuous covariate are displayed in Table 26. Table 27 has the results when Z_i follows a Bernoulli distribution. Since the influence of the covariates on T_i is the main focus of the analysis, we only present the information for β_t in order to reduce potential confusion, and make the table easier to digest. Here we can see that these simulations had a mixed outcome. It seems that the estimation procedure is performing well only for certain parameter combinations. This could possibly be explained by the fact that the proportional hazards model and additive hazards model are similar for some values. Nevertheless, the results become significantly worse as β_t gets larger. Also, the omitted data for β_u and β_w indicate that the model performs

Table 23 Case II full results for additive hazard model for T, U, and W

β_t	Continuous covariate											
	$\hat{\beta}_t$				$\hat{\beta}_u$				$\hat{\beta}_w$			
	Bias	SE	SEE	CP	Bias	SE	SEE	CP	Bias	SE	SEE	CP
0	−0.0118	0.1860	0.1906	0.957	−0.2746	0.2624	0.2518	0.792	−0.2723	0.2627	0.2523	0.793
0.5	−0.0784	0.1685	0.1855	0.953	−0.2967	0.2648	0.2521	0.751	−0.2894	0.2585	0.2523	0.783
1	−0.2424	0.1766	0.1866	0.738	−0.2916	0.2563	0.2514	0.795	−0.2652	0.2636	0.2518	0.795

β_t	Discrete covariate											
	$\hat{\beta}_t$				$\hat{\beta}_u$				$\hat{\beta}_w$			
	Bias	SE	SEE	CP	Bias	SE	SEE	CP	Bias	SE	SEE	CP
0	0.7122	0.1473	0.1469	0.001	−0.2712	0.1568	0.1514	0.540	−0.2767	0.1563	0.1511	0.555
0.5	0.2105	0.1448	0.1459	0.708	−0.2619	0.1522	0.1513	0.582	−0.2578	0.1538	0.1516	0.590
1	−0.2796	0.1441	0.1463	0.496	−0.2817	0.1492	0.1510	0.510	−0.2651	0.1512	0.1515	0.549

Table 24 Case II additive hazard model for U, and W with $\beta_t = 1$

Parameter	Continuous covariate				Discrete covariate			
	Bias	SE	SEE	CP	Bias	SE	SEE	CP
β_t	0.0118	0.1801	0.1901	0.962	0.0115	0.1485	0.1511	0.953
β_u	–0.2786	0.2667	0.2519	0.777	–0.2704	0.1528	0.1513	0.563
β_w	–0.2701	0.2468	0.2520	0.814	–0.2599	0.1524	0.1515	0.590

well in estimating these parameters. This conclusion is not too surprising since U and W were properly specified with the proportional hazards model.

The previous results motivated one final model misspecification simulation. The setup is nearly identical except the sample size is increased to $n = 400$, and the only parameter combinations examined were those that exhibited large bias in the preceding simulation. A summary of this simulation study can be found in Table 28. Increasing the sample size unfortunately did not lead to a reduction in bias. The variance estimates did get smaller but this resulted in the coverage probabilities becoming far worse.

4 Conclusions and Discussion

This chapter investigated the behavior of the methods developed for regression analysis of informative interval-censored survival data under circumstances beyond their original intended scopes. Extensive Monte-Carlo simulation studies were conducted in order to answer many questions about the flexibility of these methods. Of particular interest was how reliable these models are when the censoring is actually non-informative.

Several interesting outcomes were observed. Most importantly, for both case I and case II interval-censored data, these methods performed quite well with non-informative censoring. This suggests that these techniques can be appropriate or reasonable choices when the true nature of the censoring is unknown, provided the other assumptions hold.

The first results with non-informative censoring were positive, but overall the results were mixed. The procedures were not malleable enough to handle a number of other situations that violated their assumptions. There were issues discovered with the methods for both case I and case II interval-censored data.

For current status data, problems arose when there was a bivariate random effect and a misspecified hazard function for the observation times. In these cases, the model performed poorly both in terms of bias and empirical coverage probabilities. Consequently, a researcher would be wise to check the proportional hazards assumption on the observation times. Fortunately, the observation times are either exactly known

Table 25 Case II full results for additive hazard model for U, and W

β_t	Continuous covariate											
	$\hat{\beta}_t$				$\hat{\beta}_u$				$\hat{\beta}_w$			
	Bias	SE	SEE	CP	Bias	SE	SEE	CP	Bias	SE	SEE	CP
0	−0.0095	0.1829	0.1897	0.957	−0.2953	0.2592	0.2520	0.771	−0.2857	0.2736	0.2517	0.764
0.5	0.0248	0.1766	0.1863	0.965	−0.3123	0.2606	0.2517	0.741	−0.2864	0.2638	0.2521	0.785
1	0.0118	0.1801	0.1901	0.962	−0.2786	0.2667	0.2519	0.777	−0.2701	0.2468	0.2520	0.814
β_t	Discrete covariate											
	$\hat{\beta}_t$				$\hat{\beta}_u$				$\hat{\beta}_w$			
	Bias	SE	SEE	CP	Bias	SE	SEE	CP	Bias	SE	SEE	CP
0	−0.0026	0.1507	0.1469	0.950	−0.2731	0.1589	0.1514	0.544	−0.2578	0.1542	0.1511	0.608
0.5	0.0073	0.1414	0.1445	0.961	−0.2682	0.1561	0.1515	0.565	−0.2625	0.1594	0.1514	0.573
1	0.0115	0.1485	0.1511	0.953	−0.2704	0.1528	0.1513	0.563	−0.2599	0.1524	0.1515	0.590

Table 26 Case II additive hazard model for T with continuous covariate

β_t	β_u	β_w	Bias $\hat{\beta}_t$	SE	SEE	CP
0	0	0	–0.0115	0.1847	0.1934	0.962
		0.5	–0.0254	0.1857	0.1911	0.959
		1	–0.0143	0.1844	0.1914	0.957
	0.5	0	–0.0065	0.1851	0.1922	0.958
		0.5	–0.0264	0.1877	0.1899	0.953
		1	–0.0230	0.1837	0.1905	0.956
	1	0	–0.0249	0.1877	0.1951	0.964
		0.5	–0.0288	0.1848	0.1946	0.961
		1	–0.0139	0.1946	0.1954	0.952
0.5	0	0	–0.0728	0.1816	0.1974	0.960
		0.5	–0.0715	0.1845	0.1944	0.944
		1	–0.0769	0.1811	0.1923	0.943
	0.5	0	–0.0730	0.1843	0.1920	0.955
		0.5	–0.0746	0.1866	0.1879	0.934
		1	–0.0764	0.1789	0.1871	0.942
	1	0	–0.0798	0.1772	0.1929	0.949
		0.5	–0.0699	0.1794	0.1880	0.946
		1	–0.0731	0.1785	0.1876	0.944
1	0	0	–0.2412	0.1895	0.2061	0.793
		0.5	–0.2415	0.2078	0.2004	0.733
		1	–0.2367	0.1960	0.1985	0.763
	0.5	0	–0.2464	0.1921	0.1958	0.757
		0.5	–0.2436	0.1823	0.1911	0.754
		1	–0.2365	0.1887	0.1887	0.758
	1	0	–0.2404	0.1833	0.1933	0.767
		0.5	–0.2385	0.1763	0.1881	0.766
		1	–0.2461	0.1868	0.1848	0.734

or right-censored, and there exist many well-known techniques for model-checking with this type of data.

The procedure for case II interval-censored data also had issues with a misspecified hazard function, as might be expected. Here, generating data using the additive hazards model caused bias in the estimation of the corresponding parameters. Interestingly, the results were not all negative. Bias only seemed to exist for the parameters associated with the variables that had improperly specified hazard functions. This is somewhat surprising, and explaining why this happens remains an open question that could warrant further investigation.

Table 27 Case II additive hazard model for T with discrete covariate

β_t	β_u	β_w	Bias $\hat{\beta}_t$	SE	SEE	CP
0	0	0	–0.0095	0.1442	0.1478	0.957
		0.5	–0.0129	0.1526	0.1474	0.948
		1	–0.0105	0.1473	0.1494	0.965
	0.5	0	–0.0186	0.1479	0.1471	0.948
		0.5	–0.0192	0.1476	0.1475	0.952
		1	–0.0182	0.1497	0.1510	0.955
	1	0	–0.0219	0.1514	0.1516	0.957
		0.5	–0.0212	0.1546	0.1533	0.951
		1	–0.0340	0.1615	0.1601	0.946
0.5	0	0	–0.0789	0.1508	0.1522	0.915
		0.5	–0.0872	0.1502	0.1498	0.915
		1	–0.0840	0.1512	0.1507	0.913
	0.5	0	–0.0798	0.1474	0.1471	0.920
		0.5	–0.0934	0.1447	0.1453	0.900
		1	–0.0828	0.1456	0.1464	0.928
	1	0	–0.0871	0.1463	0.1481	0.924
		0.5	–0.0833	0.1468	0.1467	0.913
		1	–0.0962	0.1437	0.1495	0.919
1	0	0	–0.2665	0.1604	0.1614	0.599
		0.5	–0.2587	0.1577	0.1586	0.597
		1	–0.2757	0.1613	0.1572	0.578
	0.5	0	–0.2750	0.1538	0.1531	0.537
		0.5	–0.2751	0.1480	0.1489	0.525
		1	–0.2707	0.1609	0.1486	0.532
	1	0	–0.2808	0.1535	0.1508	0.516
		0.5	–0.2849	0.1465	0.1467	0.492
		1	–0.2938	0.1463	0.1461	0.460

Table 28 Case II additive hazard model for T with continuous covariate, $n = 400$

β_t	β_u	β_w	Bias $\hat{\beta}_t$	SE	SEE	CP
1	0	0	–0.2444	0.1321	0.1402	0.590
		0.5	–0.2333	0.1285	0.1363	0.595
		1	–0.2278	0.1300	0.1353	0.613
	0.5	0	–0.2416	0.1226	0.1337	0.548
		0.5	–0.2421	0.1229	0.1306	0.528
		1	–02384	0.1172	0.1289	0.540
	1	0	–0.2502	0.1202	0.1315	0.513
		0.5	–0.2529	0.1200	0.1284	0.474
		1	–0.2426	0.1249	0.1268	0.508

It is important to understand the circumstances under which a model performs as expected, and also beneficial to have an idea when a model might not be reliable. This knowledge is useful for a researcher working on a real-world data analysis problem, and can also motivate the formulation of new methodologies. Unfortunately, it is not feasible to address every possible situation one might encounter. Therefore, the work in this chapter could still be taken in a number of different directions to answer other questions that still exist. For example, here we only examined one method for each data type, and other techniques have been proposed in the literature to analyze interval-censored data. Also, we focused on the methods that utilize the proportional hazards assumption. These same unknowns could be investigated for procedures assuming a different modeling framework for the hazard functions. Regardless of what comes next, it is certain that Monte-Carlo simulations studies can play a key role in finding answers to these questions.

Finally, it is worth noting that in this chapter, we only touched a few methods developed in the literature for regression analysis of interval-censored failure time data with informative censoring, and more recently several new methods have been proposed for the same topic. For example, Ma et al. (2015, 2016) developed some sieve maximum likelihood approaches with the use of copula models for cases I and II interval-censored data, respectively, arising from the proportional hazards model. In contrast, Zhao et al. (2015a, b) gave some similar estimation procedures for the data arising from the additive hazards model. Furthermore, Wang et al. (2016) discussed regression analysis of case K interval-censored failure time data in the presence of informative interval censoring, and Liu et al. (2016) investigated the same problem but also with the presence of a cured subgroup.

References

Anderen, P. K., & Ronn, B. B. (1995). A nonparametric test for comparing two samples where all observations are either left-or right-censored. *Biometrics,* 323–329.

Betensky, R. A., & Finkelstein, D. (2002). Testing for dependence between failure time and visit compliance with interval-censored data. *Biometrics*, *58*, 58–63.

Betensky, R. A., Rabinowitz, D., & Tsiatis, A. (2001). Computationally simple accelerated failure time regression for interval-censored data. *Biometrika*, *88*, 703–711.

Chen, D., Sun, J., & Peace, K. (Eds.). (2012). *Interval-censored time-to-event data: methods and applications*. Boca Rotan: Chapman & Hall/CRC.

Dewanji, A., & Kalbfleisch, J. D. (1986). Nonparametric methods for survival/sacrifice experiments. *Biometrics* 325–341.

Finkelstein, D. (1986). A proportional hazards model for interval-censored failure time data. *Biometrics* 845–854.

Finkelstein, D., Goggins, W. B., & Schoenfeld, D. A. (2002). Analysis of failure time data with dependent interval censoring. *Biometrics*, *58*, 298–304.

Groeneboom, P., & Wellner, J. A. (1992). *Information bounds and nonparametric maximum likelihood estimation* (Vol. 19). Basel: Springer.

Huang, J. (1996). Efficient estimation for the proportional hazards model with interval censoring. *Annals of Statistics*, *24*, 540–568.

Huang, J., & Wellner, J. A. (1997). Interval-censored survival data: A review of recent progress. In D. Y. Lin & T. R. Fleming (Eds.), *Proceedings of the First Seattle Symposium in Biostatistics: Survival Analysis* (pp. 123–169). New York: Springer.

Huang, X., & Wolfe, R. A. (2002). A frailty model for informative censoring. *Biometrics*, *58*, 510–520.

Lagakos, S. W., & Louis, T. A. (1988). Use of tumour lethality to interpret tumorigenicity experiments lacking cause-of-death data. *Applied Statistics* 169–179.

Lin, D. Y., Oakes, D., & Ying, Z. (1998). Additive hazards regression with current status data. *Biometrika*, *85*, 289–298.

Liu, Y., Hu, T., & Sun, J. (2016) Regression analysis of current status data in the presence of cured subgroup and dependent censoring. *Lifetime Data Analysis* (in press)

Ma, L., Hu, T., & Sun, J. (2015). Sieve maximum likelihood regression analysis of dependent current status data. *Biometrika*, *102*, 731–738.

Ma, L., Hu, T., & Sun, J. (2016). Cox regression analysis of dependent interval-censored failure time data. *Computational Statistics and Data Analysis*, *103*, 79–90.

Peto, R. (1973). Experimental survival curves for interval-censored data. *Applied Statistics* 86–91.

Rossini, A. J., & Tsiatis, A. A. (1996). A semiparametric proportional odds regression model for the analysis of current status data. *Journal of the American Statistical Association*, *91*, 713–721.

Sun, J. (1999). A nonparametric test for current status data with unequal censoring. *Journal of the Royal Statistical Society: Series B (Statistical Methodology)*, *61*, 243–250.

Sun, J. (2006). *The statistical analysis of interval-censored failure time data*. New York: Springer.

Turnbull, B. (1976). The empirical distribution function with arbitrarily grouped, censored and truncated data. *Journal of the Royal Statistical Society: Series B (Statistical Methodology)* 290–295.

Wang, P., Zhao, H., & Sun, J. (2016). Regression analysis of case K interval-censored failure time data in the presence of informative censoring. Biometrics (in press).

Wang, Y., & Taylor, J. (2001). Jointly modeling longitudinal and event time data with application to acquired immunodeficiency syndrome. *Journal of the American Statistical Association*, *96*, 895–905.

Zhang, Z., Sun, J., & Sun, L. (2005). Statistical analysis of current status data with informative observation times. *Statistics in Medicine*, *24*, 1399–1407.

Zhang, Z., Sun, L., Sun, J., & Finkelstein, D. (2007). Regression analysis of failure time data with informative interval censoring. *Statistics in Medicine*, *26*, 2533–2546.

Zhao, S., Hu, T., Ma, L., Wang, P., & Sun, J. (2015). Regression analysis of informative current status data with the additive hazards model. *Lifetime Data Analysis*, *21*, 241–258.

Zhao, S., Hu, T., Ma, L., Wang, P., & Sun, J. (2015). regression analysis of interval-censored failure time data with the additive hazards model in the presence of informative censoring. *Statistics and Its Interface*, *8*, 367–377.

Robust Bayesian Hierarchical Model Using Monte-Carlo Simulation

Geng Chen and Sheng Luo

Abstract Many clinical trials collect information on multiple longitudinal outcomes. Depending on the nature of the disease and its symptoms, the longitudinal outcomes can be of mixed types, e.g., binary, ordinal and continuous. Clinical studies on Parkinson's disease (PD) are good examples of this case. Due to the multidimensional nature of PD, it is difficult to identify a single outcome to represent the overall disease status and severity. Thus, clinical studies that search for treatments for PD usually collect multiple outcomes at different visits. In this chapter, we will introduce the multilevel item response theory (MLIRT) models that account for all the information from multiple longitudinal outcomes and provide valid inference for the overall treatment effects. We will also introduce the normal/independent (NI) distributions, which can be easily implemented into the MLIRT model hierarchically, to handle the outlier and heavy tails problems to produce robust inference. Other data features such as dependent censoring and skewness will also be discussed under the MLIRT framework.

Keywords Clinical trial · Item-response theory · Latent variable · MCMC · Outliers · Joint model · Robust distribution · Multivariate longitudinal data

Many clinical trials collect information on multiple longitudinal outcomes. Depending on the nature of the disease and its symptoms, the longitudinal outcomes can be of mixed types, e.g., binary, ordinal and continuous. In these clinical studies, the multiple outcomes could be used as co-primary endpoints or a mixture of primary and secondary endpoints. The methods that handles the multiple outcomes may vary depending on the studies and nature of the diseases. Researchers may analyze the outcomes individually, combine them using certain rules or algorithms as a composite outcome, or analyze them simultaneously and draw inferences on the overall

G. Chen (✉)
Clinical Statistics, GlaxoSmithKline, 1250 South Collegeville Road,
Collegeville, PA 19426, USA
e-mail: geng.3.chen@gsk.com; geng.chen@live.com

S. Luo
Department of Biostatistics, The University of Texas Health Science Center at Houston,
1200 Pressler St., Houston, TX, USA
e-mail: sheng.t.luo@uth.tmc.edu

D.-G. Chen and J.D. Chen (eds.), *Monte-Carlo Simulation-Based Statistical Modeling*, ICSA Book Series in Statistics, DOI 10.1007/978-981-10-3307-0_16

treatment effects. Each method have their own advantages and disadvantages. While analysing multiple outcomes individually could be a straightforward approach, problems may appear when the conclusions from individual outcomes do not agree with each other, especially when the outcomes serve as co-primary endpoints. Combining outcomes together as a composite outcome may fixed this issue, however, how to combine the outcomes and the interpretation of the composite outcome could be potential debating points. In this chapter, we will introduce a statistical method that handles multivariate longitudinal outcomes and draw inference on the overall treatment effects using latent variables. First, we will set the scene using Parkinson's disease as an example. Then we will introduce our method, the multilevel item response theory (MLIRT) model, and explain how it works for the multivariate longitudinal data of mixed types. In addition, outliers exist in almost every clinical trial data and sometimes we cannot delete them just because they are influential points. In the following sections, we will introduce the normal/independent (NI) distribution family and demonstrate how the NI distributions are incorporated into the MLIRT model and how they tackle the outliers issues. Bayesian inference and some model selection criteria will be summarized. Brief simulation and data analysis results will be discussed. Other data features, such as dependent censoring and skewness, will also be discussed in the extended modeling section. Now let's start with the contexts that motivated the method development.

1 Parkinson's Disease as an Example

Clinical trials studies are often conducted by pharmaceutical companies and research institutes with the goal of finding treatments for certain diseases. Statistics play key roles in these research activities, from design, monitoring, interim decision making, to the final analysis and reporting. Normally, a clinical trial study will enroll multiple patients (or subjects), and patients will receive randomized treatments and will be followed for a period of time. Longitudinal information will be collected over time at each visit. Depending on the nature of the disease and its symptoms, many clinical trial studies collect information on multiple longitudinal outcomes. In addition, the multiple longitudinal outcomes can be of mixed types, e.g., binary, ordinal and continuous. Parkinson's disease (PD), for example, is a good representation of this case.

Parkinsons disease (PD) is a chronic neurodegenerative disease with multidimensional impairments. It is multidimensional because the symptoms of PD are manifested in many ways, e.g. tremors, stiffness, slowness of movements, and loss of cognitive functions (Cummings 1992; Fahn et al. 2004). Measurements such as Quality of life (QoL), Unified Parkinsons Disease Rating Scale (UPDRS) total score, Hoehn and Yahr scale (HY), and Schwab and England activities of daily living (SEADL) etc. are often used. QoL measures patients' activities of daily life and motor and non-motor syndrome (Luo et al. 2013). The UPDRS total score evaluates patients' mentation, behavior, and activities of daily life (Bushnell and Martin 1999). HY measures the disability level in daily activities (Müller et al. 2000). SEADL assesses patients' daily activities and functional impairment (McRae et al. 2000). Each outcome measurement evaluate some aspects of PD but it is difficult to identify a single measure to represent the overall disease severity (Huang et al. 2005). As a result, clinical trial studies that search for treatment to slow down the progression of PD symptoms usually collect information on multiple outcomes across visits. For example, Deprenyl and tocopherol antioxidative therapy of parkinsonism (DATATOP) study (Parkinson Study Group 1989), Tolvaptan Efficacy and Safety in Management of Autosomal Dominant Polycystic Kidney Disease and its Outcomes (TEMPO) study (Parkinson Study Group 2002), Earlier versus Later Levodopa Therapy in Parkinson Disease (ELLDOPA) study (Fahn et al. 2004) and Neuroprotection Exploratory Trials in Parkinson's Disease (NET-PD) study (Elm and The NINDS NET-PD Investigators 2012).

The multiple outcome measures collected at different visits lead to a multivariate longitudinal data structure, which contains three sources of correlations within and between outcomes for a patient: (i) different outcome measures at the same visit (intersource), (ii) same outcome measures at different visits (longitudinal) and (iii) different outcome measures at different visits (cross correlation). Analysis approaches that aim to provide inference on the overall progression of the PD need to take account for all three sources of correlations.

Many approaches have been developed to analyze multivariate longitudinal data in clinical trials. In this chapter, we are not trying elaborate each approaches nor to discuss the advantage and disadvantage of the existing methods, but rather we introduce a Monte Carlo simulation based Bayesian hierarchical model which models the overall PD progression by using a latent variable.

2 MLIRT Model

The multilevel item response theory (MLIRT) models have been increasingly used in many diseases such as PD disability, Alzheimer's disease, Huntington's disease, and dementia, to to analyze multivariate longitudinal data (Weisscher et al. 2010; Luo et al. 2012; Snitz et al. 2012; Vaccarino et al. 2011; Miller et al. 2012). The model consists two levels and the two levels of models are linked via latent variables. The first level model describes the relationship between the outcome measures and the latent variable, while the second level model describes the relationship between the latent variable and the covariate of interest. If we use PD as an example, the outcome measures in the first level model would be QoL, UDPRS etc., the latent variable would be the unobserved overall disease severity, and the covariate of interest in the second level model could be the treatment assignment as well as disease duration and time.

Now let's define the model in statistical language. Let y_{ijk} be the observed outcome k $(k = 1, \ldots, K)$ for patient i $(i = 1, \ldots, N)$ at visit j $(j = 1, \ldots, J_i)$, where $j = 1$ is baseline. y_{ijk} can be binary, ordinal and continuous. Let $\boldsymbol{y}_{ij} = (y_{ij1}, \ldots, y_{ijk}, \ldots, y_{ijK})'$ be the vector of observation for patient i at visit j and let $\boldsymbol{y}_i = (\boldsymbol{y}_{i1}, \ldots, \boldsymbol{y}_{iK})'$ be the outcome vector across visits. Let θ_{ij} be the continuous latent variable that denote the unobserved PD disease severity for patient i at visit j with higher value representing more severe disease status. In the first level model, we model the outcome measures as a function of the subject-specific latent variable θ_{ij} and the outcome-specific parameters. Specifically, we model the binary outcomes using a two-parameter sub-model (Fox 2010), the cumulative probabilities of ordinal outcomes using a graded response sub-model (Samejima 1997), and the continues outcomes using a common factor sub-model (Lord et al. 1968).

$$\text{logit}\{\text{p}(y_{ijk} = 1|\theta_{ij})\} = a_k + b_k\theta_{ij}, \tag{1}$$
$$\text{logit}\{\text{p}(y_{ijk} \leq l|\theta_{ij})\} = a_{kl} - b_k\theta_{ij}, \text{ with } l = 1, 2, \ldots, n_k - 1, \tag{2}$$
$$y_{ijk} = a_k + b_k\theta_{ij} + \epsilon_{ijk}, \tag{3}$$

where a_k and b_k are the outcome-specific parameters. a_k is the "difficulty" parameter and b_k is the "discriminating" parameter that is always positive and describes the ability that outcome k discriminates between patients with latent disease severity θ_{ij}. Moreover, for ordinal outcome in model (2), suppose outcome k has n_k categories and $n_k - 1$ thresholds $a_{k1}, \ldots, a_{kl}, \ldots, a_{kn_{k-1}}$ that satisfy the order constraint $a_{k1} < \cdots < a_{kl} < \cdots < a_{kn_{k-1}}$. The probability that patient i being in category l on outcome k at visit j is $\text{p}(y_{ijk} = l|\theta_{ij}) = \text{p}(y_{ijk} \leq l|\theta_{ij}) - \text{p}(y_{ijk} \leq l - 1|\theta_{ij})$. For continuous outcome in model (3), the random error $\epsilon_{ijk} \sim \text{N}(0, \sigma_k^2)$ with σ_k^2 denote variance of continuous outcome k. The first level model provides great flexibility of handling the three types of outcomes. But it is not necessary to have all three types of outcomes in your study to use the model.

In the second level model, the latent variable θ_{ij}, which represent the overall PD disease severity, is regressed on predictors of interest (e.g., treatment, disease duration, and time) and subject-specific random effects.

$$\theta_{ij} = \boldsymbol{X}_{i0}\boldsymbol{\beta}_0 + u_{i0} + (\boldsymbol{X}_{i1}\boldsymbol{\beta}_1 + u_{i1})t_{ij}, \tag{4}$$

where $\boldsymbol{X}_{i0}$ and $\boldsymbol{X}_{i1}$ are the design matrix that contain the covariate of interests, they may share all or part of covariates depending on the content and purpose of the studies. u_{i0} and u_{i1} are the subject-specific random intercept and random slope, respectively. In the context of PD, u_{i0} is the random intercept which determines the subject-specific PD disease severity and u_{i1} is the random slope which determines the subject-specific PD disease progression rate.

To better understand second level model (4), we use a simple example to further explain the model. Suppose there is no covariate in $\boldsymbol{X}_{i0}$ and only one variable, treatment assignment, is included in $\boldsymbol{X}_{i1}$, then model (4) simplifies to

$$\theta_{ij} = u_{i0} + [\beta_{10} + \beta_{11}I_i(\text{trt}) + u_{i1}]t_{ij}, \tag{5}$$

where $I(\cdot)$ is an indicator function (1 if treatment and 0 otherwise). In the simplified second level model (5), β_{10} and $\beta_{10} + \beta_{11}$ denote the PD disease progression rates for placebo and treatment patients, respectively. And β_{11} represent the change in disease progression rate due to treatment. If we are testing the null hypothesis of no overall treatment effect, that is $H_0 : \beta_{11} = 0$, the significant negative coefficient β_{11} would indicate that the treatment slows down the disease progression.

It is well-known that the item-response models are over-parameterized (Samejima 1997) and some constraints need to be imposed to make the models identifiable. One way we may handle this issue is to set $\mathrm{Var}(u_{i0}) = 1$ to ensure model identifiability. Other approaches, such as setting the $a_{kl} = 0$, was also explored (Luo and Wang 2014). Let the random effects vector $\boldsymbol{u}_i = (u_{i0}, u_{i1})'$ and assume $u_{i0} \sim \mathrm{N}(0, 1)$, $u_{i1} \sim \mathrm{N}(0, \sigma_u^2)$, and $\mathrm{corr}(u_{i0}, u_{i1}) = \rho$. Under the local independence assumption (i.e., conditional on the random effects vector $\boldsymbol{u}_i$, all outcome measures for each patient are independent) (Fox 2010), the full likelihood of patient i across all visits is

$$\mathrm{L}(\boldsymbol{y}_i, \boldsymbol{u}_i) = \left[\prod_{j=1}^{J_i} \prod_{k=1}^{K} \mathrm{p}(y_{ijk} | \boldsymbol{u}_i) \right] \mathrm{p}(\boldsymbol{u}_i). \tag{6}$$

To this end, we have introduced the "plain" version of the MLIRT model for the analysis of multivariate longitudinal data and we refer the model as Indep-N model. We called the current model as the "plain" version because we can add more features to it to account for more advanced data structures. In the following Sects. 3 and 7, we will discuss the outliers and dependent censoring as well as skewness and how to incorporate the data features into the MLIRT model under Bayesian framework. Due to the limited space, we will focus on the simulation and analysis results that address the outlier issues as an example of how the work was performed.

3 MLIRT Model with NI Distribution

Normal distributions are usually assumed for the continuous outcomes as well as the random effects in MLIRT model. However, the parameter estimation may be biased due to outliers and heavy tails in the continuous outcomes and/or random effects. Many approaches have been developed to handle outliers and heavy tails. For example, detection and elimination of influential data points and data transformation are two methods that are often used. However, in clinical trial studies, especially for primary efficacy analysis, the intent-to-treat (ITT) principle is often required to follow. Under the ITT principle, the analysis has to include all randomized individuals which exclude the method that eliminating the outliers from the analysis data. Data transformation methods (e.g., log, square-root and Box-Cox) might generate distributions close to normality but the disadvantages are also clear: (1) the interpretation of the results on the transformed scale may be difficult for some studies; (2) the transformation scheme usually is not universal and need to be tailed to each variables and datasets; (3) the outlier issue is further complicated when the random effects are also "departure from normality" (Lachos et al. 2011), and the transformation of the random effects may not be straightforward. In this section, we introduce a robust distribution, normal/independent distribution, to account for the outliers and heavy tailed situation on both continuous outcomes and random effects.

3.1 NI Distribution

The normal/independent (NI) distribution is a family of symmetric distributions with heavier tails. Extensive discussion about the NI distributions can be found in the literatures (Lange and Sinsheimer 1993; Liu 1996; Rosa et al. 2003; Lachos et al. 2011; Luo et al. 2013; Baghfalaki et al. 2013). Since our purpose is to incorporate the NI to the MLIRT model, without emphasizing too much on the densities and properties of NI distribution, here we use the univariate version of NI as an example.

An element of the univariate NI family y is defined as the distribution of random variable

$$y = \mu + e/\sqrt{\omega},$$

where μ is a location vector, e is random error and is normally distributed with mean zero and variance σ^2, ω is a positive weight variable which has density function $\mathrm{p}(\omega|\nu)$ with tuning parameter ν and is independent of random error e. The key of the NI distribution is the wight ω, and ω can be estimated during modeling and can be further used for outlier identification. How the NI distribution works for the outliers is that the impact of the outliers to the overall inference is control by stochastically assigning lower weights ω to the influencing points (Lange and Sinsheimer 1993). If the estimate of ω is close to 0, it indicates that the corresponding observation can be a potential outlier.

Given ω, y follows a normal distribution $\mathrm{N}(\mu,\ \omega^{-1}\sigma^2)$. The marginal density of y is given by $\mathrm{NI}(y|\mu, \sigma^2, \nu) = \int \mathrm{p}(y|\mu, \sigma^2, \omega)\mathrm{p}(\omega|\nu)d\omega$. In addition, when $\omega \to 1$ (or equivalently when $\nu \to \infty$), $\mathrm{NI}(y|\mu, \sigma^2, \nu) \to \mathrm{N}(\mu,\ \sigma^2)$ (Lange and Sinsheimer 1993; Rosa et al. 2003).

Now back to our MLIRT model. We keep the univariate version of the NI and use the continuous outcome y_{ijk} in model (3) as an example. When a univariate NI distribution is applied to model (3), we now have

$$y_{ijk} = a_k + b_k\theta_{ij} + \epsilon'_{ijk},$$

where $\epsilon'_{ijk} = \epsilon_{ijk}/\sqrt{\omega_i}$ with $\epsilon_{ijk} \sim \mathrm{N}(0, \sigma_k^2)$. The weight variable ω_i has density function $\mathrm{p}(\omega_i|\nu)$ with positive tuning parameter ν.

The NI distributions provide a family of symmetric heavy-tailed distributions with different density specification for ω_i:

- ϵ'_{ijk} follows a Student's-t distribution when $\omega_i \sim \mathrm{Gamma}(\nu/2,\ \nu/2)$ with parameter ν being the degree of freedom,
- ϵ'_{ijk} follows a slash distribution when $\omega_i \sim \mathrm{Beta}(\nu, 1)$ with tuning parameter ν,
- ϵ'_{ijk} follows a contaminated normal (CN) distribution when ω_i takes one of the two discrete values with pdf $\mathrm{p}(\omega_i|\nu) = \nu I_{(\omega_i=\gamma)} + (1-\nu)I_{(\omega_i=1)}$, where $0 < \nu \leq 1$ and $0 < \gamma \leq 1$.

The specifications for Student's-t distribution and slash distribution are straightforward. For contaminated normal (CN) distribution, ν can be viewed as the proportion of contamination, in another words, the percentage of outliers deviating from the normal distribution; γ can be viewed as the scale of contamination, that is, how severe the outliers deviate from the normal distribution with smaller γ denoting stronger deviation. When $\omega_i = 1$, then $p(\omega_i|\nu) = 1 - \nu$ and $\epsilon'_{ijk} \sim N(0, \sigma_k^2)$ with probability $1 - \nu$; when $\omega_i = \gamma$, then $p(\omega_i|\nu) = \nu$ and ϵ'_{ijk} is contaminated with probability ν and $\epsilon'_{ijk} \sim N(0, \sigma_k^2/\gamma)$ (Lange and Sinsheimer 1993; Rosa et al. 2003). Essentially, the CN distribution ϵ'_{ijk} follows is a two-component mixture distribution with pdf $p(\epsilon'_{ijk}) = \nu N(0, \sigma_k^2/\gamma) + (1 - \nu)N(0, \sigma_k^2)$.

3.2 NI Distribution in MLIRT Model

Consider the general scenario that we have outliers in both continuous outcome and random effects. In this section, we apply the NI distributions to the random error ϵ_{ijk} for continuous outcome in model (3) and the random effects vector $\boldsymbol{u}_i = (u_{i0}, u_{i1})'$ in model (4).

There are two ways of applying NI to the MLIRT models: (i) assume shared weight ω_i between the continuous outcomes and random effects as suggested by Lachos et al. (2011, 2013). and (ii) assume different weights between the continuous outcomes and random effects.

We first introduce the method using shared weight ω_i. Assume that $(\boldsymbol{u}_i, \epsilon_i) \sim NI(\boldsymbol{0}, \{(\boldsymbol{\Sigma}_u, 0), (0, \sigma^2)\}, \omega_i)$, where $\boldsymbol{\Sigma}_u$ is the variance-covariance matrix for random effects vector $\boldsymbol{u}_i$, and ϵ_i is the random error of a continuous outcome for patient i for $i = 1, \ldots, N$. Please note that $\boldsymbol{u}_i$ and ϵ_i are conditionally independent given ω_i, but not marginally independent.

Specifying the model with NI directly may not be easy. But one of the advantages in Bayesian framework is that they can be specified the parameters hierarchically. We may specify $\boldsymbol{u}_i$, ϵ_{ijk} and ω_i as:

$$\begin{aligned}
\boldsymbol{u}_i|\omega_i &\sim N(\boldsymbol{0},\ \omega_i^{-1}\Sigma_u),\\
\epsilon_{ijk}|\omega_i &\sim N(\boldsymbol{0},\ \omega_i^{-1}\sigma_k^2),\\
\omega_i &\sim p(\omega_i|\nu).
\end{aligned}$$

As introduced in Sect. 3.1, by changing the distribution of ω_i, we can now have Student's-t, slash and contaminated normal distribution for both $\boldsymbol{u}_i$ and ϵ_{ijk}. And now the continuous outcome y_{ijk} follows $y_{ijk}|\boldsymbol{u}_i, \omega_i \sim N(a_k + b_k\theta_{ij},\ \omega_i^{-1}\sigma_k^2)$. The full likelihood of patient i across all visits is

$$L(\boldsymbol{y}_i, \omega_i, \boldsymbol{u}_i) = \left[\prod_{j=1}^{J_i}\prod_{k=1}^{K} p(y_{ijk}|\boldsymbol{u}_i, \omega_i)\right] p(\omega_i)p(\boldsymbol{u}_i). \tag{7}$$

We refer to this model as model Dep-NI because $\boldsymbol{u}_i$ and ϵ_{ijk} are marginally dependent on ω_i.

In the second method, we assume different weights for continuous variable and random effects. It is generally a more flexible method since we do not apply any assumptions to the weights. The assumption that continuous variable and random effects vector share the same level of outliers and heavy tails might be reasonable for some studies, it may not always be true and may have negative impact to the model inference. Here we assume that $\boldsymbol{u}_i$ and ϵ_{ijk} are scaled by different weight variables: $\boldsymbol{u}_i \sim \text{NI}(\mathbf{0},\ \Sigma_u,\ \omega_{1i})$,and $\epsilon_{ijk} \sim \text{NI}(0,\ \sigma_k^2,\ \omega_{2ijk})$, where ω_{1i} is a subject-specific weight variable for $\boldsymbol{u}_i$, and ω_{2ijk} is a patient-visit-outcome specific weight variable for continuous outcome y_{ijk}, and ω_{1i} and ω_{2ijk} are independent.

Similarly we use hierarchical specifications and apply the NI distribution to the MLIRT model. The specification of $\boldsymbol{u}_i$, ϵ_{ijk}, ω_{1i}, and ω_{2ijk} are

$$\begin{aligned}
\boldsymbol{u}_i|\omega_{1i} &\sim \text{N}(\mathbf{0},\ \omega_{1i}^{-1}\Sigma_u),\\
\epsilon_{ijk}|\omega_{2ijk} &\sim \text{N}(\mathbf{0},\ \omega_{2ijk}^{-1}\sigma_k^2),\\
\omega_{1i} &\sim \text{p}(\omega_{1i}|\nu_1),\\
\omega_{2ijk} &\sim \text{p}(\omega_{2ijk}|\nu_2).
\end{aligned}$$

Then the continuous outcome y_{ijk} follows $y_{ijk}|\boldsymbol{u}_i, \omega_{2ijk} \sim \text{N}(a_k + b_k\theta_{ij},\ \omega_{2ijk}^{-1}\sigma_k^2)$. Let $\boldsymbol{\omega}_i$=$(\omega_{1i}, \boldsymbol{\omega}_{2i})$, where $\boldsymbol{\omega}_{2i} = \{\omega_{2ijk}\}$ for $j = 1, \ldots, J_i$ and $k = 1, \ldots, K$. The full likelihood of patient i is

$$L(\boldsymbol{y}_i, \boldsymbol{\omega}_i, \boldsymbol{u}_i) = \prod_{j=1}^{J_i}\left[\prod_{k=1}^{K}\text{p}(y_{ijk}|\boldsymbol{u}_i, \omega_{2ijk})\text{p}(\omega_{2ijk})\right]\text{p}(\omega_{1i})\text{p}(\boldsymbol{u}_i). \tag{8}$$

We refer to this model as model Indep-NI because $\boldsymbol{u}_i$ and ϵ_{ijk} are marginally independent.

4 Bayesian Inference and Model Selection Criteria

Bayesian approach based on MCMC posterior simulations was used to analyze the multivariate longitudinal data using MLIRT models. The fully Bayesian inference has many advantages. First, MCMC algorithms can be used to estimate exact posterior distributions of the parameters, while likelihood-based estimation only produces a point estimate of the parameters, with asymptotic standard errors (David 2007). Second, Bayesian inference provides better performance in small samples compared to likelihood-based estimation (Lee and Song 2004). In addition, it is more straightforward to deal with more complicated models using Bayesian inference via MCMC. The model fitting is conducted using the `BUGS` language implemented in `OpenBUGS` (`OpenBUGS` version 3.2.3).

We assume vague priors for all the parameters of interests. For example, the prior distribution of all parameters in $\boldsymbol{\beta}$ is N(0, 100), we use the prior distribution

Gamma(0.001, 0.001) for σ_u and for all components in $\boldsymbol{b}$ to ensure positivity, and use Uniform$[-1, 1]$ for ρ. Multiple chains with dispersed initial values are run. To assess convergence, we use the history plots to ensure there are no appearance of trend for all parameters. In addition, we use Gelman-Rubin diagnostic statistics to ensure the scale reduction $\hat{R}$ of all parameters are smaller than 1.1 (Gelman et al. 2013). To facilitate easy reading and implementation of the proposed models, a sample `BUGS` code for fitting model Indep-NI when using contaminated normal distribution will be available with the book.

There are various model selection methods available in Bayesian inference, for example, the log pseudo-marginal likelihood (LPML), the deviance information criterion (DIC), the expected Akaike information criterion (EAIC), the expected Bayesian information criterion (EBIC) and Bayes factor (BF) to assess model performance.

The computation of LPML is based on Conditional predictive ordinate (CPO). CPO is a cross-validation predictive method that evaluates the predictive distribution of the model conditioning on the data but with single data point deleted (Geisser 1993; Carlin and Louis 2011; Lachos et al. 2009; Chen et al. 2000). Let $\boldsymbol{y}$ be the full observed data and $\boldsymbol{y}_{(i)}$ be the data with subject i deleted. Then the CPO for subject i is defined as $CPO_i = p(\boldsymbol{y_i}|\boldsymbol{y}_{(i)}) = \int p(\boldsymbol{y_i}|\boldsymbol{\theta})p(\boldsymbol{\theta}|\boldsymbol{y}_{(i)})d\boldsymbol{\theta}$. Large CPO implies that the data for subject i can be well predicted by the model based on the data from all the other subjects. Thus lager CPO means a better fit for the model. Because there is no close form for CPO_i in MLIRT model, a Monte Carlo estimation can be obtained from the posterior distribution $p(\boldsymbol{\theta}|\boldsymbol{y})$. Since the function of CPO_i can be further derived as $CPO_i = p(\boldsymbol{y_i}|\boldsymbol{y}_{(i)}) = p(\boldsymbol{y})/p(\boldsymbol{y}_{(i)}) = 1/\int p(\boldsymbol{\theta}|\boldsymbol{y})/p(\boldsymbol{y_i}|\boldsymbol{y}_{(i)}, \boldsymbol{\theta})d\boldsymbol{\theta}$, let M be the total number of post burn-in samples, then a harmonic-mean approximation of CPO_i is $\widehat{CPO_i} = (\frac{1}{M}\sum_{t=1}^{M}\frac{1}{p(\boldsymbol{y_i}|\boldsymbol{y}_{(i)}, \boldsymbol{\theta}^{(t)})})^{-1} = (\frac{1}{M}\sum_{t=1}^{M}\frac{1}{p(\boldsymbol{y_i}|\boldsymbol{\theta}^{(t)})})^{-1}$ (Luo et al. 2013; Chen et al. 2000). A summary statistics of CPO_i is log pseudo-marginal likelihood (LPML), defined as $LPML = \sum_{i=1}^{N}\log(\widehat{CPO_i})$. A larger value of LPML implies a better model fitting.

The deviance information criterion (DIC) assesses model fittings based on the posterior mean of the deviance and a penalty on the model complexity (Spiegelhalter et al. 2002). The deviance statistics is defined as $D(\boldsymbol{\theta}) = -2\log f(\boldsymbol{y}|\boldsymbol{\theta}) + 2\log h(\boldsymbol{y})$, where $f(\boldsymbol{y}|\boldsymbol{\theta})$ is the likelihood of the observed data $\boldsymbol{y}$ given parameter vector $\boldsymbol{\theta}$, $h(\boldsymbol{y})$ is a standardized function of the data alone and have no impact on the assessment of the model fitting. Let $\bar{D}(\boldsymbol{\theta}) = E_{\boldsymbol{\theta}|\boldsymbol{y}}[D(\boldsymbol{\theta})]$ be the posterior mean of the deviance and let $D(\bar{\boldsymbol{\theta}}) = D(E_{\boldsymbol{\theta}|\boldsymbol{y}}[\boldsymbol{\theta}])$ be the deviance evaluated at the posterior mean of the parameter vector $\boldsymbol{\theta}$. The DIC is defined as $DIC = \bar{D}(\boldsymbol{\theta}) + p_D$, where $p_D = \bar{D}(\boldsymbol{\theta}) - D(\bar{\boldsymbol{\theta}})$ is the number of effective parameters and it captures the complexity of the model. A smaller value of DIC implies better fit of the model. Moreover, the expected Akaike information criterion (EAIC) and the expected Bayesian information criterion (EBIC) (Carlin and Louis 2011) are defined as $EAIC = \bar{D}(\boldsymbol{\theta}) + 2p$ and $EBIC = \bar{D}(\boldsymbol{\theta}) + p\log N$, respectively, where p is the total number of parameters used in the model and N is the sample size. Smaller values of EAIC and EBIC imply better fit of the model.

Furthermore, Bayes factor (BF) is a standard Bayesian solution (an alternative to p value) to the hypothesis testing for competing models. The BF quantifies the the degree to which whether the observed data support a hypothesis (Lavine and Schervish 1999; Lewis and Raftery 1997). Let two competing models be M_1 and M_2. Then for observed data $\boldsymbol{y}$, BF in favor of model M_1 over M_2 is defined as

$$\text{BF}(M_1; M_2) = \frac{f(\boldsymbol{y}|M_1)}{f(\boldsymbol{y}|M_2)} = \frac{\int f(\boldsymbol{y}|\boldsymbol{\theta_1}, M_1)f(\boldsymbol{\theta_1}|M_1)d\boldsymbol{\theta_1}}{\int f(\boldsymbol{y}|\boldsymbol{\theta_2}, M_2)f(\boldsymbol{\theta_2}|M_2)d\boldsymbol{\theta_2}},$$

where $\boldsymbol{\theta}_i$ is the parameter vectors for model M_i for $i = 1, 2$; $f(\boldsymbol{y}|\boldsymbol{\theta}_i, M_i)$ is the likelihood of model M_i; and $f(\boldsymbol{\theta}_i|M_i)$ is the posterior density of $\boldsymbol{\theta}_i$ for model M_i (Lewis and Raftery 1997; Gelman et al. 2013). The direct computation of the integration involved in the BF is not straightforward, so the Laplace-Metropolis estimator based on normal distribution is used to approximate the marginal likelihood $f(\boldsymbol{y}|M_i)$ (Lewis and Raftery 1997). Specifically, the $f(\boldsymbol{y}|M_i) \approx (2\pi)^{d_i/2}|\boldsymbol{\Sigma}_i|^{1/2}f(\boldsymbol{y}|\bar{\boldsymbol{\theta}}_i, M_i)f(\bar{\boldsymbol{\theta}}_i|M_i)$, where d_i is the number of parameters in $\boldsymbol{\theta}_i$, $\boldsymbol{\Sigma}_i$ is the posterior covariance matrix of $\boldsymbol{\theta}_i$, $\bar{\boldsymbol{\theta}}_i$ is the posterior mean of $\boldsymbol{\theta}_i$, $f(\bar{\boldsymbol{\theta}}_i|M_i)$ is the prior probability of parameters evaluated at $\bar{\theta}_i$, and $f(\boldsymbol{y}|\bar{\boldsymbol{\theta}}_i, M_i)$ is the likelihood evaluated at the posterior mean $\bar{\theta}_i$ (He and Luo 2013; Lewis and Raftery 1997). The interpretation of the BF is summarized by Kass and Raftery (1995). In paticular, when BF is greater than 100, decisive evidence is shown in favor of Model M_1 over M_2.

5 Monte Carlo Simulation Scheme and Some Results

Up until this section, we have been introducing all the theoretical works. Often times in Bayesian analysis, simulation studies are conducted to demonstrate the advantages/benefits of the proposed methods. For methods that have potential applications to clinical trial studies, ideally the simulation studies should be close to the real life settings or similar to some historical trial data. Here we summarize the simulation scheme for MLIRT modeling using NI distribution as an example. The complete simulation settings can be found in Chen and Luo (2016).

Recall that we refered the "plain" version of the MLIRT model as Indep-N model (Sect. 2), the model with shared weight NI distribution for continuous outcomes and random effects as Dep-NI model (Sect. 3.2) and the model with independent weights as Indep-NI model (Sect. 3.2). The purpose of the simulation study is to show that the performance of the MLIRT model with NI is better than the the "plain" model with Indep-NI performs better than Dep-NI. So the simulation basically contains three parts. In the first part, we assume there is no outliers in the continuous outcomes and random effects, that is the Indep-N model is the true model. The simulation results in this part demonstrates that under the normality assumptions, the performance of the MLIRT models with NI distributions (Indep-NI and Dep-NI) is very similar to the true model (Indep-N) despite the extra parameters from the NI distributions. The first part of the simulation could sometime get ignored or forgotten, because the simulation does not show advantages of the proposed methods. However, it is very important to show the proposed methods work well under regular assumptions not only under special cases (data with outliers in this case). In part two of the simulation, 5% outliers were generated for both continuous outcomes and the random effects. The results show that Indep-NI and Dep-NI model performs better over Indep-N in terms of bias, SD, SE and coverage probability (CP) on the parameter estimations. Furthermore, in the third part of the simulation, the results desmonstrated that among the three NI distributions (Student's t, slash and contaminated normal), the contaminated normal distribution works best for both Indep-NI and Dep-NI model with Indep-NI performs better than Dep-NI.

Overall, the simulation study conveyed the message that Indep-NI model is a more general model that work under both regular conditions (continuous outcomes and random effects follow normality assumptions) and cases when there are outliers present in both the outcomes and random effects. Note that, since under simulation settings, we know the underlining true values for each parameters, we may compare the methods directly using bias and coverage probabilities. For the analysis using real trial data, we will need to use the model selection criteria to help make decisions and conclusions.

6 Application to Trial Study Data

Analysis method innovations and improvements in the medical or biomedical field are often motivated by clinical trial study data. The introduced MLIRT models with NI distributions were motivated by the PD trial study DATATOP. The DATATOP study was a double-blind, placebo-controlled, multi-center clinical trial. A 2×2 factorial design was used to test the hypothesis that patients with early Parkinson's disease with deprenyl 10 mg/d and/or tocopherol (vitamin E) 2000 IU/d will delay the time until the application of levodopa therapy. Eight hundred eligible patients were enrolled and randomized to one of the four treatment arms: active deprenyl alone, active tocopherol alone, both active deprenyl and tocopherol, and double placebo. Longitudinal outcomes such as Unified Parkinson's Disease Rating Scale (UPDRS), Hoehn and Yahr scale (HY), and Schwab and England activities of daily living (SEADL) were collected at baseline and months 1, 3, 6, 9, 12, 15, 18, 21, and 24. In the DATATOP study, only deprenyl was found to be effective in delaying the time until the need of levodopa therapy (Parkinson Study Group 1989, 1993). The levodopa therapy provided temporary relief of PD symptoms and may significantly change the outcomes for a short period.

One of the best ways to understand the results is through data visualization methods, e.g. plots and figures. Without replicate the analysis from Chen and Luo (2016), we would like to cite a few figures to re-emphasise how the MLIRT model with NI distributions work for the DATATOP data (Chen and Luo 2016. Copyright Wiley-VCH Verlag GmbH & Co. KGaA. Reproduced with permission). Figure 1 is a spaghetti

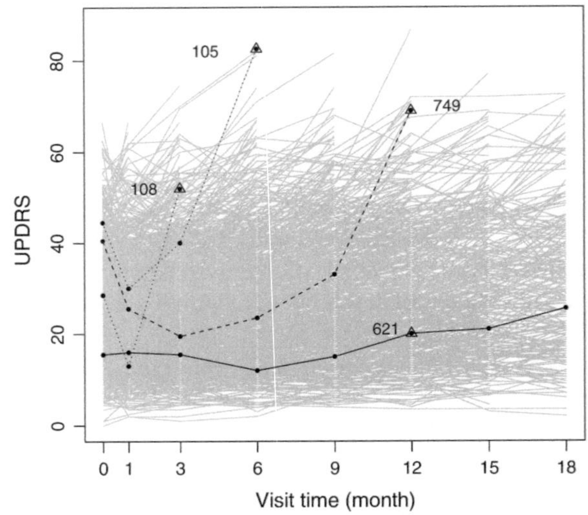

Fig. 1 Longitudinal profile of the outcome UPDRS. Numbers 105, 108, 621, and 749 denote four patients (Reproduced with permission from Chen and Luo (2016))

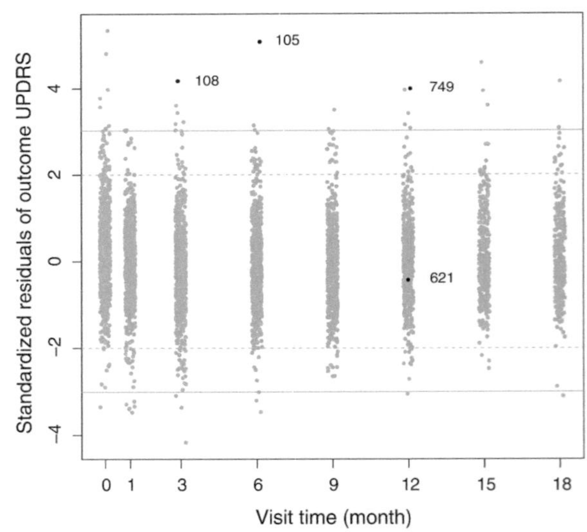

Fig. 2 Standardized residuals of the UPDRS measurements for all patients at each visit when fitting model Indep-N. The *dashed lines* are *horizontal lines* at −2 and 2 and the *solid lines* are *horizontal lines* at −3 and 3. Numbers 105, 108, 621, and 749 denote four patients (Reproduced with permission from Chen and Luo (2016))

plot that shows the longitudinal profile of the observed outcome UPDRS for all patients in DATATOP before any analysis is performed. The highlighted numbers 105, 108, 621 and 749 represent the UDPRS profile for four patients. As mentioned before, PD is a slow progression disease, UPDRS profile such as patient 621 is often observed. UPDRS profile with unexpected sudden changes, such as patient 105, 108 and 749 at visit 3, 6 and 12 month, respectively, could be potential outliers.

Figure 2 shows the standardized residuals (SRs) of UPDRS for all patients at each visit, after applying the "plain" version MLIRT model, Indep-N. For the same four patients in Fig. 1, while patient 621 maintains low SRs value, patient 105, 108 and 749

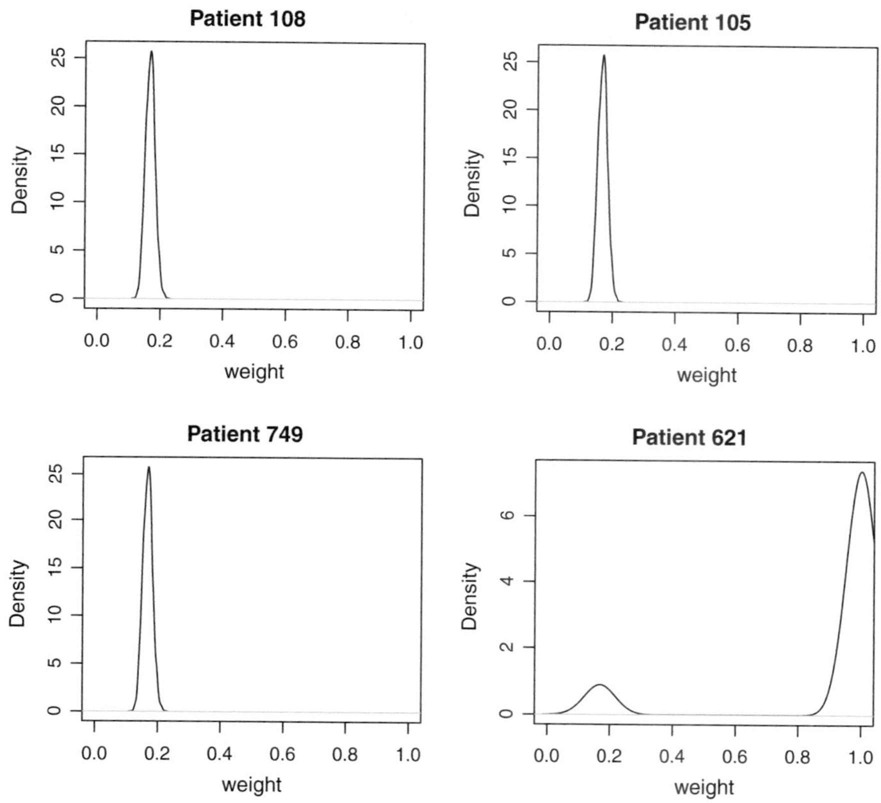

Fig. 3 Estimates of the weight variable ω_{ijk} for patients 105, 108, 621, and 749 at certain visits from model Indep-CN (Reproduced with permission from Chen and Luo (2016))

have hight SRs values (with absolute value larger than 3) at visit 3, 6 and 12 month, respectively, indicating that they are outliers in the Indep-N model. To account for the outliers, the Indep-NI model using contaminated normal distributions (Indep-CN) is applied to the DATATOP data and the results are demonstrated in Fig. 3. Clearly, low weights (around 0.19) are estimated and assigned for patient 105, 108 and 749, while high weight (0.91) is estimated for patient 621. The impact of the outliers are attenuated by the stochastically assigned lower weights during modeling. If we look at the overall performance of the models for DATATOP using the model selection criteria, not surprisingly, the the overall performance of the Indep-CN model (Indep-NI model using CN distribution) is better than the Indep-N model as shown in Table 1. The Indep-CN model had the best fit in terms of LPML and BF values. The BFs of Indep-CN over models Dep-CN and Indep-N are much larger than 100, suggesting decisive evidence in favor of the Indep-CN model.

Table 1 Model comparison statistics for the DATATOP dataset from models Indep-N, Dep-CN, and Indep-CN. The best fitting model is highlighted in bold

	LPML	BF
Indep-N	−27312.50	≫100
Dep-CN	−26930.44	≫100
Indep-CN	**−26873.84**	Ref

Reproduced with permission from Chen and Luo (2016)

7 More Extended Modeling

Besides outliers and heavy tails, other data feature such as dependent censoring and skewness, can also be considered and incorporated into the MLIRT model framework. In the following Sect. 7.1, we will briefly introduce the joint MLIRT model that accounts for the dependent censoring. In Sect. 7.2, we introduce a nice extension to the NI distribution family, the SNI distribution, which accounts for both skewness and heavy tails.

7.1 Joint MLIRT Model

In addition to outlier and heavy tail problem, the scheduled visits or follow-up of patients in longitudinal clinical studies may be stopped by terminal events. The terminal events could be noninformative such as study termination or informative such as death or dropout due to disease progression. When the terminal events are related to patients' disease conditions, the unobserved outcomes are non-ignorable. The dependent terminal event time is usually named as informative censoring or dependent censoring. It has been shown that ignoring dependent censoring leads to biased estimates (Henderson et al. 2000).

Figure 4 shows the mean UPDRS values for DATATOP patients with follow-up time less than 9 month (solid line), 9–15 month (dashed line), and more than 15 month (dotted line), respectively. Patients with shorter follow-up time have higher (worse) UPDRS values, suggesting that there is a strong association between the longitudinal outcomes and time to the initiation of levodopa therapy.

Joint modeling of the dependent terminal event time and the longitudinal outcomes provide consistent estimates (Henderson et al. 2000). In the MLIRT modeling framework, Wang et al. proposed a joint model to analyze multiple-item ordinal quality of life data in the presence of death (Wang et al. 2002). He and Luo developed a joint model for multiple longitudinal outcomes of mixed types, subject to outcome-dependent terminal events (He and Luo 2013). Luo further relaxed the proportional hazard (PH) assumption and developed a joint modeling framework replacing the PH model by various parametric accelerated failure time (AFT) models (Luo 2014).

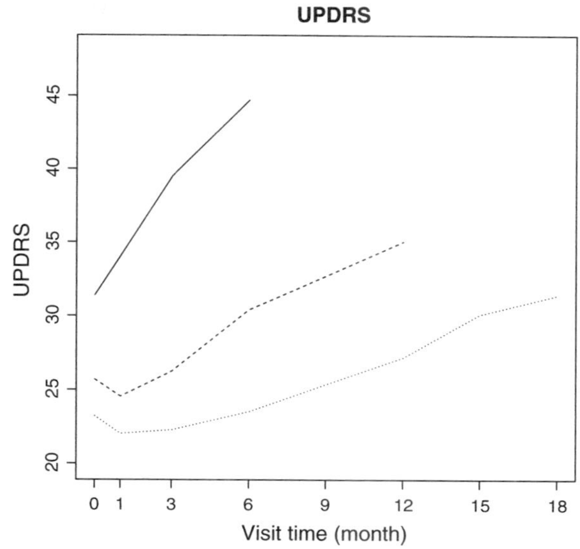

Fig. 4 Mean UPDRS values for patients with follow-up time less than 9 months (*solid line*), 9–15 months (*dashed line*), and more than 15 months (*dotted line*)

Without exhaust all the existing methods, here we introduce the joint model using the Cox proportional hazard model for dependent terminal event time.

Let t_i be the time to a terminal event ζ_i for subject i. Let $\mathbf{X}_i$ denote the vector of potential risk factors. Then the Cox proportional hazard model can be written as

$$h(t_i) = h_0(t_i)\exp(\mathbf{X}_i\gamma + \eta_0 u_{i0} + \eta_1 u_{i1}), \tag{9}$$

where γ is the unknown parameter for the potential risk factors $\mathbf{X}_i$, η_0 and η_1 measure the association between the Cox proportional hazard model and the MLIRT model. The dependent censoring issue is addressed by jointly modeling the MLIRT model and the survival model. The shared random effects u_{i0} and u_{i1} account for the correlation between the survival time and longitudinal outcomes. To allow different baseline hazard rates at different time periods, we adopt the piecewise constant function to approximate the baseline hazard function $h_0(t)$ and assume that the hazard rate is constant within each given time period. Given a set of time points $0 = \tau_0 < \tau_1 < \cdots < \tau_m$, the baseline hazard $h_0(t)$ and the baseline hazard vector $\mathbf{g} = g_0, g_1, \ldots, g_{m-1}$, the piesewise constant hazard function can be defined as $h_0(t) = \sum_{l=0}^{m-1} g_l I_l(t)$, where $I_l(t)$ is a indicator function with $I_l(t) = 1$ if $\tau_l \le t < \tau_{l+1}$ and 0 otherwise.

Under the local independence assumption (i.e., conditional on the random effects vector $\boldsymbol{u}_i$, all outcome measures for each patient are independent) (Fox 2010), the full likelihood of subject i across all visits is

$$L(\boldsymbol{y}_i, \boldsymbol{u}_i) = \left[\prod_{j=1}^{J_i} \prod_{k=1}^{K} P(y_{ijk} | \boldsymbol{u}_i) \right] \cdot h(t_i)^{\zeta_i} S(t_i) \cdot P(\boldsymbol{u}_i), \tag{10}$$

where the survival function $S(t_i) = \exp[\int_0^{t_i} h(s)ds]$ and $p(\boldsymbol{u}_i)$ is the density function for random effects $\boldsymbol{u}_i$.

7.2 MLIRT Model with Skew-Normal/Independent (SNI) Distributions

In previous sections, we have considered the outlier issues. But the outlier, or in general, departure form normality may due to skewness, outliers, or both. The skew-normal/independent (SNI) distribution is an nice extension to the NI distribution family. SNI is a class of asymmetric heavy-tailed distributions that includes skew-t (ST), skew-slash, and skew-contaminated normal distributions. In this section we focus on the ST distribution, while the application of other SNI distributions is straightforward. Similar to the properties of NI distributions, the ST distribution reduces to the skew-normal distribution (SN, asymmetric but not heavy-tailed) when degree of freedom is large and further reduce to normal distribution when the skewness approaches to zero. Here we consider the ST and SN distributions introduced by Sahu et al. (2003), which have the stochastic representation and are suitable for a Bayesian computation.

For simplicity, we illustrate the implementation of the univariate ST distribution to the continuous outcome k in model (3). Let δ be the skewness parameter for continuous outcome y_{ijk}, and ν be the degree of freedom for the ST distribution, following Sahu et al. (2003), the stochastic representation of the ST distribution for y_{ijk} is given by

$$\begin{aligned} y_{ijk} | a_k, b_k, \theta_{ij}, \sigma_k^2, \omega_{ijk}, z_{ik}, \delta, \nu &\sim N(a_k + b_k\theta_{ij} + \delta z_{ik},\ \sigma_k^2/\omega_{ijk}), \\ z_{ik} &\sim N(0, 1)I(z_{ik} > 0), \\ \delta &\sim N(0, \Gamma), \\ \omega_{ijk} &\sim \text{Gamma}(\nu/2, \nu/2), \end{aligned}$$

where the weight variable ω_{ijk} is a positive random variable with density $p(\omega_{ijk}|\nu)$, with $\nu > 0$ represents degree of freedom, z_{ik} is a subject-specific variable for outcome k that follows a truncated standard normal distribution. The skewness parameter δ indicates the skewness of outcome k, with positive δ representing a right skewed distribution and negative δ representing a left skewed distribution. The parameter Γ determines the prior variance information for δ. It is worth to mention that when the degree of freedom $\nu \to \infty$, the distribution Gamma$(\nu/2, \nu/2)$ degenerates to 1, i.e., $\omega_{ijk} \equiv 1$. In this case, the ST distribution reduces to the SN distribution. Moreover, when the skewness parameter $\delta = 0$, the ST distribution reduces to the symmetric and heavy-tailed student-t distribution. The parameters ν and δ can be estimated from the data during modeling, and small ν and large δ in absolute value are indications of heavy tails (outliers) and skewness, respectively.

After incorporating the ST distribution to the likelihood in (6), the full likelihood becomes

$$L(\boldsymbol{y}_i, \boldsymbol{u}_i, \omega_{ijk}, z_{ik}, \delta) = \prod_{j=1}^{J_i} \left[\prod_{k=1}^{K} p(y_{ijk}|\boldsymbol{u}_i, \omega_{ijk}, z_{ik}, \delta) p(\omega_{ijk}) p(z_{ik}) \right] p(\boldsymbol{u}_i). \tag{11}$$

8 Discussions

In this chapter, we introduced the application of latent variable based multilevel item response theory (MLIRT) models to clinical trial studies. The characteristics of the MLIRT model make it a great fit for analyzing longitudinal data with multiple endpoints of mixed types. The advantages of the MLIRT model include but not limited: (1) it uses the full longitudinal information and accounts for the three sources of correlations within subject via the subject-specific random effects; (2) it has a better reflection to the multilevel data structure; and (3) it simultaneously estimates the measurement-specific parameters, the covariate effects, as well as the subject-specific disease progression characteristics (Maier 2001; Kamata 2001; He and Luo 2013). The ability of the MLIRT model being able to provide overall treatment inference through multiple outcome measures lead to a great potential that the model can be applied to a wide range of therapeutic areas in clinical studies.

The underlying linear disease progression assumption in model (4) can be further relaxed by adding quadratic or higher-order term of time t to accommodate the possible disease fluctuation over time, for example, $\theta_{ij} = \boldsymbol{X}_{i0}\boldsymbol{\beta}_0 + u_{i0} + (\boldsymbol{X}_{i1}\boldsymbol{\beta}_1 + u_{i1})t_{ij} + (\boldsymbol{X}_{i2}\boldsymbol{\beta}_2 + u_{i2})t_{ij}^2$, where $\boldsymbol{X}_{i0}$, $\boldsymbol{X}_{i1}$ and $\boldsymbol{X}_{i2}$ may contain the same or different sets of covariates of interest. The choice of the form of model (4) could be determined by (a) the natural history or characteristics of the disease or (b) statistics techniques such as goodness-of-fit or Bayes factor (BF) in Bayesian framework. In addition, given a distribution assumption for the latent variables θ_{ij}, the MLIRT models are equivalent to nonlinear mixed models (Rijmen et al. 2003).

In sum, due to the flexibility of the MLIRT model and the fairly straightforward hierarchical specifications under Bayesian framework, we introduced the normal/independent(NI) distributions and incorporated the NI distribution to the MLIRT model framework to tackle the commonly-encountered outliers and heavy tails problems. We demonstrated how the NI distribution account for the outliers and attenuate their impact to the overall inference by assigning lower weights to those outlier patients. It is noteworthy that models Dep-NI and Indep-NI when using CN distributions are Bayesian mixture models because the CN distribution is a two-component mixture distribution. Gelman (2013) and Fox (2010) have discussed the identifiability issues called label switching in Bayesian mixture models. Basically the label switching problem is that the posterior distribution is invariant to permutations in the labeling of mixture components. Jasra et al. (2005) offered an excellent review of the label switching problem and provided some solutions. However, the MLIRT models using CN distribution do not have the identifiability issue because of the increasing order of the variances of the two component distributions, i.e., $\sigma_k^2/\gamma \geq \sigma_k^2$ as $0 < \nu \leq 1$. The fact that all parameters can be successfully recovered in all these models suggest that the proposed models are identifiable.

As discussed in Sect. 7, data features such as dependent censoring and skewness can also be considered and incorporated into the MLIRT model. In addition, many trials, especially phase 3 trials, often use multiple research center or institute to recruit and enroll patients. There is a potential center or clustering effect due to the care provider, environment and the population around it. Luo and Wang has developed methods to take into account the center effects (Luo and Wang 2014). As for joint models approach, besides Cox proportional model, the accelerated failure time (AFT) model is also consider (Luo 2014). We may also consider developing a nonparametric model within the MLIRT model framework to define and estimate the time dependent treatment effect. Furthermore, for the analysis methods introduced, the posterior Bayesian inference is only drawn for current data. If we have large data to properly train the model, the predictive inference could be another area we could explore further.

It might seem that the models are getting more complicated and convoluted, but remember that making complex models is not the purpose but rather a way to better fit the data structure with more flexible model assumptions.

References

Baghfalaki, T., Ganjali, M., & Berridge, D. (2013). Robust joint modeling of longitudinal measurements and time to event data using normal/independent distributions: A Bayesian approach. *Biometrical Journal, 55*(6), 844–865.

Bushnell, D. M., & Martin, M. L. (1999). Quality of life and Parkinson's disease: Translation and validation of the US Parkinson's disease questionnaire (PDQ-39). *Quality of Life Research, 8*(4), 345–350.

Carlin, B. P., & Louis, T. A. (2011). *Bayesian methods for data analysis*. Boca Raton, FL: Chapman & Hall.

Chen, G., & Luo, S. (2016). Robust Bayesian hierarchical model using normal/independent distributions. *Biometrical Journal, 58*(4), 831–851.

Chen, M. H., Shao, Q. M., & Ibrahim, J. G. (2000). *Monte Carlo methods in Bayesian computation*. New York: Springer Series in Statistics.

Cummings, J. L. (1992). Depression and Parkinson's disease: A review. *The American Journal of Psychiatry, 149*(4), 443–454.

David, D. (2007). Dunson. Bayesian methods for latent trait modelling of longitudinal data. *Statistical Methods in Medical Research, 16*(5), 399–415.

Elm, J. J., & The NINDS NET-PD Investigators. (2012). Design innovations and baseline findings in a long-term Parkinson's trial: The National Institute of Neurological Disorders and Stroke exploratory trials in Parkinson's Disease Long-Term study-1. *Movement Disorders, 27*(12), 1513–1521.

Fahn, S., Oakes, D., Shoulson, I., Kieburtz, K., Rudolph, A., Lang, A., et al. (2004). Levodopa and the progression of Parkinson's disease. *The New England Journal of Medicine, 351*(24), 2498–2508.

Fox, J. P. (2010). *Bayesian item response modeling: Theory and applications*. New York: Springer.

Geisser, S. (1993). *Predictive inference: An introduction* (Vol. 55). Boca Raton, FL: CRC Press.

Gelman, A., Carlin, J. B., Stern, H. S., Dunson, D. B., Vehtari, A., & Rubin, D. B. (2013). *Bayesian data analysis*. CRC Press.

He, B., & Luo, S. (2013). Joint modeling of multivariate longitudinal measurements and survival data with applications to Parkinson's disease. *Statistical Methods in Medical Research.*

Henderson, R., Diggle, P., & Dobson, A. (2000). Joint modelling of longitudinal measurements and event time data. *Biostatistics*, *1*(4), 465–480.

Huang, P., Tilley, B. C., Woolson, R. F., & Lipsitz, S. (2005). Adjusting O'Brien's test to control type I error for the generalized nonparametric Behrens-Fisher problem. *Biometrics, 61*(2), 532–539.

Jasra, A., Holmes, C. C., & Stephens, D. A. (2005). Markov chain Monte Carlo methods and the label switching problem in Bayesian mixture modeling. *Statistical Science*, *20*(1), 50–67.

Kamata, A. (2001). Item analysis by the hierarchical generalized linear model. *Journal of Educational Measurement*, *38*(1), 79–93.

Kass, R. E., & Raftery, A. E. (1995). Bayes factors. *Journal of the American Statistical Association, 90*(430), 773–795.

Lachos, V. H., Bandyopadhyay, D., & Dey, D. K. (2011). Linear and nonlinear mixed-effects models for censored HIV viral loads using normal/independent distributions. *Biometrics, 67*(4), 1594–1604.

Lachos, V. H., Castro, L. M., & Dey, D. K. (2013). Bayesian inference in nonlinear mixed-effects models using normal independent distributions. *Computational Statistics & Data Analysis, 64*, 237–252.

Lachos, V. H., Dey, D. K., & Cancho, V. G. (2009). Robust linear mixed models with skew-normal independent distributions from a Bayesian perspective. *Journal of Statistical Planning and Inference, 139*(12), 4098–4110.

Lange, K., & Sinsheimer, J. S. (1993). Normal/independent distributions and their applications in robust regression. *Journal of Computational and Graphical Statistics, 2*(2), 175–198.

Lavine, M., & Schervish, M. J. (1999). Bayes factors: What they are and what they are not. *The American Statistician, 53*(2), 119–122.

Lee, S.-Y., & Song, X.-Y. (2004). Evaluation of the bayesian and maximum likelihood approaches in analyzing structural equation models with small sample sizes. *Multivariate Behavioral Research, 39*(4), 653–686.

Lewis, S. M., & Raftery, A. E. (1997). Estimating Bayes factors via posterior simulation with the Laplace-Metropolis estimator. *Journal of the American Statistical Association, 92*(438), 648–655.

Liu, C. (1996). Bayesian robust multivariate linear regression with incomplete data. *Journal of the American Statistical Association*, *91*(435), 1219–1227.

Lord, F. M., Novick, M. R., & Birnbaum, A. (1968). *Statistical theories of mental test scores.* Boston, MA: Addison-Wesley.

Luo, S. (2014). A Bayesian approach to joint analysis of multivariate longitudinal data and parametric accelerated failure time. *Statistics in Medicine, 33*(4), 580–594.

Luo, S., & Wang, J. (2014). Bayesian hierarchical model for multiple repeated measures and survival data: An application to parkinson's disease. *Statistics in Medicine, 33*(24), 4279–4291.

Luo, S., Lawson, A. B., He, B., Elm, J. J.,& Tilley, B. C. (2012). Bayesian multiple imputation for missing multivariate longitudinal data from a Parkinson's disease clinical trial. *Statistical Methods in Medical Research.*

Luo, S., Ma, J., & Kieburtz, K. D. (2013). Robust Bayesian inference for multivariate longitudinal data by using normal/independent distributions. *Statistics in Medicine, 32*(22), 3812–3828.

Maier, K. S. (2001). A rasch hierarchical measurement model. *Journal of Educational and Behavioral Statistics, 26*(3), 307–330.

McRae, C., Diem, G., Vo, A., O'Brien, C., & Seeberger, Lauren. (2000). Schwab & England: Standardization of administration. *Movement Disorders, 15*(2), 335–336.
Miller, T. M., Balsis, S., Lowe, D. A., Benge, J. F., & Doody, R. S. (2012). Item response theory reveals variability of functional impairment within clinical dementia rating scale stages. *Dementia and Geriatric Cognitive Disorders, 32*(5), 362–366.
Müller, J., Wenning, G. K., Jellinger, K., McKee, A., Poewe, W., & Litvan, I. (2000). Progression of Hoehn and Yahr stages in Parkinsonian disorders: A clinicopathologic study. *Neurology, 55*(6), 888–891.
Parkinson Study Group. (1989). DATATOP: A multicenter controlled clinical trial in early Parkinson's disease. *Archives of Neurology, 46*(10), 1052–1060.
Parkinson Study Group. (1993). Effects of tocopherol and deprenyl on the progression of disability in early Parkinson's disease. *The New England Journal of Medicine, 328*(3), 176–183.
Parkinson Study Group. (2002). A controlled trial of rasagiline in early Parkinson disease: The TEMPO study. *Archives of Neurology, 59*(12), 1937.
Rijmen, F., Tuerlinckx, F., De Boeck, P., & Kuppens, P. (2003). A nonlinear mixed model framework for item response theory. *Psychological methods*, *8*(2), 185.
Rosa, G. J. M., Padovani, C. R., & Gianola, D. (2003). Robust linear mixed models with normal/independent distributions and Bayesian MCMC implementation. *Biometrical Journal, 45*(5), 573–590.
Sahu, S. K., Dey, D. K., & Branco, M. D. (2003). A new class of multivariate skew distributions with applications to Bayesian regression models. *Canadian Journal of Statistics, 31*(2), 129–150.
Samejima, F. (1997). *Graded response model*. New York: Springer.
Snitz, B. E., Yu, L., Crane, P. K., Chang, C.-C. H., Hughes, T. F., & Ganguli, M.(2012). Subjective cognitive complaints of older adults at the population level: An item response theory analysis. *Alzheimer Disease & Associated Disorders, 26*(4), 344–351.
Spiegelhalter, D. J., Best, N. G., Carlin, B. P., & Van Der Linde, A. (2002). Bayesian measures of model complexity and fit. *Journal of the Royal Statistical Society: Series B (Statistical Methodology), 64*(4), 583–639.
Vaccarino, A. L., Anderson, K., Borowsky, B., Duff, K., Joseph, G., Mark, G., et al. (2011). An item response analysis of the motor and behavioral subscales of the unified Huntington's disease rating scale in Huntington disease gene expansion carriers. *Movement Disorders, 26*(5), 877–884.
Wang, C., Douglas, J., & Anderson, S. (2002). Item response models for joint analysis of quality of life and survival. *Statistics in Medicine, 21*(1), 129–142.
Weisscher, N., Glas, C. A., Vermeulen, M., & De Haan, R. J. (2010). The use of an item response theory-based disability item bank across diseases: accounting for differential item functioning. *Journal of Clinical Epidemiology, 63*(5), 543–549.

A Comparison of Bootstrap Confidence Intervals for Multi-level Longitudinal Data Using Monte-Carlo Simulation

Mark Reiser, Lanlan Yao, Xiao Wang, Jeanne Wilcox and Shelley Gray

Abstract Longitudinal investigations, where subjects are followed over time, have played an increasingly prominent role in medicine, health, and psychology in the last decades. This chapter will address inference for a two-level mixed model for a longitudinal study where observational units are clustered at both levels. Bootstrap confidence intervals for model parameters are investigated under the issues of non-normality and limited sample size of the original data. A one stage case-resampling bootstrap will be established for constructing confidence intervals by sampling clusters with replacement at the higher level. A two-stage case-resampling bootstrap will be developed by sampling clusters with replacement at the higher level and then sampling with replacement at the lower level also. Monte-Carlo simulations will be utilized to evaluate the effectiveness of these bootstrap methods with various size clusters for the mixed-effects model in terms of bias, standard deviation and confidence interval coverage for the fixed effects as well as for variance components of the random effects. The results show that the parametric bootstrap and cluster bootstrap at the higher level perform better than the two-stage bootstrap. The bootstrap methods will be applied to a longitudinal study of preschool children nested within classrooms.

M. Reiser (✉) · L. Yao
School of Mathematical and Statistical Science, Arizona State University,
Tempe, AZ 85287, USA
e-mail: mark.reiser@asu.edu

L. Yao
e-mail: 610yll@gmail.com

X. Wang
Statistics and Data Corporation, 21 East 6th Street, Tempe, AZ 85281, USA
e-mail: xwang@asu.edu

J. Wilcox
Division of Educational Leadership and Innovation, Arizona State University,
Tempe, AZ 85287, USA
e-mail: mjwilcox@asu.edu

S. Gray
Speech and Hearing Science, Arizona State University, Tempe, AZ 85287, USA
e-mail: Shelley.Gray@asu.edu

D.-G. Chen and J.D. Chen (eds.), *Monte-Carlo Simulation-Based Statistical Modeling*, ICSA Book Series in Statistics, DOI 10.1007/978-981-10-3307-0_17

Keywords Cluster bootstrap · Two-stage bootstrap · Parametric bootstrap · Monte-Carlo bootstrap · Mixed-effects linear model · Hierarchical linear model · Repeated measures · Nested design · Classroom-based study

1 Introduction

Longitudinal study refers to an investigation where participant outcomes and possibly treatments or exposures are collected at multiple follow-up times. This type of study has played an increasingly prominent role in medicine, health and psychology over the past few decades (Fitzmaurice and Laird 2004; Carpenter 2003). The study of longitudinal data is to characterize the change of response over time and to characterize the factors that will affect the response, which would provide a model to assess how within-individual patterns change over time. Since a single individual has repeated observations, there will be some correlation among the measurements taken on the same unit. Many standard statistical techniques have been proposed to analyze longitudinal studies. For example, the full multivariate model (Beale and Little 1975) can be applied if the design is balanced and a two-stage random effects models can be applied to the unbalanced situation (Laird and Ware 1982).

In this study of longitudinal data, the linear mixed-effect model will be investigated to fit repeated measurements on level-one units which are clustered in level-two units. Hence, clusters are nested within clusters. Units within the same level-two cluster are also correlated due to some common effect such as children nested under the same teacher or patients nested under the same physician. The mixed effects model makes specific assumptions about the variation in observations attributable to the variation within a subject and to variation among subjects. The model takes into account the fixed effects, which are parameters related to the entire population, and random effects, which are associated with clusters sampled from a population (Pinheiro and Bates 2000). The parameter estimator in the linear mixed-effects model is obtained by maximum likelihood (ML) method or the restricted maximum likelihood (REML) method. The differences between ML and REML is small in a large sample, but there is slightly more bias and poorer coverage rate with ML for the estimation of the variance components in a sparse design (Patterson and Thompson 1971; Thai 2013), which may due to the feature that REML takes into account the loss of the degrees of freedom in the correlated data (Verbeke and Molenberghs 2000). The REML method will be explored in our study.

Re-sampling can be an approach to investigate the properties of any statistic of interest when the sample size is limited and some distributional assumptions are not satisfied. Many bootstrap techniques (Efron and Tibshirani 1993; Scherman 1997) have been proposed since 1979 when Efron first introduced the bootstrap method (Efron 1986) for independent and identically distributed observations. It is implemented by re-sampling the observed data repeatedly to obtain resampled data and then fitting the model to the resampled data to obtain an empirical distribution of the estimates. It is more complicated to bootstrap correlated data because the standard bootstrap method relies on the *iid* assumption. In this chapter, the parametric bootstrap, a case bootstrap at higher level, and a two-stage bootstrap for two levels will be compared. The case bootstrap consists of re-sampling clusters with replacement and keeps the entire observations within the cluster unchanged (Shieh 2002). The parametric bootstrap is realized by obtaining the residuals from the parametric distribution, and the parameters are estimated from the original data (Benton 2002). The two-stage bootstrap consists of case-resampling bootstrap at two levels. In our study, we first sample at the upper level with replacement and then within these clusters, we sample with replacement at the lower level to obtain the bootstrap sample. The mixed-effects model is then fit to each bootstrap sample.

In this chapter, we examine different bootstrap approaches to the linear mixed effects model. We will have a brief description of the two-level random effects model in Sect. 2. In Sect. 3, the details of different bootstrap methods will be introduced, including the parametric bootstrap, case bootstrap and two-stage bootstrap. The simulation settings and the results of the simulation study will be shown in Sect. 4. In Sect. 5, we will apply the bootstrap methods to a longitudinal study of preschool children nested within classrooms.

2 Linear Mixed Effects Model

2.1 *Statistical Models*

The linear mixed effects model can be applied to repeated measures or longitudinal studies where data are clustered. The linear mixed model is considered to be an extension of the classical linear model. This model takes into account the correlation among units or observations within clusters and considers the clusters as a random sample from the common population, which may be more realistic in many applications. In this section, we utilize the ideas introduced by Harville (1997) and the work of Laird and Ware (1982) to define the model for the linear mixed effects model.

First, consider a multi-level design such as students nested within classrooms. In a multi-level model with repeated measures, there are at least two types of cluster effects. For example, repeated measures on the level-one unit (students) constitute a cluster of correlated observations, and then responses from different students in the same classroom (the level-two unit) constitute another cluster of correlated observations. Let $n_{i(k\ell)}$ represent the number of repeated observations on level-one unit i nested within second-level cluster k within treatment ℓ, $i = 1, 2, \ldots, n_k$, where n_k is the number of level-one units in level-two cluster k; $k = 1, 2, \ldots, K_1$ in treatment level 1, $k = 1, 2, \ldots, K_2$ in treatment level 2, and $K = K_1 + K_2$, where K is the total number of level-two clusters; j, $j = 1, 2, \ldots, n_i$ is an index for time point, t_{ij}; and ℓ, $\ell = 1, 2$, is an index for treatment condition, i.e., control versus treatment. $N = \sum_k \sum_\ell n_{k(\ell)}$ is the number of level one clusters, and $H = \sum_i \sum_k \sum_l n_{i(k\ell)}$ is the total number of observations. Let Y be the response variable, then the linear mixed-effects model can be stated as

$$Y_{ijk\ell} = \beta_0 + \beta_1 X_{1kl} + \beta_2 t_{ij} + \beta_3 t_{ij}^2 + \beta_4 X_{1kl} t_{ij} + \beta_5 X_{1k\ell} t_{ij}^2 + b_{0i} + u_{kl} + b_{1i} t_{ij} + \varepsilon_{ijk\ell} \quad (1)$$

where X_1 is a dummy variable for the treatment, b_{0i} denotes the random intercept for student i, b_{1i} is the random slope for student i, $u_{k\ell}$ is the random intercept for classroom k within treatment l. And

$$\begin{pmatrix} b_{0i} \\ b_{1i} \\ u_{k\ell} \end{pmatrix} \sim MVN_3\left(\mathbf{0}, \begin{pmatrix} \sigma_{b_0}^2 & \sigma_{b_0 b_1} & 0 \\ \sigma_{b_0 b_1} & \sigma_{b_1}^2 & 0 \\ 0 & 0 & \sigma_c^2 \end{pmatrix}\right). \quad (2)$$

Let $\boldsymbol{Y}_i = (Y_{i1}, Y_{i2}, \ldots, Y_{in_i})'$ represent the vector of measurements at level one. The linear mixed effects model can be stated in matrix notation as

$$\begin{cases} \boldsymbol{Y}_i = \boldsymbol{X}_i \boldsymbol{\beta} + \boldsymbol{Z}_i \boldsymbol{b}_i + \boldsymbol{\varepsilon}_i \\ \boldsymbol{b}_i \sim N(0, \boldsymbol{G}_i) \\ \boldsymbol{\varepsilon}_i \sim N(0, \sigma^2 \boldsymbol{I}_{n_i}) \end{cases} \quad (3)$$

$\boldsymbol{X}_i$ is a $(n_i \times p)$ design matrix of the p fixed effect variables. $\boldsymbol{Z}_i$, with size $(n_i \times q)$, is the design matrix for the q random effects. $\boldsymbol{\beta}$ is a $(p \times 1)$ column vector of the fixed effects regression parameters, and $\boldsymbol{b}_i$ is a vector containing the random effects with size $(q \times 1)$. Here we assume that the random effects are normally distributed with mean zero and covariance matrix $\boldsymbol{G}_i$. $\boldsymbol{\varepsilon}_i$ refers to residual errors for cluster i, and are assumed to be normal with mean zero and variance $\boldsymbol{R}_i = \sigma^2 I_{n_i}$. For simplicity, we assume that random effects and residual errors are uncorrelated with $cov(\boldsymbol{b}_i, \boldsymbol{\varepsilon}_i) = \mathbf{0}$ for all i. $\boldsymbol{V}_i$, the $n_i \times n_i$ covariance matrix of the n_i observations on level-one unit i, has diagonal elements $\sigma_{jj} = \sigma_{b_0}^2 + 2t_{ij}\sigma_{b_0 b_1} + t_{ij}^2 \sigma_{b_1}^2 + \sigma_c^2$ and off-diagonal elements $\sigma_{jj'} = \sigma_{b_0}^2 + (t_{ij} + t_{ij'})\sigma_{b_0 b_1} + t_{ij} t_{ij'} \sigma_{b_1}^2 + \sigma_c^2$.

Under the model given above, all observations within a level-two unit, such as a classroom, are correlated due the $u_{k\ell}$ random effect. So $\boldsymbol{Y}_i$ and $\boldsymbol{Y}_{i'}$ within level-two unit $k(\ell)$ are correlated. Let $\boldsymbol{Z}_k$ represent the model matrix for random effects at level two, the classroom level:

$$\mathbf{Z}_k = \begin{pmatrix} \mathbf{1}\, \boldsymbol{t}\, \mathbf{0}\, \mathbf{0}\, \mathbf{0}\, \mathbf{0} \cdots \mathbf{1} \\ \mathbf{0}\, \mathbf{0}\, \mathbf{1}\, \boldsymbol{t}\, \mathbf{0}\, \mathbf{0} \cdots \mathbf{1} \\ \vdots\ \vdots\ \vdots\ \vdots\ \vdots\ \vdots\ \ddots\ \vdots \\ \mathbf{0}\, \mathbf{0}\, \mathbf{0}\, \mathbf{0}\, \mathbf{1}\, \boldsymbol{t} \cdots \mathbf{1} \end{pmatrix}, \quad (4)$$

where $\mathbf{0}$, $\mathbf{1}$, and $\boldsymbol{t}$ are vectors of length n_i, and let

$$\boldsymbol{\Sigma}_b = \begin{pmatrix} \sigma_{b_0}^2 & \sigma_{b_0 b_1} \\ \sigma_{b_0 b_1} & \sigma_{b_1}^2 \end{pmatrix}, \quad (5)$$

then $\boldsymbol{G}_k = (\boldsymbol{I} \otimes \boldsymbol{\Sigma}_b) + \sigma_c^2 \boldsymbol{e}\boldsymbol{e}'$, where $\otimes$ is Kronecker's product and $\boldsymbol{e} = (0, 0, \ldots, 0, 1)'$. The parameter space is $\boldsymbol{\theta}$, $\boldsymbol{\theta} = \{\boldsymbol{\theta}\colon \beta \in R^M, \sigma_\varepsilon^2 > 0,$ and $\boldsymbol{G}_k$ non-negative definite$\}$. The covariance matrix for the observations within a cluster at the second level is given by $\boldsymbol{V}_k = \boldsymbol{Z}_k \boldsymbol{G}_k \boldsymbol{Z}_k' + \boldsymbol{R}_k$, where $\boldsymbol{R}_k = \sigma_\varepsilon^2 \boldsymbol{I}$ and $\boldsymbol{I}$ is has dimension $n_i n_k$. $\boldsymbol{V}_k$ has diagonal block $\boldsymbol{V}_i$, which are $n_i \times n_i$ and off-diagonal blocks $\sigma_c^2 \boldsymbol{J}$, where $\boldsymbol{J}$ is $n_i \times n_{i'}$. The H-dimensional vector $\boldsymbol{Y}$ is assumed to have a multivariate normal distribution with covariance matrix $\boldsymbol{V}_Y$, which has diagonal blocks $\boldsymbol{V}_k$ and, due to the independence of the level-two units, off-diagonal blocks $\boldsymbol{0}$.

2.2 Estimation Methods

The parameters of the linear mixed effects models can be estimated by the maximum likelihood method (ML) and the restricted maximum likelihood method (REML). The REML estimator is derived from the estimator of ML to correct the loss of the degrees of freedom involved in estimating the fixed effects. If we consider the model in the general balanced case, we assume that the scores of the students Y_k at the classroom level within treatment ℓ have a multivariate normal distribution. Let $\boldsymbol{\alpha} = (\sigma_{b_0}^2, \sigma_{b_1}^2, \sigma_{b_0 b_1}, \sigma_c^2, \sigma_\varepsilon^2)'$, then the REML function is defined as

$$L_{REML}(\boldsymbol{\alpha}) = \prod_{k=1}^{K} \{(2\pi)^{-0.5} |\boldsymbol{V}_k|^{-0.5} |\boldsymbol{X}_k' \boldsymbol{V}_k^{-1} \boldsymbol{X}_k|^{-0.5} \exp(-\frac{1}{2}(\boldsymbol{Y}_k - \boldsymbol{X}_k \boldsymbol{\beta})' \boldsymbol{V}_k^{-1} (\boldsymbol{Y}_k - \boldsymbol{X}_k \boldsymbol{\beta}))\} \tag{6}$$

$\boldsymbol{\beta}$ can be profiled out of this likelihood function, and then the estimates of variance components $\sigma_{b_0}^2$, $\sigma_{b_1}^2$, $\sigma_{b_0 b_1}$, σ_c^2 and σ_ε^2 are the solutions of REML equation using Fisher Scoring Algorithm with certain conditions. Let

$$\boldsymbol{X} = \begin{pmatrix} \boldsymbol{X}_1 \\ \cdots \\ \boldsymbol{X}_2 \\ \cdots \\ \boldsymbol{X}_3 \\ \vdots \\ \boldsymbol{X}_K \end{pmatrix} \text{ and } \boldsymbol{Z} = \begin{pmatrix} \boldsymbol{Z}_1 \\ \cdots \\ \boldsymbol{Z}_2 \\ \cdots \\ \boldsymbol{Z}_3 \\ \vdots \\ \boldsymbol{Z}_K \end{pmatrix}, \tag{7}$$

Then GLS estimator for fixed effects parameters given $\boldsymbol{V}_Y$ is given by

$$\hat{\boldsymbol{\beta}} = (\boldsymbol{X} \boldsymbol{V}_Y^{-1} \boldsymbol{X})^{-1} \boldsymbol{X}' \boldsymbol{V}_Y^{-1} \boldsymbol{Y} \tag{8}$$

When the variance components are unknown, REML estimates of the variance components are used to obtain $\widehat{\boldsymbol{G}}_k$, $\widehat{\boldsymbol{V}}_k$ and $\widehat{\boldsymbol{V}}_Y$, and then the EGLS estimator of $\boldsymbol{\beta}$ is given by

$$\hat{\boldsymbol{\beta}} = (\boldsymbol{X} \widehat{\boldsymbol{V}}_Y^{-1} \boldsymbol{X})^{-1} \boldsymbol{X}' \widehat{\boldsymbol{V}}_Y^{-1} \boldsymbol{Y} \tag{9}$$

The asymptotic distribution of $\hat{\boldsymbol{\beta}}$ is multivariate normal with covariance matrix

$$\mathrm{Cov}(\hat{\boldsymbol{\beta}}) = \left(\boldsymbol{X}'\boldsymbol{V}_{\boldsymbol{Y}}^{-1}\boldsymbol{X}\right)^{-1} \tag{10}$$

Let $\boldsymbol{P} = \boldsymbol{V}_{\boldsymbol{Y}}^{-1} - \boldsymbol{V}_{\boldsymbol{Y}}^{-1}\boldsymbol{X}(\boldsymbol{X}'\boldsymbol{V}_{\boldsymbol{Y}}^{-1}\boldsymbol{X})^{-1}\boldsymbol{X}'\boldsymbol{V}_{\boldsymbol{Y}}^{-1}$ and let $\tilde{\boldsymbol{\alpha}}$ be a solution to the REML equations, then the asymptotic covariance matrix of $\tilde{\boldsymbol{\alpha}}$ is $2\boldsymbol{\Omega}^{-1}$, where element (f, g) of the $q \times q$ matrix $\boldsymbol{\Omega}$ is given by $\mathrm{tr}(\boldsymbol{P}\boldsymbol{Z}_f\boldsymbol{Z}_f'\boldsymbol{P}\boldsymbol{Z}_g\boldsymbol{Z}_g')$, and where $\boldsymbol{Z}_f$ and $\boldsymbol{Z}_g$ are columns of $\boldsymbol{Z}$ for $f = 1, 2, \ldots, q$ and $g = 1, 2, \ldots, q$.

To construct parametric confidence intervals for $\boldsymbol{\beta}$, the REML estimates of variance components are used to obtain $\widehat{\boldsymbol{V}}_{\boldsymbol{Y}}$ which can be substituted for $\boldsymbol{V}_{\boldsymbol{Y}}$ to obtain $\widehat{\mathrm{Cov}(\hat{\boldsymbol{\beta}})}$, and then appropriate multiples of the estimated asymptotic standard error can be used to form the interval. Other methods such as profile-likelihood intervals are available (Demidenko 2013). There is a large literature on confidence intervals for variance components. The Wald method is available, using estimated asymptotic standard errors from $\widehat{\mathrm{Cov}(\tilde{\boldsymbol{\alpha}})}$, but it is well known that the performance of this method is not very good for intervals on variance components. Searle et al. (1992) present a number of methods, including ANOVA and maximum likelihood methods. Burdick and Graybill (1988) review several methods for intervals on variance components.

Parametric confidence intervals for the parameters of a linear mixed model perform well in large samples even if $\boldsymbol{Y}$ does not follow a multivariate normal distribution. In the remainder of this chapter, we examine the performance of bootstrap confidence intervals for a multi-level mixed model when the sample size is not large and when $\boldsymbol{Y}$ may or may not be distributed multivariate normal.

3 Bootstrap Methods

Consider a generic scalar parameter θ. The bootstrap principle is simple (Efron 1986; Boos 2003). In the real world, $\boldsymbol{Y} = (Y_1, \ldots, Y_n)$ is an observed random sample which is sampled from an unknown probability distribution $F(\cdot)$, and the statistic of interest is a function of Y: $\hat{\theta} = s(\boldsymbol{Y})$. In the bootstrap world, $\boldsymbol{Y}^* = (Y_1^*, \ldots, Y_n^*)$ is an observed bootstrap sample which is sampled from the empirical distribution $\hat{F}(\cdot)$, where $\hat{F}(A) = \frac{1}{n}\sum_{i=1}^{n} 1_A(Y_i)$, for $A \subseteq R$, and the statistic is $\widehat{\theta^*} = s(\boldsymbol{Y}^*)$. Instead of evaluating the statistical properties (bias, standard errors, etc.) of $\hat{\theta}$ based on the sampling distribution of $\hat{\theta}$, we mimic this process by evaluating these properties of $\widehat{\theta^*}$ based on the bootstrap sampling distribution of $\widehat{\theta^*}$. The benefit of doing so is we do not actually need to compute the exact bootstrap sampling distribution of $\widehat{\theta^*}$, we can use Monte-Carlo methods to obtain an approximation: draw B independent bootstrap samples $\boldsymbol{Y}^{*(1)}, \ldots, \boldsymbol{Y}^{*(B)}$ from $\hat{F}$, compute $\widehat{\theta^*}$ for each bootstrap sample, and finally compute the estimated bias, standard errors and confidence intervals from $\widehat{\theta^{*(1)}}, \ldots, \widehat{\theta^{*(B)}}$. An important assumption is that resampled cases in the bootstrap samples are *i.i.d*.

The bootstrap principle relies on two asymptotic results: (1) The empirical distribution from the sample, $\hat{F}(\cdot)$, converges almost surely to $F(\cdot)$, the unknown theoretical distribution, assuming $i.i.d$ observations, and (2) the distribution of $\hat{\theta}^*$, from the bootstrap distribution around $\widehat{\theta}$ converges to the distribution of $\hat{\theta}$ around θ.

For longitudinal data, bootstrapping will be more complicated due to the clustering of observations. In this chapter, we will explore the parametric bootstrap, the cluster bootstrap and a two-stage bootstrap.

3.1 Bootstrap Estimates

As in the previous section, the statistic of interest is a function of Y: $\hat{\theta} = s(\boldsymbol{Y})$ to which we will assign an estimated standard error. Let $\sigma(F)$ be the standard error of $\hat{\theta}$, indicating a function of the distribution F. Then $\sigma(F) = (\text{Var}_F(s(\boldsymbol{Y})))^{\frac{1}{2}}$, and we define the bootstrap estimate of standard error as $\hat{\sigma}_B = \sigma(\hat{F})$.

In most cases, it is difficult to calculate the function $\sigma(\hat{F})$; however, since we notice that a bootstrap sample is just the random sample of size n drawn with replacement from the original data $\boldsymbol{Y} = (Y_1, \ldots, Y_n)$, we can use a Monte-Carlo algorithm approach to $\sigma(\hat{F})$. Assume the statistics calculated from each bootstrap sample are $\widehat{\theta*^{(1)}}, \ldots, \widehat{\theta*^{(B)}}$ the sample standard deviation of the $\widehat{\theta*^{(b)}}$, $b = 1, 2, \ldots, B$ is

$$\widehat{\sigma}_B^* = \sqrt{\frac{1}{B-1}\sum_{b=1}^{B}\{\hat{\theta}^{*(b)} - \hat{\theta}^{*(\cdot)}\}^2} \tag{11}$$

where

$$\hat{\theta}^{*(\cdot)} = \frac{1}{B}\sum_{b=1}^{B}\hat{\theta}^{*(b)} \tag{12}$$

It is easy to see that as $B \to \infty$, $\widehat{\sigma_B^*}$ gets closer to $\widehat{\sigma^*} = \sigma(\hat{F})$, and it has been shown by Efron that the difference between $\widehat{\sigma_B^*}$ and $\widehat{\sigma^*}$ can be ignored once B is adequately large (50–200).

For the non-bootstrap case, based on the definition of bias, the bias of the statistic of interest $\hat{\theta} = s(\boldsymbol{Y})$ for estimating parameter θ is $\text{Bias}(\hat{\theta}) = E_F(s(\boldsymbol{Y}) - \theta(F)) = E_F(s(\boldsymbol{Y}) - \theta(F)$, where E represents expectation with respect to the probability distribution F. For the bootstrap case, the bootstrap estimate of bias is $\widehat{\text{Bias}}(\hat{\theta}^*) = E_{\hat{F}}(s(\boldsymbol{Y}^*) - \theta(F)) = E_{\hat{F}}(s(\boldsymbol{Y}^*) - \theta(F)$, where E represents expectation with respect to the probability distribution $\hat{F}$.

As for the bootstrap estimate of standard error, we can apply a Monte-Carlo algorithm approach to obtain an estimate of the bias:

$$\widehat{\text{Bias}}(\hat{\theta}^*) = \frac{1}{B}\sum_{b=1}^{B} \hat{\theta}^{*(b)} - \theta(\hat{F}) \tag{13}$$

As $B \to \infty$, $\widehat{\text{Bias}}(\hat{\theta}^*) \to \widehat{\text{Bias}}(\hat{\theta})$.

3.1.1 Parametric Bootstrap

The parametric bootstrap requires strong assumptions about the model and the distributions of the errors because it depends on the model and the distribution of the errors. First for the the model of interest, which is the mixed-effects model in Sect. 2, parameter estimates $(\hat{\boldsymbol{\beta}},\ \hat{\sigma}_\varepsilon^2,\ \widehat{\boldsymbol{G}}_k)$ are obtained by a chosen method of estimation. For the model of Sect. 2, REML estimation will be used. The parametric bootstrap then proceeds as follows:

1. Simulate $\varepsilon^*_{ijk\ell}$ from $N(0, \hat{\sigma}^2)$ and simulate the random effects from the distribution $MNV(0, \hat{G}_k)$.
2. Generate bootstrap sample data by setting

$$Y^*_{ijk\ell} = \hat{\beta}_0 + \hat{\beta}_1 X_{1kl} + \hat{\beta}_2 t_{ij} + \hat{\beta}_3 t^2_{ij} + \hat{\beta}_4 X_{1k\ell} t_{ij} + \hat{\beta}_5 X_{1k\ell} t^2_{ij} + b^*_{0i} + u^*_{kl} + b^*_{1i} t_{ij} + \varepsilon^*_{ijk\ell} \tag{14}$$

3. Refit the model of interest to the new bootstrap sample from step 2, and obtain bootstrap estimates $(\hat{\boldsymbol{\beta}}^*,\ \hat{\sigma}_\varepsilon^{2*},\ \widehat{\boldsymbol{G}}_k^*)$
4. Repeat steps 1–3 B times to obtain B sets of bootstrap estimates

Then calculate bootstrap standard errors, bias, and confidence intervals from the bootstrap distribution. In some literature, including the R package *lme4*, there is a distinction between a parametric bootstrap where only $\varepsilon^*_{ijk\ell}$ is simulated and a parametric bootstrap where all random effects are simulated. The former is sometimes called the parametric residual bootstrap.

3.1.2 Cluster Bootstrap

The cluster bootstrap is a nonparametric method which does not depend on assumptions about the distribution of the response variable Y. The cluster bootstrap has been studied by Field (2007). With repeated measures in a multi-level model, there are two (or more) cluster levels, and clusters are nested within clusters. The simple cluster bootstrap would take a cluster at only one level as the resampling unit. In an application where children are nested within classrooms, for example, with repeated measurements on the children, case-resampling at the child level is a type of cluster bootstrap, but it completely ignores the correlation among children within a classroom and thus does not meet the assumption of resampling *i.i.d* units. Resampling at the level-two unit, which would be the classroom, would satisfy the assumption of *i.i.d* units, but it would not take advantage of resampling also at the level-one unit.

Re-sampling level-two clusters instead of case-resampling level-one observations can preserve the dependence of the data within each cluster and independence of observations at the cluster level. The actual steps for the cluster bootstrap are the same as described at the introduction to this section, but the units to be resampled are the clusters: $(\boldsymbol{Y}_k, \boldsymbol{X}_k, \boldsymbol{Z}_k)$. The bootstrap sample will have the same number of level-two units as the original sample, but if the study design is unbalanced, then the total number of observations in the bootstrap sample may not match the total number of observations in the original sample unless bootstrap sampling is stratified by cluster size. Stratified sampling by cluster size may be impractical.

3.1.3 Two-Stage Bootstrap

The Monte-Carlo two-stage bootstrap adds another step to the algorithm described at the beginning of this section in order to include resampling at the level-one unit in addition to resampling at the level-two unit. The multi-level bootstrap has been studied by Field and Welsch (2008). Under the mixed-effects model in Sect. 2, level-one units within the same level-two unit are not independent, but they are conditionally independent given the cluster. Some two-stage algorithms use residual resampling at the lower-level unit. In the results to be reported below, we use case-resampling within the level-two unit. Hence for each level-two cluster sampled with replacement, $(\boldsymbol{Y}_k, \boldsymbol{X}_k, \boldsymbol{Z}_k)$, we sample n_k level-one clusters, $(\boldsymbol{Y}_i, \boldsymbol{X}_i, \boldsymbol{Z}_i)$, with replacement. If the application is children nested within classrooms, for each classroom sampled with replacement, we sample n_k children within that classroom with replacement. If the cluster sizes, n_k are small, then resampling within a cluster may not be beneficial, and could be detrimental, since Monte-Carlo bootstrap is known to not work well with small samples. Due to resampling with small n_k, degenerate bootstrap samples for which estimation of the mixed-effects model in Sect. 2 will not converge becomes a larger issue.

3.2 Bootstrap Confidence Intervals

Three bootstrap confidence interval methods are explored in this chapter, which are the percentile method, Bias Corrected Accelerated (BCa) method and Bootstrap-t method (Efron 1987; Thomas and Efron 1996). For the percentile method, suppose B bootstrap replications of $\hat{\theta}^*$, are denoted by $(\hat{\theta}^{*1}, \hat{\theta}^{*2}, \ldots, \hat{\theta}^{*B})$. After ranking from the bottom to the top, the empirical order statistics are written as $(\hat{\theta}^{*(1)}, \hat{\theta}^{*(2)}, \ldots, \hat{\theta}^{*(B)})$. Then the bootstrap percentile method with 95% level of confidence interval is given by

$$\hat{\theta}^{*(0.025*B)} \leq \theta \leq \hat{\theta}^{*(0.975*B)} \tag{15}$$

Then

$$\hat{P}(\hat{\theta}^* \leq \theta_{L(percentile)}) = \frac{1}{B}\sum_{b=1}^{B} 1\{\widehat{\theta^{*(b)}} \leq \theta_{L(percentile)}\} \approx \frac{1}{2}\alpha \tag{16}$$

and

$$\hat{P}(\hat{\theta}^* \geq \theta_{U(percentile)}) = \frac{1}{B}\sum_{b=1}^{B} 1\{\widehat{\theta^{*(b)}} \geq \theta_{U(percentile)}\} \approx \frac{1}{2}\alpha \tag{17}$$

The interval is not necessarily symmetric around $\hat{\theta}$.

The BCa method is an improvement over the percentile method because it attempts to shift and scale the percentile bootstrap confidence interval to compensate for bias and nonconstant variance $\eta\ \hat{\theta}$. There are two parameters involved in calculating the BCa confidence interval. One parameter, z_0, attempts to correct bias and another, a, is the acceleration factor. z_0 is defined as

$$z_0 = \Phi^{-1}(G(\hat{\theta})) = \Phi^{-1}\{\frac{\#(\hat{\theta}_b^* < \hat{\theta})}{B}\}. \tag{18}$$

G is the cumulative bootstrap sample distribution of statistic $\hat{\theta}^*$, that is $G(t) = (\#(\hat{\theta}_b^* < \hat{\theta})) \div B$, $(b = 1, 2, \ldots, B)$, $G^{-1}(\cdot)$ is the corresponding quantile function from the bootstrap distribution. $\Phi^{-1}(\cdot)$ is the quantile function from the standard normal distribution. The acceleration factor is a jackknife estimate obtained by deleting observation X_i from the original data $(X_1, X_2, \ldots, X_n)$ and then fitting the model to produce the deleted statistics $\hat{\theta}_{-i}$, where n is the number of observations. Then the acceleration factor is

$$a = \frac{\sum_{i=1}^{n}(\tilde{\theta} - \hat{\theta}_{-i})^3}{6(\sum_{i=1}^{n}(\tilde{\theta} - \hat{\theta}_{-i})^2)^{1.5}} \tag{19}$$

where

$$\tilde{\theta} = \frac{\sum_{i=1}^{n}\hat{\theta}_{-i}}{n} \tag{20}$$

Therefore, the lower and upper limit of the confidence interval can be stated as follows:

$$\theta_{LBCa} = G^{-1}\{\Phi(z_0 + \frac{z_0 + z_{\alpha/2}}{1 - a(z_0 + z_{\alpha/2})})\} \tag{21}$$

and

$$\theta_{UBCa} = G^{-1}\{\Phi(z_0 + \frac{z_0 + z_{1-\alpha/2}}{1 - a(z_0 + z_{1-\alpha/2})})\} \tag{22}$$

where $\hat{\theta}$ is the estimate from the original data. If $a = 0$, the confidence interval is called bias corrected percentile confidence interval (BC). And if $a = 0$ and $z_0 = 0$, the interval is the same as percentile confidence interval.

Finally, the bootstrap-t confidence interval is also considered in our study. The bootstrap-t takes the same form as the normal confidence interval except that instead of using the quantile from a t-distribution (or a normal distribution), a bootstrapped t-distribution is constructed from which the quantiles are computed:

$$t_b^* = \frac{\hat{\theta}^{*(b)} - \hat{\theta}}{s.e.(\hat{\theta}^{*(b)})} \tag{23}$$

where the standard error of the bootstrap estimate $s.e.(\hat{\theta}^{*(b)})$ are computed on each bootstrap sample. Then based on these quantiles, the 95% bootstrap-t percentile confidence interval is

$$\hat{\theta} - t_{0.975}^* * S \leq \theta \leq \hat{\theta} - t_{0.025}^* * S \tag{24}$$

where $t_{0.025}^*$ and $t_{0.975}^*$ are percentile of the $t^{*(b)}$ distribution and S is the estimated standard deviation for $\hat{\theta}$.

There is also a percentile-t confidence interval:

$$\hat{P}(t_b^* \leq t_{L(percentile)}) = \frac{1}{B}\sum_{b=1}^{B} 1\{t_b^* \leq t_{L(percentile)}\} \approx \frac{1}{2}\alpha \tag{25}$$

and

$$\hat{P}(t_b^* \geq t_{U(percentile)}) = \frac{1}{B}\sum_{b=1}^{B} 1\{t_b^* \geq t_{U(percentile)}\} \approx \frac{1}{2}\alpha \tag{26}$$

The multilevel bootstrap is more complicated than the ordinary bootstrap. In the next section, we present simulation results using the parametric bootstrap and case bootstrap at one level and two levels. The parametric bootstrap generates the bootstrap samples after modeling, which assumes that the model is correctly specified. The case-resampling bootstrap draws bootstrap samples before the model is fit, which is more robust to model misspecification.

4 Monte-Carlo Simulation Study

Monte-Carlo simulations are used to model two-level longitudinal data to examine the property of the bootstrap methods applied for this setting.

4.1 The Simulation Design

The simulation design is inspired from an original study where preschool children were nested within classrooms. Classroom is the level-two unit, and child is the level-one unit. The original study was conducted to assess a treatment for developmental delay in speech and language, and classrooms were randomly allocated to either treatment or control. The original study was not quite balanced at the classroom level, and we replicate this imbalance in our simulations. The number of children per classroom was quite small, with about $n_k = 5$ children per classroom on average. Simulation results are reported below for five children per classroom and also, for comparison, 15 children per classroom. First we generate 500 Monte-Carlo data sets with 92 classrooms where each classroom has either five or 15 students and each student has six observations at time points 1, 2, 3.5, 6, 7.3, and 8.8 (weeks). Pseudo data are generated under the model given in expression (1) For the generated data, the random effects are standard normal, and the random error is standard normal. The fixed effects are treatment condition, time and quadratic time. Parameter estimates obtained from fitting the linear mixed-model to the original study data were used as the parameter values for the simulations. The values are $\boldsymbol{\beta} = (4.87,\ -0.98,\ 0.92,\ -0.015,\ 0.8989,\ -0.015)'$, $\sigma_{\beta_0} = 6.63$, $\sigma_{\beta_1} = 0.78$, $\rho_{\beta_0\beta_1} = -0.27$, $\sigma_c = 3.02$, and $\sigma_\varepsilon = 1.685$.

In order to run the simulations effectively, parallel computing is essential. For example, to perform a simulation using 500 Monte-Carlo samples with five children per classroom and 1000 bootstrap samples per Monte-Carlo sample, nearly 18 h is needed when parallel computing is run on 8 cores, about 47 h for 15 students per classroom, but almost 135 h is needed to run the simulation with five students in each classroom running on a single core. Simulations were run using the *lme4* Bates et al. (2015) package in the R system with the *foreach* and *dorng* packages for parallel computing.

In addition to the three bootstrap methods described above, parametric bootstrap, cluster bootstrap at classroom level and two-stage bootstrap, we also include a simulation without bootstrapping. As mentioned above, we also vary the classroom size. In Fig. 1, which are the figures of data sets, we find some response values are negative, because the random variables and random errors are generated from the normal

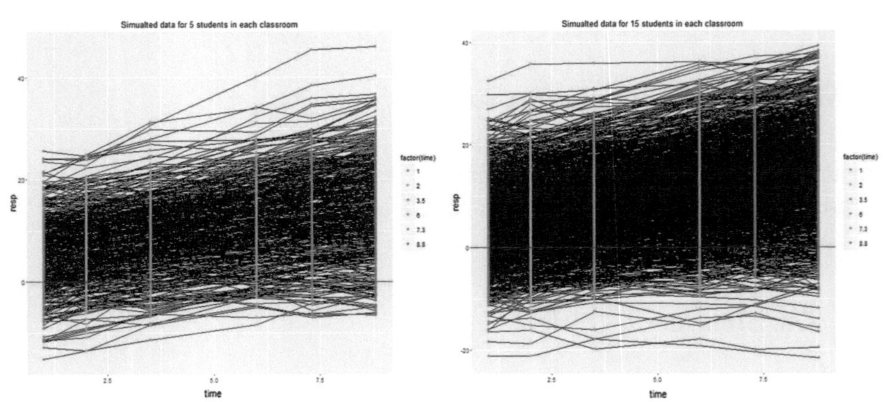

Fig. 1 Simulated data for five students (*left*) and 15 students (*right*)

distribution, which can result in good performance of the simulated data. These values will be kept in the simulation, because the simulations are not to provide realistic results but to evaluate the bootstrap methods. The value for the intercept can be set to a larger value to produce fewer negative response. The bootstrap methods will be compared with respect to the bias, standard deviation and coverage rate of confidence intervals. The coverage rate of the 95% bootstrap confidence interval is the percentage of intervals that contain the true value. The coverage rate of the 95% confidence interval will be considered to be good (90–100%), low (80–90%), or poor (<80%).

4.2 Simulation Results Five Students per Classroom

We first present simulation results with respect to the mean of bias, the mean of standard deviation and the coverage rate for five students per classroom. In the following tables, β_0 stands for the intercept parameter, β_1 for condition, β_2 for linear time trend, β_3 for quadratic time trend, β_4 for interaction of condition and linear time trend, β_5 for interaction of condition and quadratic time trend. σ_ε is the standard deviation of the residual term, σ_{β_0} and σ_{β_1} are the standard deviation of the random intercept and random slope for at the child level, respectively. $\rho_{\beta_0\beta_1}$ is the correlation of random intercept and random slope, and σ_c is the standard deviation of the random intercept at the classroom level. TID is a reference to the classroom (teacher) level. In Table 1, the coverage rates of bootstrap-t are not available for the random effects. From the table, we find that the coverage rate for the fixed effects parameters for these four methods are all near to 95%, but for the standard deviations of the random intercept at the child level (σ_{β_0}) and the random intercept at the classroom level (σ_c), the coverage rates are lower than 90% for the two-stage bootstrap with percentile and BCa confidence intervals. This may be in the two-stage bootstrap, where observations at the student level are resampled, the classroom sample size is too small, which affects the random effects more than the fixed effects. To investigate the reason for these results, we look at the bias and standard error of the parameter estimates. Here we use the mean of the bias to compare the methods. From Table 2, the bias for the fixed effects are small, which is in accordance of the coverage rate around 95%. It is interesting to note that the parametric bootstrap can achieve the smallest bias among the three methods, especially for the random effects, which suggests that the model is a good fit for the data. The bias of random effects for the cluster bootstrap at classroom level are larger than the parametric bootstrap, which may be due to re-sampling producing some degenerated bootstrap samples with only five students per classroom. The bias of the cluster bootstrap at the classroom level are smaller than the two-stage bootstrap. The two-stage bootstrap always has the largest bias among the three methods for the random effects. Furthermore, the bias of the standard deviations of the random intercept at the student level (σ_{β_0}) and random intercept at the classroom level (σ_c) are quite large in the two-stage bootstrap, which may contribute to the low coverage rate of confidence intervals. Table 3 is the average standard deviations of the parameter estimates. It is interesting to find that the standard

Table 1 Coverage rates of confidence intervals with five students per classroom

Methods	Coverage rate (%)										
	β_0	β_1	β_2	β_3	β_4	β_5	σ_ε	σ_{β_0}	$\rho_{\beta_0\beta_1}$	σ_{β_1}	σ_c
No bootstrap	97.2	96.2	96.2	96.8	94.6	96	96.2	94.4	95.6	94.2	94.6
Parametric (perc)	96	96.6	96.4	96.8	95	96	96.2	91.6	95.2	94.8	91.2
Parametric (BCa)	95.2	96.4	96	96	94.6	95.6	96.6	91.8	95	93.8	91.8
Cluster TID (perc)	95.4	95.6	96.2	95	93.8	94.4	95.6	96.6	94.8	92.8	93.8
Cluster TID (BCa)	95.6	95.8	96.4	94.8	93.8	94.2	94.6	91.6	93.6	93.6	91.2
Cluster TID (bt)	96.4	96.4	96.6	95.8	94.4	94.4	NA	NA	NA	NA	NA
Two-stage (perc)	98.8	99	99.6	99.6	99.6	99.4	99.2	*77.8*	98.2	98.4	*82.2*
Two-stage (BCa)	98.8	98.8	99.8	99.8	99.6	99.6	99.2	*69.8*	99.2	98.6	*67.6*
Two-stage (bt)	97	96.4	99.6	99.4	99.6	99.4	NA	NA	NA	NA	NA

Table 2 Bootstrap bias of parameter estimates with five students per classroom

Methods	Average bias of parameter estimates										
	β_0	β_1	β_2	β_3	β_4	β_5	σ_{β_0}	σ_{β_1}	$\rho_{\beta_0\beta_1}$	σ_c	σ_ε
No boot	0.01293	−0.0343	−0.00143	0.00009	0.00761	−0.00091	0.026992	−0.00092	−0.00024	−0.14647	−0.00054
Parametric	−0.00012	0.00009	−0.00002	−0.000007	−0.00013	0.000019	−0.00008	0.00027	0.000271	−0.01281	−0.00081
Cluster TID	0.00117	0.00066	−0.00005	0.000005	−0.00002	0.000002	−0.00033	−0.0022	0.001258	−0.130047	−0.0009
Two-stage	0.00007	−0.001	−0.00003	0.000006	−0.00036	0.000008	*−0.65341*	−0.004	−0.02664	*1.10471*	−0.0018

Table 3 Bootstrap standard deviation of parameter estimates with five students per classroom

Methods	Average of SDs of parameter estimates										
	β_0	β_1	β_2	β_3	β_4	β_5	σ_{β_0}	σ_{β_1}	$\rho_{\beta_0\beta_1}$	σ_c	σ_ε
No boot	0.592431	0.831524	0.076586	0.006417	0.118086	0.009409	0.283532	0.028045	0.047708	0.72092	0.023021
Parametric	0.611053	0.893888	0.081872	0.00658	0.119819	0.009632	0.249921	0.027583	0.047534	0.468162	0.023489
Cluster TID	0.60967	0.889384	0.08073	0.006442	0.118116	0.009437	0.282421	0.027116	0.047392	0.763778	0.023108
Two-stage	*0.72724*	*1.0610*	*0.109223*	*0.008728*	*0.159406*	*0.012757*	*0.374048*	*0.036588*	*0.068787*	*0.79873*	*0.031249*

deviations of the cluster bootstrap at the classroom level and parametric bootstrap are similar. Furthermore, the two-stage bootstrap has the largest standard deviation, which may due to the resampling the small number of observations at the student level. We also find that the standard deviation σ_{β_0} and σ_c have the largest difference compared to other parameter estimates for the three methods, which is the reason why the coverage rates of σ_{β_0} and σ_c are relative smaller than other random effects. Thus in terms of bias, standard deviation and coverage rate, we find that the parametric bootstrap and cluster bootstrap at classroom level perform better than the two-stage bootstrap.

4.3 Simulation Results 15 Students per Classroom

In this section, we present results with respects to coverage rate, bias and standard deviation when the simulation design has 15 students per classroom. Other conditions for the simulations match the conditions for the simulations with five students per classroom. Performance of the two-stage bootstrap my improve with 15 students per classroom. In Table 4, we find almost the same pattern of results as in the simulations for five students per classroom. The coverage rates for fixed effects are around 95% and the random effects are all good except for σ_ε and σ_c. The coverage rates for σ_{β_0} and σ_c in the two-stage method improved compared to five students per classroom. The bootstrap bias with 15 students shown in Table 5 are fairly small for all the parameter estimates except σ_{β_0} and σ_c for the two-stage bootstrap, but the bias for the two-stage bootstrap has improved compared to five students per classroom, which is the reason why the coverage rates for σ_{β_0} and σ_c with 15 students per classroom are greater compared to five students per classroom. For the bootstrap standard deviation results with 15 students per classroom in Table 6, the standard deviations of the cluster bootstrap at classroom level and parametric bootstrap provide similar results, while the two-stage bootstrap shows the largest standard deviations. All of the standard deviations are smaller compared to five students per classroom.

4.4 Comparison of Simulation Results for Five Students per Classroom and 15 Students per Classroom

We can see some distinctions for the two conditions of classroom cluster size from the tables of coverage rate, bias and standard deviations. To make the comparison more straightforward, we present some plots to compare the bias and standard deviation of the parameter estimates from two designs. In these plots, sdcor1 is σ_{β_0}, sdcor3 is σ_{β_1}, sdcor2 is $\rho_{\beta_0\beta_1}$, sdcor4 is σ_c, and sdcor is σ_ε.

Table 4 Confidence interval coverage rate 15 students per classroom

Methods	Coverage rate of confidence intervals (%)										
	β_0	β_1	β_2	β_3	β_4	β_5	σ_ε	σ_{β_0}	$\rho_{\beta_0\beta_1}$	σ_{β_1}	σ_c
No boot	95.6	95.4	94.8	95	96	95.6	96.2	95.8	93.6	94.4	94.2
Parametric (perc)	95.6	95.6	94.4	94.6	96	95.2	96.2	95	93.6	95	93.2
Parametric (BCa)	95.4	95.4	93.6	94.2	95.8	95.4	96	94.6	93.2	94.2	94.2
Cluster TID (perc)	95.2	94	93.6	94	95	94.6	95.6	96.2	93	93.2	90.8
Cluster TID (BCa)	95.2	94.4	93.2	93.6	95.6	95.2	96	94.8	93.8	92.6	96
Cluster TID (bt)	95.8	95	94	94	95.2	95.8	NA	NA	NA	NA	NA
Two-stage (perc)	97.8	97.2	99.2	99.4	99	99.6	99.4	*86*	99.4	99	*86.4*
Two-stage (BCa)	97.6	97	99.2	99.2	98.8	99.4	99.4	*85.2*	99.4	98.8	*72.2*
Two-stage (bt)	95.6	95.2	98.8	98.8	98.8	99.6	NA	NA	NA	NA	NA

Table 5 Bootstrap bias of parameter estimates 15 students per classroom

Methods	Average bias of parameter estimates										
	β_0	β_1	β_2	β_3	β_4	β_5	σ_{β_0}	σ_{β_1}	$\rho_{\beta_0\beta_1}$	σ_c	σ_ε
No boot	0.00269	−0.00504	−0.00021	−0.00007	−0.00445	0.00035	0.01157	0.00079	0.002104	−0.03376	−0.00017
Parametric	−0.00053	0.00083	−0.00006	0.000002	0.00005	−0.000005	0.0006	0.00011	0.00009	−0.01127	−0.00024
Cluster TID	0.00055	0.00108	0.00009	−0.00009	−0.00008	0.000013	−0.00162	−0.00073	0.00039	−0.06034	−0.0003
Two-stage	0.00081	0.00076	−0.00008	0.000003	0.00007	−0.000002	*−0.21708*	−0.00151	−0.00819	*0.39536*	−0.00063

Table 6 Bootstrap standard deviation of parameter estimates 15 students per classroom

Methods	Average SD of parameter estimates										
	β_0	β_1	β_2	β_3	β_4	β_5	σ_{β_0}	σ_{β_1}	$\rho_{\beta_0\beta_1}$	σ_c	σ_ε
No boot	0.4828	0.7248	0.04818	0.003771	0.068089	0.005443	0.14797	0.01635	0.02673	0.40326	0.013189
Parametric	0.4975	0.7287	0.0473	0.003797	0.06913	0.00555	0.1347	0.01592	0.02667	0.2957	0.01354
Cluster TID	0.4941	0.72207	0.04692	0.00375	0.068398	0.00551	0.14497	0.01565	0.02649	0.39855	0.013418
Two-stage	*0.5534*	*0.81003*	*0.06564*	*0.005254*	*0.095722*	*0.007692*	*0.18762*	*0.02189*	*0.03779*	*0.36133*	*0.018765*

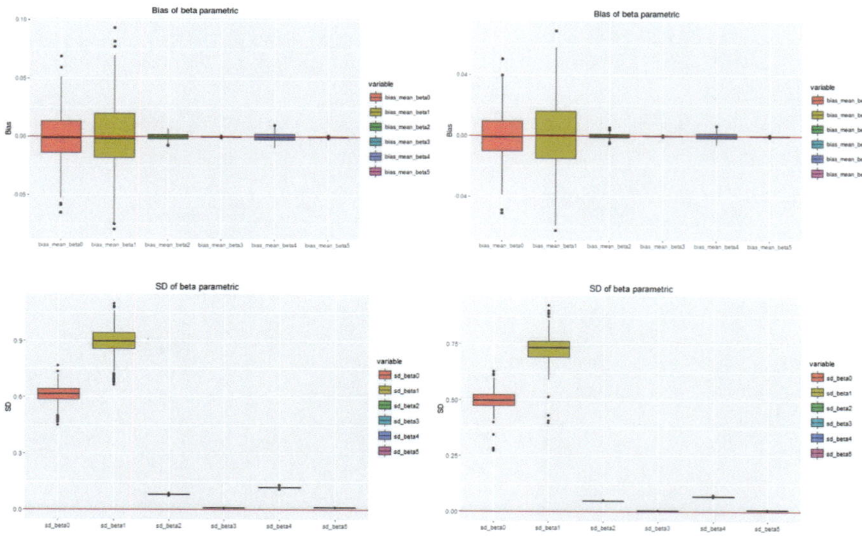

Fig. 2 Comparison of cluster size effect for parametric bootstrap (*Left* 5 students per classroom, *Right* 15 students per classroom) for fixed effect estimates

4.4.1 Comparisons of Results for Cluster Size for the Parametric Bootstrap

Simulation results for both 5 students and 15 students per classroom show the coverage rate about 95% for confidence intervals for all parameters. To analyze the difference, we present some figures to show the difference between the two conditions with respect to bootstrap bias and standard deviation.

In the simulation results, we see the bias for β_2, β_3, β_4, β_5 are very small in Fig. 2, but for β_0 and β_1, the bias and standard deviation are larger than for other fixed effect estimates. The bias in the condition with 5 students per classroom have wider distribution than in the condition with 15 students per classroom, which shows the same pattern as the standard deviations. Also, the bootstrap standard deviations of the parameter estimates with 5 students per classroom are larger than with 15 students per classroom, even though some are not noticeable in the figure. So increasing the number of level-one units in each level-two cluster can reduce the bootstrap standard error and bias of the fixed effect estimators in the parametric bootstrap method.

For the bias and standard deviation of variance component estimates in Fig. 3, the bias of σ_{β_1}, $\rho_{\beta_0\beta_1}$ and σ_ε are close to zero, but the values of σ_{β_0} and σ_c are highly skewed and have outliers far away from zero. The condition with 5 students has higher bias and standard deviation than the condition with 15 students. So more level-one units in each level-two cluster can reduce the bootstrap bias and SD of the parameter estimates.

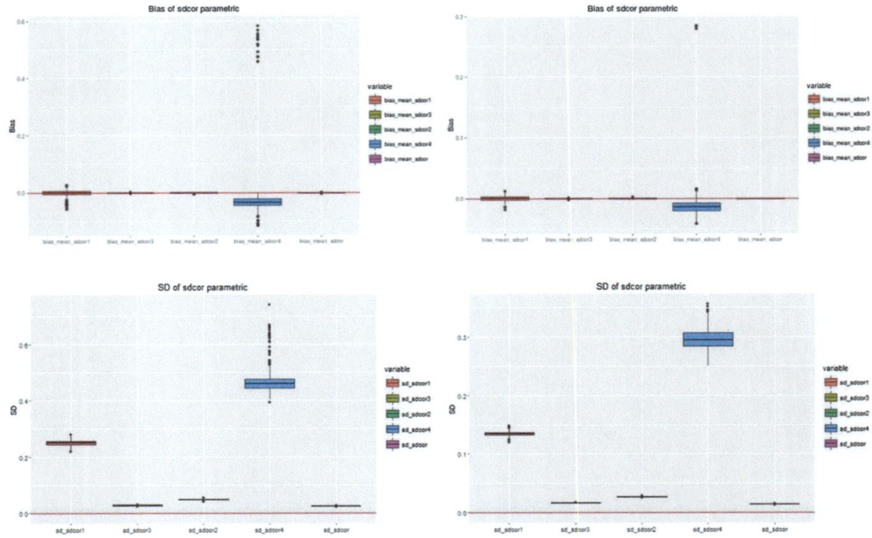

Fig. 3 Comparison of cluster size effect for parametric bootstrap (*Left* 5 students per classroom, *Right* 15 Students per classroom) for variance component estimates

4.4.2 Comparisons of Results for Cluster Size for Cluster Bootstrap at the Classroom Level

For the cluster bootstrap at the classroom level in Fig. 4, we have the same patterns for the fixed effect parameter estimates when comparing the condition with 5 Students per classroom and 15 students per classroom. The condition with 5 students per classroom has more bias and standard deviation than the condition with 15 students in each classroom, especially for β_0 and β_1 estimates. Also, the median bias with 15 students per cluster is closer to zero than 5 students in the figure. The results are theoretically plausible, as the sample size increases, the biases and standard deviations become smaller. For the variance components in Fig. 5, the condition with 15 students per classroom has less outliers, less bias and smaller standard deviation when compared with the condition with 5 students in each classroom. So from the aspects of the bias and standard deviation, more level-one units per cluster is beneficial.

4.4.3 Comparisons of Results for Cluster Size for Two-Stage Bootstrap

In Fig. 6, effects estimates in the condition with 5 students per classroom are larger compared to 15 students, which shows the same pattern as parametric and classroom bootstrap. So the estimator for intercept and condition have more bias and standard deviation than the other fixed-effect parameter estimators.

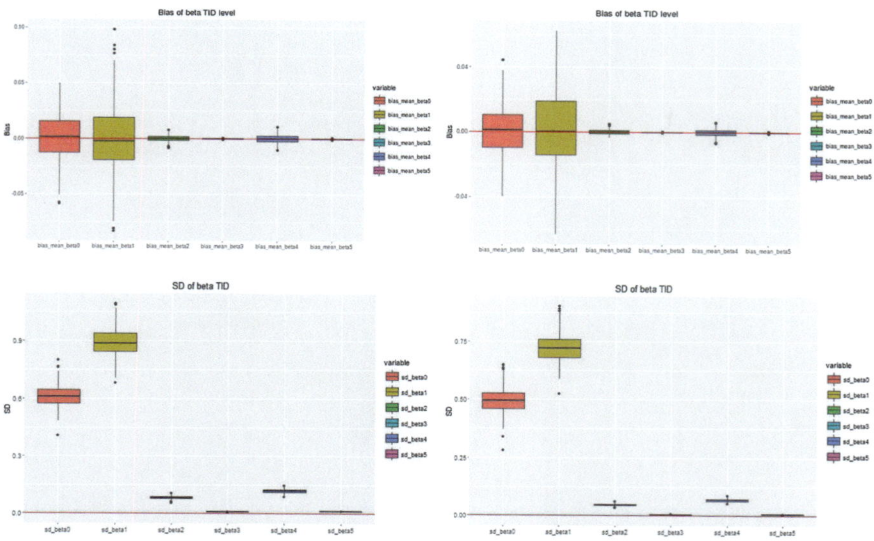

Fig. 4 Comparison of results for cluster size for cluster bootstrap (*Left* 5 students per classroom, *Right*: 15 students per classroom) for variance component estimates

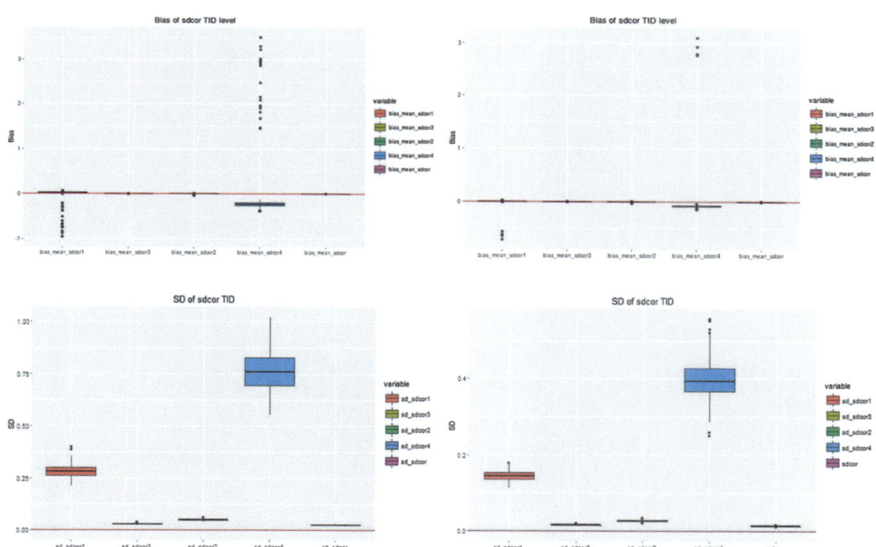

Fig. 5 Comparison of results for cluster size for cluster bootstrap (*Left* 5 students per classroom, *Right* 15 students per classroom) in terms of variance component estimates

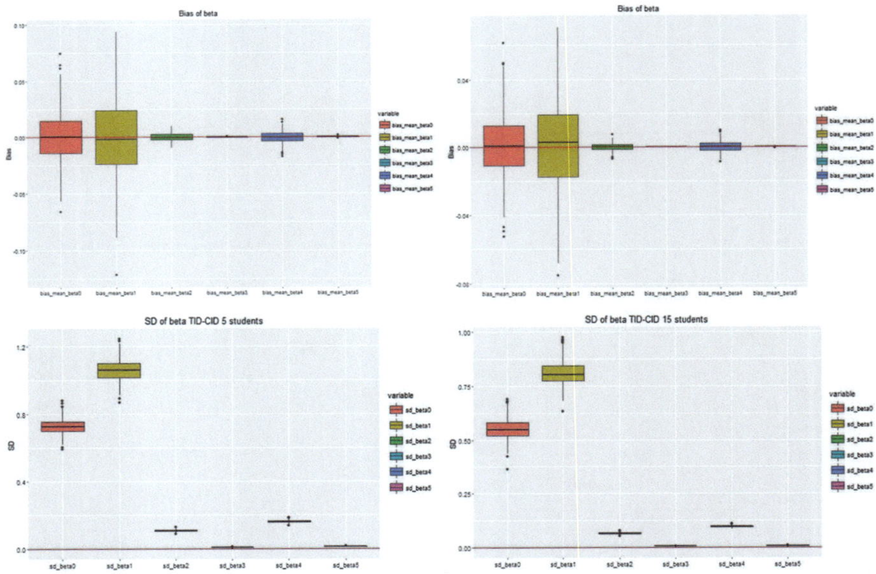

Fig. 6 Comparison of results for cluster size for two-stage bootstrap (*Left* 5 students per classroom, *Right* 15 students per classroom) for fixed-effect estimates

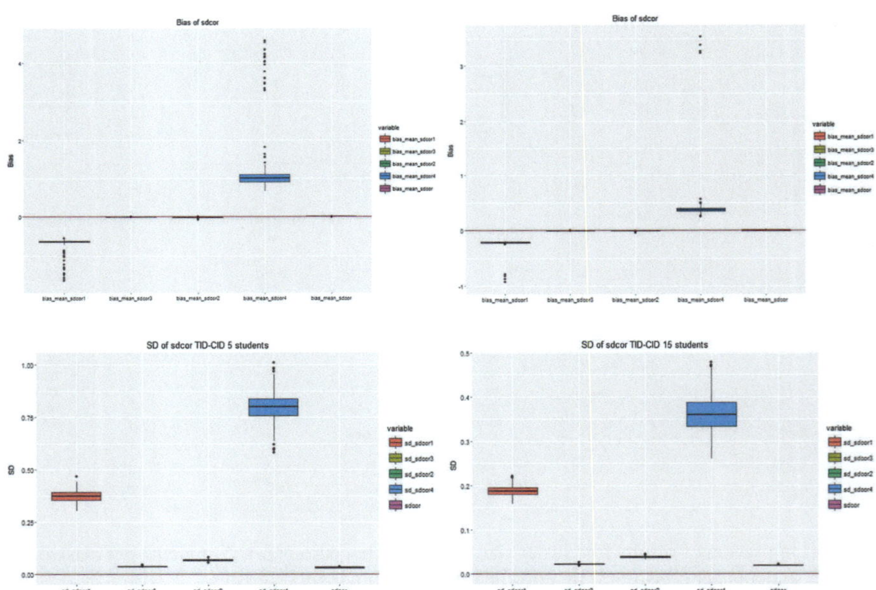

Fig. 7 Comparison of results for cluster size for two-stage bootstrap (*Left* 5 students per classroom, *Right* 15 students per classroom) for variance component estimates

In Fig. 7, the comparison of variance component estimation for the two cluster size conditions are presented. As before, the standard deviation of random intercept at the student level (σ_{β_0}) and random intercept at the classroom level (σ_c) show high bias and high bootstrap standard deviation. The bias in the two-stage bootstrap are much larger than the bias in the parametric and classroom-cluster bootstrap. For example, the median bias of σ_c for 5 students per classroom in the two-stage bootstrap is approximate 0.5, whereas it is about 0.25 for classroom-level bootstrap. So based on the biases, standard deviations and the coverage rates for the parameter estimates, the parametric bootstrap and the cluster bootstrap at the classroom level perform better than the two-stage bootstrap with cluster sizes five and 15.

5 Application

The mixed-effects model and cluster bootstrap methods were applied to a randomized study of the TELL curriculum for speech and language delayed preschool children (Wilcox and Gray 2011). This study was designed to assess a treatment of speech and language delay in preschool children. Classrooms were randomly assigned to treatment or control, and on average there were 5.4 children with developmental delay per classroom. Measurements were taken at six time points across the preschool year on several skills, and for this application we are looking at a skill called Letter Sound Identification. The scores range from zero to 26.

Before applying the bootstrap methods to the scores, we analyzed the distribution of the Letter Sound Identification responses. There may be floor and ceiling effects in the scores, and we found heavy tails in the original data set, as Fig. 8 shows. Then,the distribution of the empirical residuals from fitting the mixed-effects model were examined and a non-normal distribution was confirmed by the Shapiro-Wilk normality test with p-value less than 2.2e-16. We can also see some heavy-tail outliers from the QQ plot of the residuals in Fig. 8. So from these aspects, we know that

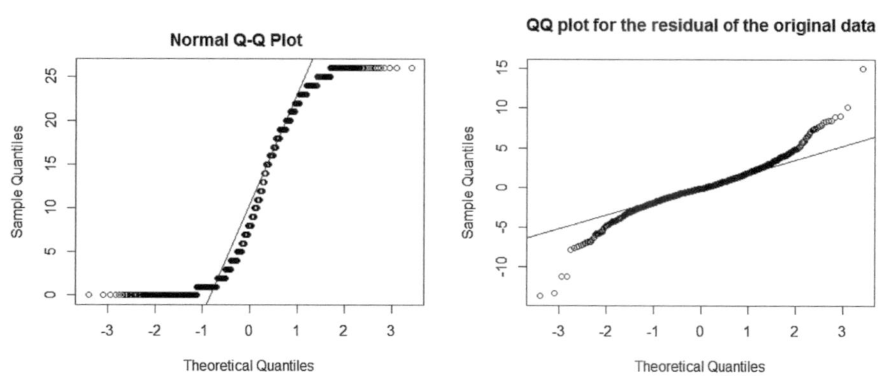

Fig. 8 *Left* QQ plot for the original data. *Right* QQ plot of the empirical residuals

the original data are not normally distributed and have several outliers. To fit the model, the bootstrap method can be a good candidate to find the proper model in this situation. So we will compare parametric confidence intervals obtained from fitting the linear mixed-effects model to confidence intervals obtained by using the parametric bootstrap, cluster bootstrap at classroom level and the two-stage bootstrap.

The original data has some missing observations, but the proportion missing was low, and we assume MAR. We fitted the model with the original data, and obtained confidence intervals by the three bootstrap methods discussed earlier. We examine the results in terms of parameter estimates, bias and standard deviation. In Tables 7, 8 and 9, the parameter estimates and the standard deviations obtained by the parametric bootstrap provide similar values as the parametric linear mixed-effects model on original data. However, there are some differences for the estimation of standard deviation among the three methods. The standard deviation from parametric bootstrap always gives smallest value, while the two-stage bootstrap has highest standard deviation and highest bias except for σ_c. It is interesting to see that the two-stage bootstrap produces the smallest standard deviation but highest bias for σ_c, which maybe due to the small cluster size.

Heat maps of the absolute value of bias and standard deviations for the parameter estimates for the three methods are shown in Figs. 9 and 10, where the individual values are represented as colors. In Fig. 9, the bias of the parameter estimates are small, and for each bootstrap the estimates for intercept and treatment condition have more bias than other fixed effects. Also, the bias of standard deviation of the variance component for intercept at the classroom level (σ_c) and random intercept at the student level (σ_{β_0}) are larger than other variance components. Overall, the two-stage bootstrap has more bias than the other methods, and the bias of σ_{β_0} and σ_c are larger than for other parameter estimates.

Heatmaps of standard deviations for the parameter estimates are shown in Fig. 10. The plot shows that the quadratic time trend, the interaction of condition and quadratic time trend, random slope for student, the correlation of random intercept and random slope, as well as the residual term have relative small standard deviation. The standard deviation of parameter estimates for intercept and the treatment condition have relative large standard deviations. In addition, the two-stage bootstrap presents larger standard deviations than the other methods. The special case is the standard deviation of random intercept from the bootstrap of level-two units is larger than the two-stage bootstrap.

Figures 11, 12 and 13, show histograms of the parameter estimates for the three bootstrap methods, where the red lines are the parameter estimates obtained from the linear mixed-effects model on the original data. From the plots, most of the estimates are centered around the red lines except σ_{β_0} and σ_c in the two-stage bootstrap. Also, the distribution of estimates for the two-stage bootstrap are wider than the parametric and cluster bootstrap at classroom level, except for σ_c, which is the reason the standard deviations for the two-stage bootstrap are larger than the other methods. However the standard deviation of the random intercept for classroom is smaller than the standard deviation in parametric bootstrap and the cluster bootstrap at classroom level. These features are in accordance with the result in Tables 7, 8 and 9.

Table 7 Estimates, bias and standard deviation of fixed effects for scores of preschool students

Methods	β_0			β_1			β_2		
	Est.	Bias	SD	Est.	Bias	SD	Est.	Bias	SD
No bootstrap	4.98	NA	0.82751	−1.12	NA	1.17161	0.914	NA	0.20908
Parametric	4.9759	−0.00419	0.821235	−1.0863	0.028432	1.165328	0.9127	−0.00182	0.209221
Cluster TID	4.9188	−0.0613	0.865659	−1.0412	0.073573	1.077025	0.9231	0.008603	0.35634
Two-stage	4.8344	*−0.1457*	*1.023809*	−0.9798	*0.134969*	*1.339487*	0.9018	*−0.01266*	*0.386505*

Table 8 Estimates, bias and standard deviation of fixed effect estimates for scores of preschool students (continued)

Methods	β_3			β_4			β_5		
	Est.	Bias	SD	Est.	Bias	SD	Est.	Bias	SD
No bootstrap	−0.0144	NA	0.01957	0.861	NA	0.28157	−0.0101	NA	0.02655
Parametric	−0.014	0.000328	0.019531	0.8595	−0.00139	0.271527	−0.0095	0.000591	0.026457
Cluster TID	−0.0152	−0.00084	0.03192	0.8408	−0.02008	0.486667	−0.0084	0.001695	0.043482
Two-stage	−00133	*0.001072*	*0.035742*	0.8759	0.015091	*0.545405*	−0.0118	−0.00163	*0.050381*

Table 9 Estimates, bias and standard deviation of variance component estimates for scores of preschool students

Methods	σ_ε			σ_{β_0}			$\rho_{\beta_0\beta_1}$			σ_{β_1}			σ_c		
	Est.	Bias	SD	Est.	Bias	SD	Est.	Bias	SD	Est.	Bias	SD	Est.	Bias	SD
No bootstrap	2.809	NA	NA	6.913	NA	NA	−0.259	NA	NA	0.788	NA	NA	2.804	NA	NA
Parametric	2.806	−0.0025	0.063	6.918	0.005	0.3761	−0.2562	0.0027	0.0739	0.787	−0.00135	0.04601	2.806	−0.103	*0.7821*
Cluster TID	2.79	−0.0221	0.155	6.89	−0.024	0.4244	−0.255	0.00445	0.0694	0.783	−0.00498	0.045997	2.599	−0.0205	*0.8381*
Two-stage	2.783	−0.0265	*0.193*	6.03	−0.883	*0.55*	−0.318	−0.0587	*0.1071*	0.779	−0.00899	*0.05871*	4.409	1.605	0.6308

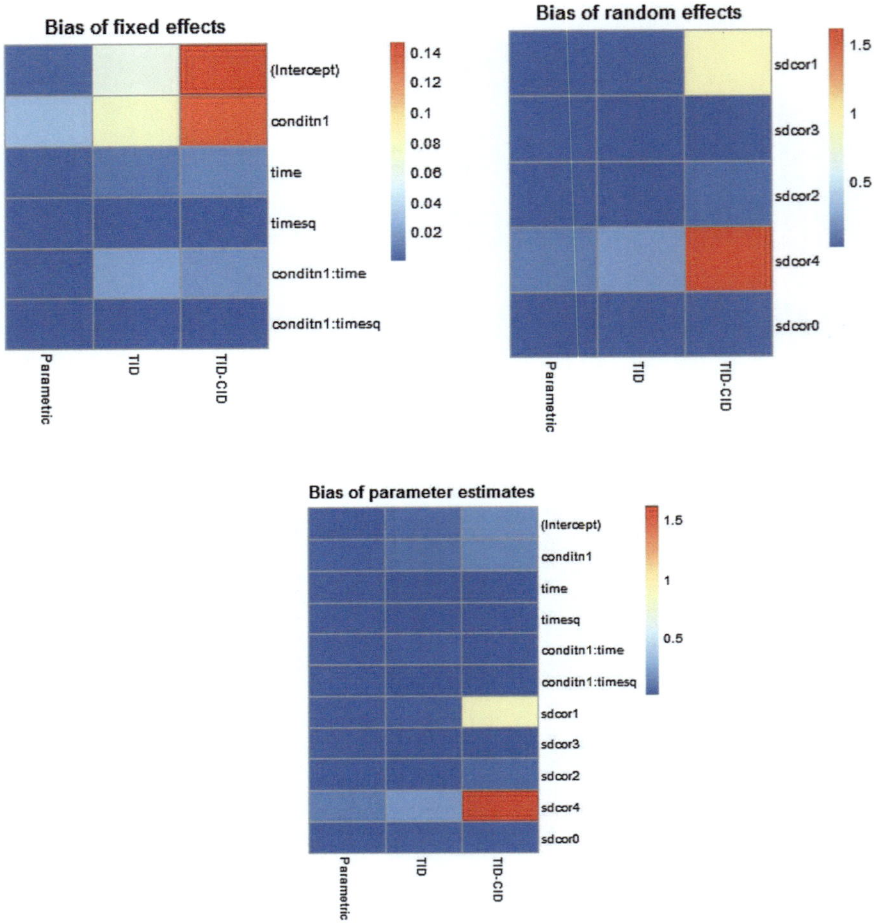

Fig. 9 Heatmap of absolute value of bias for parameter estimates

Confidence intervals to evaluate the model are presented in Table 10. Under the null hypothesis H_0: $\beta_i = 0$ for the fixed effects, we would reject the null hypothesis if zero is not within the confidence interval from the bootstrap methods. Different bootstrap method and confidence intervals provide similar results in this study; for example, there is significant linear time trend in the model and insignificant quadratic time trend. While the linear mixed-model and parametric bootstrap suggest significant interaction of condition and time, the cluster bootstrap at classroom level and the two-stage bootstrap indicate the interaction is not significant by the percentile confidence interval and BCa confidence interval.

For the parameter estimates and confidence intervals for variance components, different confidence intervals from the methods show similar results, and all the covariance estimates are significant. However, the confidence intervals of σ_{β_0} and σ_c

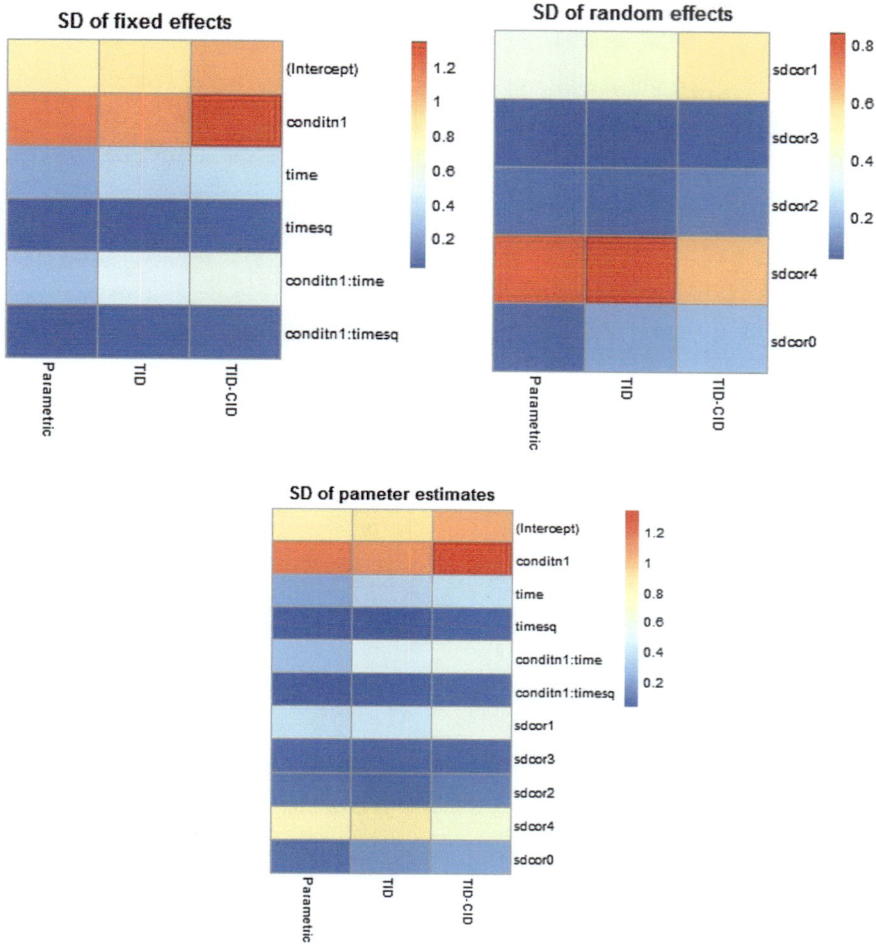

Fig. 10 Heatmap of standard deviations of parameter estimates

for the two-stage bootstrap are wider and the covariance estimates are largely different from the other two methods. It is also interesting to see that the BCa confidence intervals of σ_{β_0} and σ_c do not contain the maximum likelihood parameter estimates, maybe because the parameter estimates for σ_{β_0} and σ_c have large bias, which can explain why the coverage rates of σ_{β_0} and σ_c for the two-stage bootstrap are less than 95%. But our data set is small size, not balanced, and the data are not normally distributed, so we have higher probability to find the maximum likelihood parameter estimates are not included in the bootstrap confidence interval (Tables 10 and 11).

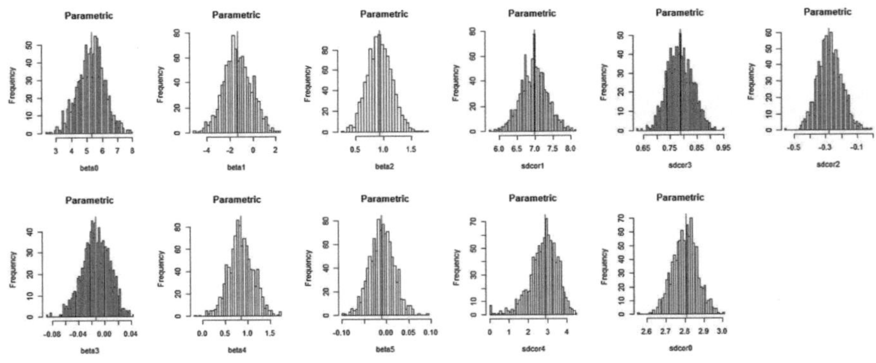

Fig. 11 Histograms of parameter estimates from parametric bootstrap

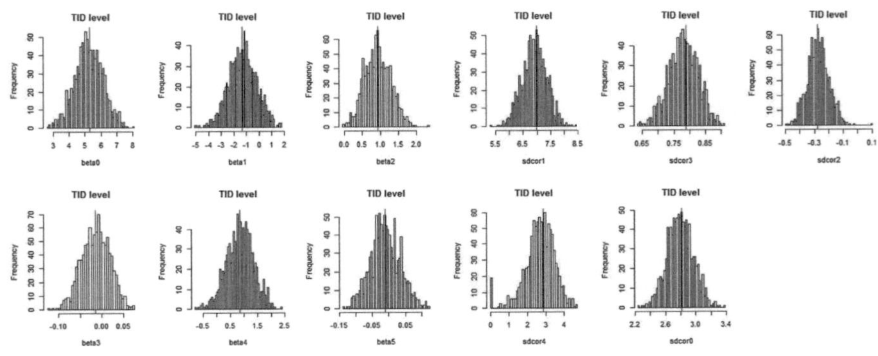

Fig. 12 Histograms of parameter estimates from cluster bootstrap

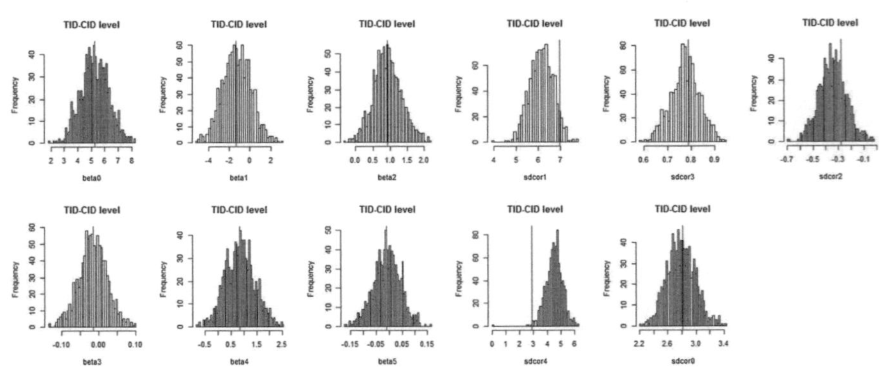

Fig. 13 Histograms of parameter estimates from two-stage bootstrap

Table 10 Estimates and confidence intervals for fixed effect parameters

Methods	Estimates and confidence intervals for the educational sample								
	(Intercept)			Treatment			Time		
	Est.	Low.	Upp.	Est.	Low.	Upp.	Est.	Low.	Upp.
No bootstrap	4.9801	3.3561	6.597	−1.1147	−3.403	1.1833	0.9145	0.5054	1.3246
Parametric (perc)	4.9759	3.3374	6.5767	−1.0863	−3.4047	1.167	0.9127	0.5241	1.3336
Parametric (BCa)	4.9759	3.3098	6.5638	−1.0863	−3.4118	1.1406	0.9127	0.5372	1.3425
Cluster TID (Perc)	4.9188	3.2484	6.5545	−1.0412	−3.2734	1.1381	0.9231	0.2857	1.6541
Cluster TID (BCa)	4.9188	3.4022	6.7693	−1.0412	−3.2955	0.8746	0.9231	0.3093	1.6803
Cluster TID (bt)	−0.09401	−2.2245	1.9183	0.068456	−1.8166	1.9781	0.008873	−3.086	3.3257
Two-stage (perc)	4.8344	2.9376	6.8847	−0.9798	−3.6348	1.616	0.9018	0.1851	1.7024
Two-stage (BCa)	4.8344	3.2606	7.2055	−0.9798	−4.0003	1.3567	0.9018	0.2491	1.7829
Two-stage (bt)	−0.187031	−2.4365	1.974	0.110393	−1.9062	2.1367	−0.09719	−3.68	3.6563

(continued)

Table 10 (continued)

Methods	Estimates and confidence intervals for the educational sample								
	Timesq			Treatment : time			Treatment : timesq		
	Est.	Low.	Upp.	Est.	Low.	Upp.	Est.	Low.	Upp.
No bootstrap	−0.01437	−0.0528	0.02394	0.8608	0.3086	1.4117	−0.0101	−0.0621	0.04199
Parametric (perc)	−0.014	−0.0534	0.02284	0.8595	0.3172	1.3718	−0.00952	−0.0594	0.0442
Parametric (BCa)	−0.014	−0.05404	0.02174	0.8595	0.2938	1.3448	−0.00952	−0.0585	0.04424
Cluster TID (Perc)	−0.01521	−0.08122	0.04528	0.8408	*−0.1372*	*1.7482*	−0.00842	−0.08995	0.07981
Cluster TID (BCa)	−0.01521	−0.0812	0.04533	0.8408	*−0.1364*	*1.7492*	−0.00842	−0.0901	0.07823
Cluster TID (bt)	−0.02212	−3.2686	3.1453	−0.06359	−3.5774	3.2355	0.05711	−3.1359	3.2708
Two-stage (perc)	−0.0133	−0.08375	0.05112	0.8759	*−0.265*	*1.8987*	−0.01175	−0.1034	0.09257
Two-stage (BCa)	−0.0133	−0.0917	0.04565	0.8759	*−0.2971*	*1.8441*	−0.01175	−0.0975	0.10031
Two-stage (bt)	0.08006	−3.4919	3.4512	0.0584	−4.0198	3.7385	−0.06415	−3.6555	3.9216

Table 11 Estimates and confidence intervals for variance components

Methods	Estimates and confidence intervals for the scores of preschool students														
	σ_{β_0}			σ_{β_1}			$\rho_{\beta_0\beta_1}$			σ_c			σ_ε		
	Est.	Low.	Upp.	Est.	Low.	Upp.	Est.	Low.	Upp.	Est.	Low.	Upp.	Est.	Low.	Upp.
No bootstrap	6.913	6.234	7.67	0.781	0.7	0.879	−0.258	−0.398	−0.102	2.804	1.096	4.018	2.809	2.687	2.935
Parametric (perc)	6.918	6.173	7.665	0.787	0.697	0.883	−0.256	−0.395	−0.1	2.702	0.74	3.999	2.807	2.682	2.932
Parametric (BCa)	6.918	6.101	7.584	0.787	0.708	0.889	−0.256	−0.407	−0.115	2.702	1.476	4.353	2.807	2.698	2.951
Cluster TID (Perc)	6.889	6.071	7.697	0.783	0.691	0.871	−0.255	−0.385	−0.111	2.599	0.0002	3.925	2.787	2.512	3.101
Cluster TID (BCa)	6.889	5.999	7.628	0.783	0.697	0.881	−0.255	−0.41	−0.131	2.599	1.709	4.494	2.787	2.562	3.17
Two-stage (perc)	6.03	4.837	7.106	0.78	0.659	0.897	−0.318	−0.52	−0.099	4.409	3.117	5.558	2.783	2.415	3.164
Two-stage (BCa)	*6.03*	*6.761*	*7.679*	0.78	0.682	0.912	−0.318	−0.401	−0.015	*4.409*	*3.2E-6*	*2.599*	2.783	2.494	3.273

6 Conclusions

In this chapter, we studied several different bootstrap methods (parametric bootstrap, cluster bootstrap at level-two and two-stage bootstrap) applied to multilevel longitudinal data. Monte-Carlo simulations were used to evaluate the three bootstrap methods. In terms of the coverage rate, bias and standard deviation of the parameter estimates, the simulations suggested that the three bootstrap methods gave us similar results for the fixed effects, and the parametric bootstrap as well as the cluster bootstrap at level-two provide better intervals than the two-stage bootstrap. The two-stage bootstrap may produce large bias and large standard deviation of the parameter estimates because the re-sampling of a small number of level-one units may produce degenerate bootstrap samples. With a larger number of level-one units in each level-two cluster, we could obtain smaller bias, smaller standard deviation and larger coverage rate for the bootstrap methods. Simulation results not reported here show that resampling of the level-one units ignoring the level-two clusters produces very poor performance for confidence interval coverage. The results are not a surprise because the level-one units are not i.i.d. ignoring the level-two units. Finally, we applied the three bootstrap methods to the scores of preschool students. The application results follow the same patterns as the simulations. Due to the large bias of the two-stage bootstrap, we can sometimes obtain maximum likelihood parameter estimates outside the bootstrap confidence interval. Simulations with a larger number of level-one units per level-two cluster may show better results for the two-stage bootstrap.

Acknowledgements This research was supported by the U.S. Department of Education, Institute of Educational Sciences Grant R324A110048. The opinions expressed in this chapter are those of the authors and no official endorsement by the IES should be inferred.

References

Bates, D., Maechler, M., Bolker, B., & Walker, S. (2015). Fitting linear mixed-effects models using lme4. *Journal of Statistical Software*, *67*(1), 1–48. doi:10.18637/jss.v067.i01.

Beale, E. M. L., & Little, R. J. A. (1975). Missing value in multivariate analysis. *Journal of the Royal Statistical Society, Series B*, *37*, 129–145.

Benton, D. (2002). Performance of the parametric bootstrap method in small sample interval estimates. *Advances and Applications in Statistics*, *2*(3), 269–285.

Boos, D. (2003). Introduction to the bootstrap world. *Statistical Science*, *18*(2), 168–174.

Burdick, R. K., & Graybill, F. A. (1988). The present status of confidence interval estimation on variance components in balanced and unbalanced random models. *Communications in Statistics—Theory and Methods (Special Issue on Analysis of the Unbalanced Mixed Model)*, *17*, 1165–1195.

Carpenter, J. R. (2003). A novel bootstrap procedure for assessing the relationship between class size and achievement. *Applied Statistics*, 431–443.

Demidenko, E. (2013). *Mixed models: Theory and applications with R* (2nd ed.). New York: Wiley.

Efron, B., & Tibshirani, R. J. (1993). *An introduction to the bootstrap*. New York: Chapman and Hall.

Efron, B. (1987). Better bootstrap confidence interval. *Journal of American Statistical Association, 82*(397). Theory and Methods.

Efron, B. (1986). Bootstrap of the mean in the infinite variance case. *The Annals of Statistics*, 724–731.

Field, C. A., Pang, Z., & Welsh, A. H. (2008). Bootstrapping data with multiple levels of variation. *The Canadian Journal of Statistics*, 521–539.

Field, C. A. (2007). Bootstrapping clustered data. *Journal of the Royal Statistical Society: Series B*, 369–390.

Fitzmaurice, G. M., & Laird. N. (2004). *Applied longitudinal analysis*. New York: Wiley.

Harville, D. A. (1997). Maximum likelihood approaches to variance component estimation and to related problems. *Journal of the American Statistical Association, 72*(358), 320–338.

Laird, N. M., & Ware, J. H. (1982). Random-effects models for longitudinal data. *Biometrics, 38*(4).

Patterson, H. D., & Thompson, R. (1971). Recovery of inter-block information when block sizes are unequal. *Biometrika, 58*, 545–54.

Pinheiro, J. C., & Bates, D. (2000). *Mixed-effects models for s and s-plus*. New York: Springer.

Scherman, M., & Cessie, S. (1997). A comparison between bootstrap methods and generalized estimating equations for correlated outcomes in generalized linear models. *Communications in Statistics, Simulation and Communication, 26*, 901–925.

Searle, S., McCullough, C., & Casella, G. (1992). *Variance components*. New York: Wiley.

Shieh, Y. Y. (2002). The application of bootstrap methodology to multilevel mixed effects linear model under conditions of error term nonnormality. *Proceedings of the Joint Statistical Meetings, Biometric Section* (pp. 3191–3196). Alexandria, VA: American Statistical Association.

Thai, H. T. (2013). A comparison of bootstrap approaches for estimating uncertainty of parameters in linear mixed-effects models. *Pharmaceutical Statistics*.

Thomas, J. D., & Efron, B. (1996). Bootstrap confidence intervals. *Statistical Science, 11*(3), 189–228.

Verbeke, G., & Molenberghs, G. (2000). *Linear mixed models for longitudinal data*. New York: Springer.

Wilcox, J., & Gray, S. (2011). Efficacy of the TELL language and literacy curriculum for preschoolers with developmental speech and/or language impairment. *Early Childhood Research Quarterly, 26*, 278–294.

Bootstrap-Based LASSO-Type Selection to Build Generalized Additive Partially Linear Models for High-Dimensional Data

Xiang Liu, Tian Chen, Yuanzhang Li and Hua Liang

Abstract Generalized additive partially linear model (GAPLM) is a flexible option to model the effects of covariates on the response by allowing nonlinear effects of some covariates and linear effects of the other covariates. To address the practical needs of applying GAPLM to high-dimensional data, we propose a procedure to select variables and therefore to build a GAPLM by using the bootstrap technique with the penalized regression. We demonstrate the proposed procedure by applying it to analyze data from a breast cancer study and an HIV study. The two examples show that the procedure is useful in practice. A simulation study also shows that the proposed procedure has a better performance of variable selection than the penalized regression.

1 Introduction

Nowadays, many datasets involving a large number of measurements (such as genetic, gene expression, proteomics and other -omic data) are produced with the hope to reveal the relation between the measurements and the phenotype and consequently the disease mechanism. For example, a dataset from a breast cancer study published by van't Veer et al. (2002) contains the observations of 97 lymph-node

X. Liu (✉)
Health Informatics Institute, University of South Florida, Tampa, FL 33612, USA
e-mail: xiang.liu@epi.usf.edu

T. Chen
Department of Mathematics and Statistics, University of Toledo, Toledo, OH 43606, USA
e-mail: tian.chen@utoledo.edu

Y. Li
Division of Preventive Medicine, Walter Reed Army Institute of Research, Silver Spring, MD 20910, USA
e-mail: yuanzhang.li2.civ@mail.mil

H. Liang (✉)
Department of Statistics, George Washington University, Washington, DC 20052, USA
e-mail: hliang@email.gwu.edu

D.-G. Chen and J.D. Chen (eds.), *Monte-Carlo Simulation-Based Statistical Modeling*, ICSA Book Series in Statistics, DOI 10.1007/978-981-10-3307-0_18

negative patients, 46 among which developed distant metastases within 5 years while the rest 51 patients remained to be disease-free after a follow-up period of at least five years. In the study, gene expression levels were measured on thousands of genes. One research interest is to explore the relationship between genes and breast cancer and consequently to build a model to predict the outcome using gene expression data.

Logistic regression model, a special case of generalized linear models (GLM, McCullagh and Nelder 1989), is a popular method to analyze binary outcome. However, linearity is only an assumption of the model but not necessarily a property of the data. An initial exploration of the marginal effect of each gene on breast cancer suggests that some genes' effects are nonlinear (see Fig. 1 and more details in Sect. 4.1). Generalized additive partially linear models (GAPLM, Härdle et al. 2004) can be used for remedying in the situation. GAPLM extends GLM by modeling the effects of some predictors through a linear function and the effects of the other predictors through additively smooth functions. Fitting a GAPLM can be simplified by approximating the smoothing functions using polynomial splines and the statistical property of the estimation has been studied (Wang et al. 2011). Considering the situation of high-dimensionality in which the number of predictors p is greater than the sample size n as in the breast cancer study, fitting a GAPLM with all predictors in it is impossible and therefore one needs to apply some variable selection method. Traditional subset selection method suffers from the intensive computational cost due to the large number of predictors. Penalized regression, such as the least absolute shrinkage and selection operator (LASSO) (Tibshirani 1996), has become an important alternative to subset selection. Tibshirani proposed the LASSO (Tibshirani 1996) by adding the L1-penalty on coefficients into the objective function in a linear regression model, which shrinks some small coefficients to zero and therefore fulfills the purpose of variable selection. A tuning parameter controls the magnitude of the penalty and is chosen by cross-validation in common practice. The idea of the penalized regression has been applied to various models including GLM (Fan and Li 2001) and GAPLM (Wang et al. 2011), and various penalty functions have been proposed (Fan and Li 2001; Zou and Hastie 2005; Zou 2006). Algorithms via coordinate descent (Friedman et al. 2010) makes the penalized regression computationally fast and affordable.

We have applied the LASSO technique for a GAPLM with spline-approximation (Wang et al. 2011) to analyze the gene expression data. As a result, 27 genes are selected with 10 genes in the nonparametric part (see Table 1 and more details in Sect. 4.1). However, the number of selected genes is still too large to build a predictive GAPLM from the gene expression data. Therefore, a practically useful variable selection method to build GAPLMs for high-dimensional data is needed. In addition, we often find that the set of selected predictors from penalized regression is not stable in real data analysis. Some perturbation of the data or a slight change of the tuning parameter can result in quite a different set of selected predictors. Research (Zhao and Yu 2006; Zou 2006; Yuan and Lin 2007; Meinshausen and Bühlmann 2006) on the consistent variable selection of the LASSO shows that there are strong assumptions on the design matrix that need to be satisfied for consistent variable selection. Decaying schemes of the tuning parameter have been studied (Zhao and Yu 2006; Meinshausen and Bühlmann 2006; Bach 2008) and it is shown that under specific settings, LASSO selects the relevant variables with probability one and the irrelevant variables with a positive probability less than one, provided that the number of observations tends to infinity. If several samples were available from the underlying data distribution, irrelevant variables could be removed by simply intersecting the set of selected variables for each sample. Bach proposed the Bolasso algorithm (Bach 2008) by providing such datasets using the bootstrap method (Efron et al. 2004). The Bolasso algorithm (bootstrapped enhanced lasso) requires the relevant variable to be selected in all bootstrap runs and it leads to consistent variable selection. The algorithm had been applied in practice (Guo et al. 2015) and had also been studied for binary data using logistic regression (Strobl et al. 2012). Meinshausen and Bühlmann proposed stability selection Meinshausen and Bühlmann (2010), which also uses subsampling technique to improve the variable selection of LASSO. Stability selection generates a stability path which is the probability for each variable to be selected by LASSO when randomly subsampling from the data against the tuning parameter. If the maximum of the probabilities for a variable over the tuning region is higher than a threshold, then the variable is selected. An upper bound on the expected number of falsely selection variables has been proven for the stability selection (Meinshausen and Bühlmann 2010, Theorem 1). However, the bound appears to be too conservative for deriving a cut off for the threshold in practice (discussion by Sylvia Richardson, (Meinshausen and Bühlmann 2010, p. 448).

Motivated by the aforementioned practical demands, we propose an easy-to-implement procedure of applying the idea of bootstrapped enhanced lasso to build generalized additive partial linear models for high-dimensional data. The requirement of the relevant variables to be selected in all bootstrap runs is too strict in practice and we consider a soft version of it to reduce the set of candidate predictors. One additional round of variable selection will be performed to build a predictive GAPLM. In our procedure, the bootstrap technique is applied with the penalized regression for GAPLM. The set of candidate predictors is then narrowed down by selecting a predictor if it is chose in at least a fraction of the bootstrap replicates and the best subset selection can be conducted afterwards. We use two real data examples and a simulation study to illustrate our proposed procedure. The remainder of the article is organized as follows. In Sect. 2, we introduce the detailed framework of the procedure to build GAPLM. Section 3 gives the technical details of the spline approximation and the penalized regression for GAPLM. In Sect. 4, we apply the proposed procedure to analyze data from the breast cancer study (van't Veer et al. 2002) and an HIV study. In Sect. 5, we provide the results of a simulation study in which we have investigated the performance of the procedure. Section 6 concludes the article with a summary and discussion.

2 Framework of the Procedure to Build GAPLM

Let's denote the covariates/predictors in the nonparametric part as W, the covariates in the parametric part as X and the response as Y. The technical details of GAPLM are given in Sect. 3. A general and flexible procedure to build generalized additive partial linear models to analyze high-dimensional data is proposed and is described as follows.

1. Perform m (e.g., $m = 500$) bootstrap replications as follows
   a. Draw a bootstrap sample consisting n observations $\{(W_1^*, X_1^*, Y_1^*), (W_2^*, X_2^*, Y_2^*), \ldots, (W_n^*, X_n^*, Y_n^*)\}$ from the original dataset $\{(W_1, X_1, Y_1), (W_2, X_2, Y_2), \ldots, (W_n, X_n, Y_n)\}$ with replacement;
   b. Apply the penalized regression of GAPLM to the bootstrap sample and record the selected covariates. The details of the penalized regression for GAPLM are described in Sect. 3.
2. Calculate the selection rates of covariates $\mathbf{r} = (r_{W_1}, r_{W_2}, \ldots, r_{W_{p_1}}, r_{X_1}, r_{X_2}, \ldots, r_{X_{p_2}})$ across the B bootstrap replications and construct a reduced set of covariates whose selection rates are higher than some threshold (C)
3. Perform the best subset selection of GAPLM on the reduced set and choose the best models according to certain criterion, such as Akaike Information Criterion (AIC) or Bayesian Information Criterion (BIC)

The Bolasso algorithm requires the relevant variables to be selected in all bootstrap runs. This requirement is too strict in real data analysis and even more when a complicated semi-parametric model such as GAPLM is used. A threshold C is introduced in Step 2 to relax the requirement and to get a reduced set of covariates. An additional model selection such as best subset selection can be conducted to get a sparse model (Step 3). The value of C should be able to reduce the number of candidate covariates to make the best subset selection practically feasible. The value can be fixed (e.g., $C = 0.5$) or be data-driven based on the sample mean and the sample standard deviation of the selection rates (e.g., $C = mean(\mathbf{r}) + sd(\mathbf{r})$).

3 Generalized Additive Partial Linear Models

Let Y be the response, whose distribution belongs to the exponential family McCullagh and Nelder (1989), and its conditional expectation given the covariates $W = (W_1, \ldots, W_{p_1})^\top$ and $X = (X_1, \ldots, X_{p_2})^\top$ is defined via a known link function g by an additive partial linear function

$$\mu = E(Y|W, X) = g^{-1}\left\{\sum_{j=1}^{p_1} \alpha_j(W_j) + \sum_{j=1}^{p_2} X_j\beta_j\right\}, \tag{1}$$

where $\alpha(\cdot) = (\alpha_1(\cdot), \ldots, \alpha_{p_1}(\cdot))^\top$ and $\beta = (\beta_1, \ldots, \beta_{p_2})^\top$ are the nonparametric components and the coefficients of the parametric components, respectively. The variance is assumed to be a function of the mean given by

$$Var(Y|W, X) = \sigma^2 V(\mu), \tag{2}$$

where $V(\cdot)$ is a known function. Note that the value of σ doesn't play any role in the specification of the quasi-likelihood later and therefore we can assume $\sigma = 1$ without loss of generality. When $g^{-1}(x) = e^x/(1+e^x)$ and $V(\mu) = \mu(1-\mu)$, we obtain the additive partial linear logistic regression model. For identifiability of the models, we assume that $E\alpha_j(W_j) = 0$.

3.1 Spline Approximation

For simplicity, we assume that the covariate W_j is distributed on a compact interval $[a_j, b_j]$, $j = 1, \ldots, p_1$, and without loss of generality, we take all intervals $[a_j, b_j] = [0, 1]$, $j = 1, \ldots, p_1$. Under some smoothness assumptions (Wang et al. 2011), the nonparametric components $\alpha_j(W_j)$'s can be well-approximated by spline functions. Let $\tau_0 = 0 < \tau_1 < \cdots < \tau_{K'} < 1 = \tau_{K'+1}$ be a partition of [0, 1] into subintervals $[\tau_k, \tau_{k+1})$, $k = 0, \ldots, K'$ with K' internal knots. Equally spaced knots or data-driven knots (sample quantiles of the observed covariate values) can be used. A polynomial spline of order u is a function satisfying the following two conditions:

(i) it is a polynomial of degree $u-1$ on each of the subintervals;
(ii) for $u > 2$, it is globally $u-2$ times continuously differentiable on [0, 1].

The collection of splines with a fixed sequence of knots has a B-spline basis $\{B_1(x), \ldots, B_{\tilde{K}}(x)\}$ with $\tilde{K} = K' + u$. Because of the centering constraint $E\alpha_j(W_j) = 0$, we instead focus on the subspace of spline functions $S_j^0 := \{s : s(x) = \sum_{k=1}^{\tilde{K}} a_{jk} B_k(x), \sum_{i=1}^n s(W_{ij}) = 0\}$ with a normalized basis $\{B_{jk}(x) = \sqrt{K}(B_k(x) - \sum_{i=1}^n B_k(W_{ij})/n), k = 1, \ldots, K = \tilde{K} - 1\}$ (the subspace is $K = \tilde{K} - 1$ dimensional due to the empirical version of the constraint). Using spline expansions, the nonparametric components can be approximated as

$$\alpha_j(x) \approx \sum_k a_{jk} B_{jk}(x), \; 1 \le j \le p_1.$$

The quasi-likelihood function is defined by $Q(\mu, y) = \int_\mu^y (y-t)/V(t)\,dt$, and the quasi-likelihood of i.i.d. observations $(W_i, X_i, Y_i)(i = 1, \ldots, n, W_i = (W_{i1}, \ldots, W_{ip_1})^\top$ and $X_i = (X_{i1}, \ldots, X_{ip_2})^\top)$ is

$$\sum_{i=1}^n Q(g^{-1}(\sum_{j=1}^{p_1} \alpha_j(W_{ij}) + \sum_{j=1}^{p_2} X_{ij}\beta_j), Y_i) \approx \sum_i Q(g^{-1}(Z_i^\top a + X_i^\top \beta), Y_i),$$

where $Z_i = (B_{11}(W_{i1}), \ldots, B_{1K}(W_{i1}), \ldots, B_{p_1K}(W_{ip_1}))^\top$ and $a = (a_1^\top, \ldots, a_{p_1}^\top)^\top = (a_{11}, \ldots, a_{p_1K})^\top$.

3.2 Penalized Regression

Variable selection can be achieved by penalized regression. The penalized regression estimates the coefficients via maximizing the penalized quasi-likelihood as follows,

$$(\hat{a}, \hat{\beta}) = \arg\max_{a,\beta}\{\sum_{i=1}^{n} Q(g^{-1}(Z_i^\top a + X_i^\top \beta), Y_i) - nP_\lambda(a, \beta)\}, \tag{3}$$

where the penalty $P_\lambda(a, \beta)$ shrinks some coefficients to zero and consequently eliminates the corresponding parametric and/or nonparametric components. The LASSO penalty (Tibshirani 1996) or other penalties, such as the SCAD penalty (Fan and Li 2001; Zou 2008) and the adaptive LASSO penalty (Zou 2006), can be used for the penalty function $P_\lambda(\cdot)$. In addition, the group LASSO penalty (Yuan and Lin 2006) is applicable if we treat the coefficients from each additive component as a group.

Specifically, we here investigate the following penalized regression based on the LASSO penalized quasi-likelihood,

$$(\hat{a}, \hat{\beta}) = \arg\max_{a,\beta}\{\sum_{i=1}^{n} Q(g^{-1}(Z_i^\top a + X_i^\top \beta), Y_i) - n\lambda_1 \sum_{j=1}^{p_1}\sum_{k=1}^{K} |a_{jk}| - n\lambda_2 \sum_{j=1}^{p_2} |\beta_j|\}, \tag{4}$$

and the group LASSO penalized quasi-likelihood,

$$(\hat{a}, \hat{\beta}) = \arg\max_{a,\beta}\{\sum_{i=1}^{n} Q(g^{-1}(Z_i^\top a + X_i^\top \beta), Y_i) - n\lambda_1 \sum_{j=1}^{p_1} \sqrt{K}\|a_j\|_2 - n\lambda_2 \sum_{j=1}^{p_2} |\beta_j|\}, \tag{5}$$

where λ_1 and λ_2 are the regularization parameters controlling the shrinkage of the coefficients in the nonparametric and parametric components. In Eq. (5), the group LASSO penalty is applied on the coefficients in each nonparametric component as a group and it makes them simultaneously zero or non-zero group-wise. In comparison, the LASSO penalty is applied on individual coefficients and it doesn't have the property of group selection. The regularization parameters can be chosen by cross-validation, AIC or BIC.

4 Real Data Examples

We have applied the proposed procedure to analyze data from the breast cancer study described in Sect. 1 and an HIV study. We have conducted 500 bootstrap replications (Step 1, a) and have considered both the LASSO penalty and the group LASSO penalty in the penalized regression (Step 1, b). We have used AIC as the criterion in the best subset selection (Step 3). As a comparison, we have also applied the penalized regression (i.e., the regular penalized regression without bootstrapping).

Cubic splines (spline order $u = 4$) are used to approximate the nonparametric components and knots are put at the sample quartiles. We use the function *bs* in the splines package in *R* to calculate the spline basis of cubic splines with 2 knots. Several packages in *R* are available to fit the LASSO-penalized logistic regression (e.g., glmnet Friedman et al. 2010) and the group LASSO-penalized logistic regression (e.g., grplasso Meier et al. 2008 and SGL Simon et al. 2013) for binary outcomes. We use the *glmnet* package for the LASSO-penalized logistic regression and the *SGL* package for the group LASSO-penalized logistic regression as both use the algorithm via coordinate descent to calculate the regularization paths. The two-dimensional grid search of the regularization parameters is computational intensive. Therefore, we treat the parametric and the nonparametric components equally by making the regularization parameters the same to reduce computational cost. The tuning parameters are chosen by 10-fold cross-validation.

4.1 Breast Cancer Data

The gene expression data in the breast cancer study van't Veer et al. (2002) are pre-processed to remove genes with more than 30% missing measurements. For the remaining genes, the missing values are imputed by the median value for that gene across all samples and the 200 genes that exhibit the highest correlations with the responses were kept in the analysis. We have tested the linearity of each gene's marginal effect on the response by fitting a nonparametric logistic model using the R function `gam`. The nonlinearity of 65 genes' marginal effects are statistically significant (at the level of 0.05). Among these genes, 34 genes still have statistically significant nonlinear effect when the extreme values (i.e., minimum and maximum) of the genes are excluded. After a further check on the plots of the marginal effects, 17 genes are included in the nonparametric part and the rest 183 genes are in the parametric part of GAPLM. The marginal effects of the 17 genes on the response are plotted in Fig. 1 and the p value of the linearity test is shown on the top of each plot.

The genes being selected by the bootstrapped penalized regression and the penalized regression are marked in Table 1. Using the LASSO penalty, the bootstrapped penalized regression has selected 4 genes in the nonparametric part and 6 genes in the parametric part, while the penalized regression has selected 6 additional genes in the nonparametric part and 11 additional genes in the parametric part. Using the group LASSO penalty, the bootstrapped penalized regression has selected no genes in the nonparametric part and 11 genes in the parametric part, while the penalized regression has selected 11 additional genes in the parametric part. The genes selected by the bootstrapped penalized regression have the selection rates over the 500 bootstrap replications greater than $C = 0.5$ and the selection rates are also given in Table 1. Overall, the penalized regression with the group LASSO penalty has selected more genes in the parametric part and fewer genes in the nonparametric part than that

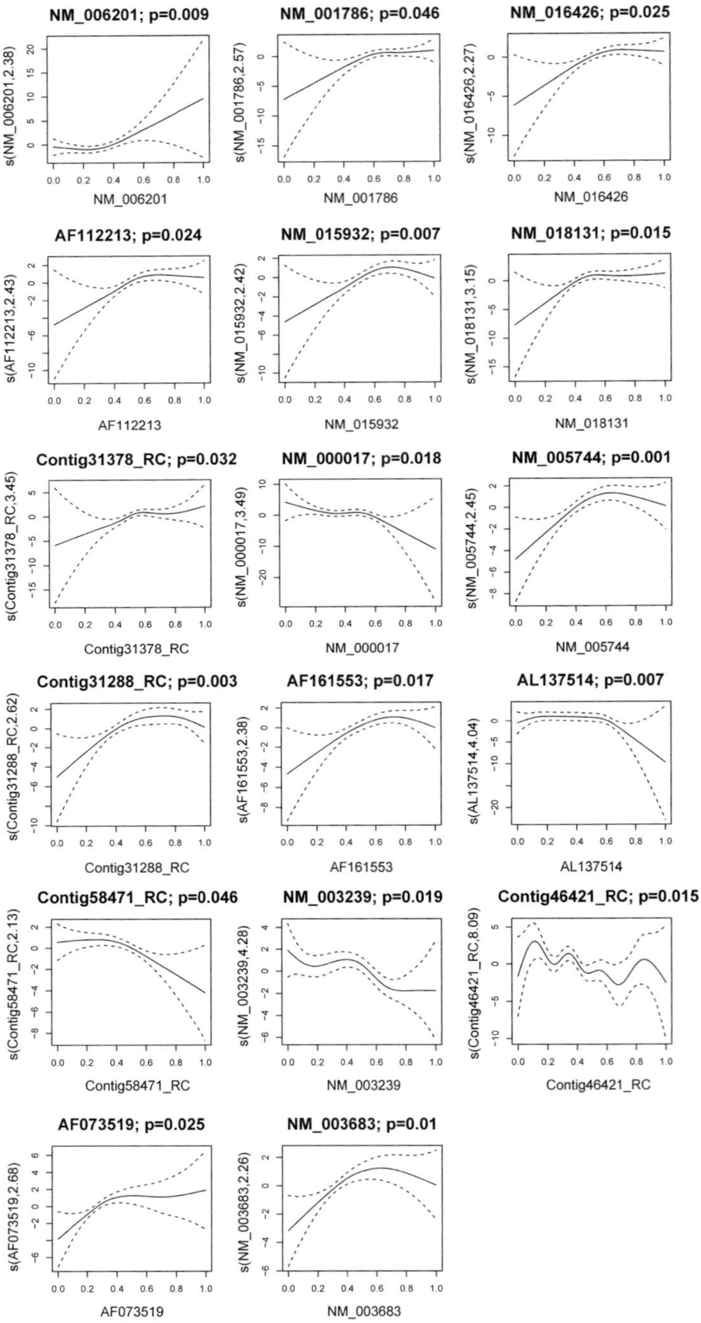

Fig. 1 The marginal effects of 17 genes estimated using the GAM analysis for the breast cancer data. The p-value of the test for linearity is displayed on the *top* of each plot

Table 1 Genes selected by the penalized regression (PR) and the bootstrapped penalized regression (Bootstrapped PR) using the LASSO penalty or the group LASSO penalty in the breast cancer example. Upper panel gives the genes in the parametric part and the middle panel gives the genes selected in nonparametric part. Symbol "X" marks the covariate that is selected and the selection rate is given in parentheses for the bootstrapped penalized regression. The total number of selected genes is in the lower panel

	LASSO penalty		Group LASSO penalty	
	(a) PR	(b) Bootstrapped PR	(c) PR	(d) Bootstrapped PR
Contig35148_RC	X		X	
AL080059	X	X(94.2%)	X	X(98.2%)
Contig47544_RC	X	X(88.6%)	X	X(92.8%)
Contig5816_RC	X			
NM_020244	X		X	X(52.4%)
NM_003147	X	X(58.0%)	X	X(64.2%)
Contig64861_RC	X		X	X(63.2%)
NM_014003	X		X	
D13540	X	X(63.8%)	X	X(68.8%)
AB018337	X	X(55.0%)	X	X(67.4%)
Contig38438_RC	X		X	X(55.8%)
NM_000127	X	X(56.2%)	X	X(61.8%)
Contig32718_RC	X		X	
Contig22253_RC	X		X	
Contig53488	X		X	X(53.8%)
Contig31839_RC	X			
Contig15355_RC	X		X	
NM_013438			X	
NM_020123			X	X(63.8%)
NM_016448			X	
Contig42563_RC			X	
Contig38726_RC			X	
NM_018313			X	
AB033032			X	
NM_006201	X			
NM_015932	X	X(74.2%)		
NM_016426	X			
NM_005744	X			
Contig31288_RC	X			
AF161553	X			
AL137514	X	X(69.8%)		
Contig58471_RC	X	X(56.4%)		
NM_003239	X	X(75.0%)		
NM_003683	X			
Total	27	10	22	11

Table 2 Models built by the proposed procedure in the breast cancer example. Model 1 is the model having the smallest AIC among all the potential subsets on List (d) in Table 1. Models 2 and 3 are the top two models with the smallest AIC among all the potential subsets on List (b) in Table 1. Upper panel gives the estimates and the p-values of the coefficients in the parametric part and the middle panel gives the p-values of the nonparametric components in the GAPLM. The lower panel gives AIC and AUC (area under the ROC curve) for each model

	Model 1		Model 2		Model 3	
	Est. (s.e)	p	Est. (s.e)	p	Est. (s.e)	p
Intercept	−20.84 (7.37)	0.005	−17.79 (5.05)	<0.001	−23.89 (6.34)	<0.001
AB018337	21.01 (7.76)	0.007	15.77 (4.86)	0.001	9.86 (3.65)	0.007
AL080059	19.18 (6.41)	0.003	14.48 (3.56)	<0.001	15.23 (4.62)	<0.001
Contig38438RC	−8.99 (4.10)	0.028				
Contig47544RC	14.58 (5.33)	0.006	11.36 (3.59)	0.002	14.97 (4.29)	<0.001
Contig53488	8.07 (4.06)	0.047				
NM_000127	14.83 (5.09)	0.004	13.57 (3.94)	<0.001	16.00 (5.11)	0.002
NM_003147	−15.05 (5.49)	0.006	−10.84 (3.90)	0.006		
Contig58471RC						0.051
NM_003239						0.050
AIC	41.30		49.63		52.73	

with the LASSO penalty. The bootstrapped penalized regression has reduced the set of genes selected by the penalized regression and made the best subset selection feasible.

Best subset selection has been conducted on the lists of genes from the bootstrapped penalized regression with the LASSO penalty and with the group LASSO penalty (Lists (b) and (d) in Table 1) respectively. All potential models under the restriction that all p-values should be smaller than 0.1 (in order to exclude overfitting) have been considered. The model having the smallest AIC (Model 1) among all potential subsets on List (d) in Table 1 and the top two models having the smallest AIC (Models 2 and 3) among all potential subsets on List (b) in Table 1 are presented in Table 2. Models 1 and 2 are nested linear logistic models and the latter is significantly better than the former (likelihood ratio test, $p = 0.002$). Model 3 is a partially linear logistic regression model with two nonparametric components (plotted in Fig. 2). The gene Contig58471RC(SLC27A1) in the nonparametric part maps to chromosome 19p13, which is a region highly susceptible to triple negative-specific breast cancer (Stevens et al. 2012). The receiver operating characteristic (ROC) curves of the three models are plotted in Fig. 3. In comparison, the ROC curve

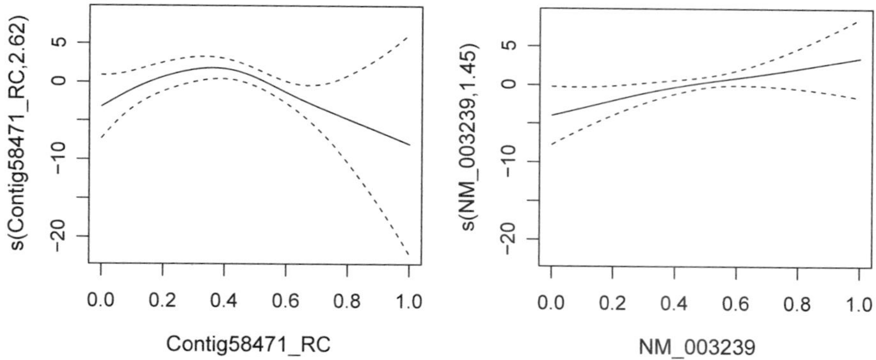

Fig. 2 Plot of the nonparametric components in Model 3 in the breast cancer example

of the model (Model 4, AIC = 95.62) fitted using the top 7 genes (NM_003862, Contig14882_RC, NM_020120, AL080059, AL050227, NM_003875 and NM_016448) according to p values of test for association with the outcome by the univariate GAM analysis is also plotted in Fig. 3. In summary, the three models built by the proposed procedure are comparable in terms of ROC (or area under the curve (AUC)) and better than Model 4.

4.2 HIV Data

The dataset includes 59 measurements on 42 HIV infected patients from a HIV study. The measurements include 51 serum neutralizing covariates on B cells, 4 serum autoreactive antibodies and 4 clinical covariates including age, time since diagnosis (years), CD4 cell count and HIV viral load (copy/per mL). The response variable is *IC50*, half maximal inhibitory concentration, which indicates how much of a particular substance is needed to inhibit a given biological process by half. When *IC50* values are below the limit of quantification (LOQ), the exact *IC50* values cannot be quantified. About 30% (14 out of 42) of the *IC50* values are below the limit of quantification and therefore the response variable is dichotomized as 0 if the *IC50* value is below or equal to LOQ and 1 otherwise. Logarithm transformation and standardization have been performed on HIV viral load, serum autoreactive antibody, CD4 cell counts and B cell counts to make their distributions less skewed. Highly correlated pairs of serum neutralizing covariates exist in the dataset and one out of each pair with correlation coefficient greater than 0.96 has been removed in the analysis. As a result, 49 out of 59 measurements/covariates have been kept for analysis.

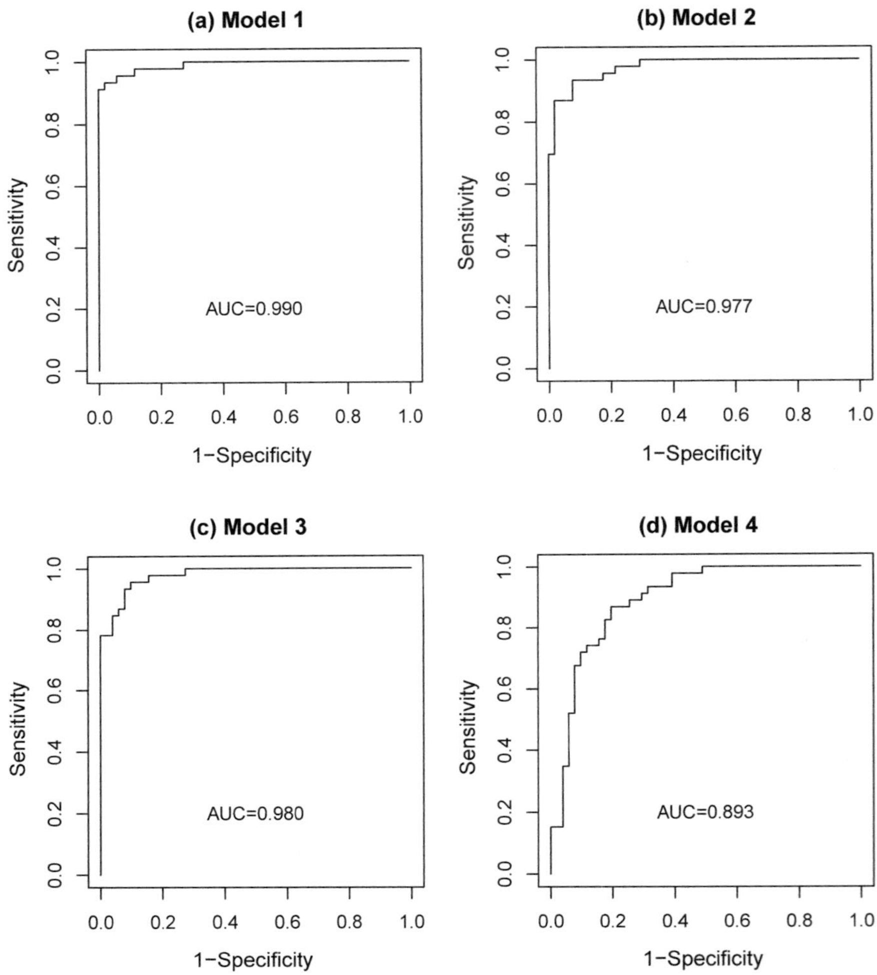

Fig. 3 ROC curves and AUCs (the Area Under the Curve) of the models built in the breast cancer example. Model 1 is the model having the smallest AIC among all the potential subsets on List (d) in Table 1. Models 2 and 3 are the top two models with the smallest AIC among all the potential subsets on List (b) in Table 1. Model 4 is the model fitted using the top 7 covariates according to p-values of test for association using the univariate GAM analysis

We first tested the linearity of each covariate's marginal effect on the response by fitting a nonparametric logistic regression using the R function `gam`. 7 covariates (time, Bcell37, Bcell38, Bcell45, antibody1, antibody2 and antibody3) have statistically significant (at the level of 0.05) nonlinear effects on the response. The effect of Bcell38 is influenced by one observation in the dataset and the nonlinearity disappears after excluding the observation. The marginal effect of antibody2 looks like piecewise linear and modeling it in a nonparametric way is not appropriate due to

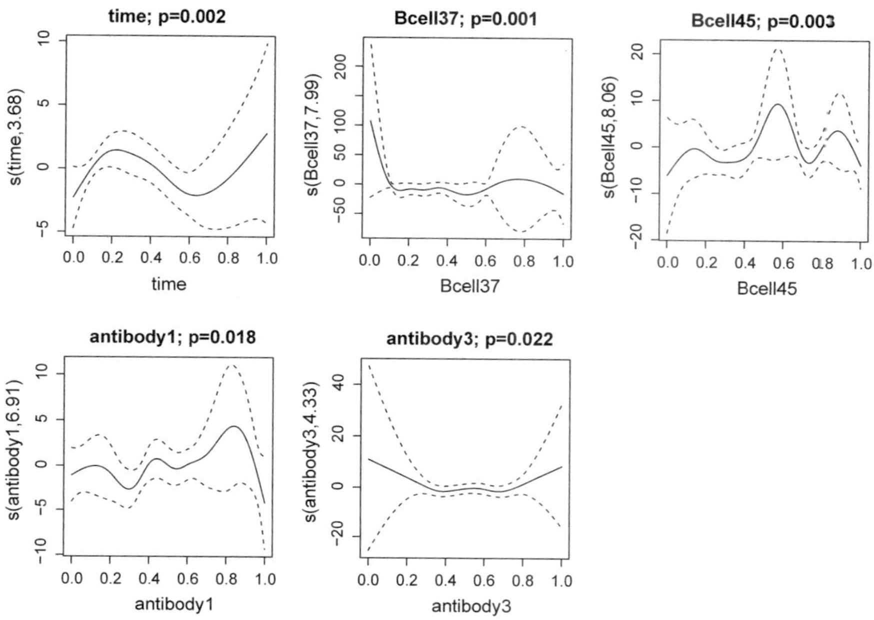

Fig. 4 The marginal effects of 5 covariates estimated using the GAM analysis in the HIV example. The p-value of the test for linearity is displayed on the top of each plot

large variance. Therefore, 5 covariates (time, Bcell37, Bcell45, antibody1 and antibody3) are included in the nonparametric part and the rest 44 covariates are included in the parametric part of the GAPLM. The marginal effects of the 5 covariates on the response are plotted in Fig. 4 and the p-value of the test for linearity is shown on the top of each plot.

The covariates being selected by the bootstrapped penalized regression and the penalized regression are marked in Table 3. Using the LASSO penalty, the bootstrapped penalized regression has selected five covariates (*age*,*antibody2* and *Bcells 20, 41, 49*) in the parametric part and three covariates (*time*, *antibody1* and *Bcell37*) in the nonparametric part, while the penalized regression has selected *Bcell20* instead of *Bcell8* in the parametric part and hasn't selected *Bcell37* in the nonparametric part. Using the group LASSO penalty, the bootstrapped penalized regression has selected almost the same covariates as the bootstrapped lasso-penalized regression has selected except for the covariate *Bcell37* in the nonparametric part. The group-lasso-penalized regression has selected five covariates (*Bcells 8, 20, 41, 49* and *antibody2*) in the parametric part and none in the nonparametric part. The covariates selected by the bootstrapped penalized regression have selection rates greater than $C = 38.7\%$ and 41.8% for the LASSO penalty and the group LASSO penalty, respectively. The selection rates are given in Table 3. The cutoff C is the sum of the mean and standard deviance of all covariates' selection rates. Overall, the bootstrapped penalized regression has expanded the set of covariates selected by the penalized

Table 3 Covariates selected from the penalized regression (PR) and the bootstrapped penalized regression (Bootstrapped PR) using the LASSO penalty or the group LASSO penalty in the HIV example. Upper panel gives the covariates in the parametric part and the middle panel gives the covariates in the nonparametric part. Symbol "X" marks the covariate that is selected and the selection rate is given in parentheses for the bootstrapped penalized regression. The total number of selected covariates is in the lower panel

	LASSO penalty		Group LASSO penalty	
	(a) PR	(b) Bootstrapped PR	(c) PR	(d) Bootstrapped PR
age	X	X(51.4%)		X(54.8%)
antibody2	X	X(76.0%)	X	X(80.0%)
Bcell8		X(41.0%)	X	X(62.2%)
Bcell20	X		X	
Bcell41	X	X(45.8%)	X	X(56.2%)
Bcell49	X	X(84.2%)	X	X(93.6%)
time	X	X(74.8%)		X(61.6%)
antibody1	X	X(48.8%)		X(46.2%)
Bcell37		X(44.0%)		
Total	7	8	5	7

regression to include more covariates except for the covariate *Bcell20*, whose selection by the penalized regression seems to be by chance as its selection rate in the bootstrapped penalized regression is low.

Best subset selection has been conducted on the lists of covariates from the bootstrapped lasso/group-lasso penalized procedure (Lists (b) and (d) in Table 3) and also the list of covariates from the lasso-penalized regression (List (a) in Table 3) as putting all the covariates in a model overfits the data (all p-values are close to 1). All potential models under the restriction that all p-values should be smaller than 0.5 (in order to exclude over-fitting) have been considered. The models with the smallest AIC from all potential subsets of the three lists are the same (Model 1) and is presented in Table 4. Model 1 is a GAPLM with one nonparametric component of the covariate *time*, which is plotted in Fig. 5. The model fitted with all the covariates in the list (List (c) in Table 3) selected by the group LASSO-penalized regression (Model 2) is also presented in Table 4. The ROC curves of the two models are plotted in Fig. 6. As a comparison, the ROC curve from a model (Model 3, AIC= 41.40) fitted using the top 5 covariates (*time* (in nonparametric part), *Bcell8*, *Bcell24*, *Bcell41* and *Bcell49*) according to the p values of test for association with the outcome from the univariate GAM analysis is also plotted in Fig. 6. In conclusion, the model built by the proposed procedure is better than the other two models considering AIC and also the area under the ROC curve (AUC).

Table 4 Models built in the HIV example. Model 1 is the model with the smallest AIC among all the potential subsets on List (d) in Table 3. Model 2 is the model fitted with the covariates selected by the of group LASSO-penalized regression (List (c) in Table 3). Upper panel gives the estimates and the p-values of the coefficients in the parametric part. The middle panel gives the p-values for the nonparametric components. The lower panel gives AIC for each model

	Model 1		Model 2	
	Est. (s.e)	p	Est. (s.e)	p
Intercept	−6.17 (4.13)	0.135	−4.62 (3.66)	0.207
age	6.87 (3.43)	0.045		
antibody2	4.76 (2.91)	0.102	2.58 (1.94)	0.183
Bcell8			−4.11 (2.63)	0.118
Bcell20			1.53 (2.35)	0.514
Bcell41	10.36 (6.51)	0.112	5.33 (3.27)	0.103
Bcell49	−6.50 (3.68)	0.078	−5.85 (2.69)	0.030
time		0.376		
antibody1				
AIC	37.95		40.26	

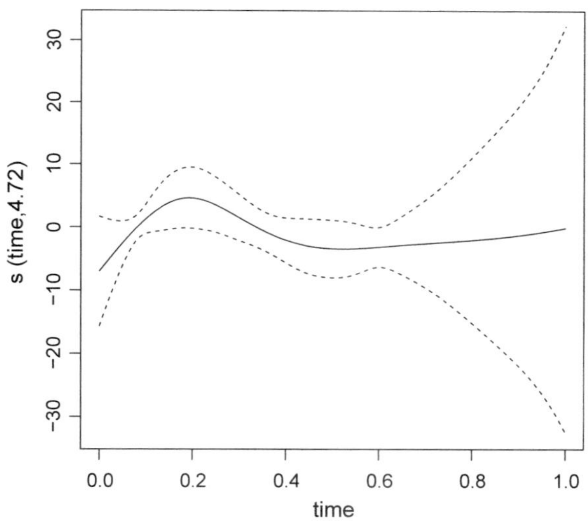

Fig. 5 Plot of the nonparametric component in Model 1 in the HIV example

5 A Simulation Study

To evaluate the performance of variable selection by the bootstrapped penalized regression and compare it to the penalized regression, we have performed a simulation study with 500 runs. We only consider the LASSO penalty in the simulation study because the computation of the bootstrapped group LASSO-penalized regression with 500 bootstrap replicates is too time-consuming. It took about 3 h when using

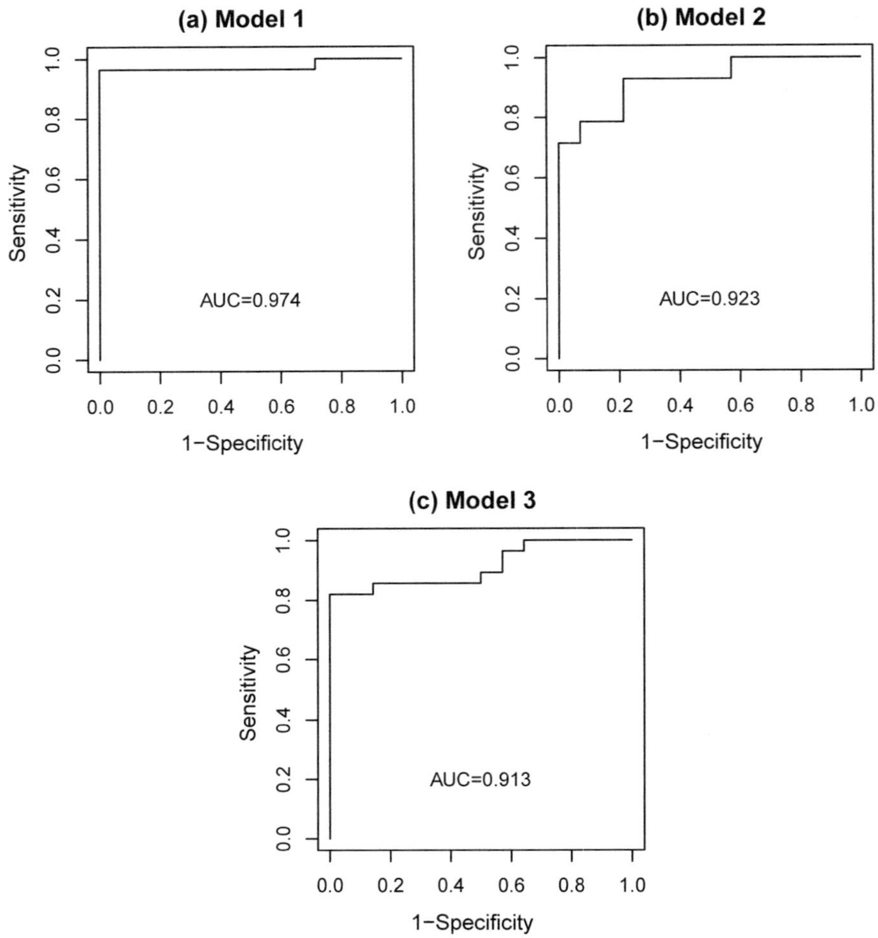

Fig. 6 ROC curves and AUCs (area under the curve) of the models in the HIV example. Model 1 is the model with the smallest AIC among all the potential subsets on List (d) in Table 3. Model 2 is the model fitted with the covariates selected by the group LASSO-penalized regression (List (c) in Table 3). Model 3 is the model fitted using the top 5 covariates according to p-values of test for association with the outcome by the univariate GAM analysis

the group LASSO penalty in the bootstrapped penalized regression while using the LASSO penalty in it only took 1 min in the real data examples.

We use the data on the covariates in the HIV dataset and Model 1 in Table 4 to generate the response. Specifically, in each simulation run, we generated the response for each observation from a Bernoulli distribution with the probability being calculated as follows,

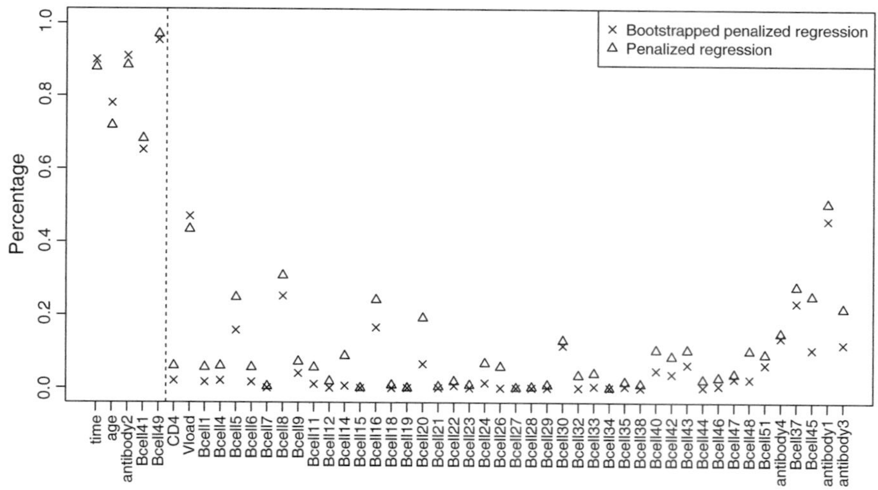

Fig. 7 Percentages of the covariates selected by the bootstrapped penalized regression and the penalized regression in the simulation study. The five covariates on the left are the covariates in the simulation model (6)

$$E(Y = 1) = \text{logit}^{-1}\{-6.17 + s(time) + 6.87 * age + 4.76 * antibody2 + 10.36 * Bcell41 - 6.50 * Bcell49\}, \qquad (6)$$

where $s(time)$ is plotted in Fig. 5. Then, we applied the bootstrapped lasso-penalized regression with 500 times of resampling on the simulated dataset. 10-fold cross-validation was used to select covariates in each bootstrapping replication and the reduced set of covariates from the bootstrapped process were those with selection rates higher than a cutoff. The cutoff was the mean plus one standard deviation of selection rates across all covariates. As a comparison, the LASSO-penalized regression was also applied on each simulated dataset and covariates were selected by 10-fold cross-validation.

The percentages of the covariates selected by the penalized regression and the bootstrapped penalized regression among the 500 simulation runs has been plotted in Fig. 7. Among the five true covariates (*time*, *age*, *antibody2*, *Bcell41* and *Bcell49*) on the left in Fig. 7, the percentages (i.e., the true positive rates) of the first three covariates being selected by the bootstrapped process (90.0%, 78.0% and 91.0% respectively) are higher than those by the LASSO-penalized regression (87.8%, 71.8% and 88.4% respectively). For the other two covariates, the true positive rates by the bootstrapped process (65.2% and 95.4% respectively) are close to but a little bit lower than those from the one round of the LASSO-penalized regression (68.2% and 97.0% respectively). For the rest covariates, which are not in the simulation model (Eq. 6), the percentages (i.e., the false positive rates) from the bootstrapped process are lower or at least equal to those from the one round LASSO-penalized regression except for one covariate *Vload*. The false selection rates of *Vload* are relatively high for both methods (43.4% by the penalized regression and 47.0% by the bootstrapped process). Overall, the bootstrapped procedure boosts 3 out of 5 true positive selection rates while controls down 43 out of 44 false positive rates comparing to the LASSO-penalized regression in the simulation study.

6 Summary and Discussion

Generalized additive partially linear model is an attractive method to analyze real data by allowing some covariates to have nonlinear effects while some others to have linear effects on the response. To address the practical need of building a GAPLM for high-dimensional data, we have proposed a procedure by using the bootstrap technique with the penalized regression to select variables for GAPLM. We have demonstrated the application of the proposed procedure to two datasets and we have applied both the lasso penalty and the group-lasso penalty. The group-lasso penalty is more strict than the lasso penalty to select nonlinear terms in the GAPLM. In addition, the bootstrapped group-LASSO regression is much more time-consuming than the bootstrapped LASSO regression. Therefore, we would suggest using the lasso penalty in real data analysis. We have also showed that the proposed procedure with the lasso penalty has a better performance of variable selection than the regular lasso-penalized regression in the simulation study.

Bootstrap is a useful technique to gauge model stability. It has been used to determine the tuning parameter in the LASSO type estimation for linear regression models (Hall et al. 2009; Chatterjee and Lahiri 2011). Recently, the technique has also been used to compute standard errors and confidence intervals of estimators in model selection Efron (2014). The stability selection (Meinshausen and Bühlmann 2010) proposed by Meinshausen and Bühlmann uses subsampling technique to estimate the structure, such as variable selection, graphical modeling or cluster analysis. The important feature of the stability selection for LASSO is the error control, where an upper bound has been provided for the expected number of falsely selection variables (Meier and Bühlmann 2007, Theorem 1). Shah and Samworth have proposed a variant of the stability selection (Shah and Samworth 2013), using complementary pairs when resampling, which is claimed to improve the error control in the stability selection. It would be interesting to investigate the theoretical property of applying the stability selection to select variables for GAPLMs, especially the theoretical aspects of the false discovery rates in the future.

Acknowledgements Material has been reviewed by the Walter Reed Army Institute of Research. There is no objection to its presentation and/or publication. The opinions or assertions contained herein are the private views of the author, and are not to be construed as official, or as reflecting true views of the Department of the Army, the Agency for Healthcare Research and Quality, the Department of Defense or the Department of Health and Human Services. Liang's research was partially supported by NSF grants DMS-1418042 and DMS-1620898, and by Award Number 11529101, made by National Natural Science Foundation of China.

References

Bach, F. R. (2008). Bolasso: Model consistent Lasso estimation though the bootstrap. In *Proceedings of the Twenty-fifth International Conference on Machine Learning (ICML)*.

Chatterjee, A., & Lahiri, S. N. (2011). Bootstrapping Lasso estimators. *Journal of the American Statistical Association*, *106*(494), 608–625.

Efron, B. (2014). Estimation and accuracy after model selection. *Journal of the American Statistical Association*, *109*(507), 991–1007.

Efron, B., Hastie, T., Johnstone, I., & Tibshirani, R. (2004). Least angle regression. *The Annals of Statistics*, *32*, 407–499.

Fan, J., & Li, R. (2001). Variable selection via nonconcave penalized likelihood and its oracle properties. *Journal of the American Statistical Association*, *96*, 1348–1360.

Friedman, J., Hastie, T., & Tibshirani, R. (2010). Regularization paths for generalized linear models via coordinate descent. *Journal of Statistical Software*, *33*(1), 1–22.

Guo, P., Zeng, F., Hu, X., Zhang, D., Zhu, S., Deng, Y., et al. (2015). Improved variable selection algorithm using a Lasso-type penalty, with an application to assessing hepatitis b infection relevant factors in community residents. *PLoS ONE*, *10*.

Hall, P., Lee, E. R., & Park, B. U. (2009). Bootstrap-based penalty choice for the Lasso, achieving oracle performance. *Statistica Sinica*, 449–471.

Härdle, W., Müller, M., Sperlich, S., & Werwatz, A. (2004). *Nonparametric and semiparametric models*. New York: Springer.

McCullagh, P., & Nelder, J. A. (1989). *Generalized linear models* (2nd ed.). London, New York: Chapman and Hall.

Meier, L., & Bühlmann, P. (2007). Smoothing l1-penalized estimators for highdimensional time-course data. *Electronic Journal of Statistics*, *1*, 597–615.

Meier, L., Geer, S. V. D., & Bhlmann, P. (2008). The group Lasso for logistic regression. *Journal of the Royal Statistical Society, Series B*, *70*(1), 53–71.

Meinshausen, N., & Bühlmann, P. (2006). High dimensional graphs and variable selection with the Lasso. *Annals of Statsitics*, *34*(3), 1436–1462.

Meinshausen, N., & Bühlmann, P. (2010). Stability selection. *Journal of Royal Statistical Society, Series B*, *72*(4), 417–473.

Shah, R. D., & Samworth, R. J. (2013). Variable selection with error control: Another look at stability selection. *Journal of the Royal Statistical Society Series B*, *75*(1), 55–80.

Simon, N., Friedman, J., Hastie, T., & Tibshirani, R. (2013). A sparse-group Lasso. *Journal of Computational and Graphical Statistics*, *22*(2), 231–245.

Stevens, K. N., Fredericksen, Z., Vachon, C. M., Wang, X., Margolin, S., Lindblom, A., et al. (2012). 19p13.1 is a triple-negative-specific breast cancer susceptibility locus. *Cancer Research*, *72*(7), 1795–1803.

Strobl, R., Grill, E., & Mansmann, U. (2012). Graphical modeling of binary data using the Lasso: A simulation study. *BMC Medical Research Methodology*, *12*(16).

Tibshirani, R. (1996). Regression shrinkage and selection via the Lasso. *Journal of the Royal Statistical Society, Series B*, *58*, 267–288.

van't Veer, L. J., Dai, H. Y., et al. (2002). Gene expression profiling predicts clinical outcome of breast cancer. *Nature*, *415*, 530–536.

Wang, L., Liu, X., Liang, H., & Carroll, R. (2011). Estimation and variable selection for generalized additive partial linear models. *The Annals of Statistics*, *39*, 1827–1851.

Yuan, M., & Lin, Y. (2006). Model selection and estimation in regression with grouped variables. *Journal of the Royal Statistical Society, Series B*, *68*, 49–67.

Yuan, M., & Lin, Y. (2007). On the non-negative garrotte estimator. *Journal of the Royal Statistical Society: Series B*, *69*(2), 143–161.

Zhao, P., & Yu, B. (2006). On model selection consistency of Lasso. *Journal of Machine Learning Research*, *7*, 2541–2563.

Zou, H. (2006). The adaptive Lasso and its oracle properties. *Journal of the American Statistical Association*, *101*, 1418–1429.

Zou, H. (2008). A note on path-based variable selection in the penalized proportional hazards model. *Biometrika*, *95*, 241–247.

Zou, H., & Hastie, T. (2005). Regularization and variable selection via the elastic net. *Journal of the Royal Statistical Society, Series B*, *67*, 301–320.

Erratum to: Monte-Carlo Simulation-Based Statistical Modeling

Ding-Geng (Din) Chen and John Dean Chen

Erratum to: D.-G. Chen and J.D. Chen (eds.), *Monte-Carlo Simulation-Based Statistical Modeling*, ICSA Book Series in Statistics, DOI 10.1007/978-981-10-3307-0

The original version of the book was inadvertently published without updating the contributor's name in the front matter. The erratum book has been updated with the change.

The updated original online version of this book can be found at http://dx.doi.org/10.1007/978-981-10-3307-0

D.-G. Chen and J.D. Chen (eds.), *Monte-Carlo Simulation-Based Statistical Modeling*, ICSA Book Series in Statistics, DOI 10.1007/978-981-10-3307-0_19

Index

D.-G. Chen and J.D. Chen (eds.), *Monte-Carlo Simulation-Based Statistical Modeling*, ICSA Book Series in Statistics, DOI 10.1007/978-981-10-3307-0

N

O

MIX
Papier aus verantwortungsvollen Quellen
Paper from responsible sources
FSC® C105338

If you have any concerns about our products, you can contact us on
ProductSafety@springernature.com

In case Publisher is established outside the EU, the EU authorized representative is:
Springer Nature Customer Service Center GmbH
Europaplatz 3, 69115 Heidelberg, Germany

Printed by Libri Plureos GmbH
in Hamburg, Germany